Performance of Light Aircraft

Performance of Light Aircraft

John T. Lowry

Flight Physics
Billings, Montana

American Institute of Aeronautics and Astronautics, Inc.
1801 Alexander Bell Drive
Reston, VA 20191

Publishers since 1930

Library of Congress Cataloging-in-Publication Data

Lowry, John T., 1935-
   Performance of light aircraft / John T. Lowry.
   p. cm. — (AIAA education series)
   Includes bibliographical references and index.
   ISBN 1-56347-330-5 (alk. paper)
   1. Airplanes – Performance. 2. Private planes. I. Title. II Series.
   TL671.4 .L68 1999
   629.133′343 – dc21

                                              99-34583
                                                  CIP

Cover design by Sara Bluestone

Copyright © 1999 by the American Institute of Aeronautics and Astronautics, Inc. All rights reserved. Printed in the United States of America. No part of this publication may be reproduced, distributed or transmitted, in any form or by any means, or stored in a data base or retrieval system, without the prior written permission of the publisher.

Data and information appearing in this book are for informational purposes only. AIAA is not responsible for any injury or damage resulting from use or reliance, nor does AIAA warrant that use or reliance will be free from privately owned rights.

Table of Contents

Preface   xv
Nomenclature   xxi

Part I. Fundamentals of Aeronautic Science

1 The Atmosphere   3

Introduction   3
Structure and Composition of the Earth's Atmosphere   4
Elementary Atmospheric Physics   6
International Standard Atmosphere   9
Example 1.1   11
Atmospheric Humidity   11
   Example 1.2   12
   Example 1.3   12
   Total Density vs "Air" Density   13
   Example 1.4   13
Viscosity   14
Speed of Sound   14
   Example 1.5   14
   Example 1.6   15
Modeling the Test-Day Atmosphere Locally   15
   Example 1.7   16
   Inchworm Method for Finding Δh 16
Reduction of Experimental Climb/Descent Data to ISA Equivalents   19
References   20

2 Atmospheric Aircraft Instruments   21

Introduction   21

vi   *Table of Contents*

Bernoulli's Law	22
Example 2.1	22
Air-Speed Indicators	23
Low-Speed ASIs	24
High-Speed ASIs	24
Example 2.2	25
Calibrating the ASI	26
Low-Altitude Speed Course Technique	27
Back and Forth with GPS Unit Technique	30
Box Pattern with GPS Unit Technique	30
Example 2.3	31
Kinetic Energy Effect	32
Example 2.4	32
Outside Air Thermometer	34
Ram Temperature Rise	34
Example 2.5	36
The Altimeter	37
Derivation of the Altimeter Setting Formula	37
Example 2.6	38
The Altimeter is a Barometer	39
Checking Altimeter Consistency	39
Example 2.7	39
Changes in Indicated Altitudes Equal Changes in Pressure Altitudes	40
General Linear Nonstandard Atmosphere	40
Example 2.8	43
Static Pressure Position Error	44
Determining Static Pressure Position Error: Technique #1	46
Determining Static Pressure Position Error: Technique #2	47
Position Error Causes Altimeter and ASI Errors	47
Example 2.9	47
References	48

3   Aircraft Performance Preliminaries — 49

Introduction	49
Units and Dimensions	49
Scalars and Vectors	51
Coordinate Systems	52
Frames of Reference	53
Linear Kinematics: Position, Velocity, Acceleration	54
Example 3.1	54
Linear Dynamics: Force, Work, Energy, and Power	55
Circular Kinematics: Angular Position, Velocity, and Acceleration	58
Example 3.2	59
Circular Dynamics: Torque, Work, Energy, and Power	59
Centrifugal and Coriolis Pseudoforces	61
Example 3.3	61

Example 3.4	62
Example 3.5	63
Technical Logic of Airplane Performance	63
Airplane	64
Atmosphere	65
Pilot	65
Quasi-Static Equations of Motion	65
References	67

4 Aerodynamic Force   69

Introduction	69
Lift, Drag, and Pitching Moment Coefficients	70
Airfoil Profiles	76
Pressure Coefficients	77
Bernoulli's Theorem	78
Times of Flight of Air Molecules Over or Under the Wing	79
Further NACA 4412, $\alpha = 8$ deg, Detail	81
Lift Coefficient from Pressure Coefficients	81
Center of Pressure	82
Pitching Moment Coefficient	82
Circulation	83
Induced AOA and Induced Drag	85
NACA 2412 Airfoil Profile	89
Finding $C_L(\alpha)$ from Section Lift Characteristics	90
Formulas for a, α_0, c_{mac}, and α_s	92
Ground Effects on Lift and Induced Drag	94
Practical Wings	97
Wing Planform Geometry	97
Aerodynamic Characteristics of Practical Wings	100
Wing Geometry: Incidence, Twist, and Body AOA	101
Flaps	103
Lift and Moment of the Entire Aircraft	110
The Plan	110
Moment Equations	110
Tail Force	112
Factors Influencing Lift	114
Drag	114
Expression for Drag D in Terms of $D_{\min} = D_{bg}$ and V_{bg}	115
Additional Drag Due to Flaps	117
References	117

5 The Engine   119

Introduction	119
Engine Power and Efficiency	123
Thermal Power	123

Table of Contents

Other Efficiency Losses	124
The Gagg–Farrar Dropoff Factor $\Phi(\sigma; C)$	126
Example 5.1	127
The Ideal Otto Air-Standard Cycle	129
Items We Have Neglected	133
Controls and Instruments	134
The Aircraft Engine Operator's Manual	135
Engine Performance Chart	138
Conclusions	142
References	143
6 Propeller Thrust	**145**
Introduction	146
Propeller Geometry	146
Details for a McCauley Propeller	148
Propeller Action and Dimensional Analysis	150
Momentum Theory	155
Physical Picture	155
Axial and Rotational Interference Factors	156
Thrust and Pressure	157
Ideal Efficiency	158
Efficiency and Power	158
Slip-stream Rotation and Torque	160
Blade Element Theory	162
Physical Picture	162
Relative Wind and Dynamic Pressure	163
Incremental Area	163
Aerodynamic Coefficients	163
Blade Element Thrust and Power Coefficient Gradings	165
Example 6.1	165
Combined Momentum and Blade Element Theory	167
Physical Picture	168
Mathematical Solution	168
Example 6.2	169
Static Thrust Case ($J = 0$)	171
Errors and Omissions	172
Example 6.3	174
Propeller Polar Diagram	175
Constant-Speed Propeller	177
General Propeller Charts	178
Example 6.4	178
Activity Factor	180
Power Adjustment Factor X	180
SDEF	182
Conclusions	183
References	183

Part II. Practical Airplane Performance

7 Introduction to the Bootstrap Approach	187
Introduction	187
Power-Available/Power Required (P_{av}/P_{re}) Analysis	190
Graphical Picture	190
Traditional Approach to P_{av} and P_{re}	192
Bootstrap Approach	194
Basic Derivation	194
Formulas for V Speeds	196
Formulas for Thrust, Drag, Rate, and Angle of Climb	199
Compendium of Bootstrap Composite Parameters	205
Sample Calculations with Composite Bootstrap Parameters	207
Bootstrap Applications 1: Variations with Weight or Altitude	207
Variation of V_x, V_{bg}, V_{md} (also V_s) and h_{md} with W or σ	209
Example 7.1	210
Variation of γ_x, γ_{bg} with W or σ	210
Example 7.2	211
Example 7.3	211
Variation of V_M, V_m, and V_y with W or σ	212
Example 7.4	212
Example 7.5	213
Variation of $\dot{h}_{max}$ with W or σ	213
Example 7.6	215
Example 7.7	216
Bootstrap Application 2: Absolute Ceiling and Speed There	216
Example 7.8	217
Bootstrap Application 3: Wind Effects on (Mostly) Full-Throttle Climbs	217
Relative Motion (and an Approximation)	219
Example 7.9	220
Graphical Analysis of Headwind/Tailwind Effects	221
Graphical Analysis of Headwinds/Tailwinds Accompanied by Updrafts/Downdrafts	223
Formula Approach to Wind Effects on Steepest Climbs	225
Example 7.10	225
Speed for Best Climb Angle V_{xhd} in Given Headwinds and Tailwinds, Given Updrafts and Downdrafts	226
Example 7.11	228
Bootstrap Application 4: MSL "Rated" Propeller Efficiency	229
Conclusions	230
Power Picture Assumptions	230
Linear Propeller Polar Assumptions	231
Evaluation	231
Advantages of the Bootstrap Approach	233
References	234

8 Maneuvering Performance — 235

Introduction	235
Review of Performance Assumptions	237
Flight Controls and Associated Variables	238
Aircraft Equations of Motion	239
Coordinated Level Turns	240
A Closer Look at the Flight Path Helix	241
Example 8.1	242
Turns, Centrifugal Force, and the Federal Aviation Administration	242
Coordinated, Slipped, and Skidded Turns	245
Example 8.2	247
Structural Limits in Level Turns	249
Extension of Bootstrap Approach to Steady Full-Throttle Maneuvering Flight	251
Banked or Unbanked Absolute Ceilings	255
Example 8.3	256
Steady Maneuvering Charts	258
A Cautionary Tale: Fatal High-Altitude Turn	267
Minimum Radius and Maximum Rate Level Turns	268
Example 8.4	270
Turn Performance of Constant-Speed Propeller Airplanes	270
Example 8.5	271
Infamous Downwind Turn	273
Where Did the Extra Kinetic Energy Come From?	274
Conclusions	276
References	276

9 Glide Performance — 277

Introduction	277
Banked or Unbanked Glides in "Exact" Theory, No Wind (Cases 1 and 2)	279
Gliding Equations of Motion (Exact)	279
Formulas for Flight Path Angle γ, Rate of Descent $\dot{h}$, and Turn Radius R, Case 1 (Exact)	280
More Facts About Helical Flight Paths	280
Formulas for Speed and Angle for Best Glide, V_{bg} and γ_{bg} (Exact)	282
Example 9.1	284
Formula for Speed for Minimum Descent Rate V_{md} (Exact)	287
Example 9.2	288
"Instant" Drag Polar Determination from Best Glide Information (Exact)	288
Banked or Unbanked Glides, Small Path Angle Approximation, No Wind (Cases 3 and 4)	289
Approximating the Force Equations (Small γ)	290
Approximating the "Exact" Solutions for γ, $\dot{h}$, and R (Small γ)	291
Approximations for V_{bg} and γ_{bg} (Small γ)	292
Approximations for V_{md} and $\dot{h}_{md}$ (Small γ)	293
Example 9.3	293

Engine Out (Return-to-Airport) Maneuver (Exact)	294
Example 9.4	296
Application: Engine-Out (Return-to-Airport) Maneuver	296
Segment 1–2, Takeoff Roll	299
Segment 2–3, Liftoff to 50 ft	299
Segment 3–4, 50 ft to 500 ft, Engine Failure	299
Segment 4–5, Getting Ready to Return to Airport	299
Zooming	300
Segment 5–6, Optimal 180-deg Turn	301
Segment 6–7, Glide Toward Runway	301
Return-to-Airport Summary	302
Speed in a Terminal Velocity Dive, V_T	303
Horizontal Wind Effects (No Updraft or Downdraft) on Wings-Level Glides (Case 8)	303
"Exact" Bootstrap Formula for $\sin(\gamma_g)$	308
Example 9.5	308
Graphical Approach	308
Formula Approach	310
Weight or Altitude Effects on γ_{bgw} in Direct Headwinds or Tailwinds	312
Downdraft and Updraft Effects on Wings-Level Glides (Small γ, Still Case 8)	314
Conclusions	315
References	316

10   Cruise and Partial-Throttle Performance   317

Introduction	317
The Bootstrap Approach Partial-Throttle Model	319
Definition of Bootstrap Power Setting Parameter Π	319
New Wine from (Mostly) Old Bootstrap Bottles	320
Example 10.1	321
Beyond the Propeller Polar	323
Brake Specific Fuel Consumption Rate	326
Calculation of GAMA-Format Cruise Performance Table	327
Example 10.2	328
Propeller Charts $C_P(J)$, $C_T(J)$, $\eta(J)$	328
Specific Endurance, Range, and V_{be}, V_{br}	330
Bootstrap Cruise Performance Table	332
Example 10.3	335
Cruise Performance Scaling Rules	336
Partial-Throttle Operation at Given Speed, Turn Rate, and Climb or Descent Rate	338
Example 10.4	338
Partial-Throttle Descent at Given Speed Along Given Approach Path	339
Example 10.5	339
Partial-Throttle Absolute Ceilings	339
Example 10.6	341
Conclusions	342
References	343

11 Takeoff Performance 345

Introduction 345
Takeoff Phases 346
    Cessna 172 Sample Takeoff Numbers 346
    Cessna 182 Sample Takeoff Numbers 347
Takeoff Forces 348
    Thrust, T 348
    Drag, D 351
        Parasite Drag 351
        Induced Drag 351
        Contaminant Drag 353
    Example 11.1 354
    Example 11.2 354
    Lift, L 355
    Weight, W 356
    Runway Reaction, N 356
    Rolling Friction, F_f 356
    Tire and Wheel Runup Neglected 356
    Net Takeoff Force (from Start to Rotation) 357
    Graphics Interlude: Takeoff Forces Parallel to Aircraft Motion 357
Physics Interlude: Dynamics When Force Depends on Speed, $F = F(V)$ 355
Application: Takeoff Phase 1, Base Case, Fixed Pitch, No Wind, No Contamination 361
    Question #1: What is $t(V)$ in general? What is $t(V_R)$ in particular? 361
    Question #2: What is $V(t)$ in general? 362
    Question #3: What is $a(t)$ in general? 362
    Question #4: What is $x(t)$ in general? 362
    Question #5: What is displacement $x(V)$ in general? 362
    Question #6: What is air speed $V(x)$ in general? 363
"Constant Average Force" Approximation 363
Calculation of Lengths of Takeoff Phases 2 and 3 364
Application: Takeoff Phase 1, Base Case, Fixed Pitch, Headwind, No Contamination 365
    Distance to Rotation in Wind, d_{Rw} 366
    Time to Rotation in Wind, t_{Rw} 368
Kinematics Under Uniform Acceleration 368
    Important Uniform Acceleration Relations 369
    The "70/50" Takeoff Rule, No Wind 369
    The "70/50" Takeoff Rule With Wind 370
Practical Perturbation Approach for Distance-to-Rotation Calculations 371
    Derivation of Perturbation Approach 372
    Example 11.3 374
    Example 11.4 374
When to Take Off Uphill Into the Wind Rather Than Downhill With the Wind 375
    Derivation of Formula for Break-Even Headwind 375
    The Break-Even Headwind Rule 376
    Approximations to Break-Even Headwind Formula 376

Example 11.5	377
Is the Optimal Break-Even Headwind Takeoff Good Enough?	377
Example 11.6	378
A "Paperless Office" Version of the Previous Calculation	378
Takeoff Power Parameter	379
Relation Between TOPP and Exponent α in $(w/\sigma)^\alpha$	380
Proper Order of Approximations	381
Derivation of Expression for d_R(TOPP)	382
Example 11.7	383
Example 11.8	385
Backengineering TOPP From Experimental or POH Takeoff Values	385
Conclusions	387
References	388

12   Landing Performance   389

Introduction	389
First Segment, d_{LA1}, from 50 ft to Just Above the Runway	391
Second Segment, d_{LA2}, from Just Above the Runway to Touchdown	000
Neglect of a Balloon and Flare	394
Third Segment, d_{LG1}, Wheel Spin-Up for 1 s	395
Experimental Determination of Wheel Moment of Inertia	396
Estimate of Aerodynamic Braking	396
Correction in Case of Runway Slope	397
Fourth Segment, d_{LG2}, Rotation for 2 s to All Wheels On the Runway	398
Fifth Segment, d_{LG3}, Hard Braking to a Full Stop (Air Speed V_{hw})	400
Braking Friction and Its Coefficients μ_B	400
How Much Load is on the Various Wheels?	402
Tag End of the Story	404
"Constant Average Force" Approximation for the Braked Landing Ground Roll	405
Wind Effect on Landing Ground Roll: Detailed and Constant Average Force Methods	407
Perturbation Approach for Landing	408
Example 12.1	409
Example 12.2	409
When to Land Downhill with a Headwind Instead of Uphill with a Tailwind	410
The Break-Even Headwind Rule for Landing	410
Approximation to Break-Even Landing Headwind Formula	410
How Gross Weight and Density Altitude Affect the Landing Roll	411
Why Landing Roll Distance d_{LG} is Proportional to W/σ	412
Conclusions	415
References	415

Appendix A   How Big Are the Error Bars?   417

Introduction	417
Sample Measurement Job	418

The Short (Necessary but Insufficient) Course on Error Theory	420
The Normal Distribution $N(\mu, \sigma)$	421
Important Questions	424
Example A.1	427
Confidence Intervals for Estimation of Mean with Dispersion Measured Through the Range	429
Estimates of Dispersion from Several Shorter Series of Measurements	431
Indirect Measurements	432
Simple Linear Regression	434
The Major Curve Fit Parameters (m and b)	437
Standard Error (S_e)	437
Coefficient of Determination (R^2)	438
Number of Observations and Degrees of Freedom	438
Scatter in the Estimates of Slope (m) and of Intercept (b)	438
Conclusions	440
References	440
Appendix B Bootstrap Approach Inputs and Outputs	**441**
Bootstrap Approach Level 1	441
Inputs for Fixed-Pitch Airplanes (Bootstrap Data Plate)	441
Inputs for Constant-Speed Airplanes (Bootstrap Data Plate)	442
Outputs for Either Fixed-Pitch or Constant-Speed Airplanes	442
Bootstrap Approach Level 2	443
Inputs for Fixed-Pitch Airplanes	443
Inputs for Constant-Speed Airplanes	443
Outputs for Fixed-Pitch or Constant-Speed Airplanes	444
Bootstrap Approach Level 3	444
Inputs for Fixed-Pitch and Constant-Speed Airplanes	444
Outputs for Fixed-Pitch and Constant-Speed Airplanes	445
Calibrated Instruments Required	445
Conclusions	446
Appendix C A Short List of Integrals	**447**
Appendix D Numerical Integration	**451**
Appendix E Derivation of Propeller Master Equation	**453**
Appendix F Flight Test for Drag Parameters	**459**
Appendix G Determination of Propeller Polar Constants m and b	**465**

Preface

While the primary purpose of this book is to introduce pilots, aviation students, and other airplane enthusiasts to quantitative methods for predicting the performance of small fixed-wing aircraft, it may also be of interest to aeronautical engineers and other aviation managers and professionals. Many of the techniques I shall describe, though fairly elementary, have heretofore only been available in research journals; some of them, not even there.

A writer steals a march on his readers' attention when he takes up an inherently interesting subject. And what could be more interesting (rhetorical question!) than airplanes and how they work? The young Wright brothers were fascinated by the toy helicopter (a design by Pénaud) their father brought home. Their intrigue expanded into the precocious realization that a scaled-up version of that toy, though carefully crafted, would not function properly.

Generations of children and adults—and especially aviators, somewhere in between—have been similarly fascinated by flight. Flight of birds, airplanes, even the erratic fall of leaves, steady tug of kites, leeward drift of balloons, and the explosive liftoff of rocket ships. Movement in that third dimension, unsupported by scaffolding or sky hook, is of interest to almost everyone, and certainly to all those with romance in their souls.

Back down to earth. This is a moderate-sized book about performance of relatively small propeller-driven aircraft. Still, a large subject. Our goal has been to clearly lay out for the reader the details, to show him or her how to answer such questions as

- If the airplane is cruising at 100 kn at 10,000 ft and the pilot decides to pull back the stick, slowing to 80 kn, and banks 20 deg, how will the airplane then move? Will it go up? Down? With what rate of climb or descent? What are the details of flight path, engine speed, turn rate?
- The airplane can take off—level concrete, 3000 lb, no wind, at sea level—in 1000 ft. What then is its distance to lift off on turf, sloped at 3 deg, when it

weighs 2500 lbf, with a 10-kn headwind, at 5000 ft? Would it be better to take off down hill?

Answers to realistic, full-fledged performance questions. Practical questions to which pilots and students and operators need answers. Historically (or so it seems to this writer), aircraft performance treatises have not been able to provide easy and accurate answers to such a full range of performance queries. That we will be able to do so in these pages is primarily due to a single conceptual advance: the bootstrap approach.

The bootstrap approach is a parametric procedure based on the demonstrably close linearity of the propeller polar diagram. Full explication of the bootstrap approach, including its extensions to maneuvering flight, partial-throttle operations, and takeoff and landing, appears here for the first time, as does explanation of the general aviation general propeller chart, which makes similar calculations practicable for constant-speed propeller airplanes.

On the other hand, we do not cover the entire aviation waterfront. No seaplanes, for instance. Little on biplanes (though those are not much stretch), and no jets. Only normally aspirated reciprocating engines, though it is not hard to rig a torque/power dropoff factor to handle turbocharged engines; some bootstrap users have done so. Subsonic speeds. None of the peculiarities of flying multiengine aircraft with an engine inoperative. And, with a few exceptions, we treat only quasi-steady-state flight. No aerobatics. Nor do stability and control issues, except for the need to fly trimmed and coordinated, appear here.

Because an airplane's performance depends on many factors—speed, weight, air density, throttle setting, flaps/gear configuration, bank angle—we can only cover the performance subject by using mathematical formulas. Only formulas (usually made more understandable and immediate by displaying corresponding graphs) can provide numerical answers to such a broad array of specific performance questions. Still, the mathematical level is only moderate. Acquaintance with calculus is occasionally assumed, but college algebra handles nine-tenths of the topics. Granted a good understanding of graphs, including slope and critical points, also helps. A few specialized topics—for example, the theory of measurement errors—are relegated to appendices.

What about computers? Most readers of technical bent, nowadays, have personal computers and an electronic spreadsheet program (e.g., 1-2-3, Quattro Pro, Excel). Data behind almost all of our graphs, and behind many numerical examples, were generated by such programs. Most of those spreadsheet templates will be available, separately, on diskette. The templates are simple—no macros, menus, or other specialized techniques—hence, readily translatable from one spreadsheet program to another. This book itself includes no computer code whatsoever. The computer is an invaluable tool when trial-and-error solutions or lengthy calculations are required, but it is only a tool. None of our attempts to transfer ideas requires the reader to have a computer handy.

The book is divided into two Parts. Part I, Fundamentals of Aeronautic Science, the first six chapters, gives the scientific and engineering foundations needed to fully understand the main performance subject. Readers mainly interested in performance "how to," and readers already having a good foundation in aerodynamics—especially one that includes propeller theory—can get through that first part quite quickly, or refer back to it from time to time as needed. Part II, Practical Airplane Performance, provides details of the performance subject proper. That is where most readers will spend the bulk of their time. Here are chapter details along with descriptions of occasional new features

Part I opens with Chapter 1, The Atmosphere. This treats Earth's gaseous envelope in somewhat greater detail than is common. It includes, for example, humidity that affects thin air to the detriment of airfoil, propeller, and especially engine performance. We also take up techniques for reducing flight test data to obtain geometric (actually geopotential) vertical intervals from corresponding pressure altitudes.

Chapter 2, Atmospheric Aircraft Instruments, is devoted mainly to air-speed indicators and altimeters. We give a new at-altitude method for calibrating the airspeed indicator using a global positioning system unit. On the altimeter side, we cover the altimeter-setting formula in detail and give two methods for determining static pressure "position" error. By the end of the chapter, the reader will be able to construct a test-day atmospheric model (one using a constant but nonstandard temperature "lapse" rate and a nonstandard sea level pressure) from data gathered during flight tests and weather-reporting stations.

Chapter 3, Aircraft Performance Preliminaries, presents several topics from physics, primarily mechanics, that bear on aircraft performance. Included topics are units and dimensions, scalars and vectors, linear and circular kinematics and dynamics, frames of reference, and pseudoforces. The chapter ends with realistic but not overwhelming quasi-static equations of motion for the airplane.

Chapter 4, Aerodynamic Force, provides what the reader will need to know about lift and drag, both parasite drag and induced drag. It focuses on building up the aerodynamics of practical wings (flapped or unflapped) from the "section" characteristics of two-dimensional profiles. It has less on the history of various wing profiles than is usual, and more on ground effect and tail force.

Chapter 5, The Engine, aims at giving the reader a modicum of both theoretical and practical understanding of gasoline-powered normally aspirated spark ignition reciprocating engines. A major aim is to present a full and clear explanation of both the sea level and the at-altitude sides of the engine performance chart. Various engine controls and instruments are tied to engine performance.

Chapter 6, Propeller Thrust, contains quite a bit of theory. In part that is because the bootstrap approach is founded in propeller theory; also because treatment of the propeller has often been a weak point in earlier books on airplane performance. Sample calculations towards propeller thrust and power coefficients are performed using first the momentum theory, then the blade element theory, and finally the two theories combined. A new and single "master" propeller equation is presented that can be solved using nothing more than a desktop spreadsheet program containing an iteration (BackSolver, SolveFor, or Solver) facility. This allows the reader, should he or she so desire, to obtain practical propeller charts *de novo* from the propeller's geometry and its aerodynamic coefficient functions. Several correction terms and techniques are considered; some are applied. On the constant-speed side, this chapter also includes a new general aviation general propeller chart. That is a recasting, using general aviation scale propeller data, of the Boeing general propeller chart developed during World War II. That early chart was developed with data for much larger (10-ft diam) propellers turned by much more powerful engines and does not work for our smaller general aviation propellers. Chapter 6 concludes the necessary background.

Part II opens with Chapter 7, Introduction to the Bootstrap Approach. This gets us (at long last!) on the path to airplane performance. For clarity and focus, this first chapter on the subject is restricted to wings-level flight, either at full throttle or gliding. Wind effects are considered here, as is the problem of absolute ceiling. The bootstrap approach is derived from its underlying assumptions (mainly, linearity of the propeller polar diagram). Bootstrap formulas for various V speeds, thrust, drag, and rates or angles of climb or descent at any speed (and at any weight, any altitude), are also derived. The nine numbers making up an airplane's "bootstrap data plate" are discussed in detail, as are the eight so-called "composite" bootstrap parameters, combinations of items from the bootstrap data plate, that are merely aids to rapid calculation.

Chapter 8, Maneuvering Performance, treats turns. Extending the bootstrap approach to allow banked flight is easy. Our previous equations of motion are also brought in. Turns of most types—coordinated, slipped, skidded—are treated and the (climbing or descending) flight path helix, it turns out, is not quite as simple as one might think. The content of a "banked ceiling" is introduced and exploited. A new "steady maneuvering chart" is described, with many examples (including that of a fatal high-altitude turn and how it could easily have been avoided). The "infamous downwind turn" paradox is disposed of, and the question of where might the extra kinetic energy come from, in the course of such a maneuver, is subjugated by calculation.

Chapter 9, Glide Performance, is not merely a reprise of the wings-level, small-flight-path treatment in Chapter 7. In some cases, our gliding equations of motion allow for exact solutions. In others—special cases of headwinds, tailwinds, updrafts, downdrafts—reasonable approximations are brought to bear. The "instant drag polar" formulas used earlier are here derived. Quite a bit of attention is paid to the engine-out-on-takeoff or return-to-airport maneuver; it is shown that the optimum bank angle for turnaround is (for general aviation aircraft) very slightly greater than 45 deg. Several special topics, including zooms and best glide speeds in wind, dot the landscape of this chapter.

Chapter 10, Cruise and Partial-Throttle Performance, further extends the bootstrap approach. But unlike the very straightforward case of maneuvering, logic behind the partial-throttle extension is somewhat subtle. Partial-throttle operations are treated as though the engine were replaced by a collection of "virtual" engines, derated engines, of diminished size. A new performance flight test—merely cruising level at several speeds—is described. From that cruise data, a strange propeller chart is obtained; it forms the basis for a new level cruise performance table that adds considerably to the standard GAMA-format one. Ours includes columns for throttle setting, thrust (therefore, drag), and (installed) propeller efficiency. (Installed propeller efficiency is about as close as one gets in airplane performance studies to the Philosopher's Stone!) Also the driving independent variable (once aircraft weight and density altitude have been decided upon) is air speed; in the GAMA table it is revolutions per minute. Moreover, the bootstrap cruise table includes very low air speeds, down through speeds for best range and best endurance. The GAMA-format table does not include those low speeds; in fact we show that it cannot. Cruise performance scaling rules, allowing one to fill out an extensive cruise performance table from data gathered at only one gross weight and altitude, are derived and explained. That time-saving technique should be especially of interest to airplane manufacturers. Examples show how a desired flight path at given weight and altitude (level, climbing, or descending; wings level or turning) can be used to find the requisite throttle setting. The chapter ends by considering some peculiar air-speed relations connected with partial-throttle absolute ceilings.

Chapter 11, Takeoff Performance, is essentially an extensive force analysis. The new forces that come into play during takeoff make for additional complications in this further extension of the bootstrap approach. Effects of different runway surface types, runway slopes, runway contamination, and wind are all considered. Ground effect is assumed throughout. Several different levels of theoretical development—more-or-less exact integrations, a constant average force assumption, a perturbation approach, and finally a lumped parametric approach, the takeoff power parameter—are considered and exemplified. A few special problems are considered. Among them is the common 70/50 takeoff rule (corrected for possible headwinds or tailwinds) and the question of when to take off uphill into the wind rather than downhill with it.

Chapter 12, Landing Performance, analyzes landings from an applied forces point of view very similarly to our treatment of takeoff. Thrust no longer, but now braking friction is added to the usual rolling friction. Again, wind, slope, and contamination effects are demonstrated. And while an integration (elementary differential equation) approach, as well as the constant average force approximation and the perturbation approximation, are treated, in this simpler case there is no need for an analog to the takeoff power parameter.

Various specialized topics and derivations make up several appendices.

It is my pleasure to thank the small band of hardy souls who helped teach me this subject and helped me put this book together. Paul Soderlind got me started with both flying and figuring the flying. Arthur Daniel was a never-ending source of encouragement; he and his wife Bonnie also provided, at their ranch near Fishtail, Montana, a convenient hideout. The late Dick Hiscocks, friendly correspondent (and lead designer of the redoubtable de Havilland Beaver), was the first aeronautical engineer to recognize the potential of the bootstrap approach. Moreover, Dick was willing to say so, out loud; I owe his memory a special debt of gratitude. Along logistical lines, the reference staff at the Billings Public Library did their best (and it was quite good) to round up arcane aerodynamics books through Interlibrary Loan. I was not having much luck getting published until Frederick Smetana put me in touch with AIAA Editor-in-Chief Paul Zarchan. From that point, Paul (and later, Megan Scheidt and Jodi Glasscock) drove me along toward completion of the draft. Andrew Bauer was the one professional aeronautical type I befriended who offered me advice, especially on propeller theory, whenever I asked for it. Sometimes even when I did not. Last and most of all, my wife Mary Ann put up with me—and with the somewhat protracted period during which I did little useful (read "remunerative") work—even though she disdains formulas. I thank them all for their support and contributions in furthering the writing and production of this book.

John T. Lowry
Billings, Montana
July 1999

Nomenclature

A	wing aspect ratio, span^2/area; area; air; aft
a	axial interference factor; speed of sound; acceleration; lift coefficient slope
a'	angular interference factor
B	wing span; cylinder bore
b	linearized propeller polar intercept; wing span
C	altitude engine power dropoff parameter
CI	confidence interval
C_D	drag coefficient
C_{Di}	induced drag coefficient
C_{D0}	parasite drag coefficient
C_L	lift coefficient
C_l	section lift coefficient
$C_{L\max}$	maximum lift coefficient
C_M	pitching moment coefficient
C_P	propeller power coefficient
C_T	propeller thrust coefficient
c	chord
c_p	specific heat at constant pressure
c_v	specific heat at constant volume
D	drag
D_i	induced drag
D_P	parasite drag
d	propeller diameter; displacement
E	composite bootstrap parameter
e	Oswald airplane efficiency factor; base of natural logarithms
e_S	saturation vapor pressure
F	composite bootstrap parameter; force; Prandtl momentum loss factor; fuel; forward

Nomenclature

f	flaps factor
G	composite bootstrap parameter
g	acceleration of gravity, 32.174 ft/s^2; ground speed
H	composite bootstrap parameter
hp	horsepower
h	altitude
h_p	pressure altitude
h_T	temperature altitude
h_ρ	density altitude
$\dot{h}$	rate of climb or descent
I	moment of inertia
J	propeller advance ratio
K	composite bootstrap parameter; Munk span factor
k	ratio of specific heats, c_p/c_v
L	lift; torque
l	length
M	engine torque; mass; Mach number
M_B	base engine torque
m	linearized propeller polar slope; slope; mass
mac	mean aerodynamic chord
mgc	mean geometric chord
N	revolutions per minute; normal force; number of tires
n	propeller revolutions per second; load factor; number
P	power; pressure
p	pressure; pitch
Q	composite bootstrap parameter; heat
q	dynamic pressure
R	radius; composite bootstrap parameter; gas constant; correlation coefficient
Re	Reynolds number
r	air composition ratio; altimeter reading; radius; range; compression ratio
S	reference wing area; Sutherland's constant; entropy; speed; cylinder stroke; standard deviation
SDEF	slowdown efficiency factor
s	altimeter setting
T	thrust; temperature; wing profile thickness
t	time; thickness
U	composite bootstrap parameter
V	air speed; volume
V_a	sound speed
V_w	wind speed
v	speed; vapor

W	gross aircraft weight; relative wind speed
W_0	standard aircraft weight
w	tire width; weight fraction
X	power adjustment factor
x	position; relative propeller blade station
Y	camber line height
y	position
Z	ratio of fuselage diameter to propeller diameter
α	angle of attack; temperature lapse rate; angular acceleration
β	propeller blade angle
Γ	circulation
γ	flight-path angle; ratio of specific heats
Δ	difference
ΔS	stall speed buffer speed
ΔV_C	pressure correction
δ	relative pressure; tire deflection
δ_f	flaps deflection angle
ϵ	ratio of molecular weights of water and air; excess angle; recovery factor; downwash angle
η	propeller efficiency; viscosity correction factor
θ	relative temperature; attitude angle; angle
Λ	sweepback factor
λ	propeller coefficient
μ	rolling friction or braking coefficient; dynamic viscosity
ν	kinematic viscosity
Π	bootstrap power-setting parameter
π	power-setting parameter
ρ	atmospheric density; radius of curvature
σ	relative atmospheric density; ground effect parameter; local solidity; standard deviation
τ	flaps effectiveness factor; wing tip correction factor
Φ	engine torque/power dropoff factor
ϕ	aircraft bank angle; propeller flow angle
φ	angle
Ω	angular speed
ω	aircraft turning rate; angular speed
′ (prime)	section quantity; per unit span
″ (doubleprime)	further modification

Subscripts and superscripts

A	finite aspect ratio
AC	absolute ceiling

Nomenclature

a	available; air
ac	aerodynamic center
av	available
α	confidence interval complement
B	base case (mean sea level, maximum gross weight); braking; body
BE	break even; blade element
be	best endurance
bg	best glide
br	best range
bta	best turnaround
C	contaminant; calibrated; cylinder clearance
c.g.	center of gravity
c	calibrated; centripetal; calm wind
D	drag; cylinder displacement
d	drag; downdraft
e	error
f	flaps; field; friction; final
G	ground
GE	ground effect
GR	ground roll
g	ground
gl	glide
h	hard case; headwind
I	in ground effect; indicated
i	induced; indicated; initial
L	lift; lower; landing; leading
LO	liftoff
l	lift
le	leading edge
M	main wheel; maximum
MT	momentum transfer
m	pitching moment; minimum
max	maximum for level flight
max ω	maximum turning rate
min	minimum for level flight
min R	minimum turning radius
md	minimum descent (rate)
N	nose wheel
n	number
n lim	load factor limit
O	out of ground effect
P	parasite; planing; power; pusher
p	parasite; piston; airplane; pressure; power

R	rotation
r	required
re	required
S	stall; specific
SC	service ceiling
s	saturation
T	thrust; total; thermal; terminal; trailing; tractor; tail
TD	touchdown
t	thrust; tailwind
te	trailing edge
U	upper
u	upper; updraft
V	volume
v	volume
W	wind; wing
WB	weight and balance
w	wind
X	power adjustment factor
x	best climb angle
xs	excess
y	best climb rate; position
0	base; rated; standard
∞	section quantity
$\oplus$	Earth

PART I. FUNDAMENTALS OF AERONAUTIC SCIENCE

1

The Atmosphere

Introduction

When not up flying, we scurry around the bottom of an ocean of air. By virtue of its protective and oxidizing chemistry, its relative transparency, and its moderate density, we enjoy or endure such atmospheric phenomena as gentle katabatic zephyrs, hurricane winds, soft rain, and seasonal lags. This blanket of air is interesting and varied, demanding its own peculiar study—meteorology—and its own texts. It's a thick blanket, of total mass 3.495×10^{17} slugs (1.1245×10^{19} lb), about 1/275 the mass of the oceans and close to one-millionth the total mass of the Earth. The atmosphere evolved, as did the oceans and the Earth's continental crust, as volcanic products.

In our study of how airplanes perform—how they claw their way up from this ocean floor into its higher reaches to cruise and maneuver, then get back down again—we usually simplify the atmosphere into a relatively static mask of its full dynamic face. Airmen cannot afford to be so eclectic. They must be ready for the real atmosphere with its occasional exceptional occurrence: a mountain wave, say, or pressure reduction caused by winds through passes, or the strikingly varied forms precipitation takes.

Airplane performance is intimately connected to the Earth's atmosphere in two ways. The word "airplane" mentions air, saying atmospheric air is the medium in which flight takes place. Though we care much more about the airplane than the air, aerodynamics shows that the *mutual* effects between air and airplane must be taken into account. Air swoops up a little just before meeting the wing and is given a downwash as the wing passes. Air swirls from the propeller. Vortices roll up off the wings tips. Aerodynamic drag ultimately heats both the airplane surfaces and the air they pass through. Those air dynamics are necessary. No air, no airplane.

Still, what the air does to the airplane—and what it allows the airplane to do—is much the more interesting story. Takeoff rolls are longer when the air is thinner. Lift is enhanced by "ground effect," air constrained by the surface. We must pay attention to winds. In this chapter we will worry over those aspects of Earth's atmosphere important to aircraft flight.

A second reason for learning about the atmosphere is narrowly technical, in the category of necessary distracting details, and will be deferred to a second chapter. Several of our cockpit instruments—altimeter, air-speed indicator, outside air thermometer, VSI—have atmospheric inputs. Those instruments come to us (no extra charge!) with an assortment of errors and artifacts. We must understand their functioning to conduct safe and efficient flight operations (e.g., stay above stall speed, get down to speed for best range V_{br}) and also because their pointers show us the way to *other* atmospheric variables we cannot know directly (e.g., true altitude differences, atmospheric density).

Here is the plan. We will look at some brute facts about the Earth's atmosphere—especially the lower 11,000 m or so, the troposphere—as regards vertical distribution of the most important atmospheric variables (temperature, pressure and density). We will explore some of the physical principles that govern air's behavior. We will examine the model international standard atmosphere (ISA) on which many of our instruments and descriptive concepts (e.g., pressure altitude, density altitude) are based. Secondary variables—humidity, viscosity, the speed of sound—will be added to the collection. We will put together what we have learned about the atmosphere to come up with schemes for figuring out, approximately, what "today's" actual atmosphere is like. Conversely, we will learn how to reduce aircraft performance data taken somewhere in the actual atmosphere to equivalent information about the airplane's behavior somewhere else in the benchmark standard atmosphere. It will be a lengthy trip.

For example, inferring vertical atmospheric intervals from cockpit indications is fairly indirect. Say you are running glide tests to find your airplane's drag parameters (parasite drag coefficient C_{D0} and airplane efficiency factor e). Even using your airport's correct current altimeter setting cranked into your precise altimeter, it seldom works to glide between an indicated 5000 ft and 4500 ft, say, and then simply tell your formulas you have floated down 500 ft. With a little extra care you can get a considerably more accurate estimate of the tapeline vertical distance. We will discuss how to take that care and get those accurate numbers.

Structure and Composition of the Earth's Atmosphere

In large-scale structure the Earth's atmosphere is a sequence of thick layers. Each layer has almost the same chemical composition as any other, but each layer has

Table 1.1 Temperature layer structure of Earth's atmosphere

Layer no.	T gradient, °F/ft	Base T, °F	Base altitude, ft	Base altitude, n mile
1	−0.00356616	59.0	0	0.00
2	0.0	−69.7	36,089	5.93
3	0.00054864	−69.7	65,617	10.78
4	0.00153619	−48.1	104,987	17.25
5	0.0	27.5	154,199	25.34
6	−0.00109728	27.5	170,604	28.03
7	−0.00219456	−5.5	200,131	32.88
8	0.0	−134.5	259,186	42.59

its own approximately constant temperature gradient (negative of its *lapse rate*). Table 1.1 gives a fuller story.

There is still some atmosphere above the eighth layer. Solar dissociation of oxygen starts at about 80 km (43.1 n mile) during the day, about 105 km (56.6 n mile) at night, and is virtually complete by altitude 120 km (64.7 n mile). Above 120 km, oxygen exists almost entirely in atomic form. Other than the split oxygen molecules, even this very rarefied atmosphere has approximately the same composition as dry dustless air at sea level.

Aircraft covered in this book cannot venture out of the lowest layer, the *troposphere*. Keep in mind that the nominal lapse rate 0.003566 F/ft is honored more in breach than observance, especially within 5000–10,000 ft of the surface. The lapse rate varies quite a bit and even (inversions) goes negative, so that temperature *increases* with altitude. (The word "lapse" implies "decreasing," hence the apparently contrary usage. It is a test.)

It is worth asking "Why does it (generally) get colder at higher altitudes?" And not so easy to answer; even a rough explanation involves at least four atmospheric facts:

1) The atmosphere is fairly transparent to sunlight. But the Earth is not and is heated by the sun. The Earth then reradiates at much longer, infrared (IR), wavelengths.[1]
2) The diatomic symmetrical molecules in air (e.g., N_2, O_2) absorb only very weakly in the IR region. But the less symmetrical triatomic molecules (CO_2 and H_2O, especially the latter) absorb quite strongly in that spectral region.[2]
3) Water vapor in the atmosphere, because it often comes from surface water or volcanoes, is more prevalent low in the atmosphere. Though the water content is quite variable with altitude and tends to layer.
4) Because of the hydrostatic relation, air pressure is higher low in the atmosphere.

Table 1.2 Major constituent molecular species in dry dust-free air

Constituent	Atoms/molecule	Molecules, %
Nitrogen	2	78.084
Oxygen	2	20.948
Argon	1	0.934
Carbon dioxide	3	0.031
Others	Weighted average 1.38	0.003
Total	Weighted average 1.99	100.000

So the idea is that lower level carbon dioxide and (especially) water vapor heat up, from the Earth, quite a bit. Higher up, however, much of the IR has already been absorbed by lower gas. Moreover, there is less such gas above the higher reaches, so that the high gas radiates out into space. That radiation is lost; the higher gas cools. The result, as we all know, is that it generally gets colder as we fly higher.

Molecular composition is much steadier than temperature. Tropospheric winds and vertical currents keep the gases well mixed, with composition as in Table 1.2. Though the true troposphere's composition may vary a fraction from this, with altitude and locality and season, we shall always ignore those tiny variations and aberrant trace molecules. Notice in particular that the constituent proportions given are those for dry dust-free air. We will later sometimes consider a small but variable humidity component. Table 1.2 gives the number of atoms in each molecular species as an aid to rationalizing the sometimes-needed ratio of specific heats, for air, as $\gamma = 1.40$.

The second layer, the *stratosphere*, is now routinely accessible to passenger jet aircraft. Not many clouds get up there; humidity is seldom a problem. Near the boundary between the two lower layers, the tropopause, fast but steady jet streams may be found. Composition of the stratosphere is essentially the same as that of the troposphere.

Elementary Atmospheric Physics

In modern college curricula, the behavior of gases is often part of a course in thermodynamics or statistical mechanics. Statistical mechanics has the advantage over thermodynamics that it is more concrete, lets us visualize gas molecules bumping into one another and into the walls of their enclosure. That goes a long way towards explaining why thermodynamics works the inscrutable way it does. Older unabashed thermodynamics, with its three laws and its Legendre transformations—switching variables around like peas in a bunko artist's shell game—

seems, on the contrary, almost theological in its formality. Derivations of thermodynamic relationships, to many, are often unsatisfying, barren.

But for the smattering of atmospheric thermal physics we need, the road from statistical mechanics to thermodynamics is far too tortuous to even look down. We will try to jolly up the subject with an occasional tangible mechanism, but some of the unmotivated substitution and exponent manipulation for which thermodynamics is infamous will remain. We warned you.

The three atmospheric variables of greatest importance to us are temperature T, pressure p, and density ρ. Now air (dry or moist) is very nearly an *ideal gas*, one in which each molecule moves independently of the others except for "occasional" well-localized molecular collisions. As such, at our relatively low pressures, it obeys the ideal gas law

$$p = Rg\rho T \qquad (1.1)$$

where p is ambient pressure (psf), R is the *specific gas constant* for air, g is the acceleration of gravity $32.174 \, \text{ft/s}^2$, ρ is mass density in slugs/ft^3, and T is temperature in absolute Fahrenheit (Rankine) units ($°R = °F + 459.7$). For dry air, with weighted average molecular weight (air) $= 28.966$, $R = 1545.5/$ mol wt (air) $= 1545.5/28.966 = 53.355$ ft/F. The 1545.5 figure is the *universal gas constant*, Boltzmann's constant k translated into engineering units.

The ideal gas law is very important because through it we discover the air's density under various conditions. Direct measurement of density is difficult: open an evacuated quart thermos in standard air and it then weighs only 1/25 oz more. Thin air. We often need to known the atmospheric density because important factors connected to airplane performance—lift, drag, thrust, engine power—depend on it. Rearranging Eq. (1.1), the standard MSL value of density for dry air is $\rho_0 = p_0/(RgT_0) = 2116.2/(53.355 \times 32.174 \times 518.7) = 0.002377 \, \text{slugs/ft}^3$, a value we use quite often.

There are other variables of interest. One of the most important is the velocity of the air. Wind is the horizontal component; updrafts and downdrafts make up the vertical component. Aviators specify wind by giving its speed with respect to the point of earth directly below and by giving the compass direction (sometimes magnetic, sometimes true) from which the wind is coming. That latter convention, by now too rusty to displace, sets vector equations on their ear; in using such relations the wind direction must often be sensibly reversed. We take note of updrafts and downdrafts to the extent that we avoid them during performance flight tests; they are very frequently the source of erroneous or unreproducible results. Turbulence is wind's higher frequency random component, in both directions, overlaid on relatively slowly shifting wind and mildly stochastic updrafts and downdrafts. We will conveniently assume our airframes hang together no matter what, so we ignore turbulence in this work. Let the structural folks and operating pilots worry about turbulence.

We will treat some additional variables later on, absolute or dynamic viscosity μ, kinematic viscosity v, the speed of sound a, and the amount of water vapor in the air, specified most often (in aviation) by citing the dewpoint temperature T_d. In a few cases—high speeds at which compressibility of the air is not entirely negligible, and to find out sonic speed itself so that we know we are indeed far subsonic—we need the ratio of specific heat of (dry) air at constant pressure, $c_p = 0.240$ Btu/(lbm°R), to its specific heat at constant volume, $c_v = 0.1715$ Btu/(lbm°R). This ratio is

$$\gamma \equiv \frac{c_p}{c_v} = 1.40 \tag{1.2}$$

The rationale for this figure is that air is mostly made up of dumbbell-shaped diatomic molecules (see Table 1.2) with 5 degrees-of-freedom (DoFs). There are 3 translational DoFs for the center-of-mass of the molecule (the only ones impacting the gas' temperature) and 2 additional rotational DoFs. A theoretical vibrational DoF is "frozen out" at the relatively low temperatures we use; the gas never (well, seldom) gets enough energy to rise above its vibrational ground state. Also absent, for slightly different reasons, is a theoretical rotational DoF about the axis of the diatomic dumbbell. What could it mean to say—because they were never a rigid body in the first place—that a swarm of bees or gnats was rotating? At any rate, elementary gas kinetic theory, or even thermodynamics, tells us that

$$\gamma = 1 + \frac{2}{\text{(active number of DoFs per molecule)}} \tag{1.3}$$

And for dry air, that number in the denominator is five. Above room temperature (about 500°R), additional DoFs actually do come into play and γ begins to decrease. At 1000°R, $\gamma = 1.38$ for air. Further atmospheric facts will be brought up as needed.

To better understand how we might get information on the vertical distribution of important variables pressure, temperature, and density, a little fluid physics is in order. Assume the atmosphere is in equilibrium, no winds or currents. At a given level, then, pressure is the same in all directions. Consider a small right circular cylinder, height dh, axis vertical. The additional pressure on the bottom face, above that on the top face, dp, is solely due to the weight of air in the cylinder: $dp = -\rho g dh$. The negative sign appears because pressure goes *down* as altitude goes *up*. This simple hydrostatic relation, together with the ideal gas law, makes it possible to take any reasonable given vertical distribution of any one of the three variables (though in most cases this will be temperature) and compute vertical distributions of the other two. Three variables, three relations; an almost done deal. To flesh the theoretical solution into usable facts, all we need is one distribution, real or assumed. One of the latter is what the model ISA gives us.

International Standard Atmosphere

The new and fuller name is the International Civil Aeronautical Organization Standard Atmosphere, and its full description would include all the layers given in Table 1.1. There are other standard atmospheres, too; the U.S. Department of Defense has finer-tuned versions for summer and winter, for tropics and the arctic, and in between. All we care about here is the ISA's lower layer, the troposphere, which is based on four major parameters: mean sea level (MSL) standard temperature $T_0 = 59°F$, standard MSL pressure $p_0 = 2116.2$ psf $= 29.921$ in. Hg), standard constant temperature lapse rate $\alpha = 0.003566$ F/ft (negative of the vertical temperature gradient), and standard acceleration of gravity $g = 32.174$ ft/s² downwards. Table 1.3 gives these defining numbers in various units.

The ISA does not pretend to accurately model the actual atmosphere above any given place on Earth at any given time, but it is very useful in at least two indirect ways. It is the benchmark to which aircraft performance numbers are referred. When we say our airplane's best rate of climb is 900 ft/min at 5000 ft, we mean at 5000 ft in the ISA. When we need a good description of the actual "here and now atmosphere," say at remote points where we cannot directly measure temperature and pressure, we assume the same mathematical *structure* as the standard atmosphere obtains, only with today's particular (and almost certainly different) values for "standard" temperature, pressure, and lapse rate: T_0', p_0' and α'. For our practical aeronautic purposes, we always take the acceleration of gravity as constant even though we know it is not quite constant. That minor prevarication gives us so-called "geopotential altitudes" instead of geometric ones. Because we actually care more about the amount of work it takes to lift an airplane from one level to another than we do about tapeline distances, the geopotential definition is not only simpler but fits our purposes better. To flesh out both these ISA uses we now delve into it more deeply, asking and answering such questions as "What are

Table 1.3 Defining parameters for ISA troposphere in British engineering and metric units

	T_0	p_0	α	g	Tropopause
British	59	2116.2	0.003566	32.174	36,089
	°F	psf	F/ft	ft/s²	ft
	518.7	29.921			
	°R	in. Hg			
Metric	15	101325	0.00650	9.80665	11,000
	°C	N/m²	C/m	m/s²	m
	288.15	760	0.001981		
	°K	mm Hg	C/ft		

the temperature and pressure at any altitude h above mean sea level?" "What about the air density at h?" Then later, "The speed of sound?" "Viscosity?"

The ISA lapse rate defines the vertical temperature distribution. At altitude h feet, standard temperature is given by

$$T_S(h) \equiv T_0 - \alpha h = 59 - 0.003566h \; °F \qquad (1.4)$$

For most computational purposes, temperatures will need to be converted to Rankine by adding 459.7.

Using the ideal gas law, Eq. (1.1), to eliminate ρ in terms of p and T, then using the first portion of Eq. (1.4) for the temperature distribution, the hydrostatic relation becomes the differential form

$$\frac{dp}{p} = -\frac{dh}{R(T_0 - \alpha h)} \qquad (1.5)$$

Integration then results in the pressure distribution

$$p_S(h) = p_0 \left(1 - \frac{\alpha h}{T_0}\right)^{1/\alpha R} = 2116.2 \left(1 - \frac{h}{145{,}457}\right)^{5.25635} \; \text{psf} \qquad (1.6)$$

It is just a matter of using the ideal gas law once again to then get

$$\rho_S(h) = \rho_0 \left(1 - \frac{\alpha h}{T_0}\right)^{(1-\alpha R)/\alpha R} = 0.002377 \left(1 - \frac{h}{145{,}457}\right)^{4.25635} \; \text{slug/ft}^3 \qquad (1.7)$$

Recall that, while the above might make it look as though temperature has a favored role in description of the ISA, or of the real atmosphere, that is not so.

The ratios of values of T, p, and ρ to their MSL standard values are known as θ (theta), δ (delta), and σ (sigma), respectively. In general, the value of any such ratio depends on the peculiarities of the actual atmosphere at the time. But because ISA values depend only on altitude (see Fig. 1.1), so do the standard ratios:

$$\theta_S(h) \equiv \frac{T_S}{T_0} = 1 - \frac{\alpha h}{T_0} = 1 - \frac{h}{145{,}457} \qquad (1.8)$$

$$\delta_S(h) \equiv \frac{p_S}{p_0} = \left(1 - \frac{\alpha h}{T_0}\right)^{1/\alpha R} = \left(1 - \frac{h}{145{,}457}\right)^{5.25635} \qquad (1.9)$$

$$\sigma_S(h) \equiv \frac{\rho_S}{\rho_0} = \left(1 - \frac{\alpha h}{T_0}\right)^{(1-\alpha R)/\alpha R} = \left(1 - \frac{h}{145{,}457}\right)^{4.25635} \qquad (1.10)$$

Aeronautical work often refers to phoney "altitudes" instead of directly to corresponding atmospheric variables. For actual pressure p', for example, the corresponding pressure altitude h_p is that altitude in the ISA with pressure p'. Similarly for density altitude and, though it is seldom used, temperature altitude.

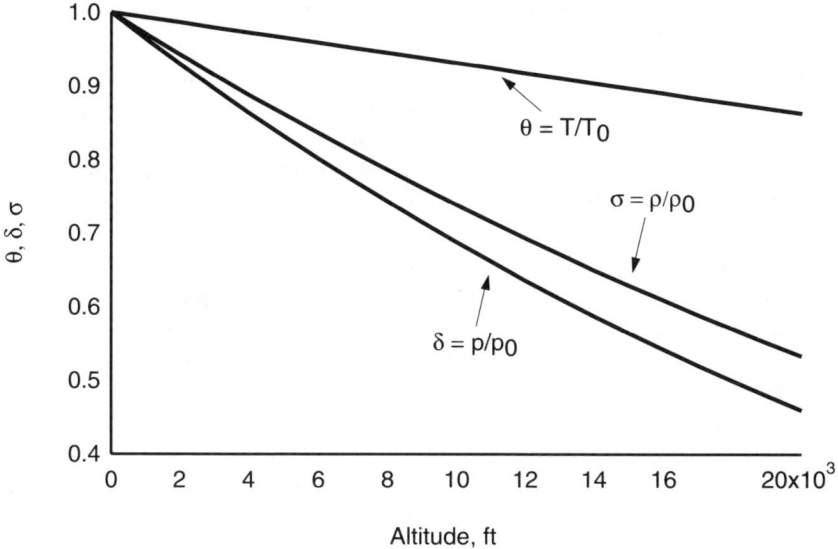

Figure 1.1 Atmospheric variables in the lower half of the troposphere.

To get operational definitions of those altitudes, one inverts Eqs. (1.8), (1.9), and (1.10). The combination T_0/α (value approximately 145,457) occurs often in atmosphere equations.

$$h_T = \frac{T_0 - T}{\alpha} = \frac{T_0}{\alpha}(1 - \theta) = 145{,}457(1 - \theta) \tag{1.11}$$

$$h_p = \frac{T_0}{\alpha}(1 - \delta^{\alpha R}) = 145{,}457(1 - \delta^{0.19025}) \tag{1.12}$$

$$h_\rho = \frac{T_0}{\alpha}(1 - \sigma^{\alpha R/(1-\alpha R)}) = 145{,}457(1 - \sigma^{0.23494}) \tag{1.13}$$

Example 1.1 When atmospheric pressure is 25 in. Hg, $\delta = 25/29.921 = 0.8355$ and $h_p = 145{,}457 \times (1 - 0.8355^{0.19025}) = 4889'$. If, in a different case, the atmospheric density is such that the density ratio $\sigma = 0.75$, then $h_\rho = 145{,}457 \times (1 - 0.75^{0.23494}) = 9506'$.

Atmospheric Humidity

Humidity makes the air feel "heavy" to us humans—cooled as we are by evaporation—but in fact humidity makes the air *less* dense. That is because the molecular weight of water (H_2O), 18.016, is considerably less than that of dry air, 28.966. The physical effect of humidity, then, is to change the gas around us from dry air to moist air, to a slightly different gas with a slightly lower average molecular weight. The bottom line is that humidity changes the specific gas constant R to a new and slightly larger value R'.

In aviation we measure humidity not by the meteorologist's "relative humidity" but rather by the "dewpoint temperature" often abbreviated T_d. The current dewpoint temperature is the temperature at which the atmosphere would be saturated with moisture, could hold no further water vapor. So T_d is the (usually lower-than-actual) temperature at which it would start to rain or at which fog would appear. Understood but not always expressed is the ground rule that the total atmospheric pressure p is assumed the same in both the actual and precipitating dewpoint cases.

Our calculation of R' from T_d and p will take two steps. First, from an empirical formula, we calculate the particular value of saturation vapor pressure e_s, which corresponds to T_d. Second, we apply a factor, involving the ratio of e_s to p and the ratio of molecular weights of water to air, $\epsilon \equiv 18.016/28.966 = 0.622$, to correct R.

An empirical formula is needed because discussion in terms of first principles would immerse us in molecular physics deeper than we care to go. Here we are a water strider insect, just denting the surface. Luckily the saturation vapor pressure depends only on temperature and not also on pressure. We will use a formula due to Tetens[3] for the job. Rewritten with T_d in degrees Fahrenheit and e_s in in. Hg, Teten's formula is

$$e_s = 0.1804375 \times 10^{(7.5T_d - 240/T_d + 395.14)} \quad (1.14)$$

Once we have e_s, we use the kind of "count them up" ratio formula you may have seen in freshman chemistry, or sophomore accounting, to get

$$R' = \frac{R}{\left(1 - \frac{e_s(1-\epsilon)}{p}\right)} = \frac{53.355}{\left(1 - \frac{0.378 e_s}{p}\right)} \quad (1.15)$$

Example 1.2 Take a fairly humid day when the dewpoint is $T_d = 70°F$ and the atmospheric pressure is 26.00 in. Hg. (or $26.00 \times 2116.2/29.921 = 1838.9$ psf). Equation (1.14) tells us the saturation vapor pressure is $e_s = 0.740$ in. Hg. Then Eq. (1.15) tells us that $R' = 53.93$.

Example 1.3 To see a practical effect of humidity, further consider that the OAT on this day is 80°F. Equation (1.1) then informs us that, for dry air, the density would be 0.001985 slugs/ft^3 or that relative density $\sigma = 0.8351$. But on this humid day we go back to Eq. (1.1) and use R' instead of R; that tells us density ρ

is actually 0.001964 slugs/ft^3 and $\sigma = 0.8261$, about 1% lower than the dry case. Not a big deal; that is why pilots *usually* ignore humidity except when carburetor icing is concerned.

Most simple aircraft performance measures depend on just σ, not on σ^2 or worse. Takeoff distance, as we shall see, is an exception. Propeller thrust, engine power output, and lift all depend approximately on σ. But then, to our advantage, so does drag! The net effect is that takeoff distance in low-powered propeller aircraft goes approximately as $1/\sigma^{2.3}$. In our case above, that would make the takeoff distance about 2.5% longer. Perhaps still negligible. But perhaps not.

Total Density vs "Air" Density

There is an important exception, or at least an additional wrinkle, to the foregoing humidity considerations. While wings and propeller blades do not care what sort of gas provides the total density in which they move, so long as it is not corrosive, engines do care. Only "air" (meaning, in fact, only air's oxygen component) combines with gasoline to provide motive power. As far as the engine is concerned, at least to first approximation, the water vapor in humid air might just as well have a molecular weight of zero. That airy dismissal of water, by the engine, gives us the clue we need to cobble together a correction. If water vapor had a molecular weight of zero, then ϵ would be zero and Eq. (1.15) would read (using a double prime for this different situation)

$$R'' = \frac{R}{\left(1 - \frac{e_s(1-0)}{p}\right)} = \frac{53.355}{\left(1 - \frac{e_s}{p}\right)} \qquad (1.16)$$

Example 1.4 Redoing our previous example, where OAT $= 80°F = 540°R$, $T_d = 70°, e_s = 0.740$ in. Hg., and the total pressure was $p = 26.00$ in. Hg $= 1838.9$ psf, we see that we get $R'' = 54.92$. Using Eq. (1.1), the density of air is then only 0.001927slug/ft^3, with relative density $\sigma_a = 0.8107$. This is almost 3% lower than the dry case, which we found would have had a relative density of 0.8351.

Using data from examples 1.3 and 1.4, one could also partition the actual total (humid case) atmospheric relative density as follows, where subscript v stands for vapor:

$$\sigma = \sigma_a + \sigma_v = 0.8261 = 0.8107 + 0.0154 \qquad (1.17)$$

Pilots in the tropics know full well that high humidity saps engine power. If you find yourself flying there, try to remember that the σ in power/torque dropoff factor $\Phi(\sigma)$ should more correctly always be σ_a.

Viscosity

For most performance purposes, we are getting down to relatively small change. About the only reason you would worry about atmospheric viscosity—internal friction, the extent to which a moving layer of air tends to make an adjacent layer move—is when you need a Reynolds number. Perhaps you would like to make corrections to lift or drag coefficients. Or perhaps you need to calculate how long it takes wing-tip vortices to die out. Still, here are the relevant facts and formulas.

For the same reasons as with humidity, viscosity needs an empirical formula. But this time there are only two constants instead of four. One is the standard ($T_0 = 59°$F or $518.7°$R) viscosity value $\mu_0 = 3.737 \times 10^{-7}$ lbf-s/ft^2. The other is "Sutherland's constant," $S = 198.7°$R. For practical purposes, absolute or dynamic viscosity depends only on temperature and follows this law:

$$\mu(T) = \mu_0 \left(\frac{T_0 + S}{T + S}\right)\left(\frac{T}{T_0}\right)^{3/2} \tag{1.18}$$

Air temperatures T and T_0 are in °R. The units of absolute viscosity are lbf-s/ft^2.

There is another flavor of viscosity, kinematic viscosity v, in aviation actually the more common. That is the ratio of the absolute or dynamic viscosity μ to density:

$$v = \frac{\mu}{\rho} \tag{1.19}$$

The units of kinematic viscosity are ft^2/s. The standard MSL value is $v_0 = 1.5723 \times 10^{-4}$ ft^2/s.

Speed of Sound

In spite of our disclaimer about confining ourselves to the "far subsonic," we include the formula for sonic speed here for two reasons: 1) how can you tell whether you are in fact far subsonic unless you know what "sonic" is? and 2) even in the propeller-driven general aviation regime, propeller tips may approach the speed of sound.

For ease of description, we are in luck here again; the sonic speed in a given gas only depends on temperature. The standard ($T_0 = 59°$F or $518.7°$R) value of the speed of sound is $a_0 = 1116.4$ ft/s. At temperature T (in °R, or some other absolute temperature scale), the sonic speed is

$$a = \sqrt{\gamma g RT} = \sqrt{\gamma p / \rho} = a_0 \sqrt{T/T_0} \tag{1.20}$$

Example 1.5 If we take the case where $T = -5°$F $= 454.7°$R, then $a = 1116.4 \times (454.7/518.7)^{1/2} = 1045.3$ ft/s. So, if flying at 100 mph $= 146.7$ ft/s, you would be at $M\,0.14$. *Really* far subsonic.

Example 1.6 For a propeller case, take a Cessna 172 propeller of 75 in diam rotating at 2700 rpm with the airplane flying at 100 mph = 146.7 ft/s. The radius at the tip is $r = 75/(2 \times 12) = 3.125'$. The angular speed is $\omega = (2\pi \times 2700)/60 = 282.74$ rad/s. The purely rotational portion of the tip speed is $v = r\omega = 883.6$ ft/s. Using the Pythagorean theorem to add in the translational portion, the resultant true air speed of the tip is $(883.6^2 + 146.7^2)^{1/2} = 895.7$ ft/s. Or, at the above $-5°$F temperature, $M\,0.857$. No wonder propellers make so much noise. A little faster than the cruise speed of a Boeing 747–400, but still subsonic.

That finishes our brief description of the atmosphere, its ISA model, and the variables of interest to us. But questions remain as to how, acting as flight test engineers, we can use this information to make proper sense of the data we gather during performance flight tests. We now turn our attention to that bevy of problems.

Modeling the Test-Day Atmosphere Locally

The kinds of questions we have in mind are "Having gone out in a real airplane into the real atmosphere, with well-calibrated instruments (altimeter, air-speed indicator, tachometer, outside air thermometer, etc.) and an accurate stop watch, running rate-of-climb tests at 85 KIAS between pressure altitudes of 4000 and 5000 ft averaging 50 s, what was our average rate of climb? Was it 1000/(50/60) = 1200 ft/min? Also, what was our true air speed?

The air-speed question is easy if we do not look closely. We recorded temperatures T_1 and T_2 at the top and bottom of the 1000-ft (pressure) altitude interval, so we get the corresponding values of relative density σ, average those to get $\langle \sigma \rangle$, and use

$$\text{KTAS} = \frac{\text{KCAS}}{\sqrt{\langle \sigma \rangle}} \qquad (1.21)$$

where we obtain the calibrated (equivalent) air-speed KCAS from the air-speed indicator calibration curve value at 85 KIAS. That much is simple. We will go into air-speed questions more deeply in the instruments chapter. We leave it, for now, after presenting a useful approximation for turning a density altitude into the corresponding relative density:

$$\sigma \doteq \left(1 - \frac{h_\rho}{70{,}000}\right)^2 \qquad (1.22)$$

This not-too-shabby approximation (only 0.06% high at 5000 ft, 1.1% high at 10,000 ft, and 1.9% low at 15,000 ft) makes it easy to find the square root of σ, the factor needed for going between calibrated and true air speeds.

The rate of climb was almost certainly not 1200 ft/min. A thousand feet of pressure altitude would, in an atmosphere described perfectly by the ISA model, imply 1000 ft of true, or at least geopotential, altitude. But such an occurrence is unlikely in the extreme. So, how can we turn two data pairs, say for example ($h_{p1} = 4000'$, $T_1 = 20°F$) and ($h_{p2} = 5000'$, $T_2 = 17.5°F$) into a good estimate of geopotential altitude difference?

Division of versions of Eqs. (1.4) and (1.6) for two positions in a nonstandard atmosphere characterized by (T_0', p_0', α') give a useful way station formula

$$\frac{T_2}{T_1} = \left(\frac{p_2}{p_1}\right)^{\alpha'R} \tag{1.23}$$

The temperatures are absolute ones, of course, and the pressures are available, from the known pressure altitudes, via Eq. (1.6). Taking the logarithm of Eq. (1.21) results in a formula for the nonstandard lapse rate

$$\alpha' = \frac{\ln(T_2/T_1)}{R\ln(p_2/p_1)} \tag{1.24}$$

Then it is easy to get

$$\Delta h \equiv h_2 - h_1 = \frac{T_1 - T_2}{\alpha'} \tag{1.25}$$

Example 1.7 Let us do a sample calculation using the two data points presented before Eq. (1.23). The two pressures—using Eq. (1.6) always with α (not α') because α is what counts in the invariable ISA tie between pressure and pressure altitude—are $p_1 = 1827.7$ psf and $p_2 = 1760.8$ psf. The two temperatures are $T_1 = 479.7°R$ and $T_2 = 477.2°R$. Equation (1.24) gives $\alpha' = 0.0026265$ F/ft. Equation (1.25) then gives $\Delta h = 951.8$ ft. In cold weather, as they say, the isobars contract. Of course it is just because cold air is denser and so the pressure difference, per tapeline foot, is greater.

The trouble with the above method is that it depends on temperature measurements considerably more accurate than we normally have. Next we derive another method that is simultaneously more sophisticated and practical.

Inchworm Method for Finding Δh

You can achieve higher accuracy if you are willing to take several measurements of pressure altitude h_p and temperature T between your lower and upper altitudes.

The Atmosphere

Derivation of the method starts with the basic atmospheric hydrostatic differential relation

$$dp = -\rho g \, dh \tag{1.26}$$

Using Eq. (1.1) for ρ,

$$\frac{-RT(p)dp}{p} = dh \tag{1.27}$$

Recall pressure altitude h_p, for pressure p, is defined as that altitude in the ISA that has p as its pressure. The virtue of pressure altitude is that it is directly measurable, in the cockpit, by simply setting the altimeter barometric subdial to 29.92 and reading off indicated altitude, altimeter "reading" r. We can get a relation for pressure altitude in terms of pressure, a more professional version of Eq. (1.12), by inverting Eq. (1.6),

$$h_p = \frac{T_0}{\alpha}\left(1 - \left(\frac{p}{p_0}\right)^{\alpha R}\right) \tag{1.28}$$

Differentiating,

$$\frac{dh_p}{dp} = -\frac{T_0 R}{p_0}\left(\frac{p}{p_0}\right)^{\alpha R - 1} \tag{1.29}$$

Use this result to rewrite our expression for dh in terms of dh_p instead of in terms of dp,

$$\frac{T(p)}{T_0}\left(\frac{p_0}{p}\right)^{\alpha R} dh_p = dh \tag{1.30}$$

A variation on Eq. (1.23) is

$$\left(\frac{p_0}{p}\right)^{\alpha R} = \frac{T_0}{T_S(h_p)} \tag{1.31}$$

so

$$dh = \frac{T(p)}{T_S(h_p)} dh_p \tag{1.32}$$

therefore,

$$\Delta h = \int_{h_{p1}}^{h_{p2}} \frac{T(p)}{T_S(h_p)} dh_p \tag{1.33}$$

One advantage of the procedure implied by Eq. (1.33) for measuring vertical intervals is that it does not assume a constant lapse rate in the air between h_{p1} and

h_{p2}. With a more varied lapse rate, one must sample a larger number of points in the atmosphere between h_{p1} and h_{p2}, true, to get accurate results. And then there is the integral to evaluate numerically. But a second advantage is that, in most cases, one need not use Eq. (1.33) in all its glory. The integrand, over the reasonable pressure altitude ranges used in performance flight testing (a few hundred to a thousand feet), usually varies quite slowly and steadily. Then it is certainly worth approximating Eq. (1.33) by using the mean of integrand values at the two ends of the interval:

$$\Delta h \doteq \left[\frac{T(h_{p1})}{T_S(h_{p1})} + \frac{T(h_{p2})}{T_S(h_{p2})}\right] \times \frac{\Delta h_p}{2} \qquad (1.34)$$

Because the ISA standard temperatures in our previous example are 479.7°R and 477.2°R, Eq. (1.34) gives us $\Delta h \doteq 951.8$ ft, the same figure as before. An even easier or lazier variation would be to run the climb or glide tests, say between pressure altitudes of 4000 and 5000 ft as before, and then simply sample the atmosphere once, at a pressure altitude of 4500 ft. Then the approximation looks like

$$\Delta h \doteq \left[\frac{T[(h_{p1} + h_{p2})/2]}{T_S[(h_{p1} + h_{p2})/2]}\right] \times \Delta h_p \qquad (1.35)$$

Assuming the temperature at the halfway point in our previous examples is the mean between extremes, $18.75°F = 478.45°R$, and using the fact that $T_s(4500') = 502.65°R$, Eq. (1.35) suggests $\Delta h = 951.9$ ft, almost the identical result given by Eq. (1.34). By splitting the difference, we imposed a constant lapse rate.

What if the experimental vertical interval you are faced with is large, or the lapse rate is quite erratic? Then take a lot of data points and use a numerical integration procedure on Eq. (1.33)—see Appendix D on using the trapezoidal rule for numeric integration. Our two-point estimate in Eq. (1.34) was essentially the roughest possible implementation of that rule. After we have carefully considered the workings of the altimeter and the way the altimeter setting works, we will return to this subject with a method for estimating a "global" nonstandard atmosphere on scant data, a way of deriving estimates for all three parameters T'_0, p'_0, and α'. However, the simple local methods described above tend to be more useful than a global model; for most aircraft performance purposes, we are only interested in the atmosphere locally.

2

Atmospheric Aircraft Instruments

Introduction

Some airplanes sport a multitude of instruments, gauges, and indicators. Others, only a few. In this chapter we are going to consider three fairly simple instruments—air-speed indicator, outside air thermometer, and altimeter—because those are most crucial to letting us evaluate airplane performance. But this is a good place to mention some of the other instruments that a typical general aviation reciprocating engine airplane is likely to have.

Regarding the engine, there will be a tachometer (measures engine or propeller circular speed) and, for the constant-speed airplane, a manifold pressure gauge (measures pressure in the engine's intake manifold, a proxy for engine torque). There often are both a cylinder head temperature gauge and an exhaust gas temperature gauge and both oil temperature and oil pressure gauges. There will be a fuel quantity gauge and, if there is a fuel pump (especially low-wing airplanes), a fuel pressure gauge and fuel flow gauge. There may even be a totalizer fuel gauge; nice to have for backfiguring accurate gross weights. There may be a carburetor deck pressure gauge. Perhaps a carburetor air temperature gauge to help detect carburetor icing danger.

The pilot has to know in what direction he or she is proceeding, so there will always be a compass and, most likely, a gyroscopic heading indicator. An attitude indicator gives him or her pitch and bank angle information. A turn-and-slip indicator or turn coordinator indicates standard turn rates and its inclinometer tells the pilot whether the airplane is heading in the direction of flight or, if not, whether it is slipping or skidding. A suction gauge tells whether the engine-driven vacuum system is operating well. An ammeter indicates whether the battery is charging or discharging and to what extent.

There is one pitot-static instrument we will not discuss: the vertical speed indicator (VSI). In performance work we carry a stopwatch and figure vertical speeds from altitude changes over time. The VSI is useful in ordinary operations but its lag, as much as 9 s, puts it at a fatal disadvantage relative to a well-calibrated thumb working a stopwatch.

Of all these, we only need to carefully understand the air-speed indicator (ASI), the altimeter, and the outside air thermometer. We need to know the inputs, outputs, and way the instrument produces the latter from the former. We need to know its common errors (random, but especially systematic, including instrument lags) and what to do to mitigate them. Our knowledge of the atmosphere will come in handy. So too will some basic gas thermodynamics, but we can get by with only a little and that largely in the very next section.

Bernoulli's Law

Bernoulli's law relates changes in pressure p, gravitational potential gh, and speed V in a fluid at different positions along the same streamline. The flow should be steady (streamlines not changing with time), irrotational, and inviscid (viscosity zero) and no external source or sink of energy should be operating. We almost always have negligible height differences and so can leave out the gravitational terms and take different, but close, streamlines. There are incompressible (low-speed) and compressible (high-speed) versions of Bernoulli's law. The low-speed version (for $M < 0.3$) is

$$p_1 + \tfrac{1}{2}\rho V_1^2 = p_2 + \tfrac{1}{2}\rho V_2^2 \qquad (2.1)$$

Applications include Venturis and air-speed indicators.

Example 2.1 Let us get ready for the air-speed indicator application by taking position #1 in air some distance ahead of the airplane's pitot tube—$V_1 = V = 100\,\text{KTAS} = 168.8\,\text{ft/s}$, $p_1 = p_0 = 2116.2\,\text{psf}$, and ρ constant (incompressible, remember?) at $\rho_0 = 0.002377\,\text{slug/ft}^3$ (the airplane under mean sea level [MSL] standard conditions). So the left-hand side of Eq. (2.1) is $(2116.2 + 33.9) = 2150.1\,\text{psf}$. Notice how small the *dynamic pressure*, the velocity term, is compared to static atmospheric pressure. Dynamic pressure is often given as q, it makes up the bulk of the formulas for lift and drag. Now take position #2 inside the pitot tube, where the air has come to rest (again, relative to the airplane). That makes $V_2 = 0$. So, using the Bernoulli relation, the pressure inside the pitot tube, $p_2 = p_1 + q$, must be 2150.1 psf. That pressure is often called *total pressure*, p_T, since it is the sum of static and dynamic pressures. There will be many more details on the ASI in the next section.

Notice from the ideal gas law, Eq. (1.1), that since $\rho_1 = \rho_2$, pressure in the Bernoulli relation is proportional to temperature and so, if $p_2 > p_1$, as it was in the example above, then $T_2 > T_1$; bringing the air in the pitot tube to a stop raises its temperature. On the other hand, when air passes over a wing or through a Venturi tube, the pressure (and hence the temperature) decreases.

We will not need the compressible form of Bernoulli's law very often because the airplanes we deal with are usually "far subsonic." But not always. For instance $M = 0.3$, at MSL, corresponds to 0.3×661.5 KTAS $= 198.4$ KTAS. Sleek fiberglass experimental aircraft could easily be beyond that somewhat artificial limit. And there will be more of those fast airplanes in the future. For that reason, we include the compressible form:

$$\left(\frac{\gamma}{\gamma-1}\right)\left(\frac{p_1}{\rho_1}\right) + \frac{1}{2}V_1^2 = \left(\frac{\gamma}{\gamma-1}\right)\left(\frac{p_2}{\rho_2}\right) + \frac{1}{2}V_2^2 \qquad (2.2)$$

Greek small gamma (γ), recall, is the ratio of specific heats, close to 1.40 for air at reasonable temperatures. The relation had to be rearranged somewhat, relative to the incompressible form, to let density ρ vary. Notice that if we tried to recast our previous example to include the possibility of compressibility, we have a problem: position #2 pressure and density are both unknown! We need another relation and find it in (approximate) knowledge that the compression took place *adiabatically*, without gain or loss of heat. Such a thermodynamic event is often called an *isentropic* process—one without entropy change—since $dQ/T \equiv dS = 0$, where S is entropy and dQ is heat transferred. One way to specify an adiabatic process, for an ideal gas, is that p/ρ^γ is constant during the process. That is the second relation we occasionally and sorely need.

Almost all aerodynamics and fluid mechanics books derive at least the incompressible form of Bernoulli's law. It turns out to be little more than a particular statement, for a volume of moving fluid, of the conservation of energy. Bernoulli's law, in whichever form (compressible or incompressible) is appropriate, is one of the main and most useful relations in aerodynamics.

Air-Speed Indicators

We shall treat ASIs of both the lower-speed type (treating air as incompressible) and higher-speed type (treating air as compressible)—though, to avoid considering shock waves, even the latter case will still be subsonic. This section will give an approximate formula that indicates when to switch from one type of ASI to the other. Fig. 2.1 is a block diagram for the ASI, showing the single pressure-difference input and the single indicated air-speed output. In theory it is a simple instrument; in practice, not so. First we consider the low-speed version.

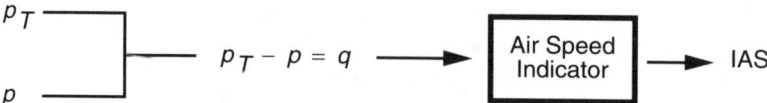

Figure 2.1 The input and output variables of an air-speed indicator.

Low-Speed ASIs

Taking Eq. (2.1) with position #1 ahead of the pitot tube, position #2 inside it, that equation specializes to

$$p + \tfrac{1}{2}\rho V^2 = p_T \tag{2.3}$$

Solving Eq. (2.3) for V, we detect a problem:

$$V = \sqrt{\frac{q}{2\rho}} \tag{2.4}$$

We cannot find V, from an ordinary ASI, without knowing air density ρ. So what we do is define *equivalent* air speed V_e as

$$V_e \equiv \sqrt{\sigma} V = \sqrt{\frac{q}{2\rho_0}} \tag{2.5}$$

Equivalent air speed, after adjusting for various systematic errors, *is* something we can read directly from an ASI. If we need to know true air speed V, we make a density altitude correction off to the side. That is as good as we can do with this device. Recall that we are, so far, at speeds low enough that there is no practical difference between equivalent and calibrated air speeds. In general aviation this speed is normally called "calibrated."

High-Speed ASIs

Equation (2.2), with position #1 ahead of the pitot tube and position #2 inside it, specializes to become

$$\left(\frac{\gamma}{\gamma-1}\right)p + \frac{1}{2}\rho V^2 = \left(\frac{\gamma}{\gamma-1}\right)\left(\frac{\rho}{\rho_T}\right)p_T \tag{2.6}$$

We also need the adiabatic or isentropic condition for this circumstance:

$$\frac{\rho}{\rho_T} = \left(\frac{p}{p_T}\right)^{1/\gamma} \tag{2.7}$$

Substituting Eq. (2.7) into Eq. (2.6) and multiplying through by $(\gamma - 1)/\gamma p$, one gets

$$1 + \frac{(\gamma - 1)\rho V^2}{2\gamma p} = \left(\frac{p}{p_T}\right)^{1-\gamma/\gamma} \qquad (2.8)$$

Using the fact that $p_T/p = q/p + 1$ in Eq. (2.8), along with inverting the pressure ratio there and subtracting unity from each side, one has

$$\frac{(\gamma - 1)\rho V^2}{2\gamma p} = \left[\left(\frac{q}{p} + 1\right)^{\gamma - 1/\gamma} - 1\right] \qquad (2.9)$$

Dividing both sides by MSL standard density and isolating the equivalent air speed, one has

$$V_e^2 = \frac{2\gamma p}{(\gamma - 1)\rho_0}\left[\left(\frac{q}{p} + 1\right)^{\gamma - 1/\gamma} - 1\right] \qquad (2.10)$$

This (compressible) time the density stratagem, going to equivalent air speed, did not solve our problem. The ASI still only has q for an input, but Eq. (2.10) requires knowledge of ambient pressure p as well. This problem is solved by constructing the ASI to read the square root of the right-hand side (RHS) of Eq. (2.10) when p is replaced by standard MSL pressure p_0. Then, of course, that new quantity is no longer quite equivalent air speed. Instead, it is called calibrated air speed, given by

$$V_c^2 = \frac{2\gamma p_0}{(\gamma - 1)\rho_0}\left[\left(\frac{q}{p_0} + 1\right)^{\gamma - 1/\gamma} - 1\right] \qquad (2.11)$$

and now an additional pressure correction, commonly called ΔV_c, must be done off to the side. Pilots of high subsonic aircraft carry a graph for that purpose, but the airplanes we deal with are slow enough ($M < 0.5$) that a surprisingly simple approximate correction, derivable from Eqs. (2.10) and (2.11) by repeated application of the binomial expansion, to various orders, is sufficient:

$$\Delta V_c \doteq \frac{-h_p V_c^3}{8RT_0 a_0^2} \qquad (2.12)$$

Variable a_0 is the MSL standard sonic speed, 1116.4 ft/s = 661.4 kn.

Example 2.2 Let us evaluate Eq. (2.12) for a sleek, high-powered craft going 300 KCAS = 506.4 ft/s at a pressure altitude of 15,000 ft. (Under standard lapse rate conditions, that speed is 378.2 KTAS.) Equation (2.12) says the correction from calibrated to equivalent air speed is −7.06 ft/s or −4.2 kn. The chart carried by jet pilots [or solving Eq. (2.11) for q and plugging that value into Eq. (2.10)] gives −5.3 kn. Close enough. The correction is negative because compressibility raises the stagnation point density and thereby overstates the ram effect, giving the inside of the ASI bellows an added boost. Our main use of Eq. (2.12) is to see whether a compressibility correction needs to be considered. In this very fast and fairly high case, correction helps.

Calibrating the ASI

Each airplane's ASI installation needs to be individually calibrated to compensate for systematic instrument errors. (In a later section, we shall consider the less common determination of position error, an error that may vary with body angle of attack (AOA) and impacts the altimeter.) Here we will describe calibration by the older, not-so-reliable measured speed course method and then by newer high-technology GPS methods.

The goal is to get an accurate graph of calibrated air speed V_c plotted against indicated air speed V_i over the entire range of your airplane's level speeds, see Fig. 2.2. If your pitot-static system, including the ASI itself, were perfect, V_i would always equal to V_c. But the system is not perfect; so, it needs to be calibrated. Once that is done, you apply corrections to V_i to get the correct value of V_c. Figure 2.3 outlines the order of application of corrections; we can almost always skip the compressibility correction.

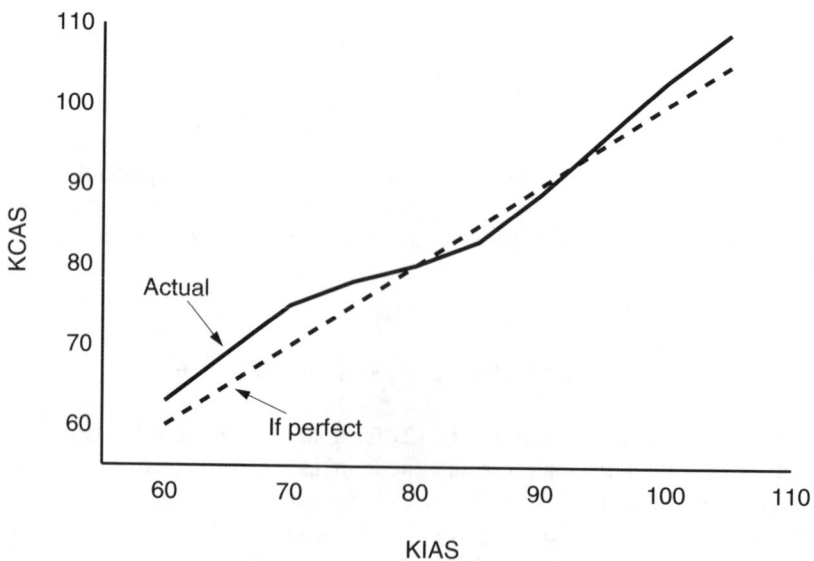

Figure 2.2 Sample ASI calibration curve.

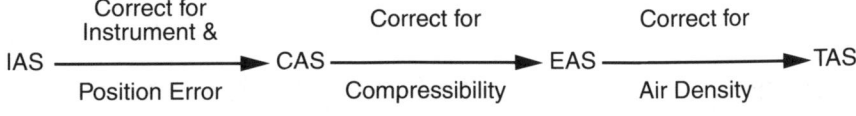

Figure 2.3   Air-speed indicator corrections.

First, general principles and cautions about the calibration procedure. You cannot do a good job on this crucial bit of instrument calibration flying unless atmospheric conditions are smooth. This almost certainly means early morning flights (late afternoons are a poor second best). A smooth atmosphere means

- Calm, or at least light and steady, winds (light enough that you can hold indicated air speed within a knot of the selected target speed) and
- Negligible vertical air mass movement (little enough that you can hold indicated altitude within 30 ft).

Before you go up, select your set of target air speeds and a somewhat random order in which to run them. Strict numeric order is poor because it amplifies effects of small systematic errors like engine temperature changes, slow atmospheric changes, or other conditions beyond your control and knowledge.

About 10 air speeds, running from somewhat above power-on stall speed to near-maximum-level flight speed, should be enough. For an airplane that stalls around 40 and can only get up to 100, this means about every five speed units. For a higher-performance airplane (say 70 to 170 KCAS), target air speeds every 10 kn or mph might be appropriate. Much depends on how smooth your final calibration curve turns out to be.

Low-Altitude Speed Course Technique

With a broad brush, you are going to record indicated air speed (IAS) and compute both true air speed (TAS) and calibrated air speed (CAS). The appropriate form is #312; see Fig. 2.4. You will make runs between parallel roads or fences at least one flight minute, and an accurately known distance, apart. If your maximum level flight speed is V_M mph, that means the roads or fences must be at least $V_M/60$ statute miles apart.

You will fly along, or parallel to, a road or fence or on a magnetic heading, perpendicular to the starting and ending landmarks. Your IAS must be well stabilized before crossing the starting line. If you have a low-wing airplane, it makes sense to start and stop the watch when those landmarks, most commonly straight roads or fences, cross under the leading edge of your wing; for a high-wing airplane, perhaps a strut is the best marker. Tell the stopwatch-wielding observer to always sit the same way in his or her seat. To get accurate start and end times you will need to be low, from 200 ft to 400 ft AGL.

Flight Physics
P.O. Box 20919
Billings, Montana 59104
(406) 248-2606

AIRSPEED INDICATOR CALIBRATION DATA SHEET – GROUND COURSE

DATE & PAGE NUMBER	AIRPLANE TAIL NUMBER	AIRPLANE MAKE, MODEL, YEAR	OWNER
4/19/95 – p 1	N 6346 D	Cessna 172P, 1982	Corporate Air

W_1, STARTING WEIGHT	W_2 ENDING WEIGHT	DISTANCE OVER COURSE	DESCRIPTION OF COURSE
2219 #		2.30 s.m. = 1.999 n.m.	Acton Rd. — Mail Box Rd.

PILOT	OBSERVER	NOTE		
J. Lowry	A. Daniel	(1) For calc'n, $\Delta t \to \Delta t \times \frac{\text{Actual IAS}}{\text{Target IAS}}$		(2) Speeds in KNOTS.

Run #	Clock Time	Magnetic Heading	Target IAS	Actual IAS	Δt	Pressure Altitude	☒F ☐C OAT	Calc'd Sigma	Calc'd GS	Calc'd <GS>	Calc'd CAS	Optional RPM
1	7:13	345	75	76	80.4	3700	23.5	0.9379	88.32	—	—	1950
2	7:18	165	75	75	99.7	3700	23.5	0.9375	72.17	80.24	77.7	1940

© Copyright 1995, Flight Physics. All rights reserved.
Form #312, 6/27/95

Figure 2.4 Filled-in Form #312.

If the airplane is crabbed a little into a steady crosswind, so be it. But crosswinds greater than 15% of your air speed should be avoided because they make for ASI errors larger than 1%. If there is too much crosswind, you may simply have to call it a day.

For each target IAS you should fly at least two back-and-forth patterns, at least four passes in all. For a dozen target air speeds and a course a couple of minutes long on the average, this means about 50 passes, 3 or 4 h of concentrated boredom. It is unlikely smooth conditions, your attention span, and your fuel capacity will all last this long. So be prepared to spend three or four mornings calibrating your ASI. You may want to work first on every other one of the target air speeds; after analyzing your initial session you should be able to tell whether the left out points are necessary.

Refer to form #312 (Fig. 2.4, form #313 is just a continuation, with less header information). Here are notes on some of the columns:

- Clock time—this is so you can later figure your gross weight. Weight changes mean different AOAs and, perhaps, different ASI position errors. We hope not. Also, recording weight (along with revolutions per minute) gives you some partial-throttle information that might be useful in the future.
- Magnetic heading—were you "back" or were you "forth"?
- Actual IAS—say your target was 70 KIAS but you started a little slow, at 68, then brought up the speed to 70 by the run's end. Then actual IAS, on the

average, would be 69 KIAS. Use that average as your "actual." If the variation is too much, or unfathomable even if small, discard the run and repeat.
- Δt—this is the run's elapsed time to the nearest tenth of a second. If actual IAS differs from target IAS, an adjustment is needed. Let us say you did a pair of back-and-forth runs with target 75 KIAS but on the first of those your actual was 76. You were going a bit faster than you intended and hence your time was a bit shorter than it should have been. The proper adjustment of your raw elapsed time is obtained by multiplying that time interval by 76/75. On the calibration curve, your KIAS along the x axis will then always be your *target* speed. Rather than mess with this adjustment, you may prefer to just throw out any ill-controlled runs.
- Pressure altitude—set your Kollsman window to 29.921 in. Hg while still on the ground. (You might also record the correct altimeter setting. You will have to land!) Say 29.921 gives you an indicated altitude of 3400 ft. And say you have decided to fly at about 300 ft AGL. That makes it $h_p = 3700$ ft for your runs.
- OAT—you need this, with pressure altitude h_p, to calculate relative atmospheric density. Be watchful for engine exhaust giving you spuriously high OAT readings. Try to place the temperature probe so that you get the same reading whether under full power or gliding level (with increasing elevator deflection).
- σ—relative density. The square root of this figure will give you the necessary correction factor.
- Calculated ground speed (GS)—this is course length divided by elapsed time. The complication comes from various possible units. If the course length is 2.3 (statute) miles, but your ASI uses knots, then you need the conversion factor 1.15076 statute miles per nautical mile. And you will need to use 3600 s/h to convert the time intervals.
- Calculated average ground speed ⟨GS⟩—this means the average for the two runs, back and forth, at the same IAS. So there will only be one of these entries for each two rows. Because back-and-forth cancels the (assumed constant!) wind effect, this single entry is the TAS for the pair. If one run (via the previous column) was at 67.1 kn and the opposite direction went at 78.8 kn, then the average is 72.95 kn. [The average wind-speed component along your course, incidentally, is half the difference or $(78.8 - 67.1)/2 = 5.85$ kn.]
- Calculated CAS—this is what we are looking for and is given by

$$\text{CAS} = \sqrt{\sigma}\text{TAS} \qquad (2.13)$$

The rearranged rewrite will be useful when we analyze flight test data:

$$\text{TAS} = \frac{\text{CAS}}{\sqrt{\sigma}} \qquad (2.14)$$

If you have half as much luck as you have put out effort, your air-speed calibration graph will look like Fig. 2.3. If different pairs of runs at the same IAS are very different, your graph fluctuates wildly, or the error is large (say over five or six units) even though relatively uniform, you need to pay remedial attention to your physical pitot-static system itself. Askue[1] has good material on solving this problem.

Back and Forth with GPS Unit Technique

As this is simply an abbreviated version of the previous technique; we will just mention differences. With the GPS unit telling you (with some small time lag) what your accurate ground speed is, you do not need such long steady runs or a stopwatch. If you use forms #312 and #313 you can forget about the distance-over-course in the header and you can neglect the column for Δt in the detail section. In addition, the column marked "Calc'd GS" should now be marked "Measured GS."

You still run back and forth along reciprocal courses (so wind effect will cancel) at a constant pressure altitude, and you still need to record OAT so you can calculate relative density σ. But if you hold IAS and pressure altitude steady for as long as 20 s, and ground speed GS fluctuates only a little (0.5 kn or less), then you are done for that run. Using a GPS unit makes the calibration work go quite a bit faster.

Box Pattern with GPS Unit Technique

This technique makes full use of the GPS and is done at altitude. Theory says that finding three unknown numbers—TAS, wind speed V_w, and wind direction θ—needs three experiments. We are only after the TAS; the rest comes with the territory.

As before, pick and record some pressure altitude h_p and record the outside air temperature periodically. Again, winds aloft must be steady; your computed results will tell you whether they are. Pick your first target IAS and maintain it throughout the three legs of your box pattern. You first fly magnetic heading 0 deg, get a "20 s" steady ground speed g_1 and record it. Then turn to heading 90 deg and repeat, getting ground speed g_2. Then to heading 180 deg and repeat, getting ground speed g_3. That is it for that first IAS, so at this point you turn back to the north to begin another cycle.

Repeat the above procedure for each of your other target indicated air speeds. You may occasionally want to go back to one of the earlier IASs to check on the wind; if the three ground speeds differ very much (more than 1 or 2 kn) between the two patterns, the wind is probably too variable for your results to be reliable. In that case, quit for the day.

Atmospheric Aircraft Instruments

Once back on the ground with good data, you have some calculating to do. Preferably in a computer spreadsheet. Here are the calculations required for each three-legged pattern:

$$p = \frac{g_1^2 + g_3^2}{2} \qquad (2.15)$$

$$\alpha = \tan^{-1}\left(g_3^2 - g_1^2, 2g_2^2 - g_1^2 - g_3^2\right) \qquad (2.16)$$

$$q = \frac{g_3^2 - g_1^2}{4\cos\alpha} \qquad (2.17)$$

$$\text{TAS} = \sqrt{\frac{p + \sqrt{p^2 - 4q^2}}{2}} \qquad (2.18)$$

$$V_w = \left|\frac{q}{\text{TAS}}\right| \qquad (2.19)$$

$$\theta = 360 - \alpha, \ \text{mod } 360 \qquad (2.20)$$

Remarks are in order.

- Variables p, α, and q are just way stations, intermediate variables, which simplify calculating the meaningful items of Eqs. (2.18–2.20).
- The ground speeds, the TAS, and the wind speed V_w are all in the same units —knots, miles per hour, feet per second, whatever; you may need to convert some of them so you end up with CAS in the same units as IAS.
- In Eq. (2.16), the inverse tangent function is the usual angle (counter-clockwise from the positive x axis) to the point (x, y); as a spreadsheet function this is written @ATAN2(x,y). Coordinates x and y will be replaced by spreadsheet cell addresses.
- If the wind is from either directly east or west, you may find that ground speeds g_1 and g_3 are *exactly* the same. Then there is a "singularity" and the equations will not compute. If that extremely unlikely case does occur, just add 0.01 unit to either g_1 or g_3.
- The vertical bars in Eq. (2.19) denote taking the absolute value of whatever is between them; as a spreadsheet function this is written @ABS(·).
- The mod (short for modulo) in Eq. (2.20) is shorthand for "the remainder when divided by" (in this case) 360; as a spreadsheet function this is written @MOD(360-α, 360) except that you will have a cell address for the argument α.

Example 2.3 Say that on your three legs (magnetic headings 0 deg, 90 deg, and 180 deg, respectively) your three ground speeds were 95.3, 130.0, and 105.9 kn. For the intermediate calculations, you will find that $p = 10,148.5$ and $\alpha =$

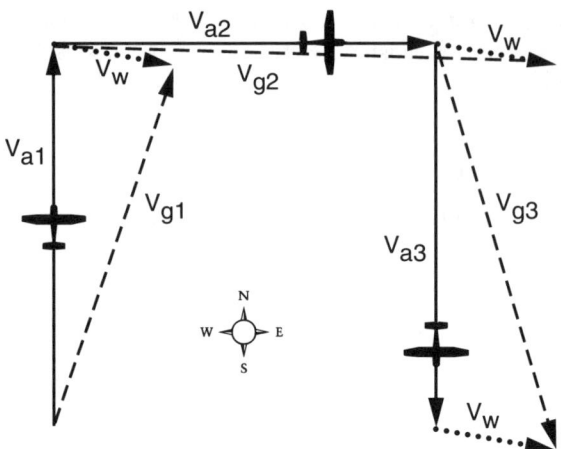

Figure 2.5 Diagram, in "air" space, of a box pattern GPS flight for ASI calibration.

81.02 deg and $q = 3417.6$. For the final calculations you will find TAS $= 93.9$, $V_w = 36.4$, and $\theta = 279.0$ deg. Figure 2.5 shows the approximate speed triangles for the three legs. Incidentally, it is unlikely an actual wind this high would be sufficiently steady for accurate ASI calibration.

Kinetic Energy Effect

We close this section on the ASI by considering the *kinetic energy effect*. The airplane experiences a real acceleration, during climbs at constant calibrated (equivalent) air speed, because of the design of the ASI. As the airplane climbs into less dense air, TAS increases. During descents at constant CAS, the effect is of course a deceleration.

Example 2.4 Consider an airplane climbing at constant CAS $V_c = 100$ KCAS $= 168.8$ ft/s. Because TAS is given by $V^2 = V_c^2/\sigma$, TAS at say density altitude 3000 ft, where $\sigma(3000\,\text{ft}) = 0.9151$, is $V(3000\,\text{ft}) = 104.54$ KTAS $= 176.44$ ft/s. At density altitude 4000 ft, where $\sigma(4000\,\text{ft}) = 0.8881$, $V(4000\,\text{ft}) = 106.11$ KTAS $= 179.10$ ft/s. In climbing from 3000 to 4000 ft at constant CAS, the airplane has in fact accelerated slightly, increasing speed by $1.57\,\text{KTAS} = 2.65$ ft/s.

So, climbing or descending, there is a small but not imperceptible acceleration which, in careful work, you need to consider. If this airplane were climbing at say

1000 ft/min, the speed increase would have taken place over 60 s and the resulting acceleration would have been $a = \Delta V/\Delta t = 2.65/60 = 0.044$ ft/s². If the airplane weighed 3000 lbf, this acceleration would correspond to an additional thrust, above that needed to overcome drag and gravity, of

$$\Delta T = \frac{W}{g} a = \frac{3000}{32.2} 0.044 = 4.1 \text{ lbf} \qquad (2.21)$$

Small, as we said, but not completely negligible.

The term "kinetic energy effect" is not particularly descriptive. The same can be said of quite a few aeronautical engineering buzz words. Granted the airplane's kinetic energy does increase between the two stations (absent unconscionably large and compensating fuel flow, which would decrease mass and thereby decrease kinetic energy), but the real acceleration is simply an artifact of ASI design.

To be able to quickly judge the extent and importance of the kinetic energy effect, we calculate the acceleration, in general, for a climb through the ISA. The descending case is the same except for a changed sign. The acceleration is

$$a_{KE} \equiv \frac{dV}{dt} = \dot{h} \frac{d\sigma}{dh} \frac{dV}{d\sigma} \qquad (2.22)$$

and we must evaluate the last two derivatives on the right.

$$\frac{dV}{d\sigma} = V_c \frac{d\sigma^{-1/2}}{d\sigma} = -\frac{1}{2} V_c \sigma^{-3/2} \qquad (2.23)$$

$$\frac{d\sigma}{dh} = \frac{d}{dh}\left(1 - \frac{\alpha h}{T_0}\right)^{1-\alpha R/\alpha R} = \frac{1 - \alpha R}{\alpha R} \sigma \left(1 - \frac{\alpha R}{T_0}\right)^{-1} \left(\frac{-\alpha}{T_0}\right) \qquad (2.24)$$

Expressions are simplest in terms of σ, so using

$$\left(1 - \frac{\alpha h}{T_0}\right)^{-1} = \sigma^{\alpha R/\alpha R - 1} \qquad (2.25)$$

and assembling results, we find

$$a_{KE} = -\frac{1}{2} \dot{h} V_c \left(\frac{\alpha R - 1}{RT_0}\right) \sigma^{2\alpha R - 1/(\alpha R - 1) - 3/2} = 1.463 \times 10^{-5} \sigma^{-0.7349} \dot{h} V_c \text{ ft/s}^2 \qquad (2.26)$$

using British engineering units throughout. Using practical aeronautical units for rate of climb and calibrated air speed, this is

$$a_{KE} = 4.1156 \times 10^{-7} \times \sigma^{-0.7349} \times (\text{ft/min}) \times (\text{KCAS}) \text{ft/s}^2 \qquad (2.27)$$

Revisiting our previous numerical example 2.4, at climb midpoint 3500 ft we had $\sigma = 0.9015$. So with $\dot{h} = 1000$ ft/min and $V_c = 100$ KCAS, Eq. (2.27) gives $a_{KE} = 0.044$ ft/s², as before.

Outside Air Thermometer

General aviation aircraft almost invariably have some sort of outside air thermometer graduated (in the U.S.) in both Fahrenheit and Celsius degrees. These are seldom either accurate enough or, more particularly, quick enough for performance flight testing. A small portable digital remote-reading thermometer is required. One such is the Universal Enterprises model DT10K (Beaverton, Oregon). The folded unit is about $3 \times 5 \times 5/8$ in. thick. The 4-in. temperature probe is at the end of a 4-ft cable, which can be hung out a cabin window. The readout is large, in both Fahrenheit and Celsius, and is to the nearest 0.1 deg. It costs approximately $50.

If you find that heat pollution from exhaust gases is a problem, it may be necessary to glide down onto the altitude you want to sample and hold altitude with elevator. These units stabilize quickly.

Ram Temperature Rise

The other problem you might have getting accurate outside air temperatures has to do with ram temperature rise. When fast-moving air is brought to a standstill, as we have seen, the pressure and density rise. So too does the temperature. Using our earlier nomenclature, say the temperature rises from actual ambient T to "total temperature" T_T. How big is the effect?

Here it is best to make use of $M = V/a$, where a is the local speed of sound given by

$$a^2 = \gamma p / \rho \qquad (2.28)$$

We have so far been eliminating temperatures in favor of pressures and densities, but the ideal gas law suggests

$$\frac{T_T}{T} = \frac{p_T \rho}{p \rho_T} \qquad (2.29)$$

Rewriting Eq. (2.8) using Eq. (2.7), (2.28), and (2.29), we (almost) have

$$T_T = T\left(1 + \frac{\gamma - 1}{2} \epsilon M^2\right) \qquad (2.30)$$

The added factor ϵ is the *recovery factor*, an empirical addition to account for the fact that the stagnation process is not perfectly adiabatic. Some heat is lost by convection, conduction, and radiation. Recovery factors range from 0.75 to 0.99 (the high value only for electrical resistance winding temperature probes), with 0.86 a good average. For a given installation, ϵ can fairly easily be determined experimentally. Running at the same altitude at two considerably different air speeds (Mach numbers), and recording total temperatures, the two statements of Eq. (2.30) result in a solution for the recovery factor:

$$\epsilon = \frac{5(T_{T2} - T_{T1})}{T_{T1}M_2^2 - T_{T2}M_1^2} \tag{2.31}$$

But there is a problem. Without ϵ we do not have a good value of temperature T. Because the speed of sound a depends solely on T, without good temperature, even if we know TAS V, we cannot know M. It is time to get organized, see where we are, what we need, and what we have.

As a preliminary step, we add a simple formula for M to our armamentarium. Reconsider Eq. (2.9). Using Eq. (2.28) in it, then isolating the square of M, one immediately gets

$$M^2 = \frac{2}{\gamma - 1}\left[\left(\frac{q}{p} + 1\right)^{\gamma - 1/\gamma} - 1\right] \tag{2.32}$$

Here then is the situation. In the cockpit you can read pressure altitude h_p, indicated (total) temperature T_T, and IAS V_i. You also have the instrument error and position error corrections (call their sum ΔV_i) needed to get calibrated air speed V_c from V_i. The question is "How would one get TAS V, true temperature T, M, and temperature recovery factor ϵ from the known information?" Here, for our subsonic case with no shock waves at the pitot tube, is how.

Step 1. Get $V_c = V_i + \Delta V_i$.

Step 2. Solve Eq. (2.11) for the compressible dynamic pressure to get

$$q = p_0\left[\left[\left(\frac{V_c}{a_0}\right)^2\frac{(\gamma - 1)}{2} + 1\right]^{\gamma/\gamma - 1} - 1\right] \tag{2.33}$$

where a_0 is the standard MSL speed of sound (1116.4 ft/s).

Step 3. Get M from Eq. (2.32).

Step 4. Write down indicated total temperature T_T and pressure altitude h_p.

Step 5. To get information that can lead to the recovery factor ϵ, fly at the same pressure altitude (near the same time, therefore at the same true temperature T) at a (considerably) different speed. Then repeat steps 1 through 4. Call the two Mach numbers M_1 and M_2, the two total temperatures T_{T1} and T_{T2}.

Step 6. Compute recovery factor ϵ from Eq. (2.31).

Step 7. Compute true temperature T by inverting Eq. (2.30):

$$T = \frac{T_T}{1 + 0.2\epsilon M^2} \tag{2.34}$$

Step 8. Compute the ambient speed of sound V_a from

$$V_a = \sqrt{\gamma g RT} \tag{2.35}$$

Step 9. Compute TAS V from

$$V = MV_a \tag{2.36}$$

Example 2.5 You are at pressure altitude $h_P = 30{,}000$ ft and hence $P = 629.7$ psf and $\delta = 0.29756$. The total (indicated) temperature is $T_T = -50°\text{F} = 409.7°\text{R}$. Your IAS $V_i = 395$ mph. This is situation number 1. Now we go through the above steps.

1) The known corrections ΔV_i give you, say, calibrated air speed $V_c = 400$ mph $= 586.7$ ft/s.
2) Equation (2.33) gives $q = 438.2$ psf.
3) Equation (2.32) gives $M = 0.9025$.
4) $T_{T1} = 409.7°\text{R}$ from the conditions of the problem. And, of course $h_P = 30{,}000$ ft.
5) To get recovery factor information, say you slow down to $M_2 = 0.40$ and find that $T_{T2} = 365.2°\text{R}$.
6) From Eq. (2.31), $\epsilon = 0.96$.
7) From Eq. (2.30), the true temperature of situation number 1 (the one we care about) is $T = 354.3°\text{R}$.
8) From Eq. (2.35), $V_a = 922.7$ ft/s $= 629.1$ mph.
9) From Eq. (2.36), $V = 832.7$ ft/s $= 567.8$ mph.

We did not say anything about equivalent air speed $V_e \equiv \sigma^{1/2} V$. From the ideal gas law $\rho = p/RgT = 629.7/(53.35 \times 32.174 \times 354.3) = 0.001035$, which makes $\sigma = \rho/\rho_0 = 0.001035/0.002377 = 0.4354$ and $\sigma^{1/2} = 0.6600$. Hence $V_e = 549.6$ ft/s $= 374.7$ mph.

We will not often need all of this compressible apparatus, but it is here when we do. Now we turn towards understanding another important instrument, one just as subtle as the ASI, one needing careful treatment even at low speeds and altitudes.

The Altimeter

In Chapter 1 we used the altimeter to find pressure altitude (height in the ISA with the same pressure as our location) as an input to estimating geometric height differences. Pressure altitude is found by turning the *altimeter setting* (which we will give variable s) to the standard MSL value 29.921 in. Hg. Though that value is the one most often employed in performance flight test work, we want the whole story. Perhaps we would need to recreate, say for forensic purposes, altitude information for an airplane with an altimeter not set to 29.92 in. So in this section we dig into the workings of the altimeter, finding out how to capitalize on its possibilities and avoid its misleading indications. Figure (2.6) gives the block diagram of the instrument, showing the altimeter's two inputs and single output.

Even if we restrict ourselves to considering somewhat idealized atmospheres, those with constant lapse rates, it takes (recall from Chapter 1) four numbers to specify any one of them: T'_0, α', p'_0, and g. There is no problem with g and, as we shall soon see, the altimeter setting knob allows us to compensate for non-standard values p'_0. But there is no possibility, with only that one adjustment, of also being able to compensate for aberrant base MSL temperatures or variant lapse rates. That is the basic problem with altimeters and one we must work around, as best we can, with whatever atmospheric information we have on hand at the time. In Chapter 1, for instance, to find accurate vertical climb or glide intervals, we used one or more temperature readings, taken within or on the boundaries, to estimate true altitude height difference Δh. Next we delve into the altimeter setting, the one adjustment we do have.

Derivation of the Altimeter Setting Formula

In the ISA, at any altitude h, Eqs. (1.8) and (1.9) tell us that

$$\frac{T}{T_0} = \left(\frac{p}{p_0}\right)^{\alpha R} \qquad (2.37)$$

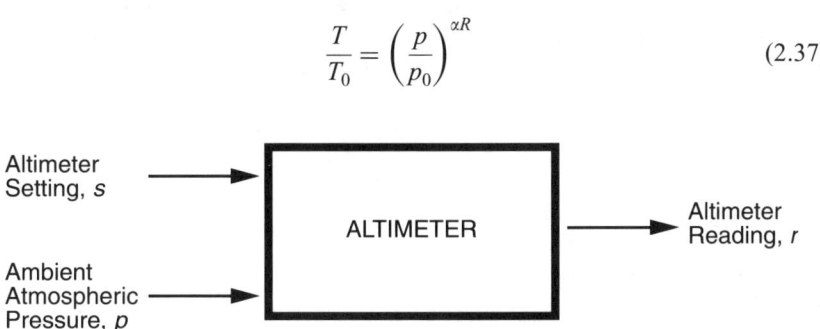

Figure 2.6   Input and output variables for the altimeter.

Now consider a second "standard" atmosphere, with the same lapse rate α and chemical composition (hence same specific gas constant R) but with a different standard temperature T'_0 and a different standard pressure p'_0. In it, we would have

$$\frac{T}{T'_0} = \left(\frac{p}{p'_0}\right)^{\alpha R} \tag{2.38}$$

Dividing Eq. (2.37) by Eq. (2.38), identifying the new standard pressure p'_0 as the altimeter setting s, and multiplying by T_0, we have

$$T'_0 = T_0 \left(\frac{s}{p_0}\right)^{\alpha R} \tag{2.39}$$

Similar to Eq. (1.6), the pressure in this nonstandard "standard" atmosphere is

$$p(h) = s\left(1 - \frac{\alpha h}{T'_0}\right)^{1/\alpha R} = s\left(1 - \frac{\alpha h}{T_0 \left(\frac{s}{p_0}\right)^{\alpha R}}\right)^{1/\alpha R} \tag{2.40}$$

When one solves for altimeter setting s, one gets

$$s = p\left(1 + \left(\frac{p_0}{p}\right)^{\alpha R} \frac{\alpha h}{T_0}\right)^{1/\alpha R} \tag{2.41}$$

and this is *almost* the formula for altimeter setting as seen in *Smithsonian Meteorological Tables*.[2] There is one further wrinkle: 0.01 in. Hg is to be subtracted from the actual pressure p to account for the approximate 10-ft vertical distance up to the cockpit location of the altimeter from the bottoms of the tires (adjusted for tread wear, presumably).

In use, to tie down all the details, the pressure p we are discussing is that at the field, hence called p_f and the altitude h is the elevation of the field barometer, hence called h_f. So the official formula for altimeter setting, in all its glory, is

$$s = (p_f - 0.01)\left(1 + \left(\frac{p_0}{p_f - 0.01}\right)^{\alpha R} \frac{\alpha h_f}{T_0}\right)^{1/\alpha R} \tag{2.42}$$

Example 2.6 Consider a field at $h_f = 4964$ ft with ambient pressure 26.00 in. Hg. Equation (2.42) then gives ($\alpha R = 0.19025$, $1/\alpha R = 5.25635$, and $T_0/\alpha = 145{,}457$) altimeter setting $s = 31.15$ in. Hg. In our performance uses we often ignore the 0.01 in. Hg offset. Doing so here would have given 31.16 in. Hg.

The Altimeter is a Barometer

Everyone says that the altimeter is just a barometer, but hardly anyone knows how to walk out onto the tarmac, glance into an airplane's cockpit, and accurately report the current ambient barometric pressure. But, with the altimeter setting formula in hand, that is easy to do. Simply solve Eq. (2.42) for pressure p, interpreting field elevation h_f as instrument reading r, to get

$$p = \left(s^{\alpha R} - p_0^{\alpha R}\frac{\alpha r}{T_0}\right)^{1/\alpha R} + 0.01$$

$$= (s^{0.19025} - 1.3124 \times 10^{-5} r)^{5.25635} + 0.01 \text{ in. Hg} \qquad (2.43)$$

Using figures from example 2.6, Eq. (2.43) does indeed give $p = 26.00$ in. Hg.

Checking Altimeter Consistency

These altimeter manipulations assume the instrument is properly calibrated. A quick way to check at least consistency, though not accuracy, is outlined in the *Federal Air Regulations* (*FAR*).[3] The technician sets $s = 29.92$ in. Hg and records the corresponding altimeter face reading r as r_0. Then the technician dials in seven other altimeter setting values, the left-hand column of Table 2.1, and sees whether the face readings are within 25 ft of those values suggested in the right-hand column. If so, the altimeter passes this mild muster.

Example 2.7 Assume, for instance, that indicated pressure altitude r_0 turns out to be 3000 ft. We can check to see whether the *FAR* table is correct by first using Eq. (2.43) to find the ambient pressure p, then solving Eq. (2.43) for altimeter reading r and checking a tabular point or two.

Table 2.1 Table IV in Appendix E of *FAR* Part 43, used to test altimeters

Altimeter setting s, in. Hg	Altimeter reading r, ft
28.10	$r_0 - 1727$
28.50	$r_0 - 1340$
29.00	$r_0 - 863$
29.50	$r_0 - 392$
29.92	r_0
30.50	$r_0 + 531$
30.90	$r_0 + 893$
30.99	$r_0 + 974$

Pressure p turns out to be 26.83 in. Hg. Solving Eq. (2.43) for r gives

$$r = \frac{T_0}{\alpha} \frac{(s^{\alpha R} - (p - 0.01)^{\alpha R})}{p_0^{\alpha R}} = 76,195.6 \times (s^{0.19025} - (p - 0.01)^{0.19025}) \quad (2.44)$$

Checking for $s = 28.50$ in. Hg with Eq. (2.44) gives $r = 1656\text{ft} = 3000\text{ft} - 1344\text{ft}$. The table is only 4 ft off.

With Eqs. (2.42 – 2.44), we have the full altimeter story, the ability to solve for any one of the three variables pressure p, altimeter setting s, and altimeter reading r when given values of the other two variables.

Changes in Indicated Altitudes Equal Changes in Pressure Altitudes

In Chapter 1 we stressed techniques for estimating true altitude differences from pressure altitude differences and selected temperature measurements. It is therefore gratifying to find that, if your altimeter does *not* happen to be set to 29.92 in. Hg, you can still easily find pressure altitude differences. No matter what altimeter setting you have, as long as you do not change it between the bottom and top of your altitude interval, the difference in altimeter readings Δr is the same as the difference in pressure altitudes Δh_p. We now show why this is the case.

We do neglect the 0.01 in. Hg offset. So Eq. (2.44) now reads

$$r = \frac{T_0}{\alpha} \frac{(s^{\alpha R} - p^{\alpha R})}{p_0^{\alpha R}} \quad (2.45)$$

If you look back into Chapter 1 for the formula for pressure altitude, you find it can be written

$$h_p = \frac{T_0}{\alpha} \frac{(p_0^{\alpha R} - p^{\alpha R})}{p_0^{\alpha R}} \quad (2.46)$$

It is clear from the forms of Eqs. (2.45) and (2.46) that differences of either r or h_p will eliminate the different constants in the parentheses of each form. So the result is that (for any constant value of altimeter setting s used in the r measurements)

$$\Delta r = \Delta h_p \quad (2.47)$$

General Linear Nonstandard Atmosphere

It often happens, in performance work, that you need an overall or "global" estimate of a particular day's atmosphere above a particular locality. With only scanty data, that may have to be a rough estimate, but the more accurate the better. Perhaps airplane X took off from airport Y and climbed at full throttle, at speeds for best rate of climb, up to cruise at indicated altitude Z. There something happened, resulting in an accident, and your job in some legal litigation support is to re-create performance numbers for that flight. To do that, you need estimates of atmospheric parameters.

We want to construct a particular nonstandard atmosphere, one with the same mathematical structure as the ISA but based on three different defining constants T'_0, p'_0, and α'. To find those three constants requires three pieces of experimental atmospheric information. We assume specific gas constant R, and of course the acceleration of gravity g, are the same as in the standard atmosphere. (If humidity is high, you can simply substitute for R the appropriately higher value R' as described in Chapter 1.) The most common situation is that we have both temperature and pressure measurements in two locations—on the ground (at the departure airport) and aloft. That is four pieces. We need four because the altitude aloft is a fourth unknown. Let us start from the beginning.

Temperature and pressure distributions, in this nonstandard atmosphere, are given by

$$T(h) = T'_0 - \alpha' h \tag{2.48}$$

and

$$p(h) = p'_0 \left(1 - \frac{\alpha' h}{T'_0}\right)^{1/\alpha' R} \tag{2.49}$$

Using subscript f to refer to the field and subscript a to refer to aloft, specializing Eqs. (2.48) and (2.49) give us these four relations:

1) Known temperature T_f on the field at known elevation h_f

$$T'_0 - \alpha' h_f = T_f \tag{2.50}$$

2) Known (or knowable) pressure p_f on the field. This may need to be backengineered from values for the then-current altimeter setting and field elevation, using Eq. (2.43). When that has been done, the relevant relation is

$$p'_0 \left(1 - \frac{\alpha' h_f}{T'_0}\right)^{1/\alpha' R} = p_f \tag{2.51}$$

3) Known temperature T_a at unknown altitude h_a. Use Eq. (2.48) again.

$$T'_0 - \alpha' h_a = T_a \tag{2.52}$$

4) Known or knowable pressure p_a at altitude h_a. Use Eq. (2.43), with altimeter setting value s and recorded at-altitude altimeter reading r, to

get p_a; then use Eq. (2.49).

$$p'_0\left(1 - \frac{\alpha' h_a}{T'_0}\right)^{1/\alpha' R} = p_a \qquad (2.53)$$

The four unknowns in Eqs. (2.50–2.53) are α', T'_0, p'_0, and h_a; everything else is measurable or calculable. Here is the stepwise solution.

Substitute Eq. (2.49) into (2.50) and Eq. (2.51) into (2.52). Divide those results, take logarithms, and solve for α':

$$\alpha' = \frac{1}{R}\frac{\ln(T_f/T_a)}{\ln(p_f/p_a)} \qquad (2.54)$$

With α' determined, use it in Eq. (2.50) to get

$$T'_0 = T_f + \alpha' h_f \qquad (2.55)$$

With both α' and T'_0, determined, use them in Eq. (2.51) to get

$$p'_0 = p_f\left(1 - \frac{\alpha' h_f}{T'_0}\right)^{-1/\alpha' R} \qquad (2.56)$$

Now h_a can be calculated from Eq. (2.52):

$$h_a = \frac{T'_0 - T_a}{\alpha'} \qquad (2.57)$$

That is not quite all there is to the story. We have good values for today's atmosphere: α', T'_0, and p'_0, and a good value for h_a. But we do not want to infer other altitude values h from temperature relations similar to Eq. (2.57)! We can only read a thermometer to about $\pm 0.5°F$. At usual values of α', this is equivalent to an altitude error of about $\pm 0.5/0.003566 = \pm 140$ ft. In addition, thermometers lag worse than altimeters; many of our flight tests will not allow sufficient time at constant altitude for the thermometer to stabilize. But do not forget we still have our altimeter even though we now know we have to read it with a calculational grain of salt. To use this more precise pressure-measuring device—± 10 ft should be possible—we use a combination of Eqs. (2.49) and (2.43):

$$p'_0\left(1 - \frac{\alpha' h}{T'_0}\right)^{1/\alpha' R} = s\left(1 - \left(\frac{p_0}{s}\right)^{\alpha R}\frac{\alpha r}{T_0}\right)^{1/\alpha R} \qquad (2.58)$$

then solve this for h:

$$h = \frac{T'_0}{\alpha'}\left[1 - \left(\frac{s}{p'_0}\right)^{\alpha' R}\left(1 - \left(\frac{p_0}{s}\right)^{\alpha R}\frac{\alpha r}{T_0}\right)^{\alpha'/\alpha}\right] \qquad (2.59)$$

where α', T'_0, and p'_0 are given by our previous results, Eqs. (2.54–2.56). For full generality, we have left altimeter setting s completely arbitrary; you can simplify Eq. (2.59) somewhat by either setting $s = p_0$ or $s = p'_0$. As are the solutions for

α', T_0', p_0', and h_a, Eq. (2.59) for h is exact within the limits of experimental errors.

Example 2.8 Your home base, Waypoint Field, elevation (and weather station elevation) 2000 ft, announces current OAT $= T_f = -25°F = 434.7°R$ and current altimeter setting $s_f = 29.00$ in. Hg. You intend to run glide tests at about 7000 ft.

You set your altimeter's barometric subdial to $s = s_f = 29.00$ and leave it there. You take off and climb to indicated altitude $r = 7500$ ft; there you fly level for a minute and find that $T_{h'} = -35°F = 424.7°R$. Then you climb up another 1000 ft, cut your throttle, and stabilize to your first target air speed. You time your glide down from $r = 7500$ ft to 6500 ft. What was your actual geometric altitude loss Δh for that indicated 1000 ft descent?

1) Find the field-level pressure from Eq. (2.43):

$$p_f = (29.00^{0.19025} - 0.000013124 \times 2000)^{5.25635} + 0.01$$
$$= 26.965 \text{ in. Hg} \tag{2.60}$$

2) Find the first at-altitude pressure, also from Eq. (2.43):

$$p_a = (29.00^{0.19025} - 0.000013124 \times 7500)^{5.25635} + 0.01$$
$$= 21.930 \text{ in. Hg} \tag{2.61}$$

3) Find α' from Eq. (2.54):

$$\alpha' = \frac{1}{53.35} \frac{\ln(434.7/424.7)}{\ln(26.965/21.930)}$$
$$= 0.0021106 \text{ F/ft} \tag{2.62}$$

4) Find T_0' from Eq. (2.55):

$$T_0' = 434.7 + 0.0021106 \times 2000$$
$$= 438.9°R = -20.8°F \tag{2.63}$$

5) Find p_0' from Eq. (2.56):

$$p_0' = 26.965\left(1 - \frac{0.0021106 \times 2000}{438.9}\right)^{-1/(0.0021106 \times 53.35)}$$
$$= 29.382 \text{ in. Hg} \tag{2.64}$$

Note that p_0' is *not* equal to s_f.

6) Find the first true altitude h_a from Eq. (2.57):

$$h_a(r=7500) = \frac{438.9 - 424.7}{0.0021106}$$

$$= 6727.9 \text{ ft} \qquad (2.65)$$

7) Find the second true altitude from Eq. (2.59):

$$h(r=6500) = \frac{438.9}{0.0021106}\left[1 - \left(\frac{29.00}{29.382}\right)^{0.0021106 \times 53.35}\right.$$

$$\left. \times \left(1 - \left(\frac{29.92}{29.00}\right)^{0.19025} \frac{0.003566 \times 6500}{518.7}\right)^{0.0021106/0.003566}\right]$$

$$= 5882.6 \text{ ft} \qquad (2.66)$$

So our -1000 ft nominal Δh is actually

$$\Delta h = h(r=6500) - h_a(r=7500)$$

$$= 5882.6 - 6727.9$$

$$= -845.3 \text{ ft} \qquad (2.67)$$

We avoided, in this cold case, a 15% error in Δh. It took work to go through all of these steps, but it does not take much more work to set up a computer spreadsheet that will give all such answers as fast as you can type in the raw data. It is worth the effort to avoid errors in numbers as important and ubiquitous as parasite drag coefficient C_{D0} and airplane efficiency factor e.

From the terrain clearance point of view, you were about 700 ft lower than your indicated altitudes. "Cold weather contracts the isobars." The altimeter is a *pressure*-measuring instrument.

Static Pressure Position Error

In our block diagrams of the ASI (Fig. 2.1) and the altimeter (Fig. 2.6), we neglected one fairly important aspect. There may be errors (see Figs. 2.7 and 2.8). In this subsection we consider the most important such error—systemic error due to not finding a *position* along the fuselage where static pressure is the same as that in the free stream at the same altitude.

Most atmospheric instrument errors, for our subsonic general aviation purposes, are either easily correctable or essentially negligible. Among the first are the brute instrument errors, for which calibration cards can be prepared during periodic instrument shop checks, and ASI calibration error, which we have treated

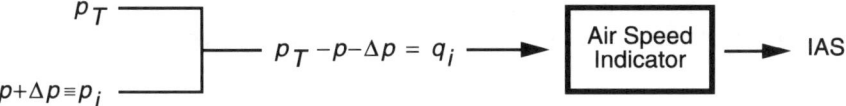

Figure 2.7 Air-speed indicator position error.

in detail. Among the almost negligible errors is lag error due to the length and small diameter of tubing between the pitot tube and the ASI and from the static port to either instrument. We can neglect lag error, in almost all cases, because it is self-compensating. If our altimeter lags say 1 s (typical lag errors are from 0.5 to 2.0 s) behind our actually reaching target altitude of 7500 ft, no matter; it will lag an almost identical amount when we pass through a second target altitude of 6500 ft. Another is that in the pitot tube itself. Up to AOAs of about 17 deg, total pressure pitot errors are negligible.

Static pressure position error, however, is a problem. First of all, it depends on several variables: pressure altitude, air speed, AOA, flaps/gear configuration, side slip angle, Reynolds number, and Mach number. Because most of our performance concerns are quasi-steady-state concerns, low maneuvering, coordinated flight, and far subsonic, we can dispose of the latter three influences as beyond our treatment. Each configuration must almost always be accounted for by having its own position error determination. Because air speed and AOA are so intimately connected, we can combine those. So we are left with position error $\Delta p \equiv p_i - p$ as a function of true static pressure p and TAS V.

That will still be plenty to worry about. If Δp is positive, so that indicated pressure p_i (actually inferred, not directly indicated) is too high, then IAS is low ($V_i < V_c$) and indicated pressure altitude is also low ($h_{pi} < h_p$). Let us further simplify the problem and cut it down to tractable size by only determining position error at one air speed and at one pressure altitude. Expanding to other values of those base variables will then just be more of the same.

We will consider two techniques. In each treatment we will assume there is only one common static port and that the size of the position error is the same for each of the two instruments. That way we can play one off against the other. Because there is only one position error, a single veridical value (in one case

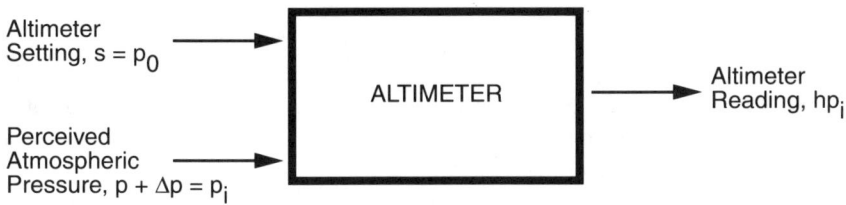

Figure 2.8 Position error gives the altimeter a false input.

CAS, in the other case pressure altitude) will suffice to fix the error. The static pressure correction, recall, is simply the negative of the static pressure error.

Determining Static Pressure Position Error: Technique #1

What is required is steady measurement—in essence, simultaneous—of indicated pressure altitude h_{pi}, IAS V_i, TAS V, and (to get to CAS V_c through density ρ) temperature T. The position error to be calculated is indicated pressure minus true pressure: $\Delta p = p_i - p$. Indicated pressure is no problem. To get it, one simply used the standard relationship, Eq. (1.6). So we need an expression for true pressure p in terms of measured quantities. Our method is to get two expressions for CAS V_c, so that can be eliminated, and then also two expressions for total pressure p_T, so that can be eliminated. We will then be left with only measurable items.

Because we are subsonic, the definition of dynamic pressure q and the ideal air speed indicator construction immediately give us

$$V_c^2 = \frac{2(p_T - p)}{\rho_0} \tag{2.68}$$

while the relation between calibrated and TAS and the ideal gas law give us

$$V_c^2 = \frac{pV^2}{RgT\rho_0} \tag{2.69}$$

Putting Eqs. (2.68) and (2.69) together and solving for total pressure then gives

$$p_T = p\left(1 + \frac{V^2}{2RgT}\right) \tag{2.70}$$

On the other hand, the realistic ASI relationship

$$V_i^2 = \frac{2(p_T - p_i)}{\rho_0} \tag{2.71}$$

gives us an alternative formula for total pressure:

$$p_T = \tfrac{1}{2}\rho_0 V_i^2 + p_i \tag{2.72}$$

Using Eqs. (2.70) and (2.72), and again using the ideal gas law and a bit of algebra, gives us the required relation:

$$\Delta p \equiv p_i - p = \frac{p_i V^2 - p_0 V_i^2 T/T_0}{2RgT + V^2} \tag{2.73}$$

Atmospheric Aircraft Instruments

Determining Static Pressure Position Error: Technique #2

This is the tower fly-by (or pacer aircraft) technique in which an accurate value of pressure altitude is independently determined. In this case, the expression for static pressure position error is immediate:

$$\Delta p \equiv p_i - p = p_0 \left[\left(1 - \frac{\alpha h_{pi}}{T_0}\right)^{1/\alpha R} - \left(1 - \frac{\alpha h_p}{T_0}\right)^{1/\alpha R} \right] \qquad (2.74)$$

Position Error Causes Altimeter and ASI Errors

We are not quite finished. Finding the position error is not an end in itself. Assuming the error is fairly small—a few hundredths of an inch of mercury—we can avoid some onerous algebra by simply taking partial derivatives of the variables of interest, V_c and h_p, with respect to pressure p (or vice versa) and using those as differential expressions. For CAS,

$$\frac{\partial V_c}{\partial p} = \frac{-1}{\rho_0 V_c} \qquad (2.75)$$

and so

$$\Delta V_c = \frac{-\Delta p}{\rho_0 V_c} \qquad (2.76)$$

In practice, one would use known IAS V_i, instead of V_c, on the RHS of Eq. (2.76).

In the pressure altitude case, it is convenient to use Eq. (1.6) and take this derivative:

$$\frac{\partial p}{\partial h_p} = \frac{-p_0}{RT_0} \left(1 - \frac{\alpha h_p}{T_0}\right)^{1/\alpha R - 1} \qquad (2.77)$$

Inverting and using the ideal gas law then gives us

$$\Delta h_p \doteq \frac{-\left(1 - \dfrac{\alpha h_p}{T_0}\right)^{1 - 1/\alpha R} \Delta p}{g \rho_0} \qquad (2.78)$$

Again, one would use known indicated pressure altitude h_{pi} instead of h_p on the RHS of Eq. (2.78).

Example 2.9 Let us use technique #2. The airplane, with altimeter set to 29.92 in. Hg, flies by the tower at an indicated pressure altitude of 100 ft. Those in the tower, with their accurate barometer, level and sighting grid, along with a little trigonometry to make up small deviations they detect from level with respect to their barometer, see the pressure altitude as 125 ft.

Equation (2.74) gives the static pressure position error as $\Delta p = 0.0269$ in. Hg $= 1.9025$ psf. If this figure is plugged right back into Eq. (2.78), using known $h_{pi} = 100$ ft, one indeed gets $\Delta h_p = -24.95$ ft. But that is not the use of the position error that matters. The one that counts operationally is what the altimeter error is at some other altitude, when there is no one around to check. Say the same airplane happens to be at $h_{pi} = 7500$ ft at $V_i = 100$ KIAS. Then using Eq. (2.78) tells the pilot that the pressure altitude error is -19.9 ft, so that actual pressure altitude, assuming position error is the only error, is 7519.9 ft. (Recall the correction is always the negative of the error.) Also, using Eq. (2.76)—and not neglecting to convert to British engineering units—the ASI error is -4.7 kn. So the airplane is actually traveling, again assuming position error is the only ASI error, at 104.7 KCAS.

You will undoubtedly run into slightly different instrument calibration and error analysis situations that require unique combinations of the tools we have assembled in this chapter. For instance, if errors are large and not correctable (say, because they are historical or recorded from an aircraft destroyed or otherwise not available), you will need to forsake the above rough differentials for more elaborate algebraic differences. You may find it necessary to evaluate position errors at several speeds, thereby partitioning ASI calibration curve errors into position errors plus nonposition errors. But the tools you need to do these things have been laid out.

References

1. Askue, V., *Flight Testing Homebuilt Aircraft,* Iowa State University Press, Ames, IA, 1992.
2. List, R.J. (ed.), *Smithsonian Meteorological Tables,* 6th rev. ed., Smithsonian Institution Press, Washington, DC, 1984.
3. *Federal Air Regulations,* Part 43, Appendix E.

3

Aircraft Performance Preliminaries

Introduction

We are still not quite ready to plunge into the vast sea of airplane performance. We need to check off a moderate list of preliminary background information in both physics and aeronautics: some terms (most will be italicized at first mention), and ideas behind them, which will be our constant companions. Much of this may be review.

First, on the physics side, some performance items (speeds and rates of climb or of yaw) involve *rates*. Some involve *positions*; a few, *accelerations*. Accelerations are tied, through Newton's *Laws of Motion*, to *forces*. Forces are *vectors* and have a direction as well as a magnitude. The same vector may be described differently in different *frames of reference*. Other performance items are described using *angles* and *rotation speeds*; then *torques* supplant forces and *angular speeds* replace linear ones. Because our main subject is quasi-steady-state performance—certainly not the considerably more advanced theory of stability and control—we will not need much of this preliminary physics apparatus, only a nodding acquaintance to help get our bearings. As we set out what we need we will also say where those bits fit into larger pictures of perhaps future studies.

Second, on the aeronautics side, we need to come up with equations specific to aircraft performance. Also, a brief treatment of airplanes and how they are controlled.

Units and Dimensions

Physics is, or hopes to be, a mathematical model of nature. Engineering science extends the fundamentals of physics to include mathematical models of industrial machines (airplanes!) and processes (internal combustion!). Inherent in this modeling is the replacement of real things (properties of physical objects, such attributes as mass, weight, position, etc.) with numbers. The numbers measure the attributes. The correct version of "I am 6 ft tall" is ungainly: "the number that measures my height, in a system of units employing feet as the unit of length, is 6." No one can stand such pedantry for long, but it is a good idea to be able to recognize that the modeling idea is always lurking in the background.

Some physics/engineering measure numbers are simply numbers: e.g., number of propeller blades per propeller B, Mach number M, relative air density σ. These pure "dimensionless" numbers either come from simple counts (as B) or from ratios in which all units have canceled (M is the ratio of air speed to the speed of sound, σ the ratio of ambient air density to "standard" [MSL] mean sea level density ρ_0, 0.002377 slugs/ft^3). Other measure numbers do have units attached: 6 ft, 9 y, 74.6 slugs, 2400 lbf. Units that measure the same sort of thing—for instance, feet and yards are both units of length—are said to have the same *dimensions*. There are only a few fundamental dimensions, but, even so, which set to choose is somewhat arbitrary. A common choice among engineers is force F, length L, time T, and temperature Θ (Greek capital theta). A common choice among physicists replaces F with mass M. Some also count electric charge as a fundamental dimension. Angles are somewhat strange. We ordinarily use degrees, but theoreticians prefer radians (rad). Because radian measure is arrived at by taking the length of the arc subtended by the angle in some circle, then dividing that arc length by the circle's radius, radians are dimensionless; accordingly, we also take degrees to be dimensionless.

Even after the fundamental dimensions have been chosen, the choice of their *units* is still open. British engineering units, still common among aeronautical engineers in the U.S., use feet (ft) for L, seconds (s) for T, and pounds force (lbf) for F. Temperature, in practical aviation work, is variously measured in degrees Fahrenheit (°F), degrees Rankine (°R, the "absolute" version of Fahrenheit), but also in degrees Celsius (°C), or degrees Kelvin (K, absolute Celsius). A similar European system employs meters (m) for L, seconds (s) for T, and Newtons (N) for F; the European system sticks with Celsius and Kelvin for temperatures. Further details are given in Chapter 6 where so-called *dimensional analysis* is used to help unravel the propeller problem.

We shall use British engineering units almost exclusively. This includes a few other named units. Mass, for instance, is expressed in *slugs*. The slug is a "derived" unit, with 1 slug defined as 1 lbf-s^2/ft, the amount of mass that accelerates at 1 ft/s^2 when acted on by a force of 1 lbf. Using square brackets to mean "the dimensions of," $[M] = FL^{-1}T^2$. The order in that product is of no consequence. The speed variable is our major exception, or augmentation, to strict use of British engineering units. The British engineering speed unit is ft/s. While our equations will use ft/s, input and output will more often cite knots (nautical miles per hour, n mile/h) or, occasionally, (statute) miles per hour (mph). Because a U.S. nautical mile is 6076.115 ft, and an hour consists of 3600 s, we have

$$1 \text{ kn} = \frac{1 \text{ n mile}}{\text{h}} = \frac{6076.115 \text{ft}}{3600 \text{ s}} = 1.6878 \text{ ft/s} \tag{3.1}$$

Hence, a speed given in knots must be multiplied by 1.6878 to express that same speed in feet per second. The reciprocal, 0.592484, is what a given speed in feet per second must be multiplied by to obtain the correct expression of that same speed in knots. Statute miles per hour have a somewhat more easily memorized factor for conversions to and from feet per second. Sixty mph = 88 ft/s. So multiply a given speed in feet per second by 60/88 to get miles per hour; multiply a given speed in miles per hour by 88/60 to get feet per second. However you do it, it is easy to make mistakes converting from one set of units to another.

Associating units with measure numbers, while still focusing on the underlying physical objects as being paramount, leads us to mathematically strange statements such as

$$24 \text{ ft} = 8 \text{ yd} \tag{3.2}$$

Equation (3.2) is patently untrue if you hold strictly to the traditional definition of "equals" as "is the same number as." We do not. Once units get into the act, we loosen up "equals" to mean, in this case, that the lengths are the same; we understand that the measure numbers for those lengths, if we employ different units to express them, will not be numerically equal.

Scalars and Vectors

The simplest quantities we measure—number of propeller blades, mass, altitude—are expressed as single numbers. The numbers may be integers or fractions, signed or not, with or without units. We will not need to use so-called *imaginary* or *complex* numbers, though those are very useful in airfoil design and in much of advanced aerodynamics. Single number quantities are known as *scalars*.

Vectors are the next most sophisticated category of physical variables. Vectors have sizes and also directions; it takes an ordered pair or triple of numbers to express a vector and a vector is diagramed by drawing an arrow with a certain length and direction. Weight is a vector anchored to the center of mass of an object and directed downward; velocity is a vector also anchored to the center of mass, directed in the (instantaneous) direction of motion of the object, with length the speed of the object in whatever convenient system of units. Two vectors are "equal," in yet another minor modification of that term, if and only if both lengths are equal, in the same system of units, and both directions are the same.

Airplane A headed north at 100 kn in level cruise has the same *speed* as airplane B headed east at 100 kn in level cruise; but they do not have the same *velocity*.

The easiest way to add two vectors is to resolve each of them into mutually perpendicular components along the axes of the same Cartesian coordinate system. The say X component of the sum is then simply the sum of the X components of the two addends. Similarly with the other components. There are complicated formulas for doing this addition within polar coordinates, but it is usually easier, if need be, to transform to rectangular coordinates, do the addition, then transform the sum back to polar coordinates.

There are two types of products of vectors: scalar or "dot" products and vector or "cross" products. The dot product of vectors A and B is

$$\boldsymbol{A} \cdot \boldsymbol{B} \equiv AB \cos[< (\boldsymbol{A}, \boldsymbol{B})] \tag{3.3}$$

where scalars A and B are the lengths of the eponymous vectors and the argument of the cosine function is the angle between the two vectors. Because the cosine function is symmetric, it does not matter whether one considers the angle from A to B or the one from B to A. The most convenient way to calculate dot products is often this relation:

$$\boldsymbol{A} \cdot \boldsymbol{B} = A_x B_x + A_y B_y + A_z B_z \tag{3.4}$$

where the subscripted quantities are the components of the two vectors along the indicated axes.

The cross product of two vectors A and B is considerably more complicated. $A \times B$ is a vector, perpendicular to both A and B, pointed in the direction of advance of the right-hand screw that would "turn" A into the direction of B, the shorter way, if A were welded to the screw head. (You may know what I mean, but probably not through the agency of my description.) The size of $A \times B$ is the product of their lengths, AB, multiplied by the sine of the angle between them. This is the area of the parallelogram formed by A and B. The computation of $A \times B$ is usually done by expanding a certain determinant made up of the three unit vectors and components of A and B:

$$\boldsymbol{A} \times \boldsymbol{B} = \begin{vmatrix} \boldsymbol{i} & \boldsymbol{j} & \boldsymbol{k} \\ A_x & A_y & A_z \\ B_x & B_y & B_z \end{vmatrix} \tag{3.5}$$

With the exception of a brief foray into the Coriolis pseudoforce, we won't go deeply enough into mechanics to need to do this.

Coordinate Systems

The same physical vector may be expressed differently in two coordinate systems. The only two kinds of coordinate systems we will need here are 1) ordinary right-handed rectilinear Cartesian ones, with three mutually perpendicular axes x, y, and z (these axes oriented, respectively, as the thumb, forefinger, and middle finger of a right hand) and 2) circular or cylindrical polar coordinates, with radial coordinate r (in two dimensions) or ρ (in three dimensions), azimuthal or angular coordinate θ (increasing counterclockwise starting from the positive x axis), and height coordinate again z. Trigonometry comes to the rescue when we need to express a vector, given in one of these coordinate systems, in terms of the other coordinate system. In the two-dimensional polar case,

$$x = r\cos\theta, \quad y = r\sin\theta \tag{3.6}$$

or, inversely,

$$r = \sqrt{x^2 + y^2}, \quad \theta = \tan^{-1}(y/x) \tag{3.7}$$

We will skip over the problem of what to do when $x = 0$; in practical cases, continuity will take care of that ambiguity. Equations (3.6) and (3.7) are for the two coordinate systems with a common origin, common z axis, and oriented as mentioned in the description of polar coordinate θ.

Frames of Reference

Coordinate systems can also be in motion with respect to one another. The major such case we will consider involves describing aircraft motion alternatively in a "fixed" Earth-based system and in a "moving" air-mass-based system. The air mass is assumed to move, with respect to the Earth below it, with constant speed in a constant direction; no accelerations. We will say something about rotating coordinate systems after we have introduced circular motion. In this easier situation, there is a simple algebraic rule for getting this coordinate transformation done properly. If we denote vectors as bolded quantities, and use the symbol $\oplus$ to denote the Earth, a for the air mass, and p for the (air)plane, with for instance $V_{\oplus p}$ meaning velocity of the plane with respect to the Earth, then

$$V_{\oplus p} = V_{\oplus a} + V_{ap} \tag{3.8}$$

The three objects featured in Eq. (3.8)—Earth, air mass, and airplane—might be any other three; those are what we are interested in here. Or they could be those same three, but permuted. It is only the algebraic relationship, with the "collapse" of the internal symbol a on the right side, that matters. In use, one often writes a relation like Eq. (3.8) in any convenient form he or she can get a handle on, then rewrites it at leisure using the usual rules of algebra (adding the same quantity to each side or multiplying a term by minus one, as $V_{\oplus p} = -V_{p\oplus}$) to get precisely the relation needed.

Wind, the motion of an air mass, is a somewhat strange case. Air movement is almost always reported with respect to the Earth. Wind velocity V_w is velocity of the air mass with respect to the Earth, $V_{\oplus a}$. We report this, however, in a uniquely confusing polar convention—wind is currently 20 kn from 250 deg, for example—giving wind speed and the direction the wind is from instead of the direction in which it is moving. Well, in Minneapolis, Minnesota, we speak of individuals from Mobile, Alabama, as "Southerners," emphasizing where they come from, and what they bring with them, rather than stressing their direction of flight. Similarly, the sea breeze brings moisture, the norther (headed south) brings, in the northern hemisphere, arctic chill. In fact our description of a wind velocity, if you ignore the word "from," is a description of $-V_w = V_{a\oplus}$, the velocity of the Earth with respect to the air mass. Practically? The wind velocity vector, which we need in equations, comes from simply tacking a minus sign onto the conventional description.

Linear Kinematics: Position, Velocity, Acceleration

To describe motion of a point—usually the center of mass or c.g. of some extended object, perhaps an airplane—one first of all needs a coordinate system. The *position* of the point at any time t is vector $x(t)$. Another symbol, other than x, may of course be used; if the motion of more than one object is being described, one will use a different symbol, or distinguishing subscripts, for the motion of each. *Displacement* of the object, between times t_1 and t_2, is the vector $[(x(t_2) - x(t_1))]$.

Velocity of the point at time t_1, $v(t_1)$, is the time derivative of x evaluated at time t_1; this derivative is conventionally written, in our culture's not-so-consistent functional notation, as $dx(t_1)/dt$. To take such a derivative, one actually uses the two or three scalar functions for the like number of components of vector x.

Acceleration of the point at time t_1, $a(t_1)$, is the time derivative of v evaluated at time t_1, written $dv(t_1)/dt$. Again one actually uses scalar functions in the mathematical work.

Example 3.1 Because we are only reviewing notation, let us see how this works in an example you already know about. You know that if you drop an object out of an airplane into a frictionless atmosphere, it suffers a constant acceleration of size $g = 32.174\,\text{ft/s}^2$ downwards. Let us describe this idealized motion (flat Earth, no variation of gravitational acceleration with altitude, no atmosphere).

First of all, to give the numbers something to refer to, we need a coordinate system. Take a rectilinear Cartesian system with origin at the position of the center of mass of the object at the instant of release. Take the x axis in the direction of flight, the y axis horizontally to the right as the pilot faces, and the z

axis straight down, all of these pertaining to the instant of release. We have picked a particularly simple coordinate system, for this problem, so that we will get a particularly simple description of the motion. One final simplification: the object is at rest with respect to the coordinate system just described at the instant of release. As is often done, let us take vectors of unit length in the directions of the x, y, and z axes to be, respectively, $\boldsymbol{i}$, $\boldsymbol{j}$, and $\boldsymbol{k}$.

In this problem, we start off knowing acceleration is the constant vector $g\boldsymbol{k}$. The velocity vector, when differentiated with respect to time, that will give this is $\boldsymbol{v}(t) = gt\boldsymbol{k}$. The position vector that, when differentiated, will give $\boldsymbol{v}$ is $z(t) = \frac{1}{2}gt^2\boldsymbol{k}$. Because the motion is in the z direction, we switched from x to z for our position variable. As in many problems of this type, the proof is in the "check." Indeed, when you differentiate $z(t)$ with respect to t you do get $\boldsymbol{v}(t)$. And a further differentiation of $\boldsymbol{v}(t)$ indeed does give $\boldsymbol{a}(t)$. All of this is true only for times t for which $z(t)$ is less than the absolute altitude of the airplane at $t = 0$.

One aspect of this example that made things simple is that the coordinate system we chose had unit vectors $\boldsymbol{i}, \boldsymbol{j}, \boldsymbol{k}$ which themselves did not vary in time. If one has to deal with rotating coordinate systems (or other accelerated, so-called "non-inertial" frames of reference), such is not the case and description of motion (kinematics) gets messier. If you study stability and control or flight dynamics (even if only for flights still within the atmosphere), you will consider moving coordinate systems in detail and in abundance. We will touch on one application below.

Linear Dynamics: Force, Work, Energy and Power

But what creates the accelerations of objects that change their velocities, changing their positions? Forces. Sir Isaac Newton encapsulated this subject in Newton's Laws of Motion:

1) An object not acted on by forces, if at rest, remains at rest. If it is moving, it maintains that same speed and moves in a straight line (i.e., no force, no acceleration; therefore, constant speed).
2) An object of mass m, acted on by a force $\boldsymbol{F}$, undergoes acceleration in the direction of that force of size F/m (i.e., $\boldsymbol{F} = m\boldsymbol{a}$).
3) If object A exerts force $\boldsymbol{F}$ on object B, then object B exerts force $-\boldsymbol{F}$ on object A (i.e., each action is accompanied by an equal-sized, but oppositely directed, reaction).

The first law is just a special case of the second. The second law is essentially a definition telling us how to quantify the size of a force (see how big is the acceleration it produces on a mass of size m). The third law is not true for all electromagnetic phenomena but will be for interactions within our purview. The second law is the big one.

When the mass of an object experiencing a force is changing in time, the second law must be expanded to read the following: the time rate of change of *momentum* $p \equiv mv$ is given by force F (i.e., $F = dp/dt = d(mv)/dt$). This formulation is needed for rocket problems in which mass is continuously being ejected from the rear nozzle. All the above is of course only true for "classical" or "nonrelativistic" problems, problems in which all speeds, with respect to any pertinent coordinate system, are very small compared with the velocity of light (about 186,000 mile/s). Accurate time measurements required by global positioning system satellites use relativistic corrections, but that is about the only current Earthly aeronautical application where those are needed.

Work and energy are two more important mechanics concepts we will occasionally use. *Work W* done on an object by a force is measured as the product of the size of the force, F, and the distance through which the object was moved *in the direction of the force*:

$$W \equiv F \cdot d \qquad (3.9)$$

where vector d is the displacement of the body while acted on by the force. Work is a scalar, as is any quantity defined as a dot product of two vectors. British engineering work units are ft-lbf. Equation (3.9) works only for simple cases, when the force is constant. In more complicated cases, in which either the force or the displacement changes with time (e.g., the infamous "downwind turn"), an integral version of Eq. (3.9) is employed. We will wait until we need that.

Doing work on an object adds to its *energy* (as does heating it). Energy exists in many different forms. We will certainly be concerned with kinetic energy and gravitational potential energy, possibly with chemical energy and heat energy; seldom will our energy net be cast so wide as to bring in radiation energy (though it did in explaining the atmosphere's temperature lapse rate) or the inherent mass energy of nuclear reactors or explosions. Energy is a scalar and has the same units as does work.

Kinetic energy (*KE*) is the energy a massive body has by virtue of its motion. The kinetic energy of a body of mass m, in a frame of reference in which its speed is v, is

$$KE = \tfrac{1}{2} mv^2 \qquad (3.10)$$

In the example of the previous section, the body falling from the airplane in a vacuum, say from height h and for time interval t_f (*f* for "final"), it is easy to calculate the work done on the body by gravity and the corresponding final kinetic energy:

$$\begin{aligned} \mathbf{F} \cdot \mathbf{d} &= m g \mathbf{k} \cdot h \mathbf{k} \\ &= mgh \\ &= \tfrac{1}{2} m g^2 t_f^2 \\ &= \tfrac{1}{2} m v^2(t_f) \end{aligned} \qquad (3.11)$$

The general statement of what happens to a body's kinetic energy if you accelerate it with force $\mathbf{F}$ between positions r_1 and r_2 is

$$\begin{aligned} \int_{r_1}^{r_2} \mathbf{F} \cdot d\mathbf{r} &= m \int_{r_1}^{r_2} \frac{d\mathbf{v}}{dt} \cdot d\mathbf{r} \\ &= m \int_{v_1}^{v_2} \mathbf{v} \cdot d\mathbf{v} \\ &= \tfrac{1}{2} m v_2^2 - \tfrac{1}{2} m v_1^2 \\ &= KE_2 - KE_1 \end{aligned} \qquad (3.12)$$

Potential energy (PE) is energy a body has by virtue of its position in a force field. Of the various types of potential energy—electrical, magnetic, chemical, mechanical, and gravitational—we will only be interested in the gravitational variety. Only potential energy *differences* are defined, but we often implicitly assume the starting point or datum is at MSL, or on the Earth directly beneath the airplane, or at some standard altitude above or below the airplane's current position. If we take our previous falling object example and calculate the potential energy difference between having the object at ground level and placing it in the airplane at absolute altitude h, then

$$PE = mgh = Wh \qquad (3.13)$$

This is just the work—using our previous variables, $(-\mathbf{F} \cdot -\mathbf{d}) = (\mathbf{F} \cdot \mathbf{d}) = mgh$—it would take to lift the body from the *PE* datum (ground level), to the airplane's altitude, against the force of gravity. Notice, in Eq. (3.11), how naturally that amount of work appeared in the kinetic energy calculation.

Energy is conserved. That means that the total energy of a fixed set of material objects, confined within an energyproof barrier (no particle or heat or radiation leaks), never changes. In the falling object case, the body lost gravitational PE as it was falling and gained an equivalent amount of *KE*. Once it hit the ground, the body's splendid isolation was broken; then its *KE* was converted into heat energy, within itself and the neighboring ground, and in mechanical potential energy in its dents and ground compression.

In force/motion/mechanics problems, one often has a choice between considering forces in detail or, alternatively, taking an energy approach in which kinetic and potential energies are interchanged. You can feel free to take whichever path seems most obvious and simplest. But do not combine them; once potential energy is introduced you must not consider the force field responsible for that potential energy.

Power is the time rate of doing work. The strict British engineering units are, therefore, units of work divided by units of time (ft-lbf/s), but we most often fall back on the familiar preindustrial unit *horsepower* (hp), which is 550 ft-lbf/s. Using P for power, common relations in which it appears (assuming force is constant in time) are

$$P = \frac{dW}{dt} = \frac{d(\boldsymbol{F} \cdot \boldsymbol{d})}{dt} = \boldsymbol{F} \cdot \boldsymbol{v} \qquad (3.14)$$

Circular Kinematics: Angular Position, Velocity, and Acceleration

We describe the angular "position" or orientation of an object by saying how many degrees or radians a given line, embedded in the object, has rotated, in a certain direction, relative to a given reference position of that line. In the airplane case, the given line is usually taken to be one down the length of the fuselage, likely passing through the airplane's c.g.: the "body axis." We often pretend the body axis is parallel to the thrust axis, the symmetry axis of the propeller shaft, but those may differ by as much as a couple of degrees, perhaps down a bit, perhaps off a bit to the right (for propellers that rotate clockwise as seen from behind). That small off-axis possibility is only a minor annoyance because, for a given airplane, it is a fixed quantity. We usually ignore it.

The angle units used in science and mathematics are *radians*, previously defined as the ratio of arc subtended to radius, where the angle is considered a sector of any circle. The common angle unit, of course, is the *degree*. Because there are 2π radians, and 360 deg, in a full circle, a radian is $180/\pi \doteq 57.2958 \doteq 57.3$ deg. As is common in engineering work, we will use radians in equations, because that makes them simpler, and use degrees in common parlance. Conversion factors of $180/\pi$, therefore, or of its reciprocal $\pi/180$, abound. Degrees are sometimes broken into 60 min *of arc*, which in turn are broken into 60 s *of arc*, but modern usage, to which we will adhere, prefers the simplicity of decimal fractions of a degree. An angular position, or the angular displacement (*rotation*) required to achieve one angular position from another, can be any (real) number of radians or degrees, positive or negative.

For our fairly elementary purposes, in which rotations take place around one fixed and well-defined axis at a time, or very close to that, angular positions and displacements can be considered scalars. In the full generality of three-dimen-

Aircraft Performance Preliminaries 59

sional space, with rotations about axes that are not fixed, correct treatment of rotations involves describing them at least as vectors and, most generally, as pseudo-vectors or axial vectors, actually antisymmetric second-order tensors. You can leave that for a course in stability and control or one in orbital mechanics. We will only show angles and angular speed as vectors when it is necessary to get proper directions for such derived concepts as angular momentum or torque.

Example 3.2 An aside. Rotational kinematics and dynamics are *much* more complicated than their linear counterparts. Here is an example. My father-in-law bequeathed to me his small (about 3 in. long) two-blade pocket knife. This knife (and many of its type) has the following strange behavior. If you place it on its back on a Formica (low friction) countertop and spin it (blades closed) clockwise as seen from above, it rotates for about 8 s. It slows down, because of friction between the knife back and the countertop, and stops. It barely wobbles while it does this. If, on the other hand, you spin the knife counterclockwise, the knife wobbles (back and forth about its long axis) fairly violently, comes to a stop in only 3 or 4 s, and then *backs up* (rotates in the clockwise direction) approximately 20 to 30 deg. How can this be? Why the asymmetry? Why the extreme wobble? Why the backing up? Advanced physics mechanics books, or specialized tomes on rigid body mechanics, answer these questions. It is not a miracle—the knife is not haunted—but the causes are subtle and the proper mathematical description is quite involved.

Because angular position or displacement—for pitch angles we are likely to use θ (Greek small theta), for angles of attack we have used α (Greek small alpha), and for flight path angles we will mostly use γ (Greek small gamma)—can, in simple cases, taken to be a scalar, so too is its time rate of change, angular velocity or rotation speed, commonly denoted ω (Greek small omega). As is angular acceleration $d\omega/dt$; because we will not need it, the fact that angular acceleration is often denoted α, the same letter as angle of attack, will not confuse us. While ω usually has rad/s for units, our most common rotation rates are of propellers or crankshafts. For those we employ variables N, revolutions per minute (rpm) or n, revolutions per second (rps).

Circular Dynamics: Torque, Work, Energy, and Power

Having cut rotary mechanics down to rotations at constant speed about a fixed axis, Table 3.1, augmented with a few comments, tells the story as we need it. While one cannot always use these relations uncritically, they obtain in most straightforward cases considered in this book. In linear mechanics, it was masses that had position, moved, were forced on, and accelerated. Here we have the more

Table 3.1 Correspondence between linear and circular quantities

Concept	Linear variable	Linear units	Circular variable	Circular units
Mass analog	m	slug	I	slug-ft^2
Position	x	ft	θ	radian, or deg
Velocity	$v = \dot{x}$	ft/s	$\omega = \dot{\theta}$	rad/s, or deg/s
Acceleration	$a = \dot{v}$	ft/s^2	$\alpha = \dot{\omega}$	rad/s^2, or deg/s^2
Force analog	$F = ma$	lbf	$L \equiv r \times F = I\alpha$	ft-lbf
Momentum	$p = mv$	slug-ft/s	$r \times p = I\omega$	slug-ft^2/s
Work	$F \cdot d$	ft-lbf	$L \cdot \theta$	ft-lbf
Kinetic energy	$\frac{1}{2}mv^2$	ft-lbf	$\frac{1}{2}I\omega^2$	ft-lbf
Gravitational PE	mgh	ft-lbf	—	—
Power	$F \cdot v$	ft-lbf/s, or hp	$L \cdot \omega$	ft-lbf, or hp

complex notion of *moment of inertia I*, which depends both on the extended object's mass and on the distribution of that mass about a given rotational axis:

$$I = \sum_{i=1}^{N} m_i r_i^2 \qquad (3.15)$$

Here m_i is mass of the *i*th of N mass points and distance r_i is the (perpendicular) distance between that *i*th mass point and the axis. An airplane's moment of inertia about its longitudinal axis plays a role, for instance, in its full-aileron-deflection roll rate. We will not be considering rapid aerobatic maneuvers. Beginning physics books often have calculations of moments of inertia of such regular solids as spheres and cylinders. We will get by citing only a couple of those results: 1) for a solid cylinder, about its axis, $I = \frac{1}{2}MR^2$; 2) for a hollow cylinder of inside radius R_1, outside radius R_2, $I = \frac{1}{2}M(R_1^2 + R_2^2)$. In both of these cases, the object has uniform density and has total mass M.

In aviation, we generally associate *torque* with engines or propellers and use a different word, *moment*, for torques on the (assumed rigid) airframe. Moments of weight forces, about the c.g. of the airplane, come into play in weight and balance calculations. An airplane loaded too heavily aft, for instance, even though its gross weight is less than the maximum allowed, is a particularly dangerous situation. It might take off apparently normally but then rotate nose up, farther and faster than could be corrected by full nose-down elevator stick deflection, and stall. Even if one gets past that takeoff problem by rotating at higher-than-normal air speed, the balance problem is waiting at the other end of the trip, when the pilot slows for landing.

Uniform gravitational forces act on an extended object as though its entire mass were concentrated at the *center of mass* (c.m.). Because gravity is the pivotal body force we are interested in, the c.m. is often also called the *center of gravity*

Aircraft Performance Preliminaries

(c.g.). For a body consisting of N masses, the position of the c.m. can be found from this relation:

$$M\mathbf{r}_{CM} = \sum_{i=1}^{N} m_i \mathbf{r}_e \quad (3.16)$$

This calculation leads to the same position for c.m. (which may not even be within the material of the body itself; consider a flat washer) regardless of the coordinate system origin chosen; of course the *coordinates* of the center of mass will depend on which origin was used.

A circular motion case of particular importance—and one that confuses much of the Federal Aviation Administration (FAA)[1,2]—is that of an object moving in a circular path at constant speed. You learned in elementary physics that for this to happen there must be a *centripetal* force on the object, directed towards the center of the circle, of size

$$F_{\text{centripetal}} = \frac{mv^2}{r} = \frac{W\omega v}{g} \quad (3.17)$$

There is also a *centrifugal* force, of the same size, directed away from the center of the circle. But the centrifugal force is not a force on the airplane. The centripetal force is that of the air on the airplane, radially inwards; the centrifugal force is that of the airplane on the air, radially outwards. This is an example of Newton's third law of motion.

Centrifugal and Coriolis Pseudoforces

For the performance of *light* aircraft, this is an optional section. When one coordinate system is rotating with respect to another, strange new effects called *pseudoforces* arise. Newton's laws of motion are only true in an inertial frame of reference. *Pseudoforces* appear when one insists on looking at particle motion from a noninertial frame of reference. In the case of uniform rotation, the *pseudoforces* are the *centrifugal force* and the *Coriolis force*. (Caution! This centrifugal pseudoforce is not the same as the *actual* force called "centrifugal" in the section just above. That fact causes much confusion.)

Example 3.3 Relative to a rotating (and smooth) merry-go-round, a block of ice, if released, would slide away outwards unless restrained. That initial pseudoforce is the so-called "centrifugal" (away from the center) force. For the block of ice to remain "unaccelerated" with respect to the (noninertial) merry-go-round, a counteracting centripetal (towards the center) force, say a barrier, would have to be applied to the block. In an inertial frame, the unconstrained released block simply continues its initial motion in a straight line while the merry-go-round turns under it and "away off to the side." The motion off to the side, once the block has some palpable velocity with respect to the merry-go-round, is a two-dimensional form of "Coriolis" force.

Example 3.4 An airplane circumnavigating the Earth, 35,000 ft above the equator, headed east, has a slight advantage over its twin (which we take to be near the same place at the same time) circumnavigating westward. Making the transition to the three-dimensional rotating Earth, we must be prepared for a brief bout of heavy lifting. The usual way of describing the influence of frames of reference rotating with angular speed Ω—the origin of that frame unaccelerated with respect to the inertial frame, and in our case coincident with it—in physics texts[3-7] or in flight dynamics texts[8,9] is to argue towards and derive:

$$a = F/m = [a] + \Omega \times \Omega \times R + 2\Omega \times V + \Omega^{\cdot} \times R \qquad (3.18)$$

Here vector a is the acceleration in *FI*, the inertial frame, and vector $[a]$ is acceleration of the particle viewed in the rotating frame (for the rotating Earth case, call that rotating frame *FE*). R is the radius vector from the origin of *FI* to the particle, V is the velocity of the particle as seen in the rotating frame, and $\Omega^{\cdot}$ is the angular acceleration of the rotating frame with respect to the inertial one. Although the Earth's rotation rate is slowing (due to tidal friction), ignore that last very small term. Symbols '×' denote cross or vector products. The pseudoforces, "seen" in the accelerated frame, are expressed best by solving for $[a]$ (and, here, neglecting the angular acceleration):

$$[a] = F/m - \Omega \times \Omega \times R - 2\Omega \times V \qquad (3.19)$$

The left-hand side of Eq. (3.19) is the acceleration seen from the noninertial (Earth-based) frame. The three terms of the right-hand side of Eq. (3.19) are, in order, the (inertial) acceleration from external actually applied forces, the centrifugal pseudoforce (often called "centripetal" because of the negative sign), and the Coriolis pseudoforce.

As further simplification, we also assume the center of the Earth is unaccelerated, so that if it were not for its rotation about its axis (also ignoring precession and nutation), it would be a suitable inertial frame. That is, we ignore the orbital acceleration of the Earth about the Sun and any acceleration of the Sun as it revolves in the Milky Way and anything else. Coriolis acceleration is commonly thought of as manifesting itself (only) by forcing moving objects in the northern hemisphere to the right and those in the southern hemisphere to the left. That is not quite correct; there will turn out to be more to that story.

Now let us specialize the velocity of the particle (airplane) V to the equatorial plane. The Earth rotates "from west to east." About its axis, seen from above the North Pole, that is considered a positive rotation with angular speed $\Omega = 2\pi/(24 \times 60 \times 60) = 7.27 \times 10^{-5}$ rad/s. The radius of the Earth is about

3960 statute miles or $R = 2.091 \times 10^7$ ft. Add $h = 35,000$ to it and you get $(R + h) = 2.094 \times 10^7$ ft. $\Omega \times (R + h)$ points East and is of length 1522.3. The negative of $\Omega \times$ that previous result, the centripetal term, points inwards, towards the center of the Earth, and is of length 0.111 ft/s^2 = 0.0034 g, where $g = 32.174$ ft/s^2. Notice there is no dependence of the centripetal term on the velocities of the eastbound or westbound airplanes. The centripetal term is there, true, but it is precisely the same for the two airplanes. But we have neglected something (else) important.

The Coriolis force! Evaluated, using FE, for the eastbound airplane, this is an outward (from the Earth's center to the airplane) force of size $2\Omega V$. Not right, not left, but outward. On the westbound airplane, an inward force the same size. So the difference between the two airplanes is that the eastbound one, if it is to maintain its flight path, enjoys an extra lift of size $4\Omega V$. If one were to take different noninertial frames, say for instance the eastbound or the westbound airplane, then calculation details would change and the various pseudoforce terms would be different. But the difference between the eastbound and westbound airplanes would always be precisely $4\Omega V$.

Example 3.5 Putting numbers to the previous example, take airplanes with air speed 500 ft/s (very close to 300 KTAS). For simplicity, no wind. Then the effect difference is $4\Omega V = 4 \times 7.27 \times 10^{-5} \times 500 = 0.1454$ ft/s^2 = 0.0045 g. For a large 500,000-lbf transport, the "weight" difference would amount to 2250 lbf. For a light airplane, of course, it is a negligible effect, practically speaking, but that is not the point. And just because it is a pseudoforce does not mean it does not have real effects in some circumstances. Rivers in the northern hemisphere wear away their right banks quicker and are a little, but measurably, higher on the right side than on the left.

Technical Logic of Airplane Performance

The problem faced by a beginning student of airplane performance is that everything seems to depend on everything else. You can not calculate angle of attack because it depends on lift, which depends on flight path, which depends on air speed, which depends on angle of attack (AOA). That sort of thing. This conundrum is certainly not insuperable—mathematics and analysis evolved, after all, specifically to overcome dependency problems of this kind—but they certainly are daunting to the beginner. What follows is a positive spin on this fundamental difficulty. If not a solution then at least a reduction of frustration.

The cast of this airplane performance melodrama is made up of three personae: airplane, atmosphere, and pilot. Now follows (getting a little ahead of ourselves!) some details on the characterization of each.

Airplane

The standard fixed-wing, propeller-driven airplane consists of three parts: airframe, engine, and propeller.

1) Airframe. This is characterized by weight W (including that of the engine and propeller), reference wing area S, wing aspect ratio A, pitching moment coefficient $c_{m0}(\alpha)$, c.g. location, $C_L(\alpha)$, $C_D(\alpha)$;
2) Engine. This is characterized by rated horsepower P_0, rated revolutions per minute N_0, and engine power P (which depends in turn on engine revolutions per minute, manifold absolute pressure [MAP], and air density ρ);
3) Propeller. This is characterized by power coefficient function $C_P(J)$, thrust coefficient function $C_T(J)$, and diameter d. J is the dimensionless *propeller advance ratio*, V/nd, where V is true air speed, n is propeller revolutions per second, and d is propeller diameter. We assume the propeller has sufficient strength and rigidity to withstand the centrifugal and torque pulse forces placed on it.

It is too early for us to do much about the engine or propeller. We do know that, when the engine is turned off, there is not any thrust; we will consider that gliding case in due time (Chapter 9). We will learn lots about engine power P in Chapter 5 and much about functions C_P and C_T in Chapter 6. Here we are just making a list.

For the two important airframe *functions*, C_L and C_D, we mean that to help characterize the airplane we must know the entire functions, their values for any α we are likely to encounter, not just a particular value at some particular angle of attack α. So when I say I know $C_L(\alpha)$ I usually mean I have a table (or graph or formula) of all the ordered pairs (α, C_L) for the entire function; I do not mean that I know what numerical value C_L has in this particular case.

The mathematics education establishment, if any such thing exists, needs to come up with better functional notation. Or perhaps a nested sequence of such notations, each appropriate to a particular level of discourse. The author's pure mathematics instructors, for instance, used to draw lines under numerals (e.g., 1) to show we meant the function whose value was everywhere that constant; Professor Dirac used lots of primes on variables—x', x'', etc.—to distinguish among variable x's alternative values. Cumbersome stuff. Ordinary functional notation is often ambiguous, thereby confusing even good students, but it is all we have at the moment.

Atmosphere

The next character is the atmosphere, of which we have a good understanding. To move the Earth, all Archimedes needed was a fulcrum and a lever; to pin down its atmosphere, all we need is the vertical distribution of either pressure or of temperature. Chapter 1 has the details; with them we can readily find out enough about the atmosphere. Density ρ is normally the pivotal performance variable.

Pilot

Finally, the pilot. Actually, we catalog the controls the pilot manipulates:

1) Ailerons—these control bank angle ϕ;
2) Rudder pedals—except for cases like wing-low approaches and forward slips, these simply maintain coordinated flight;
3) Elevators—these change pitch attitude and control air speed V; and
4) Throttle—this sets MAP to control altitude (climb or descent rate) and engine torque.

Of course we simplified. The propeller functions in turn depend on the number of blades B and on the twist angle $\beta(r)$ and blade chord $c(r)$ as functions of radial station r out from the center of the shaft. The pilot may or may not have a propeller revolutions per minute control but almost always has a mixture control knob. But remember, we are only interested here in the big picture and, for the moment at least, propulsion details do not matter.

Quasi-Static Equations of Motion

It is time to tie up loose ends even if the resulting bundle is too heavy to lift. Quite often we will use the physicist's time-honored technique of analyzing forces. Those are, of course, the forces of lift L, drag D, thrust T, and weight W acting on the airplane. It was further stipulated that this airplane will be of conventional design, flown trimmed and in coordinated flight, and not subject to accelerations except the centripetal acceleration associated with banked turns. Now what are the (sufficiently general) force equations?

Mathematics is wonderful. When handed a heavy dose of equations, first you wonder what they are trying to say. After you get their drift, you wonder whether they are true. And even following their indubitable derivation, you continue to wonder: What are the solutions?

But that is for later. For now, we only want to introduce the three force equations that will be our sometime companions. They come from staring at Fig. 3.1, concentrating on each of the three indicated spatial directions in turn, and using the two important facts that drag is always parallel to the flight path while lift is always perpendicular to it.

Parallel to the flight path:

$$T \cos \alpha - D - W \sin \gamma = 0 \qquad (3.20)$$

Perpendicular to the flight path and in the vertical plane:

$$L \cos \phi - W \cos \gamma + T \sin \alpha = 0 \qquad (3.21)$$

Perpendicular to the flight path and in the horizontal plane:

$$L \sin \phi - \frac{W \omega V}{g} = 0 \qquad (3.22)$$

In these equations, α is AOA, ϕ is bank angle, γ is flight path angle (negative when descending), ω is turning rate (angular speed) in radians per second. V is true air speed of the airplane in feet per second The other symbols have their usual meanings. Figure 3.1 depicts the general situation. The attitude angle θ is also shown there; it can easily be calculated from the AOA α and the flight path

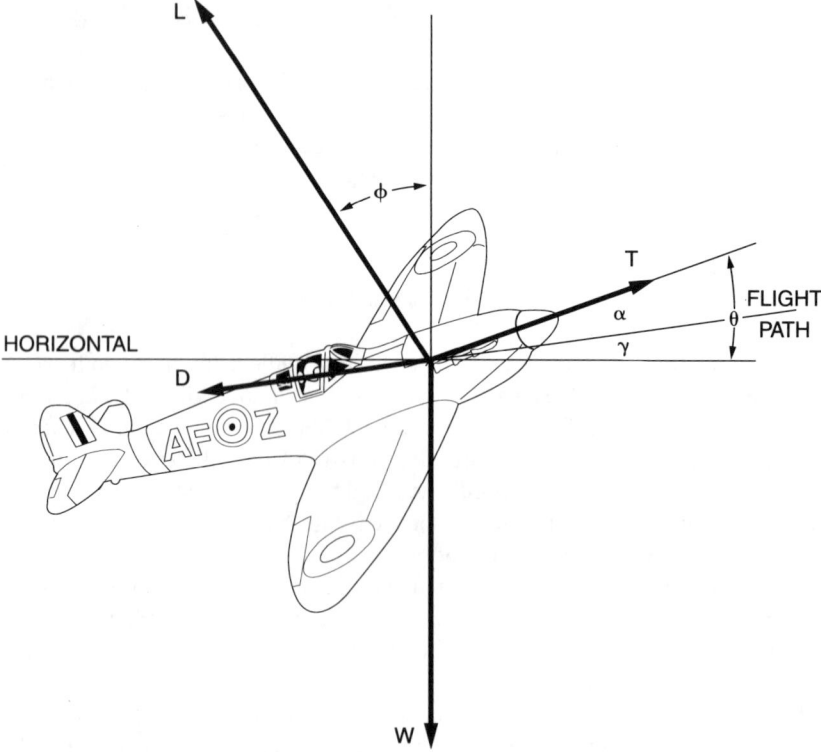

Figure 3.1 Forces and angles for the airplane in flight.

angle γ: $\theta = \gamma + \alpha$. If the propeller shaft is not aligned with the body angle of attack reference line then, as mentioned earlier, an additional constant offset angle must be added to our α.

It is a good exercise to jot down these "full" equations and then consider, one case at a time, simplified versions under several different circumstances: 1) banked glides, $T = 0$; 2) unbanked glides, $T = \phi = 0$; 3) unbanked powered flight, $\phi = 0$; 4) AOA small enough so that $\cos\alpha \doteq 1$, $\sin\alpha \doteq 0$; and 5) flight path angle small enough that $\cos\gamma \doteq 1$, $\sin\gamma \doteq 0$. Some of these cases, especially the first two, the gliding ones, we will consider in great detail in Chapter 9. The fifth case is a standard assumption of both power-available/power-required (P_{av}/P_{re}) analysis and of the bootstrap approach. Our performance appetite is whetted. But next we need to know more about the aerodynamic force.

References

1. Illman, P. E., *The Pilot's Handbook of Aeronautical Knowledge*, Tab Books, Blue Ridge Summit, PA, 1991, p. 34.
2. Federal Aviation Administration, *Flight Training Handbook*, Doubleday, New York, [undated], p. 32.
3. Goldstein, H., *Classical Mechanics*, Addison-Wesley, Reading, MA, 1959, p. 135.
4. Corbin, H. C., and Stehle, P., *Classical Mechanics*, 2nd ed., Wiley, New York, 1960, pp. 145–146.
5. Symon, K. R., *Mechanics*, Addison-Wesley, Cambridge, MA, 1953, pp. 234–242.
6. Kittel, C., Knight, W. D., and Ruderman, M. A., *Mechanics*, Berkeley Physics Course Vol. 1, McGraw-Hill, New York, 1965, pp. 84–88.
7. Sommerfeld, A., *Mechanics*, Lectures on Theoretical Physics, Vol. I, Academic Press, New York, 1952, pp. 165–167.
8. Thompson, W. T., *Introduction to Space Dynamics*, Dover, New York, 1986, pp. 20–22.
9. Etkin, B., *Dynamics of Atmospheric Flight*, Wiley, New York, 1972, pp. 121–133.

4

The Aerodynamic Force

Introduction

The *aerodynamic force* on the airplane is simply the total of all forces on the airframe due to the moving airstream. We can think of each portion of airframe surface as subjected to its own aerodynamic force increment, acting inwards or outwards or tangentially, but by restricting ourselves to rigid airframes we can simplify that complicated picture of a huge flock of arrows. Physical mechanics tells us that every system of forces on a body is equivalent to a single force, through an arbitrary point, plus a force couple. A force couple corresponds to two equal but oppositely directed forces acting at a distance from one another.

Because wings and airframes almost always have a plane of symmetry, and generally move (relative to the air) in that plane, we can simplify matters to a single aerodynamic force, acting at a *center of pressure*, plus a torque about an axis perpendicular to the plane of symmetry. In particular cases, either the force, torque, or both might be zero. For our needs, determination of quasi-static performance, we push this simplification even further, with the lines of action of propeller thrust and aircraft drag usually assumed to act in the longitudinal (approximate) symmetry axis of the fuselage and aircraft weight and lift concentrated at points on that axis but acting perpendicular to it. In this approximate theory, we ignore torques due to drag and thrust. The latter is not always reasonable. We also ignore, except in special cases, aerodynamic side forces. In practice, total torque on the airframe, in steady flight, vanishes because we trim it away by imposition of counteracting torque.

The aerodynamic force, acting in the aircraft's plane of symmetry, is split into a *lift* component L perpendicular to the free airstream direction and a *drag* component D in the direction of the free airstream. Lift is the *sine qua non* ("without which not") of heavier-than-air flight and will get most of our attention. The torque or aerodynamic *pitching moment* will be considered solely due to lift forces. Drag is split into *parasite drag*—further decomposed into pressure or profile, skin friction, cooling, trim drag, and a few specific others and a catch-all "interference" drag—and *induced drag*, also termed "drag due to lift." We will not treat parasite drag in any detail. We will confine its expression to a parasite drag coefficient C_{D0} because we are dealing with airplane performance not design and we can more easily measure the airplane's parasite drag characteristic than ponder and compute it. Induced drag, on the other hand, we will treat at length.

So much for starters and generalities. We shall proceed to define lift, drag, and pitching moment coefficients, depending (almost) exclusively on the shape and orientation relative to the free airstream of the lifting surface under consideration. A family of airfoil profiles will be described. A numerical example will show how Bernoulli's theorem (relating airstream speed to pressure) and the Kutta–Joukowski theorem (relating airstream *circulation* to lift) can be used to compute lift and pitching moment coefficients from empirical data on pressures distributed around an airfoil profile. *Pressure coefficients* will be used in that analysis. Aerodynamic differences between actual three-dimensional wings and their infinite-span "two-dimensional" constructs will be treated in detail. Induced drag (an essential concomitant of lift), associated with an *induced angle of attack*, and modified by *ground effect*, will be treated.

Moving to the aerodynamic characteristics of practical wings, we analyze their planforms (views from above) and their spanwise geometry (profiles with variable *angles of incidence*, variable profile orientation or *twist*). We will integrate lift and pitching moment characteristics tabulated for two-dimensional airfoils to approximate actual wing behavior. Extension of flaps—giving us a slightly different wing for each flaps deployment angle—will be brought into the discussion. A further lifting surface, the horizontal tail, enters from back stage when we discuss lift and moments on the entire aircraft. We will end our discussion of lift with a brief but cautionary collection—some specimens considered, some almost ignored—of practical factors.

As mentioned, drag will get fairly short shrift. But we will develop one useful approximate formula for calculating the drag force as a modification of its minimum value and then another formula for estimating the drag effect of deployed flaps.

Lift and drag and pitching moment—rediscovered by every child holding a flattened hand out the moving car window—happen when a surface and the air are in relative motion. We extend that discovery with some of the cultural and quantitative details.

Lift, Drag, and Pitching Moment Coefficients

Aerodynamic force and torque depend on several variables. To separate environmental, kinematic, and surface size considerations from those of surface shape and orientation in airflow, we use dimensionless coefficients implicitly defined by

$$L = \tfrac{1}{2}\rho V^2 S C_L(\alpha) \tag{4.1}$$

$$D = \tfrac{1}{2}\rho V^2 S C_D(\alpha) \tag{4.2}$$

$$M = \tfrac{1}{2}\rho V^2 S c(y) C_M(\alpha) \tag{4.3}$$

The lift, drag, and pitching moment coefficients are, respectively, C_L, C_D, and C_M. Variable ρ is air density, V air speed, S reference wing area, $c(y)$ the length of chord at spanwise position y (measured laterally from the aircraft centerline), and α angle of attack, the smaller angle from the direction of relative motion of the wing (with respect to the air mass) and some reference chord line associated with the wing profile (drawn from tail to nose). That reference chord line may be taken to be the longest line starting and ending inside the profile from tail to nose, along the bottom of the profile, or otherwise; the definition does not much matter but needs to be stated. Jumping ahead a little, we will also need to consider infinite-span or two-dimensional cases in which any of lift, drag, or pitching moment might be infinite. The useful concept then is lift, drag, or pitching moment per unit span. Using primes for derivatives with respect to the spanwise space variable y and using small letters for subscripts (as is conventional), these coefficients are implicitly defined by

$$L'(y) = \tfrac{1}{2}\rho V^2 c(y) C_l[\alpha(y)] \tag{4.4}$$

$$D'(y) = \tfrac{1}{2}\rho V^2 c(y) C_d[\alpha(y)] \tag{4.5}$$

$$M'(y) = \tfrac{1}{2}\rho V^2 c^2(y) C_m[\alpha(y)] \tag{4.6}$$

Coefficient values are normally graphed or tabulated for the two-dimensional case. Adjusting factors, to be discussed below, are then used to find the corresponding values of coefficients for an actual three-dimensional surface or wing.

It turns out that, besides depending crucially on angle of attack (AOA) each of the coefficients also depends to some extent on Mach number $M = V/V_S$ (where V_S = the local speed of sound), and on Reynolds number $Re = Vl/\nu$ (where l = some representative length—perhaps of the wing chord—and ν kinematic viscosity of the air). In our far subsonic cases with full-scale aircraft, neither of these additional two dimensionless factors will much matter.

One additional wrinkle in the drag situation for the three-dimensional case is that total drag D is the sum of parasite drag D_p (we prefer D_0) and induced drag D_i

$$C_D(\alpha) = C_{Dp} + C_{Di} = C_{D0} + \frac{C_L^2(\alpha)}{\pi e A} \tag{4.7}$$

There is no induced drag in the two-dimensional case. In the induced drag term, e is the *airplane* (or *Oswald*) *efficiency factor* and aspect ratio A is B^2/S (B the wing span). We shall take up induced drag in due course. Because C_{D0} is just a single number (for a given airplane in a given flaps/gear configuration in a given state of repair and cleanliness), one sees that variation in drag coefficient depends on variation in lift coefficient. Lift is the driving concept.

Actual wings are usually made up of different wing sections—profile shapes—of various sizes and orientations. When the full wing is to be considered, our usual case, we take for AOA some agreed-upon orientation of the airplane's fuselage with respect to the free airstream, the so-called "body" angle of attack, α_B. Except in confusing cases, we leave off subscript B.

Let us consider lift coefficients in greater detail. It is a convenient aeronautical circumstance that most operational aircraft wing sections are "about the same." That is, almost all of them (especially in our subsonic general aviation regime) have graphs of C_l vs α looking like Fig. 4.1, data[1] collected at $Re = 6 \times 10^6$.

At smaller AOAs, these graphs are quite close to a straight line, with slope a somewhat less than $0.11/\deg$ or $2\pi/\mathrm{rad}$. Because of wing profile camber (arched shape), C_l seldom falls to zero before some small nose-down angle of

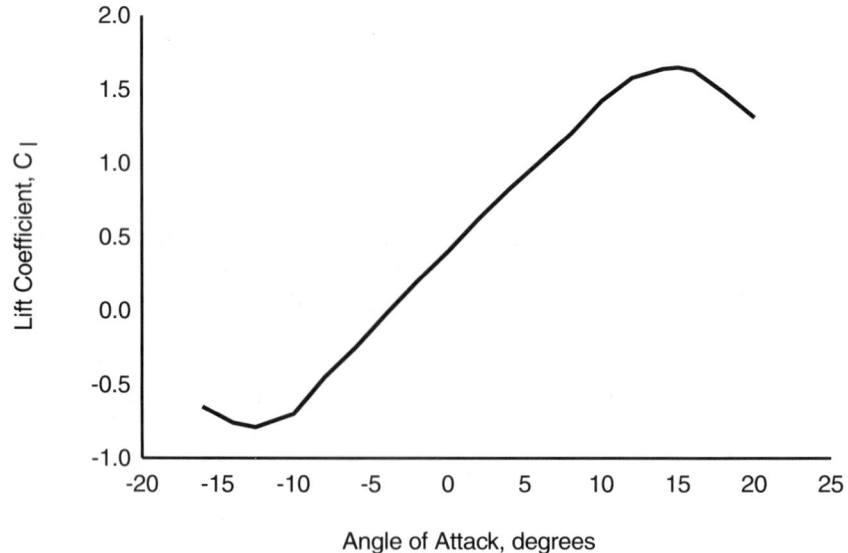

Figure 4.1 Lift coefficient C_l as a function of angle of attack α.

attack, $\alpha_0 < 0$. The lift coefficient function reaches a maximum value, $C_{l\max}$ at some sizeable stall AOA α_S. For flaps up, an α_S of about 16 deg is typical. Except for some latitude in fairing between the linear portion and the curved portion at the top, the four numerical values a, α_0, α_S, and $C_{l\max}$ pretty well complete the description of any particular wing's $C_l(\alpha)$ and, once the discussion shifts to an actual wing with finite aspect ratio, its $C_L(\alpha)$. So the wing's lift characteristics, awaiting the appearance of flaps or ailerons, is not hard to determine.

You might occasionally need an analytic polynomial expression for $C_l(\alpha)$ for values of α between the linear portion of the curve and its peak. For generality, call the point at the end of the linear portion $[\alpha_1, C_l(\alpha_1)]$ and the point at the peak, the stall point, $[\alpha_2, C_l(\alpha_2)]$. We know both heights and both slopes (a on the left, zero on the right). Four pieces of information correspond to a polynomial with four constants, a cubic:

$$y = C_l(\alpha) = a\alpha^3 + b\alpha^2 + c\alpha + d \tag{4.8}$$

Writing down the four facts we know gives the following system for coefficients a, b, c, and d:

$$\begin{bmatrix} \alpha_1^3 & \alpha_1^2 & \alpha_1 & 1 \\ \alpha_2^3 & \alpha_2^2 & \alpha_2 & 1 \\ 3\alpha_1^2 & 2\alpha_1 & 1 & 0 \\ 3\alpha_2^2 & 2\alpha_2 & 1 & 0 \end{bmatrix} \begin{bmatrix} a \\ b \\ c \\ d \end{bmatrix} = \begin{bmatrix} C_l(\alpha_1) \\ C_{l\max} \\ m \\ 0 \end{bmatrix} \tag{4.9}$$

So as not to get confused, we used m in lieu of what is often called a or a^∞ for the linear slope. The 4×4 matrix on the left side of Eq. (4.9) is an example of a so-called *confluent alternant*[2] matrix. While its special structure leads to special techniques for its inversion, we shall simply use standard methods, below, when we curve fit C_l for the NACA 2412 profile of the Cessna 172 wing. It is often good enough to restrict the spline to a parabola, crossing out all second rows and the first column in the alternant matrix in Eq. (4.9). That lets $C_{l\max}$ float to its own level, granted, but one will not be too far off. Viscosity and reality kick additional sand in the face of these *steady* or slowly achieved stall angles; compared to static wind tunnel results, actual aircraft dynamic stalls usually give the pilot a short-term break. Devolving to a parabola will not lead you far astray for another reason; like squares and circles, all parabolas have the same shape. They are predictable. If you do take your spline down to a parabola, you get this more tractable form for C_l when α is in $[\alpha_1, \alpha_2]$:

$$C_l(\alpha) \doteq \frac{-a}{2(\alpha_2 - \alpha_1)} \alpha^2 + \frac{a\alpha_2}{(\alpha_2 - \alpha_1)} \alpha + C_l(\alpha_1) + \frac{a\alpha_1(\alpha_1 - 2\alpha_2)}{2(\alpha_2 - \alpha_1)} \tag{4.10}$$

Let us take an example. For the NACA 4412 airfoil (see Fig. 4.2), for $Re = 6 \times 10^6$, the two points in question are approximately (10 deg, 1.4) and (15 deg, 1.65); the slope of the linear portion is closely 0.10/deg. Either

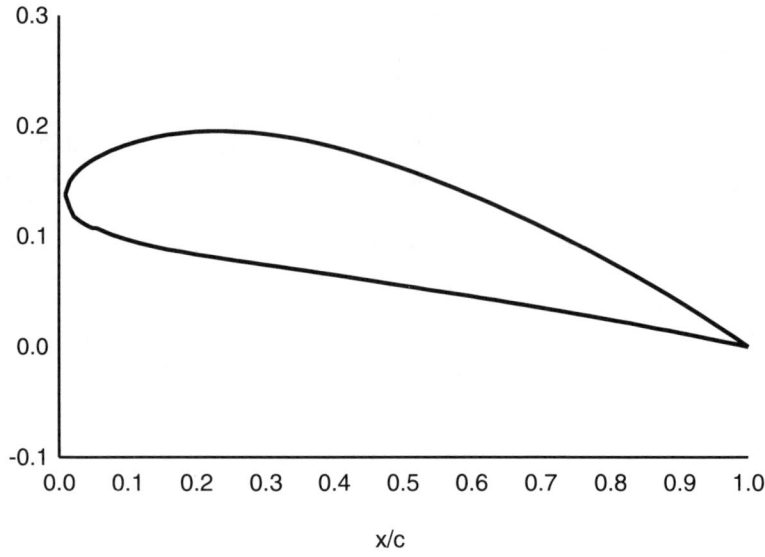

Figure 4.2 NACA 4412 profile at $\alpha = 8\,\text{deg}$.

prescription above (since the coefficient of the α^3 term is zero) then gives the result:

$$C_l^{\text{NACA 4412}}(\alpha) \doteq -0.01\alpha^2 + 0.3\alpha - 0.6, \qquad 10\,\text{deg} \leq \alpha \leq 15\,\text{deg} \quad (4.11)$$

For the straight line portion, written in terms of the zero-lift AOA $\alpha_0 = -3.8\,\text{deg}$, we find

$$C_l^{\text{NACA 4412}}(\alpha) \doteq 0.10(\alpha + 3.8), \qquad \alpha < 10\,\text{deg} \quad (4.12)$$

Figure 4.2 shows this so-called NACA 4412 profile section at an (for future purposes) 8 deg AOA.

Let us now turn briefly to pitching moments. Moments (hence the common variable M) or torques have an annoying feature not shared with forces. Because they are measured as the products of forces and distances from the line of action of the relevant force to a point or axis, the same physical situation gives different torque values depending on which center or axis one chooses. On the other hand, it is not difficult to compute a torque value about one axis when you know its value about another. We shall treat cases in which we first compute pitching moments about the wing's leading edge, then convert those (for a reason to be discussed) to moments about the quarter-chord point (one-fourth of the way back along the chord from the leading edge). A sign convention for torques also needs to be selected. In aerodynamics, positive torques are defined as those tending to raise the leading edge of the wing when in ordinary noninverted flight.

The not-very-exciting graph of $C_{mac}(\alpha)$, moments taken about the quarter-chord point, for the NACA 4412 wing, is shown in Fig. 4.3. The quarter-chord point is essentially the aerodynamic center. Again, this data[1] is for $Re = 6 \times 10^6$. For smaller values of α, $C_m(\alpha) = -0.09$ looks about right.

What about actual wings? We are not yet in a position to consider the aerodynamic details of an actual three-dimensional NACA 4412 wing, as is on the Aeronca L3, Bellanca Champion, or the Stinson Voyager. Or any other. We must first understand induced AOA and drag. But here is a simple qualitative rationale for our reticence. For two-dimensional wings, symmetrical in the span direction, air passing around the wing is deflected neither left nor right. Not so for actual wings, where higher pressure under the wing and lower pressure above it means that air flowing under the wing travels slightly outwards, not straight back. And air flowing over the wing travels slightly inwards. In simpler cases, which most of ours will be, we ignore that deflection, when not close to the wing tips, and consider the airflow as strictly parallel to the plane of symmetry of the airplane. In this dominant case, we can reconstruct the average behavior of actual wings (though not details close to the tips!) by considering them as made up of thin slices of two-dimensional wing sections, profiles of various shapes, sizes, and orientations. Then ordinary integral calculus or numerical integration gives us reasonably good approximations to the aerodynamic data on which we can base our performance calculations. We will be doing this, but let us first take up another question.

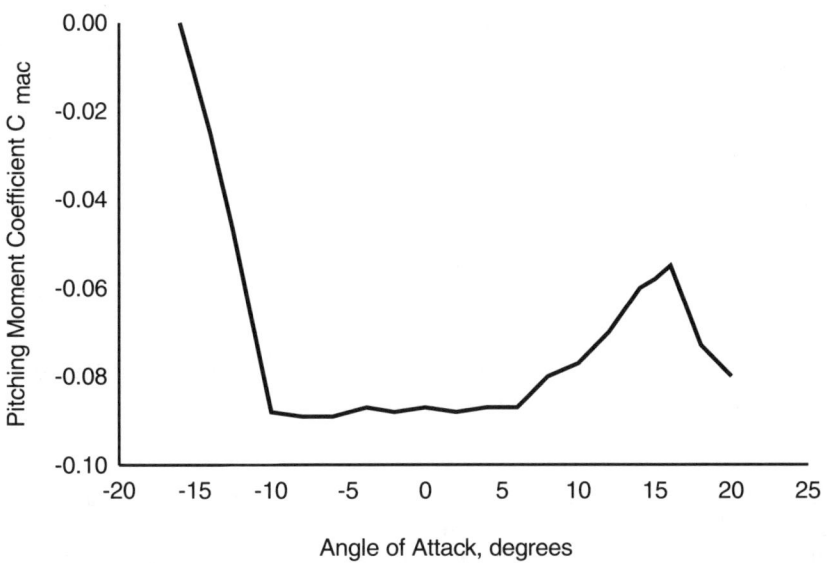

Figure 4.3 Pitching moment coefficient about the quarter-chord point for the NACA 4412 section profile.

Airfoil Profiles

We used the NACA 4412 profile, above, and will again. What is this all about? NACA stands for National Advisory Committee on Aeronautics, precursor to the current National Aeronautics and Space Administration (NASA). The NACA 4412 profile is a member of the so-called "NACA 4-digit" profile family, with the four digits partially coding that shape as follows: the first digit 4 means that the profile's maximum camber ϵc is 4% of the chord length; the second digit 4 means that the chordwise location of that maximum camber is at a point pc 40% of the way from the leading edge to the trailing edge; the final two digits 12 mean that the maximum thickness τc of this profile is 12% of the chord length (and occurs at $0.3c$ from the leading edge). Later we will make even greater use of the Cessna 172 wing, very close to a NACA 2412 profile. Hence, the Cessna wing has only one-half the maximum camber of the NACA 4412, used on say the Aeronca L3, but is otherwise the same.

The three shape parameters ϵ, p, and τ are not the whole story. To be able to draw the profile from such scant information, one needs a formula for the camber line height $Y(x)$ measured up from the chord line joining the extreme nose and tail of the profile, x the distance from the leading edge to the station, and another for the profile thickness $T(x)$, thickness measured (strangely enough) perpendicular to the camber line instead of to the chord line. For the NACA 4-digit family of profiles, those are smoothly joined parabolas

$$Y(x; \epsilon, p, c) = \begin{cases} \dfrac{\epsilon x}{p^2}\left(2p - \dfrac{x}{c}\right), & 0 \leq \dfrac{x}{c} \leq p \\ \dfrac{\epsilon(c-x)}{(1-p)^2}\left(1 + \dfrac{x}{c} - 2p\right), & p < \dfrac{x}{c} \leq 1 \end{cases} \quad (4.13)$$

and

$$T(x; \tau, c) = 10\tau c\left[0.2969\sqrt{\dfrac{x}{c}} - 0.126\,\dfrac{x}{c} - 0.3537\left(\dfrac{x}{c}\right)^2 + 0.2843\left(\dfrac{x}{c}\right)^2 - 0.1015\left(\dfrac{x}{c}\right)^4\right] \quad (4.14)$$

It turns out that the leading edge has radius of curvature $r = 1.1019\tau^2 c$. The upper and lower profiles are given (to our approximation) by

$$y_{U/L} = Y \pm T/2 \quad (4.15)$$

These NACA 4-digit details were only given to illustrate the relative complexity; we will not be using them further. NACA scientists came out with several other profile families—5-digit, modified 4- and 5-digit, and 1-, 6-, and 7-series—while aeronautical researchers in other nations and offices did likewise. Additions to profile shapes and their taxonomy continues to this day. Details of some of the standard profile families can be found scattered throughout the literature.[1,3,4]

Pressure Coefficients

We have another coefficient to consider. When cataloging the distribution of pressures on an airfoil, it is customary to use dimensionless pressure coefficients C_p defined as

$$C_p \equiv \frac{(p - p_0)}{q} \tag{4.16}$$

where p is the pressure at the airfoil surface location, p_0 is the ambient atmospheric pressure, and q is the so-called *dynamic pressure*

$$q \equiv \tfrac{1}{2}\rho V^2 \tag{4.17}$$

one could expect from the free airstream with density ρ moving at relative speed V. The dynamic pressure can also be looked at as the *coherent* kinetic energy density (energy per unit volume) of the airstream. Pressure p, on the other hand, is (for an ideal gas) two-thirds of the *random* translational kinetic energy per unit volume.

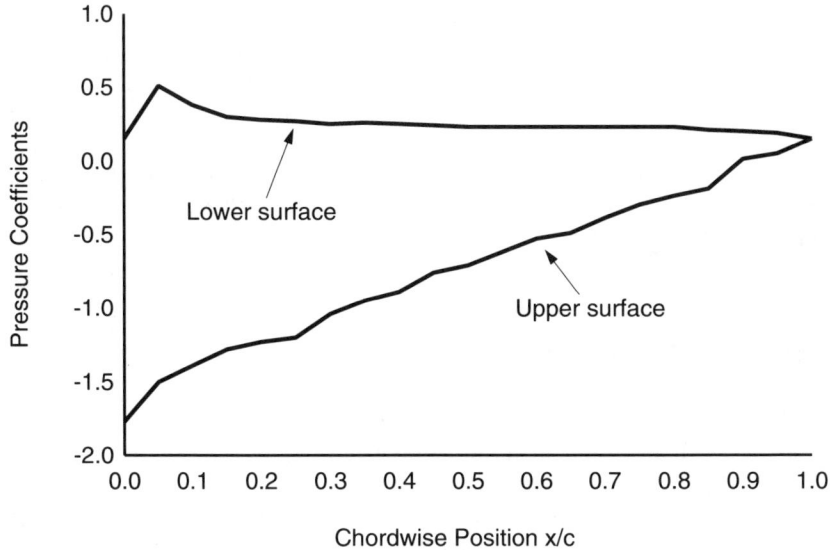

Figure 4.4 Experimental pressure coefficient values on the NACA 4412 section, 8 deg AOA.

Table 4.1 Experimental pressure coefficients at 21 chordwise positions

(1)	(2)	(3)
0.00	0.15	−1.77
0.05	0.51	−1.50
0.10	0.38	−1.39
0.15	0.30	−1.28
0.20	0.28	−1.23
0.25	0.27	−1.20
0.30	0.25	−1.04
0.35	0.26	−0.95
0.40	0.25	−0.89
0.45	0.24	−0.76
0.50	0.23	−0.71
0.55	0.23	−0.62
0.60	0.23	−0.53
0.65	0.23	−0.49
0.70	0.23	−0.39
0.75	0.23	−0.30
0.80	0.23	−0.24
0.85	0.21	−0.19
0.90	0.20	0.01
0.95	0.19	0.05
1.00	0.15	0.15

The pressure coefficient is constructed to remind us that what is important is the excess or deficit of pressure on the airframe, compared to ambient atmospheric pressure p_0, relative to the dynamic pressure. Values of the pressure coefficient, for given airfoils at given AOAs, are typically given at many locations on the upper (U) and lower (L) surfaces of the wing section. Here (Fig. 4.4 and Table 4.1) are experimental pressure coefficient data[1,5,6] for the NACA 4412 profile at an 8 deg AOA.

Bernoulli's Theorem

This theorem (or law, or equation), which we have presented before, is recast here because it is fundamental to these aerodynamic inquiries. For incompressible irrotational inviscid fluids that suffer no energy changes except those due to the work done by pressure differences along the streamline, one gets Bernoulli's theorem:

$$p_1 + \tfrac{1}{2}\rho V_1^2 = p_2 + \tfrac{1}{2}\rho V_2^2 \tag{4.18}$$

for any two positions 1 and 2 along any given streamline in the fluid. This form of Bernoulli's theorem ignores differences in gravitational potential (head) along the streamline. For gases in our practical circumstances, that approximation is perfectly acceptable. Our first use of Eq. (4.18) will be to take position 1 far out in front of our airfoil, so that $p_1 = p_0$, the ambient atmospheric pressure, and $V_1 = V$, the air speed of the wing. Various positions 2 will then be taken on the upper or lower surfaces of the section. On the upper surface, generally speaking, pressure will be lower and streaming speed v will be higher; on the lower surface, pressure will be higher than ambient and streaming speed v lower than section air speed.

Combining Bernoulli's theorem with the definition of the pressure coefficient gives us two useful results:

$$C_{pU/L} = 1 - \left(\frac{v_{U/L}}{V}\right)^2 \tag{4.19}$$

and

$$\frac{v_{U/L}}{V} = \sqrt{1 - C_{pU/L}} \tag{4.20}$$

Equation (4.20) gives us relative streaming speeds and, turning matters upside down, relative times of flight for air molecules passing either over or under the wing. Let us take a closer look.

Times of Flight of Air Molecules Over or Under the Wing

First, there is a problem calculating relative times of flight over and under the wing. "Over the wing" should mean from the forward stagnation point (at positive AOAs, somewhere on the lower surface of the wing, slightly abaft the leading edge), then up over the leading edge and on over the top surface of the wing to the sharp trailing edge, where there is an additional stagnation point. Similarly for under the wing. But molecules at stagnation points have no speed in either direction; they just hit and sit, taking infinite time to go anywhere. If we decide to clock molecules by moving slightly off the stagnation point to start, and end just before the aft stagnation point, we should be prepared to defend our choice of cutoff, how close to the stagnation points we venture. Unfortunately, there is insufficient reason for picking any particular cutoff. Still, we will implicitly or casually pick one in the calculations to follow. Our results will make illustrative sense, but they will not be precise. Another source of imprecision comes from our treating the airfoil as so thin that paths along wing surfaces are portions of chord lines. Experimental error and a coarse horizontal grid kick additional sand in the face of our numerical work. And then there is the fact that actual air is not inviscid. We are using a variant of so-called *thin airfoil theory*, and it is "well known" that thin airfoil theory is not sufficiently accurate to give good pressure coefficient values. Arguing in reverse will not fully palliate the defect. This is rough work.

In Table 4.1 we divided the wing chord into 20 equal portions $\Delta x/c = 0.05$. Time to traverse portion Δx_i is proportional to $(v_i/V)^{-1}$, the quantity we will numerically integrate to determine relative times of flight over and under the wing. For several of these calculations, we will use the simple but adequate "trapezoidal rule" of Appendix D:

$$\int_a^b f(x)\,dx \doteq \Delta x \sum_{i=0}^n f_i - (f_0 + f_n)(\Delta x/2) \qquad (4.21)$$

Table 4.2 sets out numerical facts relevant to our time-of-flight ratio and section lift coefficient calculations. In the time-of-flight case, we see that molecules passing under the wing take $1.164/0.789 = 1.48$ times as long for the trip as do molecules passing over the wing. If the wing is moving to the left, relative wind towards the right, this difference in speeds over and under amounts to clockwise circulation. Later on, we will directly calculate that circulation. Qualitatively, it is important to note that the divided airstream does *not* smoothly

Table 4.2 Data for relative time-of-flight calculations, NACA 4412, 8 deg AOA

(1) x/c	(4) V/vL	(5) V/vU
0.00	1.0847	0.6008
0.05	1.4286	0.6325
0.10	1.2700	0.6468
0.15	1.1952	0.6623
0.20	1.1785	0.6697
0.25	1.1704	0.6742
0.30	1.1547	0.7001
0.35	1.1625	0.7161
0.40	1.1547	0.7274
0.45	1.1471	0.7538
0.50	1.1396	0.7647
0.55	1.1396	0.7857
0.60	1.1396	0.8085
0.65	1.1396	0.8192
0.70	1.1396	0.8482
0.75	1.1396	0.8771
0.80	1.1396	0.8980
0.85	1.1251	0.9167
0.90	1.1180	1.0050
0.95	1.1111	1.0260
1.00	1.0847	1.0847
Sums	24.3625	16.6174
Integrals	1.1639	0.7887

Further NACA 4412, $\alpha = 8$ deg Details

Lift Coefficient from Pressure Coefficients

Even with rough data and numerical procedures, it is worthwhile to see examples of calculations showing how pressure data taken at various positions around the airfoil leads to evaluation of the lift coefficient, the center-of-pressure position CP, the pitching moment coefficient, and circulation. Table 4.3 gives some of the calculated columns we need. Column (6) integrates to give C_l; column (7) shows CP, the x/c position which evenly divides the total lift; and column (8) integrates to give the pitching moment coefficient about the leading edge.

Table 4.3 Pitching moment calculations for NACA 4412 profile, $\alpha = 8$ deg

(1) x/c	(6) $C_{pL}-C_{pU}$	(7) Cumulative sum of (6)	(8) Moment
0.00	1.92	1.92	0.000
0.05	2.01	3.93	0.100
0.10	1.77	5.70	0.177
0.15	1.58	7.28	0.237
0.20	1.51	8.79	0.302
0.25	1.47	10.26	0.368
0.30	1.29	11.55	0.387
0.35	1.21	12.76	0.424
0.40	1.14	13.90	0.456
0.45	1.00	14.90	0.450
0.50	0.94	15.84	0.470
0.55	0.85	16.69	0.468
0.60	0.76	17.45	0.456
0.65	0.72	18.17	0.468
0.70	0.62	18.79	0.434
0.75	0.53	19.32	0.398
0.80	0.47	19.79	0.376
0.85	0.40	20.19	0.340
0.90	0.19	20.38	0.171
0.95	0.14	20.52	0.133
1.00	0.00	20.52	0.000
Sum	20.52	Sum	6.6135
C_l	0.978	C_{mle}	−0.33068
		C_{mac}	−0.08618

From the definitions of lift per unit span L' [Eq. (4.4)], pressure coefficients [Eq. (4.16)], and dynamic pressure [Eq. (4.17)], with judicious addition and subtraction of ambient pressure p_0, we see that

$$C_l = \int_0^1 (C_{pL} - C_{pU}) \, d(x/c) \qquad (4.22)$$

The section lift coefficient for the NACA 4412 profile at an 8 deg AOA, we get here as 0.978. This is almost 23% below the $C_l = 1.20$ given in Abbott and von Doenhoff.[1] Our experimental data did not show us a stagnation point below the leading edge, and coarse graining (5% c slices) cost us some far-reduced pressure values just behind the leading edge. And there are other inaccuracies.

Center of Pressure

Another point. If you run a cumulative sum on column (6) in Table 4.3, you find that precisely one-half the total pressure acts forward of the quarter-chord point ($x/c = 0.25$) and one-half acts abaft that point. Thin subsonic symmetric airfoil theory claims such will always be the case, that the so-called "center of pressure" (or *CP*), the single place where the total lift force can be presumed to act, is always at $x = 0.25c$. Actual airfoils, or even theoretical thin *cambered* ones, are not so accommodating. Also note that when the airfoil is oriented to $\alpha_0 = -3.8$ deg, the AOA at which the lift coefficient is zero, there *can be* no "single place where the total lift force can be presumed to act." Because there is still a nonzero pitching moment, and that must be presumed to be due to a force couple, there are *two* places where lift acts. This complication is one reason pitching moments are used in lieu of center of pressure calculations. Center of pressure cannot always give the whole story.

Pitching Moment Coefficient

Let us further extend our example by computing the section pitching moment coefficient about the leading edge, C_{mle}. That only requires our getting back to our original pressure coefficient data and doing a numerical integration of column (8), the product of columns (1) and (6), in Table 4.3:

$$C_{mle} = -\int_0^1 (x/c)(C_{pL} - C_{pU}) \, d(x/c) \qquad (4.23)$$

The value turns out to be $C_{mle} = -0.3307$. But the leading edge of the wing, once one gets past introductory examples, is not the most convenient place about which to take moments. The so-called "aerodynamic center," which for thin airfoil theory is precisely at the quarter-chord point, has the felicitous feature that moments taken about that point are independent of AOA. The rule of thumb is that pitching moment due to wing camber has the midchord point as axis and that pitching moment due to AOA has the quarter-chord point as axis. Even for real wings, this is close to being the case. For a standard value of $Re = 6 \times 10^6$, for instance, Abbott and von Doenhoff[1] have aerodynamic center for the NACA 4412 wing at $x/c = 0.246$. Adjusting the axis—by now you may have discovered that, if distances are measured in units of c and forces measured in units of q, $L' = C_l$ and $M' = C_m$—first in a way to show the general pattern, and then specifically to the aerodynamic center ac, taken as the point $x/c = 0.25$, we have

$$C_{mac} = C_{mle} + C_l \frac{(x_{ac} - x_{le})}{c} = C_{mle} + \frac{C_l}{4} \qquad (4.24)$$

Using the first of our computed values of C_l in Eq. (4.24), we get $C_{mac} = -0.0862$. Abbott and von Doenhoff[1] make it about -0.088 while both Hiscocks[7] and Diehl[8] show it about -0.089. In either case, close to our calculated value. In view of our inaccurate C_l, somewhat lucky.

Circulation

The Kutta–Joukowski theorem presents a simple and powerful notion discovered independently by Kutta in 1902 and by Joukowski in 1906. It relates circulation Γ around a two-dimensional airfoil—essentially the integrated excess airstream flow over the wing compared with that under it—to the lift that section experiences

$$L' = \rho V \times \Gamma \qquad (4.25)$$

The lift direction is given by rotating the relative wind direction 90 deg in a direction opposite to the direction of circulation. For instance, with the nose of the airfoil to the left, circulation is clockwise and, by the right-hand rule, a vector into the paper. The airstream speed is to the right. Turning V into Γ with a screwdriver, the shortest way, would make a right-hand screw advance upwards. That is the direction of lift (per unit span) L'. The same goes for lift of actual wings with finite aspect ratio.

Circulation is the line integral of the air flows over and above, or under and below, that of the free stream. Theory says that any path around the wing profile one chooses gives the same circulation value Γ. Above the wing (at ordinary positive AOAs), where local streaming speed is higher than V, that excess is

$$v_U - V = V\left(\frac{v_U}{V} - 1\right) = V\left(\sqrt{1 - C_{pU}} - 1\right) \qquad (4.26)$$

while below the wing, reversing sign to go in the direction of the integration, we have

$$V - v_L = V\left(1 - \frac{v_L}{V}\right) = V\left(1 - \sqrt{1 - C_{pL}}\right) \qquad (4.27)$$

Putting these last two results together with the Kutta–Joukowski theorem [Eq. (4.25)] and Eq. (4.4), and solving for the section lift coefficient, gives us

$$C_l = 2 \int_0^1 \left(\sqrt{1 - C_{pU}} - \sqrt{1 - C_{pL}} \right) d(x/c) \qquad (4.28)$$

Table 4.4 has the numerical values we need to calculate section lift coefficient from the Kutta–Joukowski theorem. Circulation is integrated airstream speed excess. The relation between pressure and stream speed comes from the Bernoulli relation. Doubling the integral gives $C_l = 0.873$, even farther from the nominal 1.2 value than our earlier approximation. By now we should be used to reality going beyond our theoretical expectations, at least beyond our simpler ones.

Just a minute! Equations (4.22) and (4.28), two calculations of C_l using pressure coefficients, do not agree! But notice that if one linearizes the integrand of Eq. (4.28) with the binomial expansion $(1 + x)^{1/2} \doteq 1 + \tfrac{1}{2}x$, the two forms are equivalent. We warned this was rough work.

Table 4.4 Rough C_l calculation, NACA 4412 airfoil, 8-deg AOA

(1) x/c	(9) Square root $(1 - C_{pU})$	(10) Square root $(1 - C_{pL})$	(11) Difference, (9)–(10)
0.00	1.6643	0.9220	0.7424
0.05	1.5811	0.7000	0.8811
0.10	1.5460	0.7874	0.7586
0.15	1.5100	0.8367	0.6733
0.20	1.4933	0.8485	0.6448
0.25	1.4832	0.8544	0.6288
0.30	1.4283	0.8660	0.5623
0.35	1.3964	0.8602	0.5362
0.40	1.3748	0.8660	0.5087
0.45	1.3267	0.8718	0.4549
0.50	1.3077	0.8775	0.4302
0.55	1.2728	0.8775	0.3953
0.60	1.2369	0.8775	0.3594
0.65	1.2207	0.8775	0.3432
0.70	1.1790	0.8775	0.3015
0.75	1.1402	0.8775	0.2627
0.80	1.1136	0.8775	0.2361
0.85	1.0909	0.8888	0.2021
0.90	0.9950	0.8944	0.1006
0.95	0.9747	0.9000	0.0747
1.00	0.9220	0.9220	0.0000
		Sum	9.0967
		Integral	0.4363
		C_l	0.8725

Induced AOA and Induced Drag

Detailed treatment of an infinite-span, two-dimensional wing shows that there is an upwash in front of the aerodynamic center and a downwash behind it. We did not have to consider those because the initial upwash vanishes (but without becoming a downwash) at the two-dimensional aerodynamic center and the trailing downwash reverts to zero in the far wake. This means that two-dimensional lift is perpendicular to the freestream direction; there is neither an induced AOA nor any induced drag. See Fig. 4.5.

Not so for finite-span wings. The airstream at the aerodynamic center of a wing with finite aspect ratio has a discernible downwash velocity (say of size w). The downwash angle there, or the *induced* AOA α_i, is then $\tan^{-1}(w/V)$, where V is the air speed. See Figs. 4.6 and 4.7. With the aerodynamic force vector now tilted rearwards by amount α_i with respect to the free stream direction, a component opposing flight, induced drag, has made its appearance. In fact these angles are not the same at all spanwise locations; induced angles are greater near the wing tips. The downwash speed w and induced drag, everything else being equal, are larger for smaller aspect ratios A. That is why high-performance sailplanes have large span wings with small chord. Eventually, the downwash dissipates through viscous processes that heat the air.

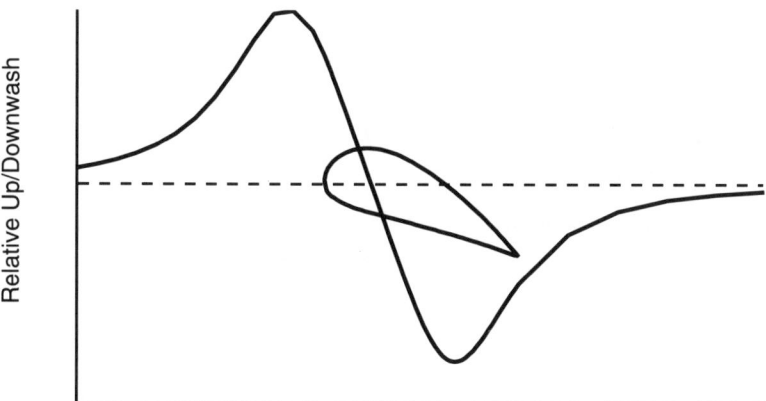

Figure 4.5 With infinite aspect ratio (section profile) case, there is no downwash in the ultimate wake.

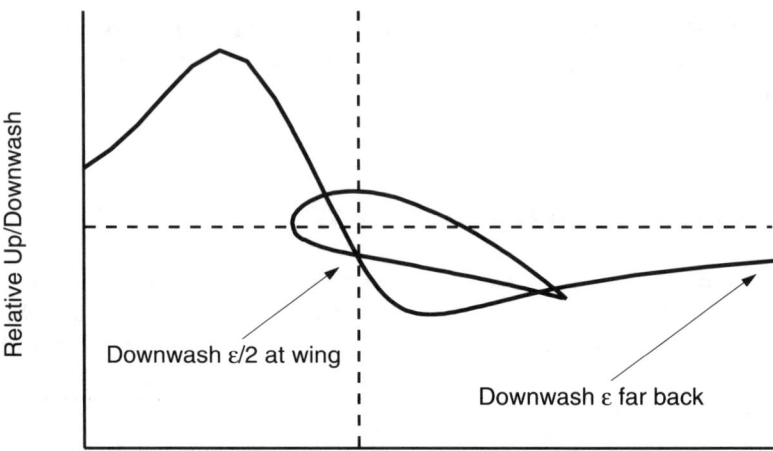

Figure 4.6 Actual wings, with finite aspect ratios, show induced AOA and downwash in the ultimate wake.

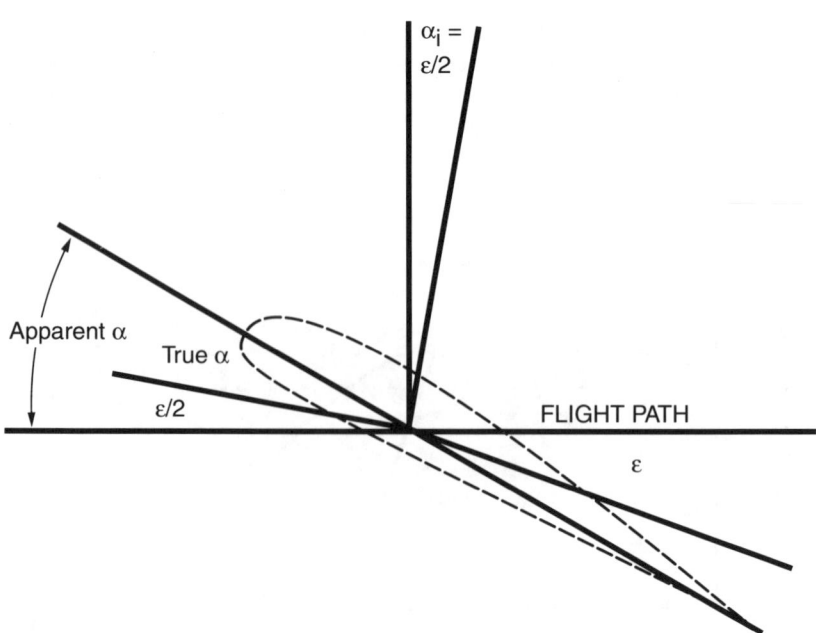

Figure 4.7 Downwash in the ultimate wake averages twice the induced AOA.

There are some minor issues to mention, if only to gloss over. One is that our focus on only the aerodynamic center of the wing comes from our implicit picture in which the system of vortices corresponding to the wing is simplified to a single horseshoe vortex of constant circulation strength. There is a vortex bound to the moving wing and two equally strong vortices, semi-infinite in extent, trailing from near each wing tip. (There is also a transverse starting vortex line and additional spanwise vortices shed whenever the wing's AOA changes.) Therefore, all lift appears to take place at that one chordwise location. Not quite right (a set of nested horseshoe vortices would be better) but good enough for our heuristic and performance purposes. A second simplification and minor error, again arising from our assumption of a single horseshoe vortex with constant circulation strength, is our assumption that all spanwise wing stations can be treated as equivalent. In fact, there is considerably greater downwash nearer the wing tips. Thirdly, the theory we present is only accurate for elliptically loaded wings; other shapes or loadings lead to minor correction factors, again especially near the wing tips, which have been tabulated. We will not bother with those so-called δ (delta) and τ (tau) correction factors, depending on both wing taper and aspect ratio; our experimental evaluation of airplane efficiency factor e will take the place of these and several other corrections. Among them is a fourth factor that is neglected here, that the airplane's wing is not the only lifting surface to consider. The horizontal tail, and in some cases even the fuselage, may contribute to lift. Still, the big if simple picture we sketch is substantially correct, good enough for our performance calculations.

Abaft the aerodynamic center ac, downwash grows to a maximum value, then diminishes. But—for thin wings with elliptical loading and close enough for practical wings—the downwash angle in the ultimate wake, ϵ, is twice the downwash angle α_i at the aerodynamic center. Why twice? Consider two positions near one of the trailing vortices—point A very near a wing tip, point B far back in the wake. The Biot–Savart law says that an element ds of a straight line vortex of constant circulation Γ induces (downwash) velocity increment dw according to

$$dw = \frac{\Gamma \sin(\theta)\, ds}{4\pi r^2} \quad (4.29)$$

where angle θ is from the direction of the vortex line to the point a distance r from the line element. Near the wing, then, the bound vortex contributes little or nothing to downwash at point A because θ is zero; the effective portion of the contribution comes from integrating Eq. (4.29) along the semi-infinite trailing vortex. At point B far back in the wake, however, that trailing vortex appears infinite, stretching out in both directions. That is how the distinguishing factor of two comes about. See Prandtl and Tietjens[9] for details of this calculation.

So at this point we only know that

$$\alpha_i^A = \frac{\epsilon^A}{2} \qquad (4.30)$$

but we do not know the size of either angle.

An easy, if approximate, way of evaluating ϵ is to consider the impulse per second given a tube of air with diameter equal to the wing span, equating that time rate of change of momentum to the lift force. Temporarily consider a fully effective wing, $e = 1$. Take a wing of span B moving through the air at V ft/s. It deflects downwards each second, by angle ϵ, the air within a circular cylinder of diameter B and length V (see Fig. 4.8). The mass of air affected (each second), the mass flux, can be found by multiplying the volume of this cylinder by the air density:

$$m = \rho A V = \frac{\rho \pi B^2 V}{4} \qquad (4.31)$$

Because ϵ is a small angle (about 2 deg for a loaded Cessna 172 at 100 kn) we can use the "small angle approximation": $\epsilon \doteq \sin\epsilon \doteq \tan\epsilon$, ϵ in radians. (For 2 deg, that would be $0.03491 \doteq 0.03490 \doteq 0.03492$.) The velocity change of the affected air mass is then $V\epsilon$. The force (change of momentum per second) on that air mass is the mass affected times its change of velocity:

$$L = \frac{\rho \pi B^2 V^2 \epsilon}{4} = \frac{\rho V^2 S C_L}{2} \qquad (4.32)$$

Solving for the downwash angle in the ultimate wake gives

$$\epsilon = \frac{2C_L}{\pi A} \qquad (4.33)$$

and so the induced angle $\alpha_i^A = \epsilon/2 = C_L/\pi A$.

It is important to have these qualitative and semiquantitative facts in mind, but our main purpose is to get a handle on the behavior of finite aspect ratio wings using wing section data, data for infinite aspect ratio wings.

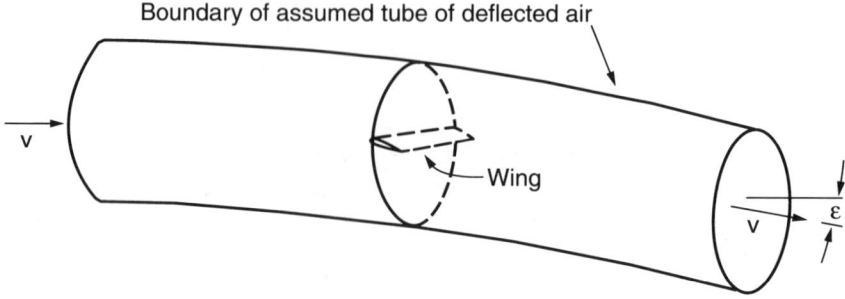

Figure 4.8 The wing stays up by the air being deflected down.

With a real wing (finite A), to get the same lift as we do for a "wing" with infinite A, under identical conditions of air speed and air density, we must have the same lift coefficient C_L. This means the *true* AOA in either situation must be the same. Because having a finite aspect ratio A reduces the true α by $\epsilon/2$, we can get back the old lift simply by increasing our apparent AOA by $\epsilon/2$. (Here and in what follows we restrict ourselves to the linear portion of the lift coefficient graph.) For that "apparent" α we actually see and fly by, our angle with the horizontal must now be

$$\alpha^A = \alpha^\infty + \alpha_i^A = \alpha^\infty + \frac{C_L}{\pi A} \text{ rad}$$
$$= \alpha^\infty + \frac{180 C_L}{\pi^2 A} \text{ deg} \qquad (4.34)$$

Now that we are moving to real wings it's convenient to take up a different sample profile, the less highly cambered NACA 2412 used on the Cessna 172 single-engine trainer.

NACA 2412 Airfoil Profile

For our first sample calculations we used the NACA 4412 profile because (at least for $\alpha = 8$ deg) pressure coefficient data for it was available. And it does appear on several airplanes. But the relative ubiquity of the Cessna 172—now back in production—prompts a change to its NACA 2412 profile. See Fig. 4.9 in which the implicit message is that your wing sees a smaller AOA than does your wingman.

But what are the NACA 2412 section lift and pitching moment characteristics? Using Abbott and von Doenhoff[1] for data for a typical $Re = 3.1 \times 10^6$, one finds the following. The linear portion of the graph of $C_l(\alpha)$ runs from $(-8 \text{ deg}, -0.60)$ to $(8 \text{ deg}, 1.05)$. That means $a^\infty = 0.1031$. The stall point is at $(15 \text{ deg}, 1.60)$; the angle for zero lift is $\alpha_0 = -2.1$ deg. The pitching moment coefficient about the quarter-chord point is fairly constant, with AOA, at $c_{mac} = -0.043$.

Solving Eq. (4.9) to get a cubic fit for the nonlinear portion, for AOAs between 8 and 15 deg, one gets

$$C_l(\alpha) = -0.0011029\alpha^3 + 0.030686\alpha^2$$
$$- 0.17612\alpha + 1.0597, \quad 8 \text{ deg} \leq \alpha \leq 15 \text{ deg} \qquad (4.35)$$

That is all the (somewhat compressed) section data we shall need for the NACA 2412 profile.

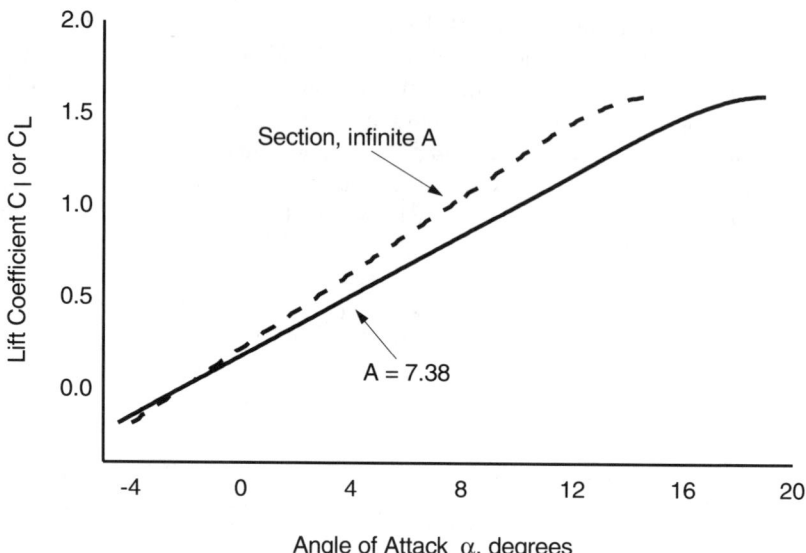

Figure 4.9 Lift coefficient as a function of AOA for the NACA 2412 "section" and for the Cessna 172 wing, aspect ratio 7.38.

Finding $C_L(\alpha)$ from Section Lift Characteristics

We now need to pause and restate our current goal. It is to take aerodynamic coefficient data for the NACA 2412 section and, from it, to construct what? To construct a sequence of formulas or graphs, one for each flaps/gear configuration, each of which gives lift coefficient $C_L(\alpha; \delta_f, \text{up/down})$, as a function of AOA α, for any specified flaps/gear configuration. In the Cessna 172, where the GUMPS approach-to-landing check always comes back with "down and welded," gear configuration is vacuous. But it might not be on some higher-powered airplane for which you might be constructing these graphs or formulas. Back to business.

We have seen that, for any given profile, a wing with finite aspect ratio A requires a higher angle of attack α^A, relative to the infinite-span "section" at AOA α^∞, to attain the same coefficient of lift. These AOAs differ by induced AOA α_i given implicitly by Eq. (4.34). But that equation reminds us that there is an exception: when $C_l = C_L = 0$, which is the case at the so-called "zero lift angle" α_0; then induced angle $\alpha_i = 0$. No lift, no drag due to lift. Working for now only in the linear region—a defect to be corrected below—we have, in general,

$$C_L(\alpha^A) \equiv a^A(\alpha^A - \alpha_0) = C_l(\alpha^\infty - \alpha_0) = a^\infty(\alpha^\infty - \alpha_0) \qquad (4.36)$$

Using the relation between the two AOAs,

$$\alpha^\infty = \alpha^A - \frac{C_L}{\pi A} \qquad (4.37)$$

we have

$$C_L(\alpha^A) = a^\infty \left(\alpha^A - \frac{C_L}{\pi A} - \alpha_0 \right) \tag{4.38}$$

Rearrangement then gives

$$C_L \left(1 + \frac{a^\infty}{\pi A} \right) = a^\infty (\alpha^A - \alpha_0) \tag{4.39}$$

or finally

$$a^A = \frac{a^\infty}{\left(1 + \frac{a^\infty}{\pi A}\right)} = \frac{1}{\frac{1}{a^\infty} + \frac{1}{\pi A}} \tag{4.40}$$

For wings with palpable sweepback angles Λ—which aircraft might also fly at nonnegligible Mach numbers M—there is further complication:

$$a^A(\Lambda, M) \doteq \frac{\cos \Lambda}{\sqrt{1 - M^2}} a^A \tag{4.41}$$

The simpler form, Eq. (4.40) suffices for our hardly swept and far subsonic aircraft. It may be easier to remember in a form analogous to the formula for the total resistance of two resistances in parallel:

$$\frac{1}{a^A} = \frac{1}{a^\infty} + \frac{1}{\pi A} \tag{4.42}$$

But what if we are not in the linear region? Matters get messier but there is no difficulty in principle. Restricting the lift coefficient function so it starts at some point with positive slope and stops at the stall point, and calling

$$C_L = f_1(\alpha^\infty) \tag{4.43}$$

then f_1 has an inverse

$$\alpha^\infty = f_1^{-1}(C_L) \tag{4.44}$$

One can similarly think of Eq. (4.34) as

$$\alpha^A = f_2(C_L) = \alpha^\infty + \alpha_i^A = \alpha^\infty + \frac{C_L}{\pi A} \tag{4.45}$$

with the inverse function, we need

$$C_L = f_2^{-1}(\alpha^A) = \left(\alpha^\infty(C_L) + \frac{C_L}{\pi A} \right)^{-1} (\alpha^A) \tag{4.46}$$

where of course exponents here mean "inverse" in the functional sense. Using our curve fit, Eq. (4.35), for the nonlinear portion, Fig. 4.9 gives the graphical picture, showing the definitely lower slope when the wing aspect ratio is finite.

Let us look at some numbers for the NACA 2412 Cessna 172 wing. Because that airplane's wing span is $B = 35.83$ ft and its reference area is $174\,\text{ft}^2$, the aspect ratio is $A = B^2/S = 7.38$. For the new slope, converting from rad^{-1} to deg^{-1}, and rounding slightly,

$$a^A = \frac{a^\infty}{\left(1 + \dfrac{180 a^\infty}{\pi^2 A}\right)} = \frac{0.103}{\left(1 + \dfrac{180 \times 0.103}{\pi^2 \times 7.38}\right)} = 0.082 \qquad (4.47)$$

with zero lift AOA

$$\alpha_0^A = \alpha_0^\infty = -2.18 \text{ deg} \qquad (4.48)$$

The only other parameter we need to reevaluate is the angle of AOA for stall:

$$\alpha_s^A = \alpha_s^\infty + \frac{180 C_{L\text{max}}}{\pi^2 A} = 15.0 \text{ deg} + \frac{180 \times 1.60}{\pi^2 \times 7.38} = 19.0 \text{ deg} \qquad (4.49)$$

The maximum lift coefficient and the pitching moment coefficient are not affected (at least not in this theory of thin elliptically loaded wings) by our taking on a finite aspect ratio. Stalling, however, is very nonlinear. So these results for α_s and for $C_{L\text{max}}$ are not sufficiently accurate. Stalling actually takes place earlier than our predicted 19.0 deg; consequently, $C_{L\text{max}}$ does not quite get up to its old section value of 1.60. Therefore, we shall use experimentally determined POH stall speeds to fix $C_{L\text{max}}$ and experimental flight tests to fix α_s.

For reasonable values, lift coefficient unity and aspect ratio 7, the downwash angle is 0.0455 rad at the aerodynamic center, about 2.61 deg, and twice this amount far behind, which might include the location of the horizontal tail. The effect is impaired by the fact that wing spans are typically much larger than horizontal tail spans, so the tail gets much less than an average share of the wing's downwash. Still, this must be included in all but very preliminary aircraft design work. It is also worth mentioning that the vertical component of airstream, when *beyond* the wing span, is briefly *upwards*. Wings are complicated.

Formulas for a, α_0, c_{mac}, and α_s

These formulas are modified from the classic works of Diehl[8] and Woods[10] and from spreadsheet regression analysis. After presenting each formula we will then, as a reality check, compare the NACA 2412-calculated formula value with the "known" value.

$$\begin{aligned}
a^\infty &= 0.1097 - 0.07\left(\frac{t}{c}\right) \\
&= 0.1097 - 0.0007 N_{34}
\end{aligned} \qquad (4.50)$$

Because in our case $t/c = 0.12$ ($N_{34} = 12$), this gives 0.101 for the slope; the section curves gave about 0.103. This is good agreement here and Eq. (4.50) generally gives good results.

$$\alpha_0 = -67\left(\frac{m}{c}\right) - 700\left(\frac{m}{c}\right)\left(\frac{L}{c}\right)$$
$$= -0.67N_1 - 0.07N_1N_2 \qquad (4.51)$$

For our NACA 2412, $N_1 = 2$ and $N_2 = 4$, so the formula gives $\alpha_0 = -1.9$ deg. The section curves gave about -2.2 deg. In general, the regression formula standard error here is 0.2 deg (precisely ours!) and R^2, a measure of how much of the total variation in the set of α_0 values is attributable to the above relationship, is 0.96. This is good.

$$c_{mac} = -1.1\left(\frac{m}{c}\right) - 3.8\left(\frac{m}{c}\right)\left(\frac{L}{c}\right) + 3.8\left(\frac{m}{c}\right)\left(\frac{t}{c}\right)$$
$$= -0.011N_1 - 0.0038N_1N_2 + 0.00038N_1N_{34} \qquad (4.52)$$

The formula gives -0.043 and this is exactly the section curve value. The standard error for this regression formula is 0.004 and $R^2 = 0.98$. Excellent agreement. And a nice formula to have; there is nothing we can easily measure, even with flight tests, to come up with a value for c_{mac}.

$$\alpha_s = 9.54 + 5.72\left(\frac{L}{c}\right) + 26.4\left(\frac{t}{c}\right)$$
$$= 9.54 + 0.572N_2 + 0.264N_{34} \qquad (4.53)$$

This formula gives 15.0 deg for the NACA 2412 airfoil; just exactly what we got from the original section curve. Nevertheless, this formula is not very accurate: the standard error is 1.5 deg, and $R^2 = 0.49$. The scatter diagram looks like an ineffective arrow shot through a swarm of gnats. You can only expect so much from statistics.

If we did come up with a statistical formula for C_{Lmax} it would suffer the same fate as that for α_s. In most cases it is best to use POH or experimentally determined stall AOAs and maximum lift coefficients. Nevertheless, the following quite limited rule may be of use. If your computed N_1 and N_2 are the same as those for a section curve you possess but your computed N_{34} is different, you can try to adjust the section C_{Lmax} with this rule:

$$C_{Lmax}(N_{34}) = C_{Lmax}(N_{34} = 12) - 0.03|N_{34} - 12| \qquad (4.54)$$

it being understood that all the other parameters are the same. You may have to use the rule twice, of course, if the section curve you have is not for $N_{34} = 12$.

At this point, you have fairly good estimates of the most important section values. Before we proceed to compensate for the vagaries of aspect ratio, wing geometry (in all three dimensions), and flaps, we detour to consider ground effects.

Ground Effects on Lift and Induced Drag

We have just seen that the lift coefficient graph for the real wing, with finite aspect ratio A, may have considerably smaller slope than does the graph for the (infinite aspect ratio) wing section. That was because finite A comes with a downwash angle that lowers the effective wing AOA. Now we look on a brighter side of this story. A wing near the runway develops less downwash because of the interference of the Earth's surface. Consequently, the lift slope is increased and lift is enhanced. This is the so-called "ground effect on lift." This same phenomenon also decreases induced drag.

While the physical picture is simple, the quantitative details are not. See Fig. 4.10, where vertical force components are lift, horizontal ones are induced drag, and depicted tilt angles are exaggerated, about 20 times realistic values. The following analysis is due originally to Prandtl (1919) and Wieselsberger (1923), independently arrived at by several others since then.

Von Mises[5] shows how Prandtl's wing theory explains interference between the two wings of a biplane. We can use his analysis by replacing our wing, at distance h above the ground, by similar biplane wings vertically separated by $2h$ with the lower wing turned upside down. Then the coefficient of interference drag, for a wing not necessarily elliptically loaded, is

$$C_{D12} = -\frac{C_L^2 \sigma}{\pi e A} \qquad (4.55)$$

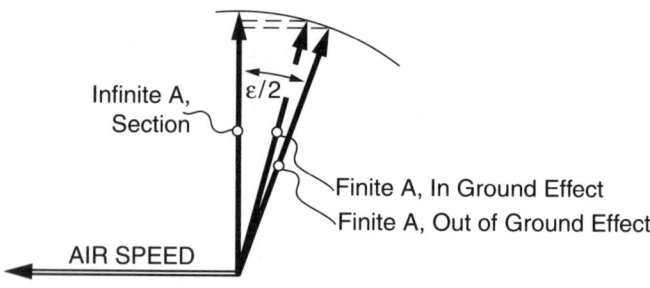

Figure 4.10 The aerodynamic force on real wings tilts back, but less so when near the surface.

where

$$\sigma \equiv \frac{b^2 I}{8\pi^2 h^2} \tag{4.56}$$

a positive number no relation to relative air density, and

$$I \equiv \int_{-1}^{1} \int_{-1}^{1} \sqrt{(1-x^2)} \sqrt{(1-y^2)} \times \frac{\left[1 - \left(\frac{b}{4h}y - \frac{b}{4h}x\right)^2\right] dx\, dy}{\left[1 + \left(\frac{b}{4h}y - \frac{b}{4h}x\right)^2\right]^2} \tag{4.57}$$

You may recognize C_{D12} as very close in appearance to the ordinary induced drag term

$$C_{Di}^O = \frac{C_L^2}{\pi e A} \tag{4.58}$$

where superscript O means "out of ground effect." I means "in ground effect." So

$$C_{Di}^I \equiv C_{Di}^O + C_{D12} = C_{Di}^O(1 - \sigma) \tag{4.59}$$

Because C_{Di} is proportional to the induced angle α_i [Eq. (4.34)] and because this is reduced by ground effect by factor $(1 - \sigma)$, that is equivalent to increasing aspect ratio A to $A/(1 - \sigma)$. Making use of our earlier discussion, that is equivalent to changing the slope of the lift coefficient curve from

$$a_O^A = \frac{a^\infty}{\left(1 + \frac{180 a^\infty}{\pi^2 A}\right)} \tag{4.60}$$

to

$$a_I^A = \frac{a^\infty}{\left(1 + \frac{180 a^\infty (1 - \sigma)}{\pi^2 A}\right)} \tag{4.61}$$

or

$$C_L^I = \frac{C_L^O}{\left(1 - \frac{\sigma}{\left(1 + \frac{\pi^2 A}{180 a^\infty}\right)}\right)} \tag{4.62}$$

Another form, one we will find useful when treating the takeoff problem, comes from inverting Eq. (4.60) and putting that result into Eq. (4.62):

$$C_L^I = \frac{C_L^O}{\left(1 - \frac{180\sigma a_O^A}{\pi^2 A}\right)} \quad (4.63)$$

Now to find σ. It depends on the ratio of wing height above ground, h, to wing span B. Figure 4.11 shows the results of numerical integration using a two-dimensional version of Simpson's rule with Cessna 172 parameters. Lift enhancement is our concern here; ground effect on lift is evidently pretty small. The "standard curve fit" there was used in the Daedulus man-powered flight project and also appears in performance reports for large jet transports. That fit is

$$\sigma \doteq e^{-4.22(h/B)^{0.768}} \quad (4.64)$$

close to and much more convenient than numerical integration of Eq. (4.57). A perhaps more accurate formula, because it includes the influence of airplane efficiency factor, is due to Suh and Ostowari,[11] cited by Hubin[12]:

$$\sigma \doteq \frac{2e}{\pi^2} \ln\left[1 + \left(\frac{\pi B}{8h}\right)^2\right] \quad (4.65)$$

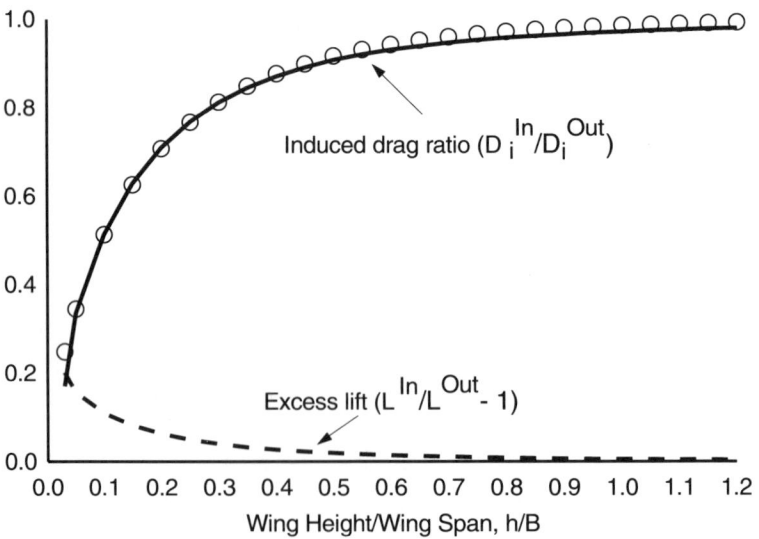

Figure 4.11   Ground effect induced drag factor and curve fit.

One or another of Eqs. (4.64) or (4.65) will be used later in computing distance to liftoff, the takeoff run. This finishes up our treatment of techniques you are likely to need for computing your airplane's lift in any operational circumstance.

Practical Wings

Wing Planform Geometry

Aircraft wings come in many sizes and shapes. The section coefficients are independent of size, or almost so (Reynolds numbers!), so we can concentrate on wing shapes: from above (planform), from in front (dihedral or anhedral), and of course in sagittal section (both the local profile shape and its nose-up or nose-down orientation, the "angle of incidence"). Planforms vary from rectangular through elliptic, tapered, swept back, and sometimes even swept forward. Because we are not dealing with stability considerations, dihedral is relatively unimportant to us. Anhedral (drooped wings in flight) is relatively rare. Figures 4.12 and 4.13 depict the planform nomenclature.

Table 4.5 gives coordinates for the geometrically important points on a Cessna 172 wing. It is generally most economical in dealing with practical wing geometry to start with the kind of data in Table 4.5, then develop formulas for the various (almost always) straight-line segments, rather than develop specialized formulas in terms of root chord, tip chord, sweepback angle, etc.; practical wings usually reveal some wrinkle eluding generalization. Except this one. Notice that in even such an apparently simple wing as for the Cessna 172, the leading edge taper does not begin at the root and the wing tip is an add-on. Even so, you may still have to approximate rounded shapes.

The next goal is to obtain formulas for the various straight-line segments. To find slope m and y intercept b of a (nonvertical) straight line, when you are given the coordinates (x_1, y_1) and (x_2, y_2) of any two points on that line, one needs

$$m = \frac{(y_2 - y_1)}{(x_2 - x_1)}$$
$$b = \frac{(x_2 y_1 - x_1 y_2)}{(x_2 - x_1)}$$
(4.66)

Although for this fairly simple wing we could get by with simple geometry, in general we need a formula for chord length as a function of spanwise station, $c(y)$. Because we know points along the leading and trailing edges (LE and TE), a first step is to get formulas for those edges, LE(y) and TE(y). Then we can get $c(y)$ by subtraction, $c(y) = $ TE$(y) - $ LE(y). After applications of Eq. (4.66), we find

$$c(y) = \begin{cases} 5.35, & 0 \leq y \leq 8.33 \\ -0.1832y + 6.877, & 8.33 < y \leq 17.28 \\ 3.71, & 17.28 < y \leq 17.92 \end{cases} \quad (4.67)$$

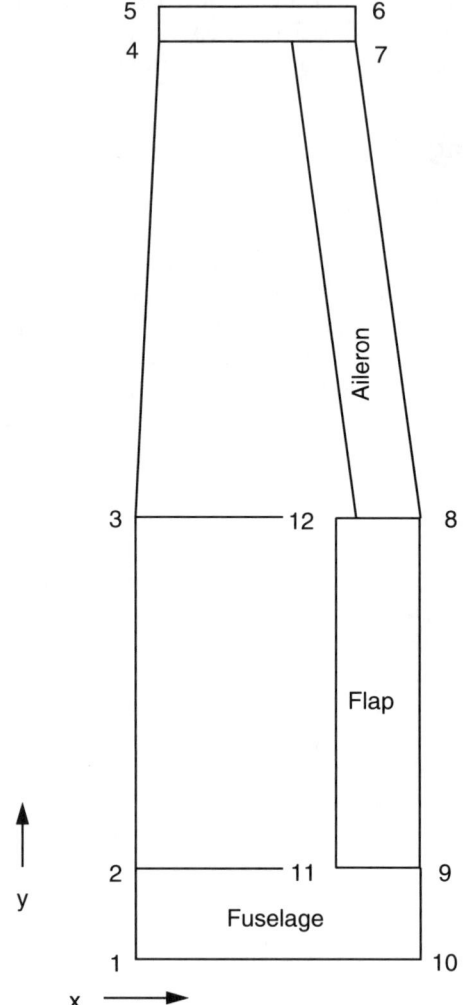

Figure 4.12 Sketch of Cessna 172 wing with numerals keyed to Table 4.5.

Calculation of total wing area S immediately suggests itself. Using wing span $B = 35.83$ ft, this is

$$S = 2 \int_0^{B/2} (y)\,dy = 175.0 \text{ ft}^2 \qquad (4.68)$$

Take off 1 ft² for the rounded corners of the wing tips, and we have the Cessna 172 POH[13] value of 174 ft². The average wing chord length, the *mean geometric*

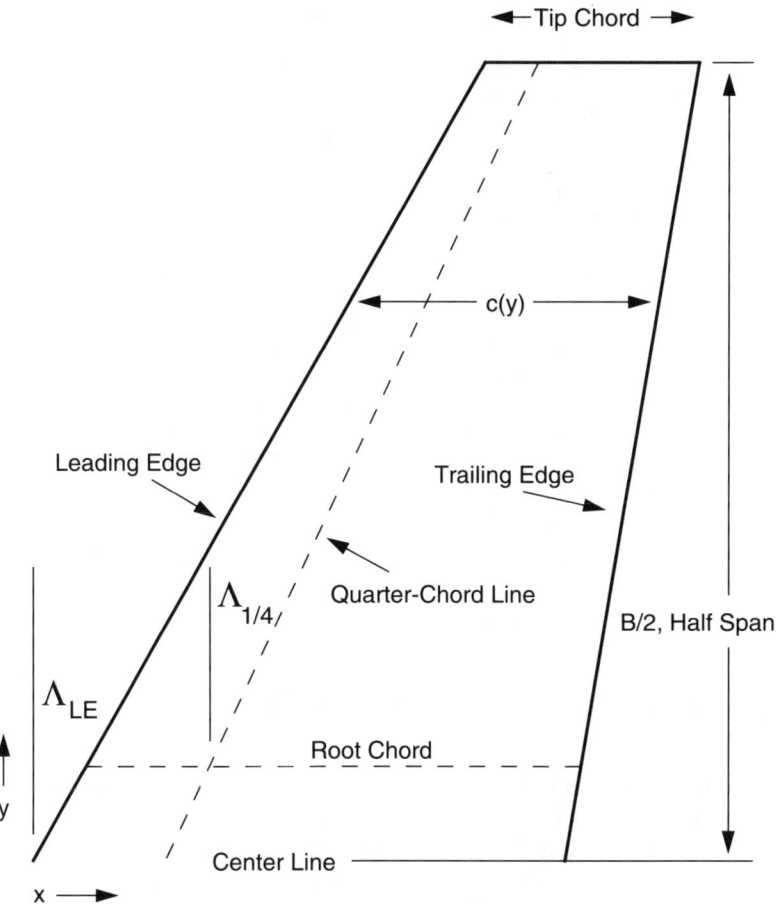

Figure 4.13 General aviation wings are only very slightly swept back, if at all.

chord, mgc, is then $\langle c \rangle = S/B = 4.85$ ft. The wing aspect ratio is the ratio of span to average chord, $A = B/\langle c \rangle = B^2/S$; for the Cessna 172 wing, $A = 7.38$.

A slightly more complex calculation would be that for the *mean aerodynamic chord*, *mac*, useful for more advanced stability considerations and as a weighting factor in composite pitching-moment calculations. The definition, and evaluation for this Cessna 172 wing, is

$$mac = \frac{2}{S} \int_0^{B/2} c^2(y) \, dy = 4.98 \text{ ft} \qquad (4.69)$$

Table 4.5 Cessna 172 right wing points and their coordinates (dimensions in feet)

Point #	Coordinates	Details
1	(0, 0)	Leading edge (LE), Center line (CL)
2	(0, 1.71)	LE, Fuselage
3	(0, 8.33)	LE, Start of taper
4	(0.42, 17.28)	LE, End of taper
5	(0.42, 17.92)	LE, Tip
6	(4.13, 17.92)	Trailing edge (TE), Tip
7	(4.13, 17.28)	TE, End of taper
8	(5.35, 8.33)	TE, Start of taper
9	(5.35, 1.71)	TE, Fuselage
10	(5.35, 0)	TE, CL
11	(3.72, 1.71)	Inboard leading flap corner
12	(3.72, 8.33)	Outboard leading flap corner

(The Cessna POH, in the weight and balance section, gives $mac = 4.90$ ft.) Because this Cessna wing is close to rectangular and has similar partial tapers front and rear, mgc and mac differ only a little. This concludes what we need to know about wing section characteristics.

Aerodynamic Characteristics of Practical Wings

Next we take a look at how real three-dimensional (finite span) wings are put together and how to combine section characteristics to get a handle on the real wing's aerodynamics. There are several problems. We have seen how wings of finite span have less steep curves $C_L(\alpha)$ than their infinite-span counterparts with the same profile. "Three-dimensional" wings (finite aspect ratio) are subjected to an "induced AOA," whereas their unrealistic cousins (infinite aspect ratio) are not. In addition, practical wings may be designed with "washout," an increasingly nose-down orientation as spanwise position nears the tips, to ensure that stalls initiate far enough inboard to avoid ailerons. And, of course, chord length may vary spanwise, generally getting somewhat shorter with progress towards the tip. Even the profile shape may vary to some extent along the wing. So it is not an altogether trivial problem—even assuming that wing sections act relatively independent of one another, with negligible lateral "crossflow"—to piece together the aerodynamic characteristics of a practical three-dimensional wing out of its two-dimensional components. But, to work our way towards considering the flight characteristics of the entire airplane, we must do that. Considering the entire airplane, the AOA most often used is the so-called *body AOA*, α_B, the angle between the free airstream and some reference line imbedded in the fuselage (sometime a top or bottom door sill, sometimes a line between two screw heads protruding slightly for just this purpose).

Wing Geometry: Incidence, Twist, and Body AOA

It would not be realistic to obtain a formula for the Cessna 172 lift coefficient function $C_L(\alpha; \delta_f = 0)$—we have appended specification of the flaps deflection angle δ_f we need to carry along—assuming an *untwisted* wing of aspect ratio 7.38 with *constant* chord length and with α measured between the relative airflow and the direction of the *chord*. The Cessna 172 wing is twisted and has a slight taper in the outboard half. And we want to measure AOAs from the relative airflow to the *fuselage*-level direction. Adjustments are needed.

According to Clarke,[14] the Cessna 172 wing is set 0.78 deg above the fuselage body-level direction for stations from the wing root (where it meets the fuselage) to station 8.33 ft. In our numerical example to follow, we will take that so-called "angle of incidence," α_I, to be the case from the airplane's center line. (Reference wing areas S usually assume such a continuation through the fuselage.) From station 8.33 ft out to station 17.28 ft the wing is progressively (we will assume linearly) tilted downwards ("washed out") from 0.78 to -2.83 deg. For the tip, from station 17.28 to 17.92 ft, we assume a constant incidence of -2.83 deg. The purpose of washout is to ensure that a stall, when it comes, does not start outboard where the ailerons are. Using our two-points-make-a-line formulas,

$$\alpha_I(y) = \begin{cases} 0.78 \text{ deg}, & 0 < y \leq 8.33 \\ -0.403y + 4.14 \text{ deg}, & 8.33 < y \leq 17.28 \\ -2.83 \text{ deg}, & 17.28 < y \leq 17.92 \end{cases} \quad (4.70)$$

What we have, in a sense, are many little wings with various chords [with lengths distributed as in Eq. (4.67)] and pointing various directions all joined together. If the level line in the fuselage is pointed to body AOA α_B, then the AOA of the winglet at station y is

$$\alpha(y) = \alpha_B + \alpha_I(y) \quad (4.71)$$

For the incremental piece of wing at y, with "span" dy, the lift is

$$dL(y) = \tfrac{1}{2} \rho V^2 C_L[\alpha(y); 0] c(y) \, dy \quad (4.72)$$

If we add these up across one-half the span and use our relation for $\alpha(y)$, we have

$$L = \tfrac{1}{2} \rho V^2 2 \int_0^{B/2} C_L[\alpha_B + \alpha_I(y); 0] c(y) \, dy$$
$$= \tfrac{1}{2} \rho V^2 S C_L(\alpha_B; 0) \quad (4.73)$$

where we have brought in the definition of the lift coefficient function we are ultimately after. Getting down to coefficients

$$C_L(\alpha_B; 0) = \frac{2}{S} \int_0^{B/2} C_L[\alpha_B + \alpha_I(y); 0] c(y) \, dy \quad (4.74)$$

The integral on the right, even though we have formulas for all of the pieces, presents two problems. The first is that part of the wing stalls before the whole wing does. Hence, to evaluate the integral we need to extend our definition of $C_L[\alpha(y); 0]$, at least for close-in ys, to the right of the stall point. This is not difficult; we simply let the curved analytic spline keep running. The second problem is that we are getting too many pieces in this puzzle: two formulas for $C_L(\alpha; 0)$, three for $\alpha_I(y)$, and three for $c(y)$. Too cumbersome a result for practical performance calculations. (Remember those? They are what we are driving toward.) We need to simplify, to approximate.

Well, integration is a linear process, $C_L(\alpha)$ is fairly linear and so is $\alpha_I(y)$. At least on the linear portion of the lift coefficient curve,

$$C_L[\alpha_I(y) + \alpha_B; 0] \doteq a\alpha_I(y) + C_L(\alpha = \alpha_B; 0) \qquad (4.75)$$

where the strange functional notation on the right reminds us that the *old* function C_L is involved there, not the new one we want to evaluate. That term does not depend on y, so we take it out of the integral, use the fact that the integral of the chord over one-half the span is one-half the wing area, and get (again, only for the linear portion),

$$C_L(\alpha_B; 0) \doteq C_L(\alpha = \alpha_B) + \frac{2}{S}\int_0^{B/2} a\alpha_I(y)c(y)\,dy \qquad (4.76)$$

The integral on the right—slightly messy because of the different formulas for $\alpha_I(y)$ and $c(y)$ at different spanwise stations—is an ideal candidate for a computer spreadsheet. Just use separate columns, somewhere off to the right, for each of $\alpha_I(y)$ and $c(y)$. Figure 4.14 shows the dependence on station of the variable part of the integrand. For $m = 0.082$, the integral evaluates to -0.715; up near the stall, taking the slope to be $0.082/2 = m = 0.041$, to one-half that, -0.358. When multiplied by $(2/S)$ these give -0.008 (rounding to -0.01) and -0.004 (rounding to 0.0), respectively.

After all this, it turns out that the constant term in our C_L formula is decreased by 0.01 in the left-hand portion and is essentially unchanged near the stall. It is an amazing coincidence that correcting the lift coefficient curve for root incidence, twist, and taper gives (for the flaps zero configuration) almost exactly the uncorrected result we started with.

Or is it? Maybe the Cessna pioneer designers intended this. They did not have computers and needed to keep things simple. We shall see a similar near-null result when we calculate the longitudinal location of the aerodynamic center of the Cessna 172 wing. Nevertheless, your own wing may not be this simple and, if it is not, you will need your own version of these calculations.

What if you do give this problem the full *cumbersome* treatment? The result is that the stall AOA is depressed a little, from 19.0 to 18.2 deg, and so is C_{Lmax}, from 1.60 to 1.576. The main parts of the curves are almost identical. Nothing to

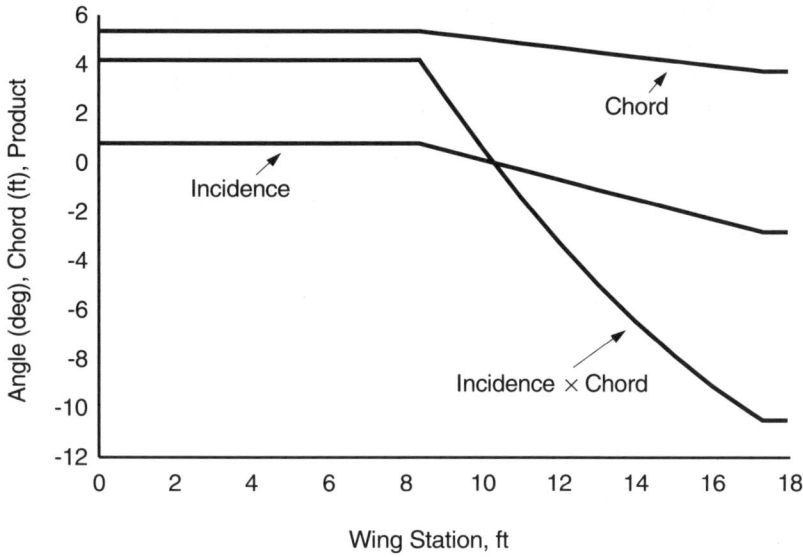

Figure 4.14 Incidence angle α_i times chord length c gives integrand for calculating $C_L(\alpha_B; 0)$.

write home about regarding the stall characteristics, especially because we most likely use alternative techniques to estimate α_S and $C_{L\max}$.

Here is a sufficiently accurate result, for the flaps zero case, using one-half the linear-portion slope when up near the stall:

$$C_L(\alpha; 0) = \begin{cases} 0.082\alpha_B + 0.16, & -16.9 \leq \alpha_B \leq 15.8 \\ 0.041\alpha_B + 0.82, & 15.8 < \alpha_B \leq 19.0 \end{cases} \quad (4.77)$$

This very slight change in intercept b means α_0 is now raised to -2.0 deg. Next we need similarly practical results for flaps variously deployed.

Flaps

It is somewhat unrealistic to associate a given airplane with only a single wing. Most have movable trailing edge flaps that can be extended to various flaps angles δ_f. Extending flaps translates the $C_L(\alpha)$ curve for the flapped region of the wing straight up (more or less) by an amount that depends on how much the flaps were extended and on the type and size of flaps installed. There is a somewhat smaller effect on the $C_L(\alpha)$ curve near the stalling point, and that flapped stall AOA α_S is slightly farther to the left. Flaps extension has negligible effect on the slope of the $C_L(\alpha)$ curve. These general features can be seen in any of the section lift curves from Abbott and von Doenhoff[1] in which split flaps extended 60 deg were investigated. See Fig. 4.15, the section lift coefficient function for the NACA 65$_2$-415, used by some Piper aircraft.

In this section, we use an assortment of derived, empirical, and curve fit formulas (most of the base data from McCormick[15]) to estimate these flaps effects on $C_L(\alpha)$ for the flapped area of the wing. "Flapped area" here means "area of the wing that has flaps" and *not* "flaps area." Effects on the entire wing are then taken to be based on the ratio of flapped area to total wing area. As simple and versatile as this procedure is, we will find we get quite respectable estimates for our sample aircraft.

Our starting point is the realization that lift *with* flaps is equal to lift *without* flaps plus an increment due to having a flapped area A_F with additional lift coefficient $\Delta_f C_L(\delta_f)$:

$$L = \tfrac{1}{2}\rho V^2[SC_L(\alpha;0) + A_F \Delta_f C_L(\delta_f)] \tag{4.78}$$

where

$$\Delta_f C_L \equiv a\tau(c_f,c)\eta(\delta_f)\delta_f \tag{4.79}$$

with functions and parameters defined as

$$\begin{aligned} a &= \text{left-portion slope,} \\ \tau &= \text{flap effectiveness factor,} \\ c_f &= \text{flap chord, and} \\ \eta &= \text{viscosity correction factor.} \end{aligned} \tag{4.80}$$

The flap effectiveness factor τ is defined as

$$\begin{aligned} \tau(c_f,c) &= 1 + \frac{\sin(\theta_f)}{\pi} - \frac{\theta_f}{180} \\ \theta_f &\equiv \cos^{-1}\left(\frac{2c_f}{c} - 1\right) \text{ deg} \end{aligned} \tag{4.81}$$

The viscosity correction factor η is complicated by having different forms for each of plain, split, and Fowler/slotted flaps types:

$$\eta_{\text{plain}} = \frac{-89.9015}{\delta_f^2} + \frac{15.8051}{\delta_f} + 0.1191 \tag{4.82}$$

$$\eta_{\text{split}} = \frac{0.4228}{\delta_f} - 0.0031\delta_f + 0.539 \tag{4.83}$$

$$\eta_{\text{slotted}} = \left(\frac{\delta_f}{35.27}\right)^5 - \left(\frac{\delta_f}{24.21}\right)^4 + \left(\frac{\delta_f}{18.03}\right)^3 - \left(\frac{\delta_f}{14.53}\right)^2 + \left(\frac{\delta_f}{17.54}\right) + 0.570 \tag{4.84}$$

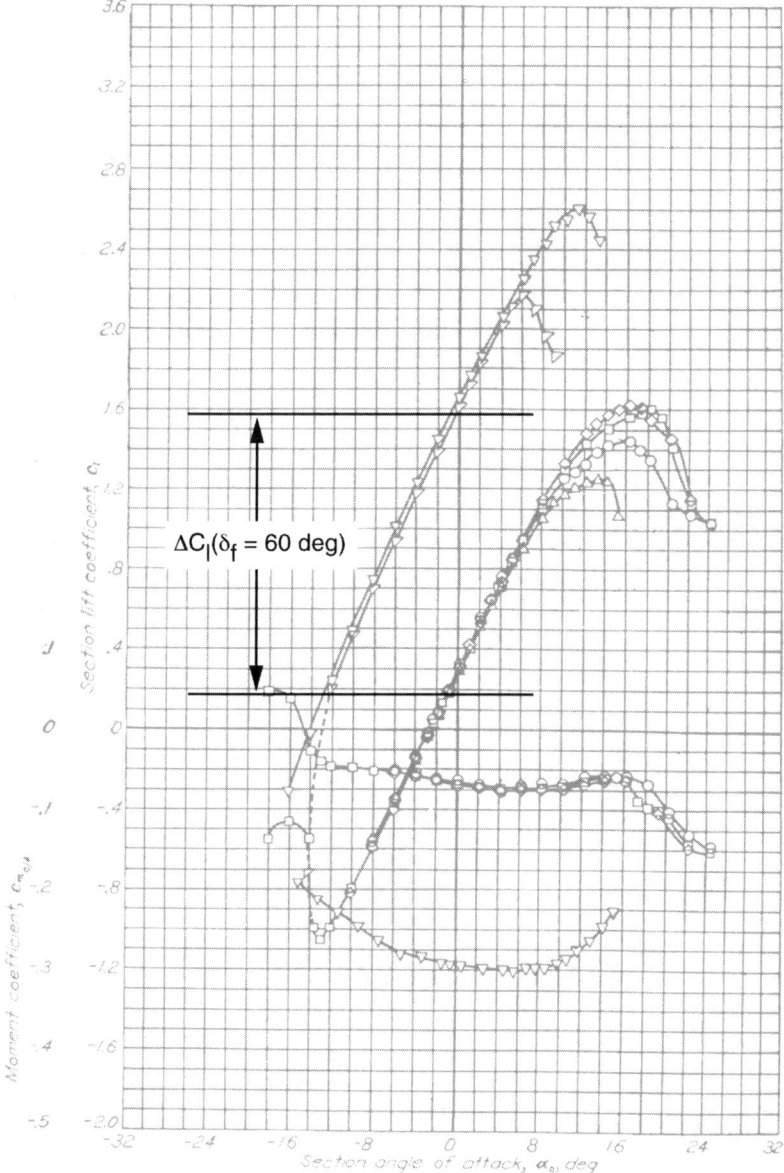

Figure 4.15 Experience (or imagination) lets you see these graphs as two pairs of parallel lines.

Figures 4.16 and 4.17 are graphs of the flap effectiveness and viscosity correction factors. Best to look at these formulas as shorthand tables of experimental results. We can use those empirical adjustments to evaluate what flaps do to lift for our sample Cessna 172. Slotted/Fowler flaps apply. Slope a is 0.082, $c_f = 1.63$ ft, $c = 5.35$ ft, $\theta_f = 113.0$ deg, $\tau = 0.6653$, and δ_f is one of 10, 20, 30, or 40 deg. The flapped region A_F is 70.8 ft^2 = 0.4071S. After turning the calculational crank, using Table 4.6, we find the results of Table 4.7.

We have yet to deal with our orphan stall points $[\alpha_s(\delta_f), C_{L\max}(\delta_f)]$. As mentioned above, this involves further correcting ΔC_L with a factor (which depends on the type flaps installed), we shall call $f_{L\max}$. Brute force polynomial curve fits to graphs found in McCormick[15] yield

$$f_{L\max}^{\text{plain/split}} = 16.667 \left(\frac{c_f}{c}\right)^4 - 6.667 \left(\frac{c_f}{c}\right)^3 \\ - 2.667 \left(\frac{c_f}{c}\right)^2 - 0.183 \left(\frac{c_f}{c}\right) + 1 \quad (4.85)$$

and

$$f_{L\max}^{\text{slotted}} = 16.667 \left(\frac{c_f}{c}\right)^4 - 25.000 \left(\frac{c_f}{c}\right)^3 \\ + 6.333 \left(\frac{c_f}{c}\right)^2 - 0.600 \left(\frac{c_f}{c}\right) + 1 \quad (4.86)$$

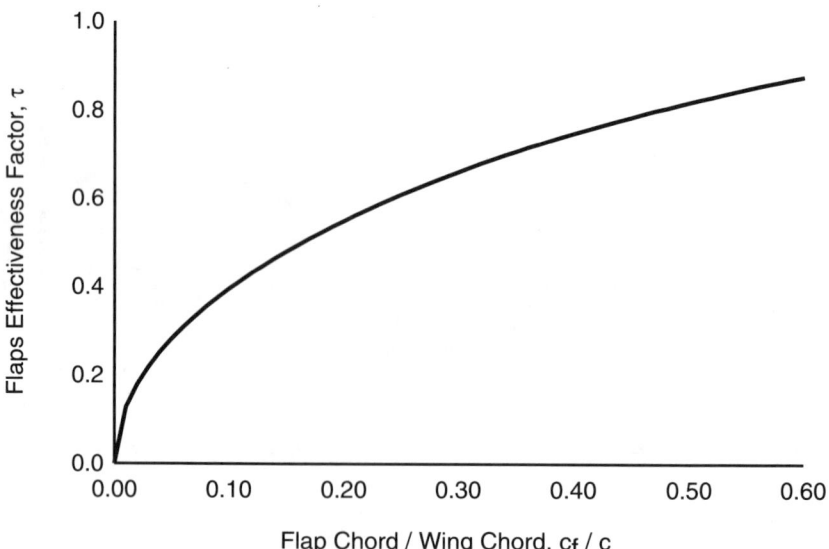

Figure 4.16 When theory fails, scientists often take refuge in "finagle factors."

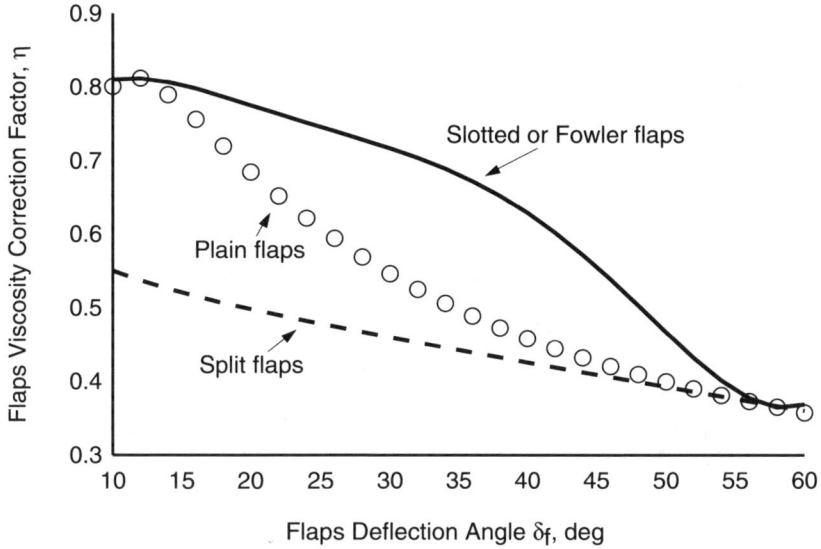

Figure 4.17 However, "polynomial curve fits" sounds better.

Straightforward application of the proper one of these formulas gives the results in Fig. 4.18.

The Cessna 172 POH[13] only cites stall speeds for flaps 0, 10, and 30. Using those values (and picking a most forward c.g.), Table 4.7 gives POH values compared with spreadsheet results. The handbook gives stall speeds to the nearest knot. That implied half-knot error implies an error of about 0.04 C_{Lmax}; so we see that our calculations are close to the mark. More importantly, we have filled in the gaps in the POH data. Even though we are not placing much credence in our approximations to C_{Lmax} itself, we have some reason to believe this calculation of *differences* in C_{Lmax}.

These Cessna data are examples of the fact that incremental flaps effectiveness *always* decreases with increasing flaps angle. The last notch adds only a little

Table 4.6 Flaps effect on lift coefficient for a Cessna 172

δ_f	$\Delta C_L(\delta_f)$
10	0.18
20	0.34
30	0.48
40	0.56

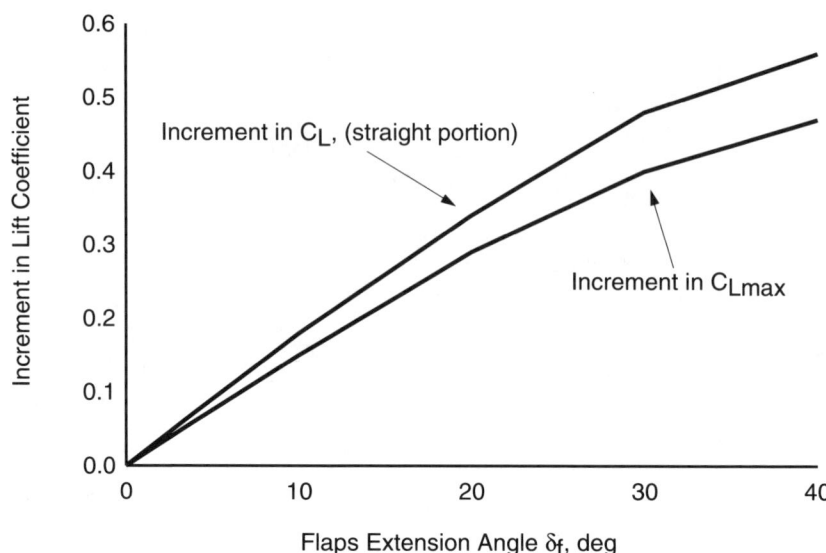

Figure 4.18 C_{Lmax} increases due to flaps are consistently smaller than C_L increases within the linear portion of the graph.

reduction in stall speed, but it adds a lot of drag. On approach to landing, especially with full flaps, watch your air speed!

Table 4.8 shows the flaps effect first described in Table 4.6. These supplement (for the main region, not close to stalling) the earlier formula for lift coefficients with retracted flaps.

The graphs corresponding to Table 4.8 appear in Fig. 4.19. Positive incremental effect of flaps lessens with increasing flaps angles. As an estimate of the decrease of stall AOA α_s with flaps extension you can use this rule of thumb: $\Delta\alpha_s = \alpha_0/4$.

Table 4.7 POH and calculated ΔC_{Lmax} due to flaps

C172 Flap	POH, KCAS stall speed	POH C_{Lmax}	POH ΔC_{Lmax}	Calculated ΔC_{Lmax}	Calculated – POH difference
0	52	1.51	0.00	0.00	0.00
10	49	1.70	0.19	0.15	−0.04
20	NA[a]	NA	NA	0.29	NA
30	46	1.93	0.42	0.40	−0.02
40	NA	NA	NA	0.47	NA

[a] Not available.

Table 4.8 Cessna 172 lift coefficient functions in the linear region

δ_f, flaps angle	$C_L(\alpha_B; \delta_F)$	α_0, deg
0	$C_L(\alpha_B; 0) = 0.082\alpha_B + 0.16$	−2.0
10	$C_L(\alpha_B; 10) = 0.082\alpha_B + 0.34$	−4.2
20	$C_L(\alpha_B; 20) = 0.082\alpha_B + 0.50$	−6.1
30	$C_L(\alpha_B; 30) = 0.082\alpha_B + 0.64$	−7.8
40	$C_L(\alpha_B; 40) = 0.082\alpha_B + 0.72$	−8.8

We have one last question to answer: What do flaps do to c_{mac}? Both Diehl[8] and McCormick[15] show experimental data, but here it pays to remember that our uses for c_{mac} are simple ones. We are not trying to investigate the dynamic stability of a supersonic transport; we are just planning to get a good estimate of the tail-down force on a trimmed general aviation aircraft during moderate maneuvers. Accordingly, we will use the following average approximation for all flaps types for their effect on the pitching-moment coefficient:

$$c_{mac}(\delta_F) = c_{mac}(0) - 0.18\Delta C_L(\delta_F) \qquad (4.87)$$

Combining this result with Table 4.6 and the base value, −0.043, gives the results in Table 4.9. After lengthy but necessary adjustments dealing with the structure of the wing, we are finally ready to tackle the entire airplane.

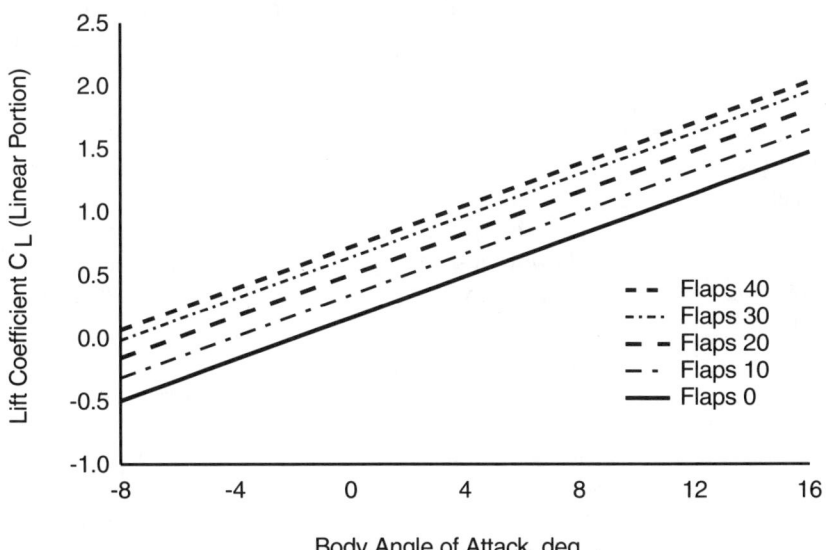

Figure 4.19 Lift coefficient curves for the Cessna 172 wing (NACA 2412) not close to stall.

Table 4.9 Pitching-moment coefficients, any flaps configuration, Cessna 172

δ_f	$c_{mac}(\delta_f)$
0	−0.043
10	−0.075
20	−0.104
30	−0.129
40	−0.144

Lift and Moment of the Entire Aircraft

The Plan

When, above, we went after the best possible $C_{L\max}$ values, we failed to take tail force into account. We now need to correct that oversimplification. If the airplane weighs 2400 lbf and the tail force is 300 lbf downwards, then the wing has to support 2700 lbf, not 2400. And if the tail force is 100 lbf upwards, then the wing only has to support 2300 lbf. If tail lift is measured the same way wing lift is—upwards forces positive—we have in general, in unaccelerated level flight,

$$L_W + L_T = W \tag{4.88}$$

To get the full lift required of the wing, we need to calculate tail force. Afterwards, we may decide to ignore it, or give it a "rule of thumb" value once and for all, but we at least need to know to how much we are giving short shrift. You may have run into hangar-flight rules of thumb: "tail-down force adds another 10 or 15% to your weight." We shall see.

Here is an outline of the plan. In steady unaccelerated flight, total forces and moments on the airplane must be zero. From knowledge of c.g., wing and tail geometry, POH data, and earlier work, a small amount of algebraic unraveling will give us expressions for lift on the wing and lift on the horizontal tail. From those we can find $C_{L\max}(F/A, \delta_f)$, where F/A stands for forward-most c.g. or aft-most c.g.

Moment Equations

For the airplane to be in equilibrium just before the stall, both the sum of the forces on it must be zero—the condition just above—and also the sum of the force moments (torques) on it must be zero. Because the moment due to the wing pitching coefficient is about x_{ac}, it is convenient to use that point of reference for all moments. There are three moments: for wing lift, weight, and tail lift. We are assuming drag and thrust lines act through or near the c.g. so that we can ignore their moments. Such is usually the case for trainer aircraft, though for seaplanes or amphibians it often is not.

$$0 = M_{LW} + M_W + M_{LT}$$
$$= \tfrac{1}{2}\rho V^2 Sc c_{macw} + (x_{c.g.} - x_{acw})W + (x_{acw} - x_{act})L_T \quad (4.89)$$

Equation (4.89) ignores small effects of nonzero body AOA and ignores the pitching-moment contribution of the horizontal tail itself. Downwash off the wings also changes airflow over the tail; we ignored that, too. If necessary, one can add in some or all these factors. In general, aircraft control issues are not within our chosen realm and can be safely skirted.

We do know or can readily figure out the constant geometric locations, the aerodynamic centers (average quarter-chord points) of the wing (x_{acw}) and tail (x_{act}). Once the airplane is loaded, settling the location of the center of gravity ($x_{c.g.}$), we know everything except L_T. These longitudinal locations and the vertical moment-producing forces we are interested in, are depicted in Fig. 4.20.

Here are some details. From a graph in the weight and balance section of the POH, the most forward c.g. position obeys

$$x_{Fc.g.}(W) = \begin{cases} 2.92, & W \leq 1950 \\ 2.92 + 0.00083(W - 1950), & W > 1950 \end{cases} \quad (4.90)$$

The farthest aft c.g. for the Cessna 172 is always at 3.94 ft. For maximum gross weight 2400 lbf, the location of the farthest forward c.g., $x_{Fc.g.}$, is 3.294 ft behind the datum, the front surface of the lower portion of the firewall.

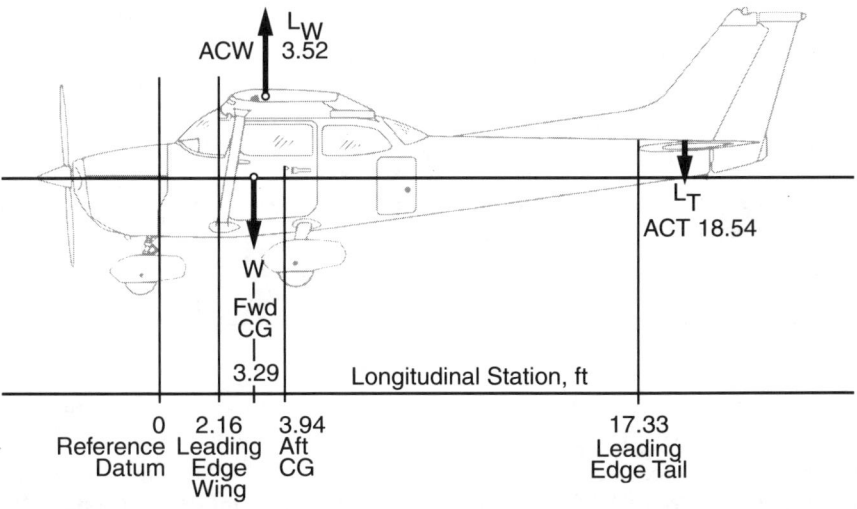

Figure 4.20 Cessna 172 alphanumeric soup of important points, stations, and forces.

Average aerodynamic centers (quarter-chord points) of the wing and tail are, strictly speaking, what you get by weighting each section's quarter-chord point by its chord. For the wing, this is the case we mentioned earlier that hardly (for *this* aircraft) makes any difference. But, because your own airplane may not be so obliging, here is the general formula

$$x_{acw} = \frac{2}{S} \int_0^{B/2} c(y) x_{ac}(y) \, dy$$
$$= \frac{2}{S} \int_0^{B/2} c(y)(\tfrac{3}{4} x_{LE}(y) + \tfrac{1}{4} x_{TE}(y)) \, dy \quad (4.91)$$

In using Eq. (4.91), do not forget that our longitudinal origin has been shifted to the firewall specified by the POH weight and balance section, 2.16 ft ahead of the leading edge of the wing where it meets the fuselage. It turns out that $x_{acw} = 3.51$ ft. Taking one-quarter of the chord length of the early part of the wing, 5.35 ft/4 = 1.34 ft, and adding it to $x_{LEW} = 2.16$ ft, one gets the very respectable estimate of 3.50 ft. If your wing is not swept back, and is not tapered much, you can do the same.

The same technique applies to finding x_{act}, with the added inducement to casualness that a couple of inches more or less, in this case, does not make any difference. From our plan drawing of the horizontal tail, and the dimensions and stations on it, we estimate $x_{act} = 17.77$ ft. The average horizontal tail chord, c, is 3.38 ft.

We have already calculated the effect of flaps extension on c_{mac} for the Cessna 172; Table 4.6 gives those results. You will have to use your section graph or statistical value of c_{mac} with the increments you found near the end of our earlier discussion. That completes preparatory work getting the needed parameters.

Tail Force

Adding the three moment expressions together, setting their sum equal to zero, and solving for L_T gives

$$L_T(x_{c.g.}, \delta_f) = \frac{\tfrac{1}{2}\rho V^2 S c c_{mac}(\delta_f) + (x_{c.g.} - x_{acw})W}{(x_{act} - x_{acw})} \quad (4.92)$$

Figures 4.21 and 4.22 show the resulting Cessna 172 tail force for a range of calibrated air speeds for the two extreme loading conditions (those specified in the POH stall speed table) and for all flaps settings. Near stalling, we see the tailforce varies from 54 lbf up (for flaps up, c.g. aft) to 96 lbf down (for flaps 40, c.g. forward).

The Aerodynamic Force 113

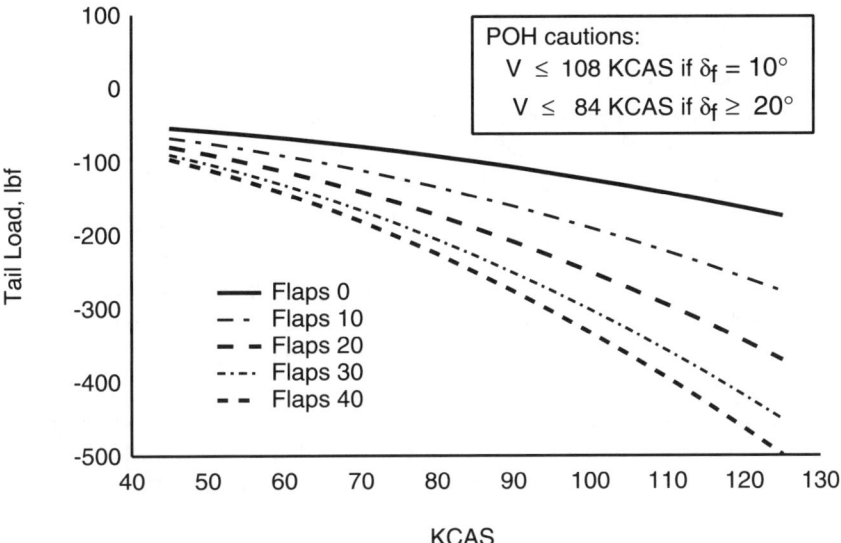

Figure 4.21   Negative (downwards) tail forces add to the burden, raise stall speed.

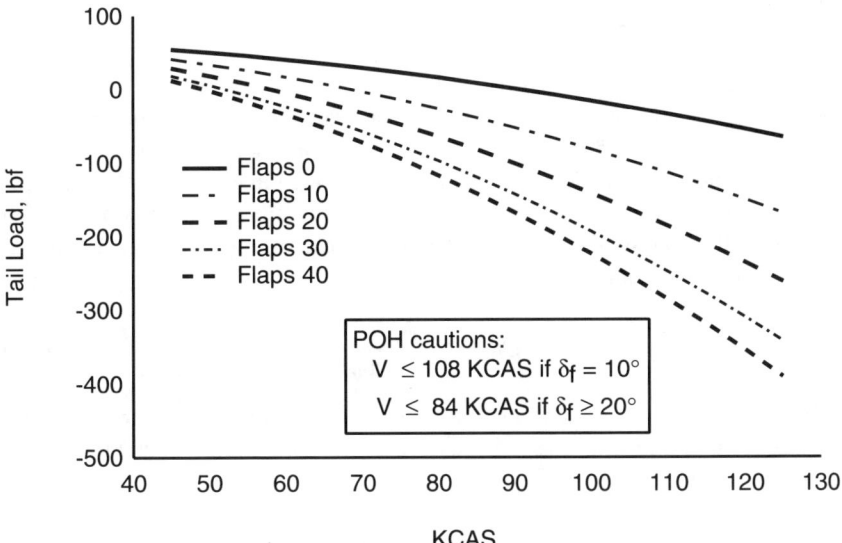

Figure 4.22   At low speeds, positive (upwards) tail forces lower the stall speed a bit.

Factors Influencing Lift

In spite of the lift details we have treated, one must not become sanguine. Recall that we are generally saved, in our low speed, low powered, quasi-steady-state regime, by the fact that flight path angles are small and, therefore (wings level), net lift is close to gross weight. Here is a lexicon of practical effects on lift in general aviation—some large, some medium, some small—to emphasize what "a complete theory" would be up against.

1) Environmental—air density (large), air viscosity or Reynolds number (small), Mach number (small), ground effect (medium), gusts (variable), centrifugal and Coriolis pseudoforces (quite small, but treated in Chapter 3).
2) The airplane—wing area (large), wing shape (medium), wing profile (medium), angle of incidence and twist (medium), sweepback (medium), dihedral (small), nonwing surfaces (variable), surface smoothness and cleanliness (variable), protuberances (small).
3) Airplane configuration—flaps extension angle (large), landing gear extension (small).
4) Operational factors—body AOA (large), air speed (large), yaw and roll angles (small), trim settings (small), throttle and/or revolutions per minute settings (medium).

Drag

In some nonaviation applications—oil droplets dropping in still air, spheres slowly rising or sinking in a resting fluid—the drag force goes as the first power of the relative speed of object and fluid. Stokes law governs. In aviation we seldom see those cases. While it is primarily a phenomenological (not fundamental) relation, in our regime we shall stick with the time-honored and experimentally supported "quadratic drag polar" describing the relation between drag and lift coefficients:

$$C_D = C_{D0} + \frac{C_L^2}{\pi e A} \qquad (4.93)$$

The drag force, of course, is then arrived at through Eq. (4.2). There are cases, especially with highly cambered airfoils, when it is convenient to bend the rule of Eq. (4.93) and proclaim

$$C_D = C_{Dmin} + \frac{(C_L - C_{Lmin\text{-}drag})^2}{\pi e A} \qquad (4.94)$$

but that only amounts to sliding the $C_L(\alpha)$ over a bit; the requisite mathematical manipulations are essentially unchanged. We will use only the simpler form.

In Eq. (4.93), we now know all about aspect ratio A. What remains to be found is the pair of constants C_{D0}, the parasite drag coefficient, and e, the airplane efficiency factor. Because, for the case we care about—the entire airplane—we will arrive at those two values by means of glide tests and we leave the details to the chapter on glide performance. But before we leave the aerodynamic force, we should familiarize ourselves with a graph of $D(V)$. It is important to keep in mind that, because C_L rises at slow speeds, the total drag curve is cup-shaped, with a definite minimum. Later we shall show that this minimum occurs at V_{bg}, the speed for best (longest) glide in still air. For now, assume that result. For the Cessna 172, which Fig. 4.23 mimics, V_{bg} is at about 72 KCAS when the airplane weighs 2400 lb.

Expression for Drag D in Terms of $D_{\min} = D_{bg}$ and V_{bg}

If one considers the airplane in wings-level (unbanked) flight and uses Eqs. (4.1), (4.2), and (4.93) along with the so-called "small flight path angle assumption"

$$L = W \cos \gamma \doteq W \qquad (4.95)$$

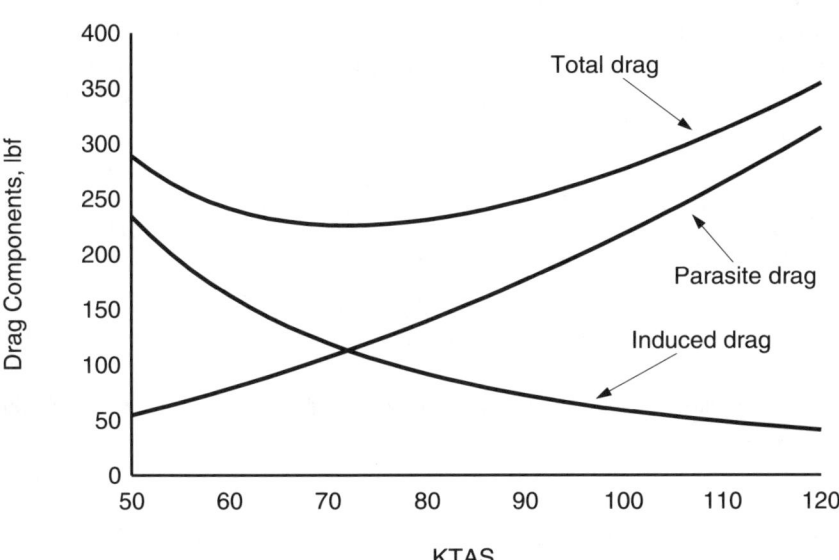

Figure 4.23 Parasite, induced, and total drag force on an airplane in level flight.

where W is the aircraft weight and γ is the flight path angle, then the total drag depicted in Fig. 4.23 is given by the formula

$$D(V) = AV^2 + B\frac{1}{V^2} \tag{4.96}$$

with

$$A \equiv \tfrac{1}{2}\rho S C_{D0} \tag{4.97}$$

and

$$B \equiv \frac{2W^2}{\rho \pi e A S} \tag{4.98}$$

A convenient property is shared by functions f of the form of Eq. (4.96):

$$f = Ag + B/g \tag{4.99}$$

where function g—for us, $g(V) = V^2$—does *not* have an extremum (minimum or maximum) in the interval of interest. Taking the derivative of f, a little algebra shows that the extremum of f occurs only when

$$g^2 = \frac{B}{A} \tag{4.100}$$

or, rearranging, the extremum of f occurs precisely when

$$Ag = B/g \tag{4.101}$$

In our case, this means the minimum value of drag occurs when parasite drag is equal to induced drag. Jumping ahead to pick nomenclature and an idea from the glide performance chapter, the speed at which this drag minimum occurs is the speed for best (longest) glide in still air, V_{bg}. So minimum drag will be twice either drag component:

$$D_{\min} = 2AV_{bg}^2 = 2B/V_{bg}^2 \tag{4.102}$$

Solving Eq. (4.102) for constants A and B (not *quite* constant, recall), we then have the very nice—even memorable—result that

$$D \doteq \tfrac{1}{2} D_{\min} \left(\frac{V^2}{V_{bg}^2} + \frac{V_{bg}^2}{V^2} \right) \tag{4.103}$$

A more careful analysis shows that this approximation depends on the smallness of

$$\cos^2 \gamma - \cos^2 \gamma_{bg} \tag{4.104}$$

Not normally a problem. Eq. (4.103) is not original (see McCormick[15]), though this derivation might be. We will use this *very approximate* (meaning *very close*) result fairly often. For our sample Cessna 172, for example, using this assumption

Table 4.10 Parasite drag coefficient C_{D0} vs flaps deflection angle δ_f

δ_f, deg	$C_{D0}(\delta_f)$
0	0.037
10	0.039
20	0.045
30	0.055
40	0.066

[Eq. (4.103)] gives a value of drag that differs from the exact formula (for all speeds between stall and level maximum) by less than 1 lbf.

Additional Drag Due to Flaps

McCormick[15] has a formula—possibly originally from Cahill—that you can use to check your own experimental runs at various flaps settings or use in case you have to deal with an airplane that does not come with drag information on each flaps setting.

$$\Delta C_D = \begin{cases} 1.7 \left(\frac{c_f}{c}\right)^{1.38} \left(\frac{A_f}{S}\right) \sin^2 \delta_f, & \text{plain or split} \\ 0.9 \left(\frac{c_f}{c}\right)^{1.38} \left(\frac{A_f}{S}\right) \sin^2 \delta_f, & \text{slotted or Fowler} \end{cases} \quad (4.105)$$

Here ΔC_D is the additional drag coefficient due to flaps deflection δ_f. Variable c_f is the chord length of the flap and A_f is the area of the flapped portion of the wing (not of the flaps themselves). Variables c and S are the ordinary wing (mean aerodynamic) chord and reference area, respectively.

Using Eq. (4.105) for the Cessna 172 gives Table 4.10. Extending flaps has negligible effect on the airplane efficiency factor e but considerable effect on both lift and drag. Successively higher flaps increments give relatively smaller lift increases but relatively larger drag increases.

After this long introduction to the aerodynamic force, we are almost ready for practical work. But first we must consider the engine and the propeller.

References

1. Abbott, I.H., and von Doenhoff, A.E., *Theory of Wing Sections*, Dover, New York, 1959, pp. 62, 478, 488.
2. Aitken, A.C., *Determinants and Matrices*, Oliver and Boyd, Edinburgh, 1964, p. 119.
3. Moran, J., *An Introduction to Theoretical and Computational Aerodynamics*, Wiley, New York, 1984.

4. Roskam, J., and Lan, C.-T.E., *Airplane Aerodynamics and Performance*, DARcorporation, Lawrence, KS, 1997.
5. Von Mises, R., *Theory of Flight*, Dover, New York, 1959, pp. 247, 272.
6. NACA, TR 563 (1936).
7. Hiscocks, R.D., *Design of Light Aircraft*, published by the author, 1995, p. 24.
8. Diehl, W.S., *Engineering Aerodynamics*, Ronald Press, New York, 1936.
9. Prandtl, L., and Tietjens, O.G., *Applied Hydro- and Aeromechanics*, Dover, New York, 1957, p. 197.
10. Woods, K.D., *Technical Aerodynamics*, McGraw-Hill, New York, 1935.
11. Suh, Y.B., and Ostowari, C., "Drag Reduction Factor Due to Ground Effect," *Journal of Aircraft*, Vol. 25, No. 11, 1988.
12. Hubin, W.N., *The Science of Flight*, Iowa State University Press, Ames, IA, 1992, pp. 157, 180.
13. Cessna Aircraft Co., *1986 Model 172P Information Manual*, Wichita, KS, 1985.
14. Clarke, B., *The Cessna 172*, Tab Books, Blue Ridge Summit, PA, 1987.
15. McCormick, B.W., *Aerodynamics, Aeronautics, and Flight Mechanics*, Wiley, New York, 1979.

5

The Engine

Introduction

To many aviators, the airplane's engine is a "black box." Fuel in, power out. Not only a black box but—with fewer and fewer "round engines" these days—a box hidden inside a cowling, yet another box. Out of sight, out of mind. Unless, of course, it seizes, sputters, or dies. Then the engine attracts our full attention!

As it will in this short chapter, a preemptive strike. We will study, in broad outline, how engines work (we limit ourselves to normally aspirated spark ignition reciprocating internal combustion gasoline engines) and their more important performance measures and variables. To some extent, how those depend on each other. Very little on coddling an engine or maintaining it and nothing on repairs; we remain focused on performance under the prosaic assumption that everything is copacetic. This background should be of value to ordinary pilots and even to aircraft accident investigators.

Modern internal combustion engines are a technological marvel. For all of their complication and intricacy and all of the violence of their pistons' thrash, these engines are extremely reliable. They are also complicated; a lot goes on behind the scenes. The chemistry is not as simple as some would have us believe; the thermodynamics is intricate and the mechanical engineering is extreme. Without an iota of hard evidence as support, the author estimates more complex modern internal combustion engines as somewhere between 1/1000 and 1/100 as complex as the simplest viruses. Just a guess.

These engines can be put to many uses and the range of constructed engine sizes is vast. MacMahon (an engineer) and Bonner[1] (a biologist) consider a list of 39 internal combustion engines. The smallest engine listed is the Webra Speedy, a single cylinder two-cycle engine weighing 4.75 oz, displacing 0.110 in.^{3}, and developing 0.45 hp at 22,000 rpm. The largest is from Burmeister & Wain, a 12-cylinder four-cycle Diesel engine weighing 225,000 lbf, displacing 726,180 in.^{3}

and developing 27,800 hp at 110 rpm. That range of engine masses spans a factor of 758,000! About one-half the engines MacMahon and Bonner list are aircraft engines, either for models or full-scale airplanes. As a hero of the motion picture *Flight of the 'Phoenix'* (the more intellectual and neurasthenic one) expressed, the distinction between model and full-scale aircraft is somewhat artificial. "The pilot for today's flight is Col. Charles Lindbergh." Or, "Col. Thom Thumb."

MacMahon and Bonner go on to propose and defend various *allometric* formulas connecting pairs of engine variables selected from among engine mass m, brake horsepower bhp, engine rotation speed N, diameter of cylinder bore B or length of stroke S, average linear piston speed S_p, and total engine displacement volume V_D. An allometric formula is one in which two variables x and y are related according to

$$y = bx^a \qquad (5.1)$$

or, equivalently, according to

$$\ln y = \ln b + a \ln x \qquad (5.2)$$

where, for any given pair of variables x and y, a and b are constants. Plotted on log–log paper, the points of variable pairs closely following an allometric formula lie close to a single straight line with slope a and y intercept (where $x = 1$, $\ln x = 0$) $\ln b$. Because parameter b can be thought of as depending only on the arbitrary units used to express variable y, parameter a is the number that holds all the important information. Table 5.1 gives the results for several engine variable pairs. It does not, however, show the extent to which the corresponding 39 points were scattered about that best-fit line. The zero exponent for S_p shows that mean piston speed (hence, maximum piston speed because those are proportional) is fairly constant over the large range of engine masses considered.

As the first and last rows in Table 5.1 explicitly show, not all of the dependent varibles y are independent of one another. For even some variables *not* given as functions of others, further analysis provides an explanation of their associated exponents a. Take displacement V_D, for instance. How might we explain that it has $a = 4/3$? As follows.

Table 5.1 Allometric parameters connecting engine variables

Variable x	Variable y	Exponent a
m	hp $\propto N \times V_D$	1
m	N (rpm)	$-1/3$
m	B (Bore)	1/3
m	S (Stroke)	1/3
m	V_D (Displacement)	4/3
m	$S_p \propto N \times S$	0

One structural limitation on internal combustion engines is the maximum strain energy per unit volume that pistons can take without breaking down. That energy density is proportional to the mass density of the piston material, approximately a constant, and the square of the average piston speed, S_p^2. So constant limiting strain means constant S_p^2 means constant S_p. Now switch gears and consider power. Power P goes up directly with the engine mass m, with exponent one. This is shown by the fact that as we move from single- to multiengine aircraft we do not appreciably gain or lose in the *specific power*, power-to-weight ratio. We can gain power, of course, at the expense of equivalent weight, and we *may* gain reliability—light twins, if not heavily loaded and if at only moderate altitudes, may continue level flight after one engine fails. We know more about power. Almost any amount of fuel can be pushed through an engine if one does not care about burning it; in earlier times, you may have had a car like that. Power is primarily limited by the speed with which the engine can breathe in combustion air, on the product of the amount of air brought in per cycle, engine displacement V_D, and the number of complete engine cycles per time, $N/2$ in the case of a four-stroke cycle. So power $P \propto V_D \times N$. Because $N \propto m^{-1/3}$ (see the second row of Table 5.1), we are left with the conclusion that the exponent of m in the expression for V_D must be $1 + 1/3 = 4/3$. We have here strayed from performance into rudiments of engine design, but is is always worthwhile to point out that there are *reasons* for (most) things.

As the recent focus on power and engine displacement illustrated, our initial descriptive snippet that the engine is a black box—fuel in, power out—is far from the mark. Though still much simplified. Fig. 5.1 gives a more realistic picture. The engine mechanism cycles, but the fuel, air, and waste products do not.

Though *power* is most often cited as the primary useful engine output, *torque*, in most productions, best plays that leading role. Power and torque (or *moment*, hence the common variable M) are related by

$$P = 2\pi n M \tag{5.3}$$

Torque M is more closely the engine's *ability to do work* (under given conditions of throttle position, a competently leaned mixture, and sufficient input air density) while power P is more closely the rate at which the engine is

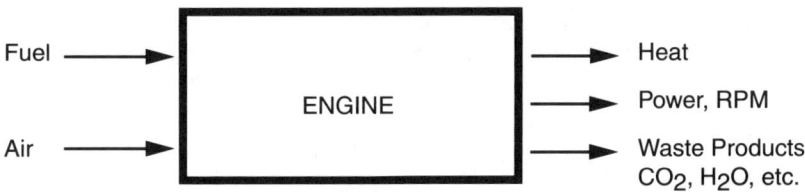

Figure 5.1 An engine that does useful work must also produce heat and entropy.

actually doing work. If an increased load is placed on the engine (leaving everything else the same) engine circular speed—here measured by revolutions per second n instead of by the more common revolutions per minute, N—decreases. In a common aviation example, this might come about by a decision of the pilot of a fixed-pitch propeller airplane to climb without changing his throttle position. Revolutions per minute would decrease, engine power output would decrease, but torque would remain almost constant. If he decided to (power) dive, on the other hand, both power and revolutions per minute would increase while, again, torque would stay fixed.

Besides wanting to know how much power or torque an engine produces, we are interested in its fuel efficiency. The type and specifications of the fuel matter, of course, but those choices are quite limited. Table 5.2 lays out the major variables for common aviation gasolines. In addition, of course, the gasoline must be very clean and uncontaminated, especially free of water. Gasoline engines are finicky and, by design, only accommodate one or two varieties of fuel. If the correct type of fuel is not available, it is easier on the engine to temporarily use fuel one step higher in octane (antiknock) rating rather than fuel a step lower. Jet-propelled aircraft generally have less delicate fuel appetites.

In Table 5.2, the appellation *lower* heating value refers to water, among the combustion products, in vapor form. If this vapor were to condense, additional heat would be liberated and the so-called *higher* heating value, about 1400 Btu/lbm higher, would be appropriate. By convention, lower heating values are used in thermal efficiency calculations of internal combustion engines. The abbreviation LL in the heading of the second column stands for "low lead."

All of these "avgas" varieties have nominal specific weights of 6.0 lbf/U.S. gall. More exactly, for 100/130 aviation gasoline

$$\text{Fuel lbf/U.S. gall} = 6.077 - 0.00409 \times °F \tag{5.4}$$

More to the point is adjustment of the mixture, the fuel-to-air ratio F:A, for maximum speed or maximum range or whatever operational goal the pilot has in mind. This involves the "leaning" we spoke of, and most airplanes have a mixture control knob for that purpose. Earth-bound automobiles enjoy smaller altitude excursions and seldom need to control the F:A. The chemically correct or *stoichiometric* ratio of fuel weight to air weight for complete combustion

Table 5.2 Common aviation gasolines

Variable	100 LL	100/130	115/145
Minimum lean/rich antiknock rating	91/98	100/130	115/145
Color	Blue	Green	Purple
Minimum lower heating value, Btu/lbm	18700	18700	18700
Maximum tetraethyl lead, ml/gal	2	3	4.6

depends slightly on the relative proportions of hydrogen to carbon in the fuel. So does the heating value. For heating value $H = 18{,}700\,\text{Btu/lbm}$ cited above, the stoichiometric F : A is $(F:A)_s = 0.067 = 1/14.9$. With 25% excess air, this drops to about $(F:A) = 0.053 = 1/18.9$. The mixture's F : A is of course a major influence on fuel economy, generally measured as *brake specific fuel consumption rate* or *bsfc*, pounds of fuel consumed per hour per horsepower. But the F : A also influences cylinder and exhaust gas temperatures (which can thereby be used to monitor and adjust the mixture) as well as the chemical composition of the exhaust gas.

Stoichiometrically, pure hydrocarbon fuel should burn to nothing but carbon dioxide, CO_2, and water, H_2O. In practice, of course, there is always some sulfur in the fuel; it should burn to sulfur dioxide, SO_2. If the mixture is lean, with an excess of oxygen, the exhaust contains that uncombined oxygen; if the mixture is rich, with an excess of fuel, the exhaust contains uncombined hydrogen, H_2, and some partially burned carbon as carbon monoxide, CO, the gas one sees as a darker plume in the inner and cooler portion of a candle flame. Because it behooves an aircraft engine to do things quickly rather than exactly, in practice there are always unburned carbon particles and some methane, CH_4, expelled in the products of combustion.

The main engine performance variables we have discussed so far are power P, torque M, engine circular speed N or n, the fuel-to-air mixture mass ratio (F : A), and fuel efficiency expressed as *bsfc*. That is only the beginning. Now it is time to tackle the details.

Engine Power and Efficiency

Of the many questions one might have on the workings of the aircraft internal combustion engine, two stand out:

- Given reasonable mechanical specifications for such an engine (including a rated revolutions per minute, which we are very far from being able to predict), what maximum power (or torque) output we can expect from it?
- Given that maximum output, how efficient is the engine in turning the chemical energy in the fuel and air into mechanical work?

Tentative answers to these questions come from looking into the engine's size and speed relationships together with its F : A and fuel/air heating value.

Thermal Power

It is not realistic to assume that the engine can gulp a full volume V_D of new air each engine cycle. V_D is the total volume displaced by all the cylinders, granted, but various pressure losses in the intake system (carburetor, intake manifold, intake valves), with temperature increases that rarefy the induction air, and the fact that some of that gas is indeed fuel vapor, lead to a finagle factor known as *volumetric efficiency*, η_v. This is a measure of how well the engine "breathes." Volumetric efficiency is normally close to 85%, but an "asthmatic" engine might not do that well.

We can use η_v, with previously defined rates and sizes, to find that thermal power P_T, the combustion energy per unit time the burned fuel should liberate, is

$$P_T = \frac{\text{heat}}{\text{fuel wt}} \times \frac{\text{fuel wt}}{\text{air wt}} \times \frac{\text{air wt}}{\text{vol}} \times \frac{\text{vol}}{\text{cycle}} \times \frac{\text{cycles}}{\text{time}}$$
$$= H \times \text{F:A} \times \rho g \times \eta_v V_D \times n/2 \qquad (5.5)$$

For our sample Cessna 172, powered by a Lycoming O-320-D2J engine with $V_D = 320 \text{ in.}^3$ total displacement, using the fuel heating value featured in Table 5.2, running "full bore" at sea level at its rated 2700 rpm = 45 rps with a stoichiometric F : A, the thermal power is

$$P_T = \frac{\left(18{,}700 \times 778 \, \dfrac{\text{ft-lbf}}{\text{Btu}}\right) \times 0.067 \times (0.002377g) \times (0.85 \times 320/12^3) \times \dfrac{45}{2}}{550 \, \dfrac{\text{ft-lbf/s}}{\text{hp}}}$$

$$= 480 \text{ hp} \qquad (5.6)$$

The explicit conversion factors in Eq. (5.6) are for the mechanical equivalent of heat and to convert to horsepower. Our rough sample answer for thermal power is three times the 160-hp rated output of the engine. But we have further losses and inefficiencies to explore.

Other Efficiency Losses

The engine transforms only about one-third of the heat of combustion into useful shaft work. Here is an approximate breakdown of the various energy loss mechanisms for a normally aspirated engine.

- Otto cycle thermodynamic efficiency, about 57.5%. Internal combustion engines of this type approximate the ideal Otto cycle. Thermodynamic analysis of the Otto cycle (adiabatic compression and power strokes, constant volume ignition and heat dumping) shows that the efficiency of the cycle depends only on the compression ratio r (8.5 : 1 for our sample Lycoming O-320-D2J) according to

$$\eta_{\text{Otto}} = 1 - r^{1-k} \qquad (5.7)$$

where k is the ratio of specific heat at constant pressure to that at constant volume, C_p/C_v. One normally takes standard air as an approximation to the "working fluid" in the gasoline engine; the nominal value of k for standard air is 1.40. In fact the working fluid is a much more complex mixture of gases and, in additional fact, k for air at the higher temperatures inside the engine is closer to 1.30.
- Volumetric efficiency, about 85%. This is the figure we used above. Depending on how much one lumps into the engine's respiratory process, 85% may have been putting a good face on it. Beside the air-passage restrictions mentioned, which should include pumping losses during the intake and exhaust strokes, and the rarefying temperature increases, there are effects due to exhaust and intake valves both being open during part of the cycle and due to the finite *clearance volume* V_C (the gas volume remaining between cylinder heads and pistons when the latter are at top dead center). From the fact that by definition

$$\frac{V_C + V_D}{V_C} = r \tag{5.8}$$

it follows that

$$V_C = \frac{V_D}{r - 1} \tag{5.9}$$

Efficiency to this point, 57.5% × 85% = 48.9%.
- Fuel-rich operation efficiency, 86%. The stoichiometric F : A is about 0.067. To achieve the most complete burning of the fuel, say for economy, one must lean the mixture to about F : A = 0.055. To get maximum power from the engine, to the other side, one must enrich the mixture, to about F : A = 0.078. Our over-rich efficiency figure comes from the fact that 0.067/0.078 = 85.9%. Efficiency to this point, 48.9% × 86% = 42.0%.
- Mechanical efficiency, 88%. This loss includes all sources of hard-surface friction: piston rings rubbing against the insides of the cylinders and on the pistons themselves, bearing friction along the crankshaft and camshaft, rods in rod guides, etc. Efficiency to this point, 42% × 88% = 37.0%.
- Accessory bleed efficiency, 95%. Almost all engines turn alternators or generators (to power the airplane's electrical devices, including vacuum pumps powering some cockpit instruments) and magnetos (for spark ignition). Some airplanes go further and siphon off engine power for cabin air conditioning or pressurization. This item is quite variable. Efficiency to this point, 37% × 95% = 35.1%.
- Efficiency due to other effects, 95%. This might include various chemical effects, energy entrapped in unburned molecular species, radiative effects, blow-by gas and fuel losses, humidity effects, anything not yet covered. Final total efficiency, 35.1% × 95% = 33.3%.

Do not be misled by this close confirmation of the calculation in Eq. (5.6). None of these efficiency figures will precisely hold true as one moves from one engine to another, and none will remain constant for any single engine subjected to changing conditions of maintenance, time before overhaul, atmosphere, load, or temperature. These numbers are in the ballpark, but only that. A somewhat different energy branching outline (adapted from Liston,[2] originally for a supercharged engine) should bring home this looseness. Here the Otto cycle efficiency appears explicitly as heat in the exhaust gas.

Fuel energy, 100%

A. Exhaust, 51.6%
  1. Heat, 47.0%
  2. Chemical, 4.6%
     a. CO, 3.1%
     b. CH_4, 1.5%
B. Other thermal, 12.2%
  1. Conduction to air, 7.2%
  2. Conduction to oil, 1.6%
  3. Radiative and miscellany, 3.4%
C. Indicated output, 36.2%
  1. Mechanical loss, 4.3%
  2. Brake horsepower output, 31.9%

Notice that the implied mechanical efficiency is $1 - 4.3/36.2 = 88\%$. This is a "creative adjustment" we imposed to ensure coincidence with bootstrap parameter C (about 0.12 for our sample engine) used in the altitude power dropoff factor $\Phi(\sigma)$.

The Gagg–Farrar Power Dropoff Factor $\Phi(\sigma; C)$

The inevitable decrease of shaft power P output with diminishing air density at altitude is expressed in the bootstrap approach as well as by most investigators as

$$P(\sigma) = \Phi(\sigma; C) \times P(\sigma = 1) \qquad (5.10)$$

where the dropoff multiplier $\Phi(\sigma; C)$ is that originally arrived at by Gagg and Farrar[3] in 1934:

$$\Phi(\sigma; C) = \frac{\sigma - C}{1 - C} \qquad (5.11)$$

where engine-specific parameter C is the ratio of power lost to friction (such loss assumed constant with altitude) to *indicated power* at sea level, $P_i(\sigma = 1) \equiv P_{i0}$:

$$C \equiv \frac{P_f}{P_{i0}} \qquad (5.12)$$

Indicated power gets its name from so-called "indicator diagrams," graphs of cylinder pressure vs volume, drawn by mechanical monitoring instruments ported into an active cylinder. We will soon derive Eq. (5.11), but let us first show how the power dropoff factor varies with (dry) density altitude (Fig. 5.2) and what the effect of humidity on engine performance can be (Table 5.3).

Example 5.1 Data in Table 5.3 were constructed under the following assumptions: the air is saturated with moisture (so that the outside air temperature T is the same as the dewpoint temperature T_d); atmospheric pressure is assumed standard, $p = p_0 = 29.921$ in. Hg $= 2116.2$ psf.

Describing variables and units as we go, let us calculate the last row of Table 5.3, the hot humid case with $T = T_d = 100°F$. Saturation vapor pressure e_s, in in. Hg, comes from Teten's formula given as Eq. (1.14). We see that water vapor is providing about 1/15 of the atmospheric pressure. The specific gas constant R'', treating vapor molecules as nonentities (which they essentially are, because they do not contribute to the chemistry of fuel combustion), is given by Eq. (1.16). $R''/R = 57.041/53.355 = 1.069$; discounting water vapor, the specific gas constant is almost 7% larger. To find σ_a, the density of (only) "air," we first use the ideal gas law given as Eq. (1.1), with R'' substituted for R. Temperature T will be $(100 + 459.7)°R = 559.7°R$. Doing that arithmetic and dividing by $\rho_0 = 0.002377$ slug/ft³, we find $\sigma_a = 0.8667$. Ordinarily, and as given by Eq.

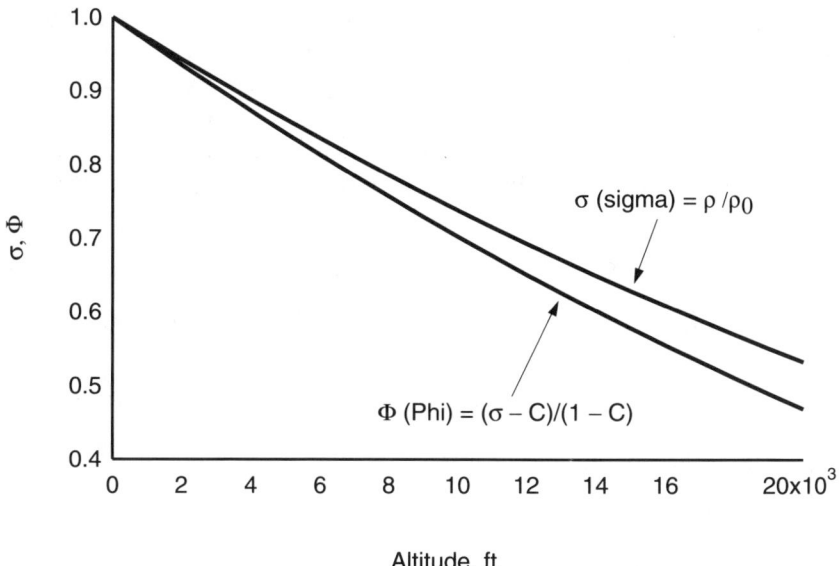

Figure 5.2 The Gagg–Farrar factor Φ drops off with density altitude a bit steeper than relative density σ.

Table 5.3 Humidity effect on engines, saturated air, standard pressure

$T = T_d$	e_s	R''	σ_a	h_{pa}	$\Phi(\sigma_a)$
0	0.045	53.435	1.1265	−4128	1.144
10	0.071	53.481	1.1015	−3343	1.115
20	0.110	53.551	1.0772	−2563	1.088
30	0.166	53.653	1.0532	−1781	1.060
40	0.248	53.801	1.0293	−989	1.033
50	0.363	54.010	1.0052	−176	1.006
60	0.522	54.302	0.9805	671	0.978
70	0.740	54.707	0.9549	1569	0.949
80	1.033	55.262	0.9278	2540	0.918
90	1.422	56.018	0.8986	3608	0.885
100	1.933	57.041	0.8667	4807	0.849

(1.13), that relative density would connote a density altitude of 4807 ft and a Gagg–Farrar dropoff factor, by Eq. (5.11), of $\Phi = 0.849$. Even though pressure is at the mean sea level (MSL) standard value, heat and humidity have "raised" the effective altitude almost 5000 ft.

For aerodynamic purposes, there will still be a humidity effect because water molecules are lighter than "air" molecules, but not as large an effect as on the engine. Calculating R' with Eq. (1.15) shows it to be 54.691, only 2.5% larger than standard. The wet atmosphere relative density turns out to be $\sigma = 0.9040$, and the aerodynamic density altitude is 3410 ft. Quite a bit rarer than the pressure figure would suggest, but not as rare as the atmosphere experienced by the engine.

Now we return to our interrupted derivation of the Gagg–Farrar dropoff factor, Eq. (5.11). We need to know that indicated power is the engine's power before friction power is subtracted:

$$P_i(\sigma) \equiv P(\sigma) + P_f \qquad (5.13)$$

and that indicated power $P_i(\sigma)$ is assumed to drop off with altitude just as does relative density σ itself

$$P_i(\sigma) = \sigma P_i(\sigma = 1) = \sigma P_{i0} \qquad (5.14)$$

Then, from Eqs. (5.12–5.14),

$$P(\sigma) = P_i(\sigma) - P_f = \sigma P_{i0} - P_f = P_{i0}(\sigma - C) \qquad (5.15)$$

On the other hand, at sea level, we have

$$P(\sigma = 1) = P_{i0} - P_f = P_{i0}(1 - C) \qquad (5.16)$$

Dividing Eq. (5.15) by Eq. (5.16),

$$\Phi(\sigma; C) \equiv \frac{P(\sigma)}{P(\sigma = 1)} = \frac{\sigma - C}{1 - C} \qquad (5.17)$$

which is Eq. (5.11).

When dealing with engine performance charts, it is sometimes useful to know the ratio of friction power P_f to MSL full-throttle power $P(\sigma = 1)$. That is easily seen to be

$$\frac{P_f}{P(\sigma = 1)} = \frac{P_f}{P_{i0} + P_f} = \left(\frac{1}{C} + 1\right)^{-1} = \frac{C}{1 + C} \qquad (5.18)$$

Gagg and Farrar published their work in 1934 but general aviation engines have not changed appreciably since then beyond lowering the value of dropoff parameter C through advances in materials, closer tolerances, and improved lubrication. Their equation was empirical, a distillation of then-available data, and was found by them to be applicable to either normally aspirated or gear-driven supercharged (though not turbosupercharged) aircraft. They also found that manifold pressure (MAP), at full throttle and constant revolutions per minute, also varies as does brake power:

$$\text{MAP}(\sigma) = \Phi(\sigma; C) \times \text{MAP}(\sigma = 1) \qquad (5.19)$$

although Scott[4] argues cogently that it makes more sense for MAP to vary as does pressure ratio δ. Pressure ratio decreases with altitude faster than does Φ (which, in turn, decreases faster with altitude than does σ).

Engine brake horsepower at constant MAP increases with altitude because exhaust backpressure decreases with altitude. This may explain why the power dropoff parameter $C = 0.12$ (apparently corresponding to mechanical efficiency $\eta_m = 0.88$) is relatively low. In general, $\eta_m = 0.85$ is considered a generous (low friction) value. In older engines, $C = 0.165$ was considered standard. Analysis of the Lycoming O-320-D2J engine manual charts suggests $C = 0.1137$ is about right, but we are now straining at quite small distinctions. The upshot is that we, along with almost everyone else dealing with normally aspirated, propeller-driven aircraft, assume and use the Gagg–Farrar relation given by Eqs. (5.10) and (5.11).

The Ideal Otto Air-Standard Cycle

So that you can work through some of this material with your own engine, at this point we take up a theoretical thermodynamic heat engine akin to the Otto cycle. To build realism we will use sample specifications. Table 5.4 gives pertinent numbers for the Lycoming O-320-D2J engine found on older Cessna 172 airplanes. The engine has four horizontally opposed cylinders (that is what the O in the model designation stands for), direct drive, and is air cooled. The

Table 5.4 Lycoming O-320-D2J specifications

Engine variable	Specification
Rated MSL power @ rpm	160 hp @ 2700 rpm
Cylinder bore, B	5.125 in.
Piston stroke, S	3.875 in.
Engine displacement, V_D	319.8 in.^{3}
Compression ratio, r	8.5 : 1
Spark timing	25 deg before TDC
Engine weight	275 lbf
Fuel type	100/100 LL

carburetor is from Marvel–Schebler, model MA-4SPA, single barrel, float type, with manual mixture control and idle cutoff. There is a fuel primer. The ignition features dual spark plugs and two magnetos.

We will display the air-standard Otto cycle pressure-volume diagram for one 80-in.^{3} cylinder of the Lycoming. Because the compression ratio is $r = 8.5$, from Eq. (5.9) the clearance volume V_C is $80/7.5 = 10.67$ in.^{3} and the cylinder volume at bottom dead center (BDC) is 90.67 in.^{3}. Our assumption is that the engine is running at sea level under standard conditions; the intake air is at 59°F and at $p_0 = 2116.2$ psf $= 14.7$ psi. We pick up the one-cycle (two crank revolutions) narrative at the start of the intake stroke, point A in Fig. 5.3.

During the intake stroke A–B, work is done on the piston by the incoming air. In fact, the cylinder is being used as a vacuum pump and some work is being done by the other cylinders, energy extracted from the flywheel and crankshaft. The theoretical work needed is simply $p_A \times (V_B - V_A) = 14.7 \times 80$ lbf-in. $= 1176$ lbf-in. $= 98$ ft-lbf. In fact, pressure in the cylinder during intake is below ambient atmospheric.

At point B, the piston is at BDC and begins the compression stroke B–C. This is assumed adiabatic. A brief review of elementary thermodynamics reminds us that, in such a process, one with no heat exchanged between the working fluid and the environment, pressure and volume are related by

$$pV^k = \text{constant} \qquad (5.20)$$

and we will take k, the ratio of specific heats, at the nominal value 1.40 for air. For a mixture of fuel and air, 1.37 would be better. The constant on the right-hand side (RHS) of Eq. (5.20) is obtained by evaluating the left-hand side (LHS) at any point; we would do that at B because we know both pressure and volume there. The result is

$$p_C = p_B \times \left(\frac{V_B}{V_C}\right)^{1.4} = 294 \text{ psi} \qquad (5.21)$$

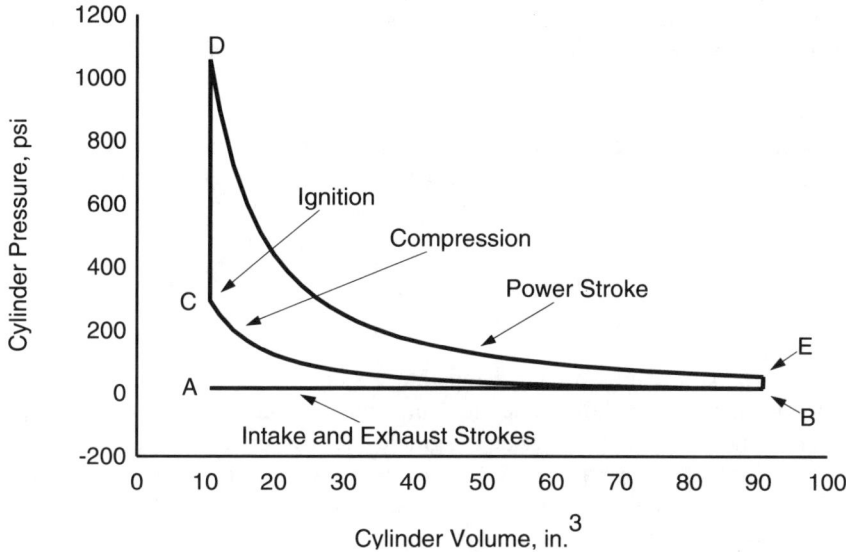

Figure 5.3 The Otto cycle is a loose model of an internal combustion engine cylinder during two rotations of the crankshaft.

At point C the breaker breaks, the spark sparks, and combustion takes place so rapidly (in our model) that the piston moves not at all. Pressure and temperature of course rise precipitously. How high?

We can find out by arranging that the power stroke, D–E, occurs along an adiabat such that the net amount of work of the entire diagram comes out to be what one would expect of one cylinder of our engine, during two crankshaft revolutions, at MSL rated power and revolutions per minute. Here are the details.

Each of the four cylinders delivers $160/4 = 40$ hp by cycling $n/2 = 22.5$ times per second. Dividing to find the energy or work delivered per engine cycle, converting that first to ft-lbf and then to in.-lbf because we are using psi (lbf/in.^{2}) as our pressure units, we find the O-320-D2J delivers 11,733.4 in.-lbf per cycle per cylinder.

Next, we use Eq. (5.20) to find that the work associated with an adiabatic interval is

$$\int_{V_1}^{V_2} p\,dV = \frac{p_1 V_1 \left(1 - \left(\frac{V_1}{V_2}\right)^{k-1}\right)}{k-1} \tag{5.22}$$

On the compression stroke, when $V_1 = V_B$ is greater than $V_2 = V_C$, this work is negative. On the power stroke, when work is done *by* the engine, it is positive. Arranging to have the net work over one engine cycle, B–C–D–E–B, equal our agreed-upon 11,733.4 in.-lbf, we find that $p_D = 1058.9$ psi. With a known situation at one end of the power stroke, it is easy enough to find that, at the other end, $p_E = 52.9$ psi.

There are two further details to wrap up. We would like to have a feeling for temperatures inside the cylinder at various times during an engine cycle, and it is useful to relate the average pressure on a piston, the so-called *brake mean effective pressure*, *bmep*, to the torque being developed by the engine. The ideal gas law from the atmosphere chapter, or even the high school chemistry version

$$\frac{p_1 V_1}{T_1} = \frac{p_2 V_2}{T_2} \tag{5.23}$$

can be arranged to give us the temperature story. Table 5.5 is a summary of important ideal air cycle variables for our model of one cylinder of our sample engine. It takes the engine over one engine cycle (two crankshaft rotations). The engine is at MSL with standard dry air as the working fluid, producing 160 hp at 2700 rpm. The model was rigged (by adjusting pressure at point D) to put out the same amount of work as does that prototype cylinder. Remember this is only a rough model of the actual engine cycle.

A given engine's *bmep* is proportional to its torque output. Writing that pressure using psi (lbf/in.^{2}) as the unit, and displacement using cubic inches, the surprisingly simple relation is

$$bmep = \frac{150.8 M}{V_D} \tag{5.24}$$

Table 5.5 Standard air cycle model of one cylinder of the Lycoming O-320-D2J engine

Cycle point or process	Description	V, in.^{3}	p, psi	T, °F	Work, ft-lbf
A	Start intake	10.67	14.7	59	
$A–B$	Intake stroke				98.0
B	Start compression	90.67	14.7	59	
$B–C$	Compression stroke				−376.0
C	Ignition	10.67	294.1	762	
$C–D$	Constant-volume combustion				0.0
D	Start power	10.67	1059	3938	
$D–E$	Power stroke				1353.7
E	Start scavenging	90.67	52.9	1407	
$E–B$	Scavenging hot gas				0.0
B	Start exhaust	90.67	14.7	59	
$B–A$	Exhaust stroke				−98.0
	Net work per cylinder per cycle				977.7

Equation (5.24) results from definitions and geometrical considerations. Returning to British engineering units, torque M is $P/2\pi n$. Average delivered power P—the engine pulsates, do not forget—is net work per time. A wrinkle is that $bmep$ is defined as the average cylinder pressure during only the *power* stroke, not the much smaller long-term average pressure. Because work is force times distance (stroke S), and force is (average) pressure times total piston face area A, the work done by all cylinders, over one cycle, is $bmep \times A \times S$. But recall that $A \times S = V_D$. Multiply that product by the number of cycles per unit time, $n/2$, to get the figure for net work per unit time, net power. Invoking the above relation between power and torque, then solving for $bmep$ and converting length units to inches, gives Eq. (5.24). Because, for this engine, $M = 311.2$ ft-lbf and $V_D = 320 \text{in.}^3$, Eq. (5.24) tells us that $bmep = 146.7$ psi.

Now what about the result of Table 5.5, that the net work per cylinder per cycle is 977.7 ft-lbf = 11,732 in.-lbf? What sort of $bmep$ does that connote? Pressure times area times stroke equals pressure times volume (80 in.^{3} per cylinder) should then be 11,732 in.-lbf. Solving for pressure $bmep$, we again have 146.7 psi. Makes sense.

Items We Have Neglected

As we have mentioned, the air-standard cycle is fairly far from the actual activity inside an engine. Here are the main items we have neglected:

- During an actual intake stroke, the cylinder is being forced to work as a vacuum pump and pressure inside it is somewhat below local atmospheric. Conversely, during an actual exhaust stroke, cylinder pressure is somewhat higher than ambient. In effect, work must be done *on* the engine by the other cylinders to make up for this "pumping loss."
- Neither the compression nor the power stroke is completely adiabatic, without transfer of heat. Heat is passing out through the cylinder walls to the cooling system (here, airflow) during each of those strokes.
- Ignition does not take place instantaneously. Except for instants at TDC and BDC, the piston is always moving. The sharp top of the cycle, illustrated in Fig. 5.3 as point D, is actually quite rounded, and lower.
- Scavenging is not so quick and efficient as we have indicated. In fact, nothing is.
- The working fluid is not air. At the beginning, it is a combination of dry air, some water vapor, and fuel as both vapor and small entrained droplets. After ignition, the working fluid is a combination of gases—the so-called "products of combustion"—of quite variable composition.
- Our theoretical engine had infinitely close tolerances; actual engines have at least some gas losses into the crankcase or otherwise.

- Our theoretical valves worked infinitely quickly and were free of leaks. In the actual engine, compromises must be made; for at least a brief period, in actual engines, both the intake and the exhaust valves are open at the same time.
- In an actual engine, complete combustion requires excess air. On the other hand, maximum power requires air amounts of less than stoichiometric proportions, a rich mixture. Fuel efficiency and high power are, in reality, mutually exclusive.
- The real engine suffers many kinds of mechanical losses, friction and wiggles and loose bearings, of which the theoretical engine is totally unconcerned. Even the theoretical engine also suffers inevitable second law losses—in the Otto cycle, related only to compression ratio—but there is more entropy gain, efficiency loss, in the real engine.

Still—and though we have so far ignored a further loss, a further efficiency of around 75 to 80%, the transformation from shaft power to thrust power (power available, $P_{av} = TV$)—those wonderful engines somehow take those wonderful flying machines from place to place well and reliably.

Controls and Instruments

How do we control the engine and how do we monitor its performance? Table 5.6 lists the more common engine controls and instruments filling those roles. For further details on the functioning and operation of these devices, and others, see Thomas.[5] Larger engines, on what are now considered "antique" aircraft, sometimes had an additional control for advancing or retarding ignition spark timing.

Engine controls for ordinary operations with fixed-pitch airplanes are limited to only a throttle and a mixture control. The throttle controls air to the cylinders and, indirectly, thereby controls fuel flow rate. The other variable of interest is the F : A. That is controlled by the mixture knob and assessed either by memorized numbers of turns out from full forward, by reading the exhaust gas temperature gauge, or by leaning until the engine runs "rough," then enrichening the mixture a turn or so. Constant-speed propeller airplanes have another "degree of freedom" in that propeller revolutions per minute can be selected (within mechanical low- and high-stop blade angle limits) independently of the air and fuel flow controlled by the throttle. The tachometer monitors revolutions per minute and a MAP gauge monitors airflow to the engine. There is no good reason, except cost, for a fixed-pitch airplane not to have a MAP gauge; having one would certainly help the pilot make sense of the engine performance chart.

Table 5.6 Engine controls and instruments

Major function	Control	Instrument	High performance	Low performance
Power output	Throttle		X	X
	Propeller pitch		X	
		Tachometer	X	X
		Manifold pressure	X	
Fuel/air mixture	Mixture control[a]		X	X
		EGT[b]	X	X
Fuel management	Fuel selector		X	X
	Boost pump[c]		X	
	Carburetor heat[d]			X
		Fuel quantity	X	X
		Fuel pressure	X	
		Fuel flow	X	
Starting	Ignition switch		X	X
	Primer[e]		X	X
Engine cooling	Cowl flaps		X	
		Oil temperature	X	X
		Oil pressure	X	X
		CHT[f]	X	

[a] The mixture control also affects engine power output.
[b] EGT = exhaust gas temperature gauge.
[c] In fuel-injected engines, this auxiliary fuel pump primes for starting. It is also used to suppress vapor lock at high altitudes and in hot weather.
[d] Carburetor heat is pulled on to prevent or cure carburetor icing when ambient humidity is high.
[e] Most fuel-injected engines have no primer. See table footnote c.
[f] CHT = cylinder head temperature gauge.

The Aircraft Engine Operator's Manual

Aircraft engine manuals—like much aviation technical material—are mixed bags of information and quasi-information. The fuel flow information is particularly sketchy and, in places, inconsistent. For instance a full-throttle fuel usage rate, in gallons per hour or pounds per hour, appears nowhere. Graphs portraying the effects of leaning or enriching the mixture on exhaust gas and cylinder head temperatures and on power and *bsfc* rate, have no numbers on the lean-rich axis and none on the *bsfc* axis. Nevertheless, we recommend you buy the manual appropriate to your engine. It does have much more information than appears in the *Pilots Operating Handbook*, more on leaning and on troubleshooting, and more on engine maintenance and disassembly. These manuals cost around $15 and, for those with Lycoming or Continental engines, are available from

Textron Lycoming
652 Oliver Street
Williamsport, PA 17701
(717)-323-6181

Teledyne Continental Motors
Attention: Publications Department
P.O. Box 90
Mobile, AL 36601

We will often be referring to the engine manual[6] for our sample Lycoming. To get the appropriate manual for other equipment, contact the factory or an overhaul facility.

Figure 5.4 represents an important graph one might find in an engine operating manual: the effect of leaning the fuel/air mixture on shaft power output. Figure 5.5 is a translation of two different engine manual graphs of brake specific fuel consumption rate and the way that number depends on shaft power; the small discrepancy between the two rightmost points is negligible. Brake specific fuel consumption rate is an important concept primarily because—except for the sizeable jump at 75% power—it is close to constant. That it not only the case for

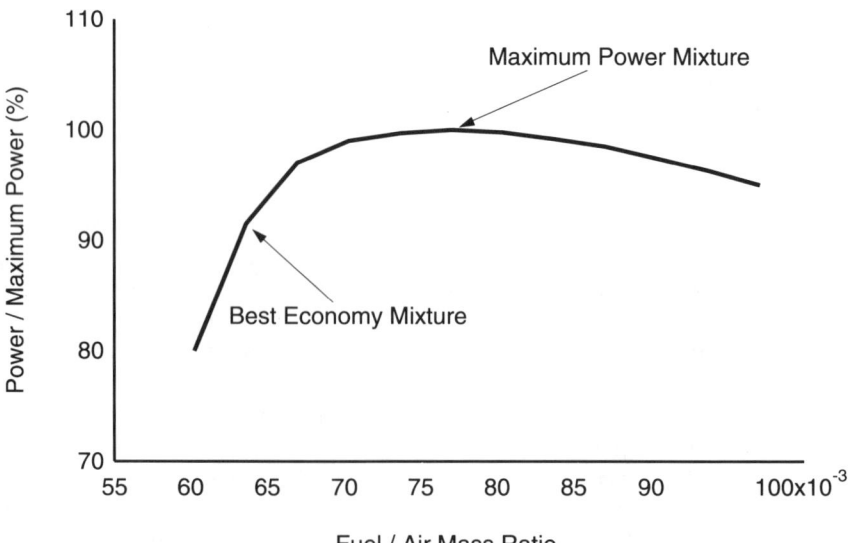

Figure 5.4 For this engine in this airplane, best economy cruise has an A : F of about 16 : 1 and best power mixture about 13 : 1.

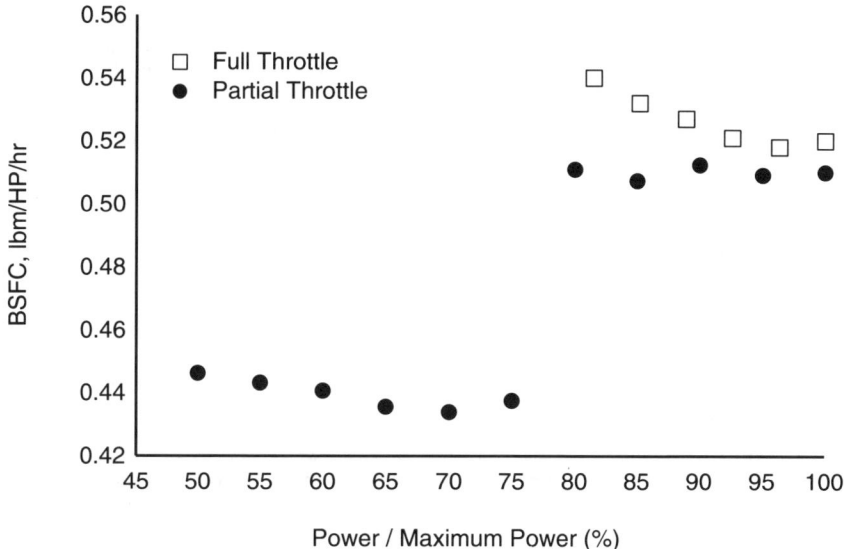

Figure 5.5 Because of the jump in *bsfc* around 75% power, most cruising is done at or below that power setting.

different power levels of the same engine but also as one moves from engine to engine.

Power as a function of engine circular speed, at full throttle and sea level, is displayed in our sample engine manual in two different graphs. Figure 5.6 recaps the data and points out that we should not pay much attention to minor variations. Power drop factor (p.d.f.) is

$$\text{p.d.f.} \equiv \frac{\text{power at 80\% of rated revolutions per minute}}{\text{rated full-throttle power (at rated revolutions per minute)}} \quad (5.25)$$

If torque were truly constant with load, as we suppose in the bootstrap approach, p.d.f. would always be 80%. The p.d.f., as a measure of power response (and constancy of torque under varying loads) has fallen out of use, but you may find it in older books. Diehl,[7] for instance, uses p.d.f. extensively. You will also find, in engine operator's manuals and elsewhere, graphs of so-called "propeller load horsepower." While Thomas[5] refers to this graph as a representation of engine behavior, in the constant-speed propeller case, when the propeller speed control is full in (high revolutions per minute, low blade pitch), with throttle varied, that is not quite right. The idea behind the propeller load curve, according to its definition, is that it follows a strict n^3 relation. From the definition of propeller coefficient, that would mean that C_P is held constant while different loads, hence varying revolutions per minute, are applied to the engine.

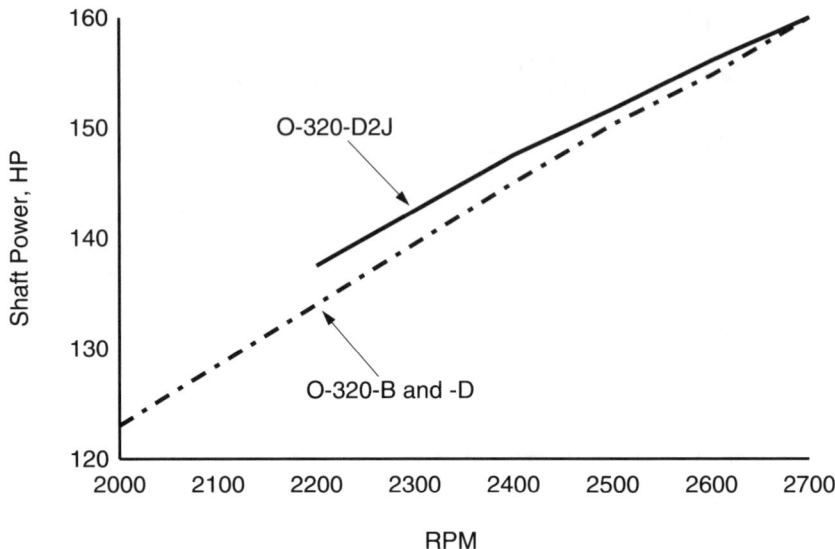

Figure 5.6 The lower graph, for the sample Cessna 172 engine, gives a p.d.f. of 0.824.

That can only be in the static, air speed $V = 0$, case. For all practical purposes, you can ignore propeller load horsepower as a source of useful information.

After these skirmishes with partial pictures of engine performance, we are ready now for our main assault, on the infamous engine performance chart itself.

Engine Performance Chart

Figure 5.7 is the engine performance chart for our sample general aviation engine, the Lycoming O-320-D2J. When the chart is entered with triples (MAP, revolutions per minute, h_p), it produces the corresponding value of brake horsepower P. (As usual, we have simplified the pair $[h_p, \text{OAT}]$ to density altitude.) At this point the reader should review the somewhat ungainly procedure involved in using the chart. Directions appear in the boxed area in the upper left; again, ignore the temperature correction because that will have already gone into the prior calculation of density altitude.

Two cautions. First, there are other ways to present engine performance and dynamometer data. This is only one method, though it is the most common presentation these days. (See Liston,[2] especially Chapter 6, for several others.) Second, one must not take these charts terribly seriously. They come from what is now the fairly distant past and they were constructed (hand drawn) from scanty

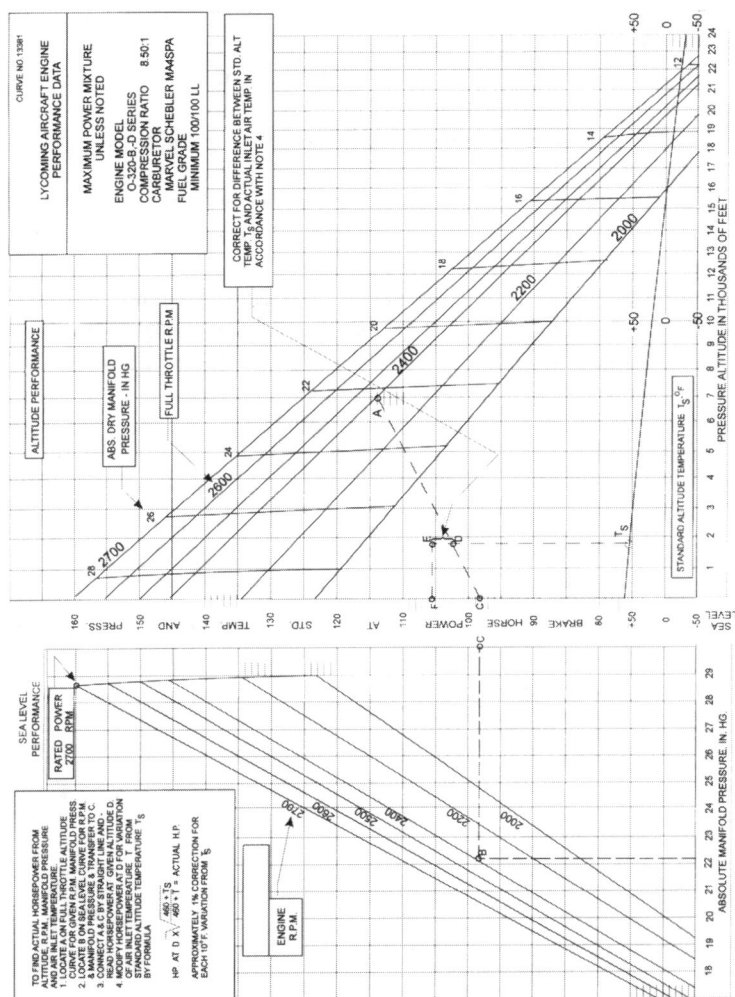

Figure 5.7 Engine performance chart for the Lycoming 0-320-D2J engine that powers older models of Cessna 172. (For informational purposes only)

data, perhaps only a dozen or two dozen actual experimental points. This means that some linearity assumptions that go into their make-up stretch credulity; there is very little reason to believe that anything as complicated (and adjusted, and compromised) as a modern internal combustion engine behaves linearly! On the other hand, these charts would not still be used if they were far from the mark. What all this means is that while we do want to analyze the charts, we do not want to turn that analysis into its own little research project. There is little value to attaining a *very accurate* set of curve–fit formulas reproducing your engine's performance chart. A rough one, yes.

Start with the LHS, the so-called sea level performance side (see Fig. 5.8). That graph results from extending the straight lines on the LHS of Fig. 5.7. That was done to see where, if anywhere, the lines for various revolutions per minute focus. They do, but instead of at MAP $= 0$, as some say, they focus, for this engine chart, at about (MAP, hp) $= (5$ in. Hg, -19 hp). Nineteen horsepower is quite close to $C \times P_0 = 0.1137 \times 160 = 18.2$ hp, what we would expect to see (from earlier considerations) as the horsepower loss due to friction, P_f. It is important to note that here manifold pressure reductions, as one moves from right to left, are due to throttling. The various revolutions per minute are due to having different loads on the engine. If one gets formulas for these six lines, say by taking values at 20 and at 28.5 in. Hg, one will get six values of line slope and intercept. Then using linear regression on those two sets of six values, and combining these results to one global formula, one finds that, approximately:

$$\text{HP}(N, \text{MAP}, \sigma = 1) \doteq 0.002277 N \times \text{MAP} + 1.4670 \, \text{MAP} - 0.01036 N - 30.41 \tag{5.26}$$

Checking a few values shows that this is quite a good fit to the "experimental" data. But perhaps we have only recreated the formula from which those experimental data were first obtained.

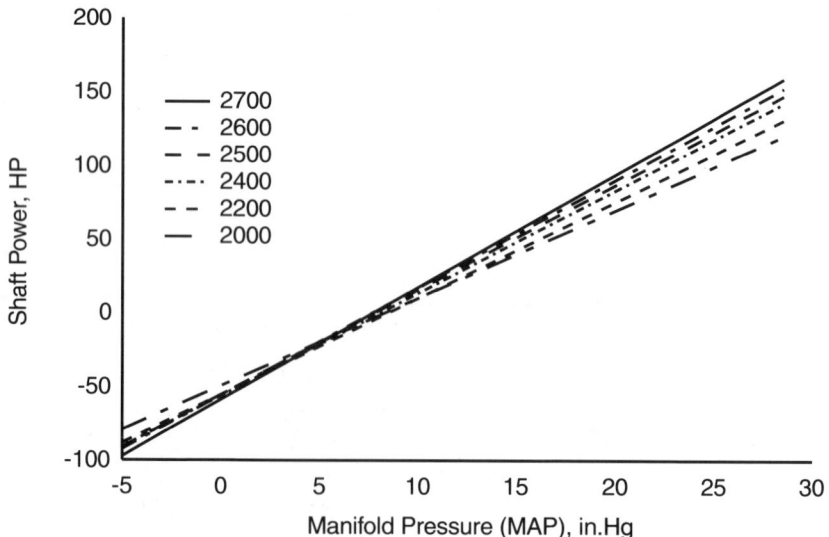

Figure 5.8 Extending the curves of engine performance chart LHS farther to the left, one sees they come to a focus of about (MAP, hp) $= (5$ in. Hg, -19 hp).

The RHS or "altitude performance" portion of the engine chart is also understandable (see Fig. 5.9). Everything on this side is understood to be at full throttle. With revolutions per minute fixed, power drops off with altitude just as does full-throttle torque, according to the Gagg–Farrar factor, Eq. (5.11). The straight lines (in fact they are not *exactly* straight) come from the chart's designer having used a nonlinear scale, along the bottom of the chart, for (density) altitude. Linear distances, starting from the left end, are in fact proportional to $1 - \Phi(\sigma)$. Those lines all come to focus at the "altitude" at which $\Phi = 0$. That will be (using the more exact figure, obtained by conducting a statistical study on this RHS) where $\sigma = C = 0.1137$. That is, at $h_\rho = 58,180$ ft. There the engine is effectively friction locked and would not turn over.

There is one further aspect of the altitude performance chart worth mentioning. Notice that when you do the sample problem built into the chart, your final output power P, under otherwise identical conditions, is *higher* if you are at *higher* altitudes. This small power boost is an effect due to reduced exhaust backpressure at higher altitudes. If you do several problems like the sample, you will find that all such slant lines running up and to the right have about the same slope. A *unit* along the abscissa or horizontal axis, is here a unit of $1 - \Phi$. Let us now find that slope and then translate it into an approximate power gain, expressed as horsepower per thousand feet increased altitude, due to reduced exhaust backpressure.

First, the slope of the upwards slanted dashed line in Fig. 5.7, the one running from point C to point A. The vertical change is easy: $\Delta y = \Delta \text{hp} = (114 - 98.3) = 15.7$ hp. The horizontal change, from MSL up to 7200 ft

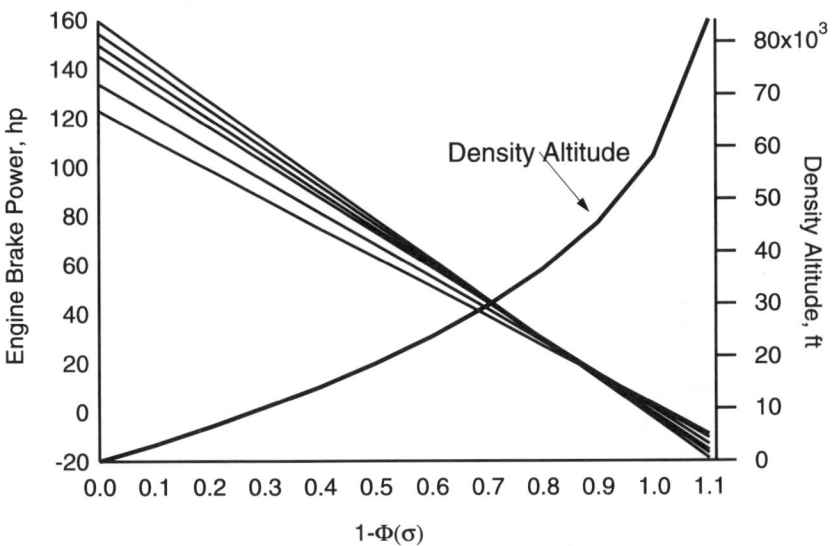

Figure 5.9 Modeling the RHS of the engine performance chart.

(density) altitude, takes a bit of work. From Eq. (1.10), $\sigma(7200\text{ ft}) = 0.8057$. Using $C = 0.1137$, then Eq. (5.11) says that $\Phi(7200\text{ ft}) = 0.7807$. Then the horizontal change is $1 - \Phi = 0.2193$. Finally, the slope is $15.7/0.2193 = 71.6\text{ hp}$.

Next we need to translate that 71.6-hp per unit into horsepower per thousand feet. The first step is to write down the calculus "chain rule"

$$\frac{d\text{ hp}}{d(h_\rho/1000)} = 1000 \frac{d\text{ hp}}{d(1-\Phi)} \frac{d(1-\Phi)}{dh_\rho} \qquad (5.27)$$

The product of the first two factors on the right we know to be about 71,600. Next we need to work with the rightmost factor in Eq. (5.27). Because, from Eq. (5.11),

$$1 - \Phi = \frac{1-\sigma}{1-C} \qquad (5.28)$$

we see that

$$\frac{d(1-\Phi)}{dh_\rho} = \frac{1}{(C-1)} \frac{d\sigma}{dh_\rho} \qquad (5.29)$$

The derivative on the right, using Eq. (1.10) again, can be written as

$$\frac{d\sigma}{dh_\rho} = \frac{(\alpha R - 1)\sigma}{RT_0} \left(1 - \frac{\alpha h_\rho}{T_0}\right)^{-1} \qquad (5.30)$$

At sea level, Eq. (5.30) evaluates to -2.926×10^{-5}; at for instance 10,000 ft, where $\sigma = 0.7385$, Eq. (5.30) evaluates to -2.320×10^{-5}. Picking up the other factors 71,600 and $(C-1)^{-1} = -1.128$, we find that reduced exhaust backpressure will lead to a horsepower increase, near sea level, of about 2.36 hp/1000 ft. Near 10,000 ft, the increase is somewhat less, about 1.87 hp/1000 ft. So, roughly, for this engine, reduced backpressure at constant MAP gives a gain of about 2 hp/1000 ft.

Using our original figures off the engine chart, 15.7 hp over 7200 ft, gives a quite comparable 2.2 hp/1000 ft. The advantage of our calculation is that it tells us how much and in what direction that horsepower gain figure changes with altitude. It decreases slightly, about 0.05 hp/1000 ft.

Conclusions

This wraps up what one needs to know, at least in simpler cases and for performance purposes, about the airplane's engine. There are only minor wrinkles when one has more than one engine, or a fuel injected engine, and considerable changes when an engine is turbocharged—more specialized than we could go into. Our rather cavalier efficiency numbers, our rule of thumb on the power boost due to reduced exhaust pressure, and even the Gagg–Farrar power dropoff factor, are approximations. Internal combustion gasoline spark ignition engines have been very thoroughly and quantitatively studied (see Taylor,[8] and Heywood[9]) and many more details of their performance are known. If one operates in the tropics, or anywhere that it is both hot and humid, certainly consider humidity effects detailed in the atmosphere chapter. Looking to the future, try to keep in mind that determining bootstrap propeller polar slope parameter m, via climb tests, will actually be a determination of the product mM; use reasonable and independent means to assess the health of your airplane's engine and its torque output M. That is especially true, as for mountain search and rescue missions, when you need good performance at high density altitude.

References

1. MacMahon, T.A., and Bonner, J.T., *On Size and Life*, Scientific American Books, New York, 1983, pp. 60–64.
2. Liston, J., *Power Plants for Aircraft*, McGraw-Hill, New York, 1953, p. 226.
3. Gagg, R.F., and Farrar, E.V., "Altitude Performance of Aircraft Engines Equipped with Gear-Driven Superchargers," *Society of Automotive Engineers Transactions*, Vol. 29, 1934, pp. 217–223.
4. Scott, A.P., *Benchmark*, Sequoia Aircraft Corp., Richmond, VA, 1990, p. A-12.
5. Thomas, K., *Aircraft Engine Operating Guide*, Belvoir Publications, Riverside, CT, 1985, pp. 21–53.
6. Anonymous, *Operator's Manual Textron Lycoming Aircraft Engines Series O-320, IO-320, AIO-320 & LIO-320*, Textron Lycoming, Williamsport, PA, 1981.
7. Diehl, W.S., *Engineering Aerodynamics*, Ronald Press, New York, 1936.
8. Taylor, C.F., *The Internal-Combustion Engine in Theory and Practice, Vol. I: Thermodynamics, Fluid Flow, Performance* and *Vol. II: Combustion, Fuels, Materials, Design*, M.I.T. Press, Cambridge, MA, 1985.
9. Heywood, J.B., *Internal Combustion Engine Fundamentals*, McGraw-Hill, New York, 1987.

6

Propeller Thrust

Introduction

While you are mingling with the crowd at the static aircraft display on Airport Appreciation Day, notice how hard it is for visitors to keep their hands off propellers. Simple object, the propeller, but attractive. We admire its sheen and its shape. But once the engine starts, the propeller's status as art object fades into a dangerous blur. Propellers are not just for decoration.

This chapter takes up propellers in detail. We will get acquainted with propeller geometry; propeller size and shape are crucial to the propulsive job. The bulk of our time will be spent exploring propeller *action*: how that action is described (thrust and power coefficients and efficiency) and how it is calculated (momentum theory, blade element theory, and a combination). Our interest is in explaining existing propellers, not so much in designing new ones, so we will sidestep such structural questions as whether the propeller material will stand up to the centrifugal or bending stresses. And we will sidestep the very interesting question of how to design an *optimal* propeller for a given job. We will leave out a lot of details. Even so, this will be the most strenuous background chapter in the book. Why?

> What at first seemed a simple problem became more complex the longer we studied it. With the machine moving forward, the air flying backward, the propellers turning sidewise, and nothing standing still, it seemed impossible to find a starting-point from which to trace the various simultaneous reactions. Contemplation of it was confusing.
>
> Orville Wright, *How We Invented the Airplane*

In spite of these difficulties, the Wrights did well with propellers. By standing on their shoulders, and with the help of other researchers before or since them, so can you. Our challenge will be to maintain clarity and focus; your challenge will be to track a lengthy procession of symbols, to keep in mind what they mean. We start with the geometrical description.

Propeller Geometry

There are B identical blades of radius R (diameter $d = 2R$) (see Fig. 6.1). At each "propeller station" r along the blade (or the corresponding "relative station" $x \equiv r/R$), as measured from the center of the hub, there is an aerodynamic section profile characterized by some name or number (see Fig. 6.3) and by blade width or *chord length* $c(x)$, maximum thickness $t(x)$, and angle $\beta(x)$. Angle β is usually measured, as in Fig. 6.3, from the plane of propeller rotation to the flat underside of the profile. (When the profile consists of two conjoined convex portions, as it does near the hub, the "flat" is taken to be that line of junction.) In some cases (e.g., Clark Y sections) the β measurement is taken to the zero lift line running forward from the trailing edge; you will need to decide, or know, which definition of β is meant. First, we will consider fixed-pitch propellers; the constant-speed type will have their say later on.

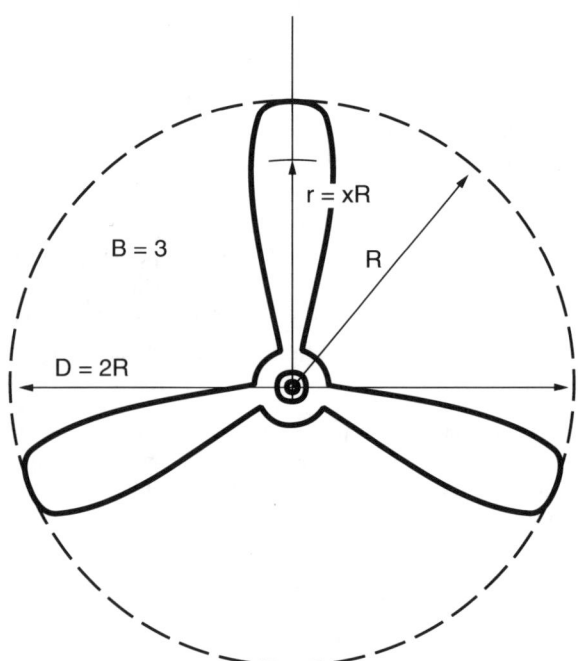

Figure 6.1   Propeller blade number, diameter, and relative station.

Propeller Thrust   147

Figure 6.2   Clark Y and RAF6 propeller airfoil sections.

Radius R (or diameter d) tells us the size of the propeller. All of the rest—B, $c(x)$, $t(x)$, $\beta(x)$, and the designation of the profile—tells us shape. We will ignore "extraneous" details such as number, placement, and diameter of mounting holes; diameter of the centering hole; boss thickness; direction of rotation; weight; and so forth.) We do not need to specify the propeller's angular position, say, in degrees one blade is from, say, vertically pointed upwards, except under conditions of asymmetrical flight. And we will not treat those.

Most general aviation propellers have either Clark Y or (more often) RAF6 profiles (the latter modified to have a flat lower surface). Saying a propeller "has RAF6 profile" actually means it has a *family* of RAF6 profiles of differing chords $c(x)$ and differing "thickness ratios" $\tau(x) \equiv t(x)/c(x)$. Table 6.1 has numerical details on the RAF6 shape. The RAF6 profile is claimed to have leading edge radius 0.10 (of the maximum thickness) and trailing edge radius 0.08.

Details for a McCauley Propeller

The 1986 Cessna 172P stock propeller, the McCauley 1C160/DTM7557, has RAF6 profiles. 1C160 is the basic planform design number. DTM gives details (if you have the propeller repair manual giving the code) on the proper type of

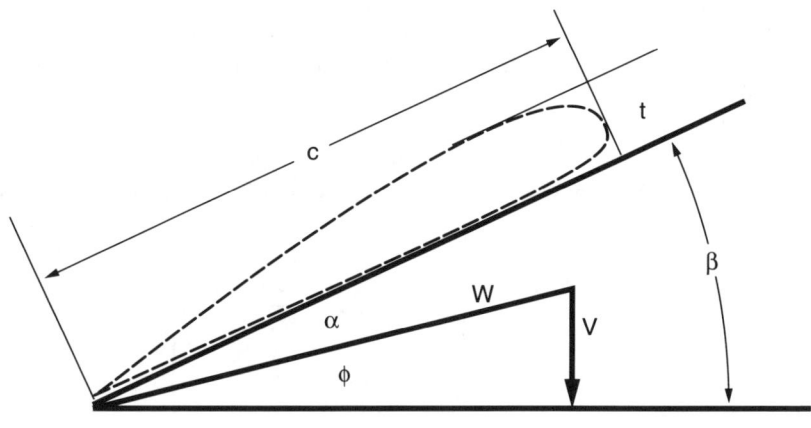

$$r\Omega = 2\pi n r = \pi n x d$$

Figure 6.3   Propeller blade section with chord c, thickness t, and resultant air speed W.

Table 6.1 RAF6 edge radii and ordinates (scaled to maximum of unit at chord fraction 1/3)

Chord fraction	Ordinate
0.025	0.41
0.05	0.59
0.10	0.79
0.20	0.95
0.30	0.998
0.40	0.99
0.50	0.95
0.60	0.87
0.70	0.74
0.80	0.56
0.90	0.35

crankshaft for the propeller, blade tip contour, etc. The 75 means the propeller diameter $d = 75$ in. $= 6.25$ ft. The final digits 57 give the nominal pitch p, the distance (again in inches) that the blade section at $x = 0.75$ would advance, in some nonslipping medium like wet sand or soft clay during one full rotation. Nominal pitch can be translated into nominal blade angle at $r = 0.75$, $\beta_{0.75}$, by visualizing that full rotation scratching the inside of a cylinder of radius $0.75R$. Cutting that cylinder open (see Fig. 6.4) shows a triangle with angle

$$\beta_{0.75} = \tan^{-1} \frac{p}{0.75\pi d} \tag{6.1}$$

Substituting our DTM7557 values gives $\beta_{0.75} = 17.88$ deg. But that is at only one station. The propeller overhaul manual gives dimensions and angles at several blade stations. See Table 6.2, where all linear dimensions are in inches. When a tolerance range was given, the larger value was selected.

Why no blade angles for the early sections? Overhauling a propeller includes possibly retwisting to correct out-of-specification angles. Close to the hub the blade is too thick to be retwisted. To get those missing figures, take one of those

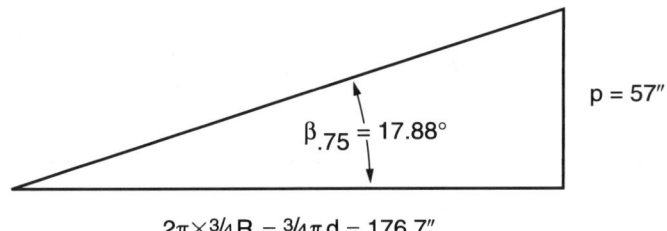

Figure 6.4 Unwrapped helix for McCauley 7557 propeller, $r = \frac{3}{4}R$.

Table 6.2 Selections from the McCauley 1C160 propeller overhaul manual

Station r	Relative station x	Chord or width c	Thickness t	Blade angle β, deg
5	0.133	5.634	2.432	NA[a]
9	0.24	5.766	1.190	NA
12	0.32	5.796	0.920	NA
15	0.40	5.826	0.760	26.83
18	0.48	5.776	0.650	24.28
24	0.64	5.326	0.510	20.00
30	0.80	4.510	0.390	17.00
33	0.88	3.910	0.335	15.80
36	0.96	2.981	0.258	14.80

[a] NA = not applicable.

many-needle, contour-copying devices (plus a good protractor and a bubble level) out to the airplane and measure. It is easier with the propeller off the airplane, on the bench.

In doing propeller calculations it is most convenient to have evenly spaced relative blade stations. And more of them. To that end, we made curve fits to the data of Table 6.2. Results appear in Fig. 6.5 and in Eqs. (6.2), (6.3), and (6.4).

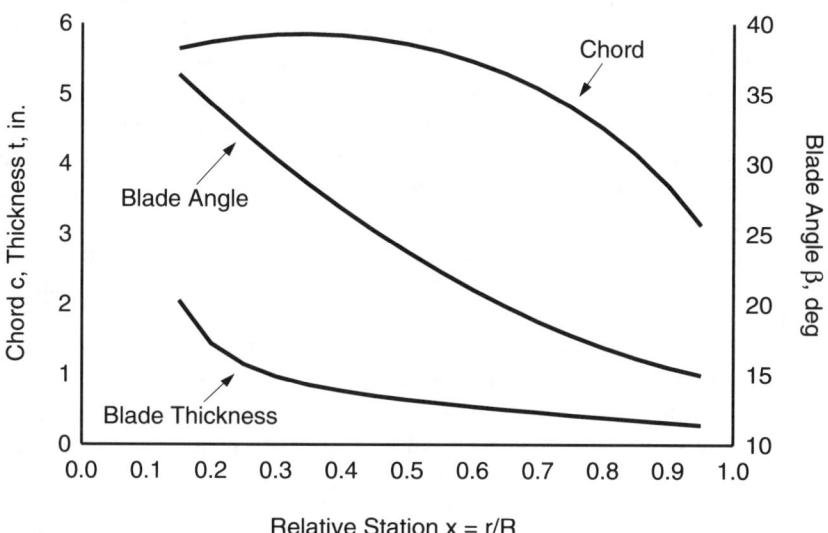

Figure 6.5 Overhaul manual and curve-fit geometry for McCauley 7557 propeller.

The units for chord $c(x)$ and thickness $t(x)$ are inches; blade angle $\beta(x)$ is in degrees.

$$c(x) = \sqrt{26.50 + 45.11x - 65.99x^2} \qquad (6.2)$$
$$t(x) = (-0.33546 + 6.7181x - 9.2040x^2 + 6.767x^3)^{-1} \qquad (6.3)$$
$$\beta(x) = 43.375 - 49.801x + 20.929x^2 \qquad (6.4)$$

So much for geometry. Now we need a little action.

Propeller Action and Dimensional Analysis

All the (fixed-pitch) propeller can do is rotate at some angular speed, n revolutions per second (rps) or $N = 60n$ rpm. From that rotation, the size and shape of the blades give us resistive torque Q (often called brake moment M), propulsive thrust T, and consumed shaft or brake power P. What else matters? Air density ρ, for one; there is no thrust in a vacuum! And the airplane's air speed V matters because V (along with r and n) influences section air speed, hence aerodynamic forces, on the propeller section at r.

Seven variables! We can simplify the problem a little by eliminating either power P or torque M; those are related by

$$P = 2\pi n M \qquad (6.5)$$

Power is more in the mainstream, so let us retain power. Now we have the shape of the propeller plus six interrelated variables: d, n, ρ, P, T, and V. What we want is some sort of procedure for calculating absorbed power P and thrust T in terms of the others. What we have is—not much!—a short list of variables. A list meant to demonstrate tidiness, commendable routine administration? Are we assiduous sergeants, "mustering the troops" after a day in the field? Not at all. There is an important principle (a theorem, in fact) that we want to demonstrate. And it does not require much more than this list of variables.

Buckingham's II Theorem[1]: if there are m physical variables defined in terms of n independent fundamental quantities, then there are $(m - n)$ independent dimensionless groups of variables.

Moreover, those $(m - n)$ groups will be related by $(m - n - 1)$ algebraic or graphical relationships. Some background for this "dimensional analysis" subject. Most physical quantities have units and dimensions. In our British engineering system, the fundamental units are feet (ft), pounds force (lbf), and seconds (s). The associated dimensions are length L, force F, and time T. Mass in this system is a derived quantity with unit slug (lbf-s^2/ft) and dimensions $FL^{-1}T^2$. That makes density (mass per unit volume) have units of slug/ft^3 and dimensions $FL^{-4}T^2$. Angle is a strange case. We ordinarily use degrees; but, for most theoretical work, radians (rad) are preferred and because radians are measured as the ratio of subtended arc to radius, are dimensionless. So we take degrees to be dimensionless too. Table 6.3 shows the units and dimensions of each of our six propeller variables.

Applying Buckingham's Π theorem to our case, $m = 6$ and, because each of the dimensions F, L, and T appears at least once somewhere in those six expressions, $n = 3$. Hence there are $(6 - 3) = 3$ independent dimensionless groups made up of our six variables and, therefore, two relations (only!) among those three quantities. If we can find those three groups we will have saved ourselves much work. Otherwise we might expect to need five relations for the six variables. Discovering those five experimentally (or theoretically) would likely be a big job. And, of course, we *will* be able to find those three dimensionless groups. Right now.

Two of the dimensionless groups can be extracted simply by close inspection of the dimensions column in Table 6.3. Look at entries in the first, second, and fourth rows and see V/nd is dimensionless. This composite variable is commonly known as the "propeller advance ratio" and given variable letter J. In 1 s, the airplane "advances" distance V while the propeller tip goes around a linear distance $n\pi d$; that ratio (ignoring constant π) gives J its name.

Looking at the last three rows, you can see that the group TV/P, commonly known as "propeller efficiency" and given symbol η, is also dimensionless.

To get a third and last independent dimensionless group (independent means not a sum, difference, product, or quotient of previous such groups), we can construct a new product by leaving out at least one of J's variables $\{V, n, d\}$ and also at least one of η's variables $\{T, V, P\}$. Let us leave out V (because it appears in both earlier groups) and P. Then the dimensions of our product must satisfy (you can always take one factor raised to the first power):

$$[d n^a \rho^b T^c] = L \times T^{-a} \times F^b L^{-4b} T^{2b} \times F^c \qquad (6.6)$$

Collecting like terms on the right gives us $L^{1-4b} T^{-a+2b} F^{b+c}$. Because the group is dimensionless, each of the three exponents must be zero. Solving those three simple equations gives us $b = \frac{1}{4}$, $a = \frac{1}{2}$, and $c = -\frac{1}{4}$. So the product

Table 6.3 The six propeller variables with their units and dimensions

Variable	Meaning	Dimensions	Units
d	Propeller diameter	L	ft
n	Angular speed	T^{-1}	s^{-1}
ρ	Air density	$FL^{-4}T^2$	slug = lbf-s^2/ft^4
V	True air speed	LT^{-1}	ft/s
P	Power	FLT^{-1}	ft-lbf/s
T	Thrust	F	lbf

$dn^{1/2}\rho^{1/4}T^{-1/4}$ is dimensionless. Clearing fractional exponents by raising everything to the fourth power, so then also the product $d^4n^2\rho/T$ is dimensionless. Insiders will recognize this last combination as the reciprocal of the dimensionless "propeller thrust coefficient,"

$$C_T \equiv \frac{T}{\rho n^2 d^4} \tag{6.7}$$

Another way of seeing that a dimensionless thrust coefficient must look *something* like Eq. (6.7)—aerodynamicists at different times and places do use slightly different definitions of these coefficients—is to think of thrust as a sort of lift, of C_T as a sort of lift coefficient. Then C_T must depend, as C_L does, on the force, air density, some reference area, and the square of an air speed. In this thrust case, one can reasonably (and that is all we can ask) take the reference area to be d^2 (total blade area might make more sense but, for blades of the same shape, blade area and d^2 are proportional). And the relevant air speed might be the linear speed at the tip, $2\pi nR = \pi nd$. Multiplying the three factors (ignoring the constant π^2) gives the denominator in Eq. (6.7).

We now have these three independent dimensionless groups of variables: J, η, and C_T. There are many alternate ways this analysis could have gone. Instead of *FLT* we could have used the more common set *MLT*. By "leaving out" different variables appearing in earlier groups, a number of alternative dimensionless coefficients could have been found. And powers of products or quotients of those would have resulted in different final "answers." But no matter what choices we made, we would always have ended with $(6 - 3) = 3$ independent dimensionless groups. And any *other* dimensionless group would always be expressible in terms of those three.

The most common set of dimensionless variables for propeller work is $\{J, C_T, C_P\}$, where C_P is the "propeller power coefficient" defined as

$$C_P \equiv \frac{P}{\rho n^3 d^5} \tag{6.8}$$

Just as we drew a parallel between lift and thrust to "explain" C_T, we can explain C_P by looking at power as force times speed; this new speed factor corresponds to another power of tip speed, another power of the product nd.

Definitions of the two propeller coefficients shows that efficiency $\eta = TV/P$ can be expressed, in terms of this new and final set, as

$$\eta = J \frac{C_T}{C_P} \tag{6.9}$$

We have not lost anything; if we have J, $C_T(J)$, and $C_P(J)$ we can still find η. Advance ratio J gets special billing as the independent variable because it is made up of only "kinematic" quantities, quantities that merely describe motion. Power and thrust, on the other hand, are "dynamic" quantities, involving forces and perhaps accelerations. We prefer to describe the complex in terms of the simple. There is nothing lost by our making that choice; given say the value of C_P for a fixed-pitch propeller, it is almost always easy, and occasionally quite useful, to invert $C_P(J)$ to find J.

Our refined goal then is to calculate, for a given propeller shape with given aerodynamic coefficients $C_l(\alpha; \tau)$ and $C_d(\alpha; \tau)$, graphs of $C_T(J)$ and $C_P(J)$, as was done by McCauley, the Cessna 172 propeller manufacturer, for Fig. 6.6. That will take some doing.

We left out something you may have noticed. What about the influence of Reynolds number (Re)? And of Mach number (M)? Good questions. It turns out that aircraft propellers differ little enough in size and speed that scale effects, Reynold's number effects, are negligible. Not so for Mach number. When the fastest part of the propeller blade, the tip, reaches speeds above M 0.85, noisy compressive losses hurt efficiency rather badly. Using swept-back blade tips, thinner blade sections, and a larger number of blades helps alleviate this problem, but the only good solution, in our non-experimental realm, is to stay clear of the transonic region. In summary, Reynolds number does not much matter but Mach number does.

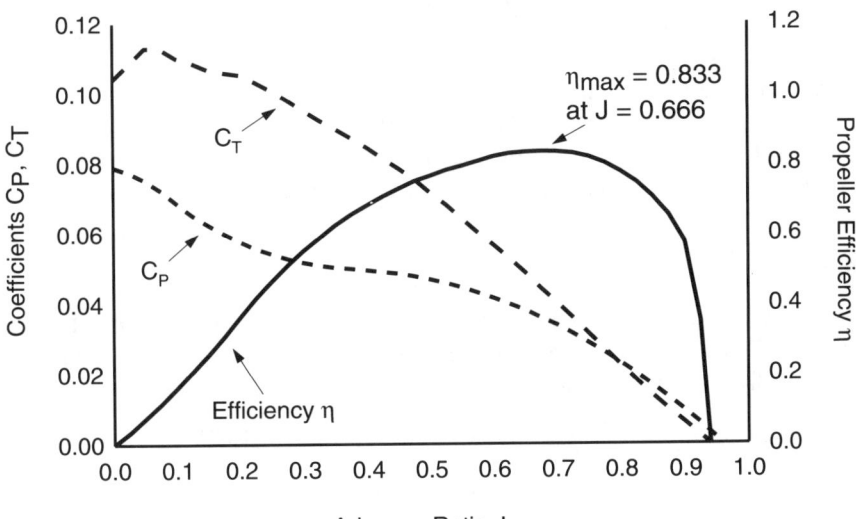

Figure 6.6   McCauley 7557 propeller characteristics.

There is a simpler way to see why propeller work calls for precisely two relations. The propeller is subjected to two types of dynamic forces: 1) tangential (torques), from rotational air resistance to the engine, described by $C_P(J)$; and 2) axial, from reaction of air the propeller pushes back on, described by $C_T(J)$. There are radial forces within the propeller structure too, of course, but these (we earnestly hope) are completely balanced and static; there is no perceptible movement in that radial direction along the length of the blade. There are forces in two directions and two coefficient functions to describe them. It is that simple. But dimensional analysis is needed for the details.

Dimensional analysis reduced the number of needed relationships, but it will never complete the whole job of finding $C_T(J)$ and $C_P(J)$. We have got an aerodynamic problem requiring an aerodynamic solution. Forces and pressures and torques and energies—the stuff of mechanics—will have to be brought to bear. We now turn to that work by considering two physically quite different theories of propeller action—the momentum theory and the blade element theory. Then by combining them we will get an amalgam stronger than either.

Momentum Theory

A propeller creates forward thrust by pushing back on the air. By Newton's third law of motion, the air pushes forward on the propeller just as hard. The stream of air passing through the propeller is also subjected to changes in kinetic energy, static pressure, and angular momentum. The momentum theory analyzes these changes—predominantly as a overall balancing picture—to come up with conditions on the resulting thrust, on power absorbed by the propeller, and with an upper limit to efficiency. It is a broad-brush theory. It replaces the propeller by an "actuator disk" of area $A = \pi d^2/4$ with an infinite number of blades. It pays scant attention to detail. Though the momentum theory may seem only a preliminary unrealistic cartoon of propeller action, it is quite a useful way station.

Physical Picture

We will take the pilot's view and consider the propeller at rest (in the longitudinal or axial direction) with undisturbed air coming at us with speed V. Figure 6.7 specifies five important axial positions:

1) Far in front of the propeller, in undisturbed ambient air;
2) Just before the propeller;
3) Halfway through the propeller;
4) Just behind the propeller; and
5) Far to the rear (the undissipated slip-stream wake).

Propeller Thrust

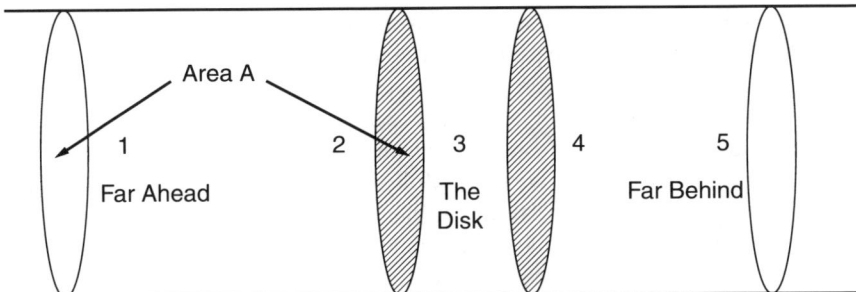

Figure 6.7 Momentum theory air stream with five important locations.

We drew the airstream as a perfect cylinder because one of the (slightly false) assumptions of momentum theory is that the slip-stream contraction, behind the propeller, is negligible. (In particular, using momentum theory to learn about static thrust, thrust with no forward air speed ($V = J = 0$), is probably very far from the mark.) Table 6.4 defines the most important variables and asks the important questions.

Axial and Rotational Interference Factors

The "axial interference factor" a is the proportional increase in air speed above free stream, from V to $V(1 + a)$, *at the propeller*. The "rotational interference factor" a' is the proportion of propeller rotation speed taken on by the airstream immediately at the propeller, from zero to $\omega = a'\Omega$, where Ω is the propeller angular speed $2\pi n$. The relative wind felt by the propeller blade is faster in the axial direction and slower in the tangential direction than without these interference effects. We will not be able to calculate values for the two interference factors until we get to the combined momentum and blade element theory, the most sophisticated propeller theory we will discuss. But we can here find ultimate

Table 6.4 Momentum theory airstream variables and locations

Position #	Air speed	Angular speed	Pressure	Disk area
1	V (free stream)	0	p (ambient)	$> A$
2	$V(1+a)$	0	$p - \Delta p_2$	A
3	$V(1+a)$	$a'\Omega$	—	A
4	$V(1+a)$	$x\Omega$	$p + \Delta p_4$	A
5	$V(1+b)$	$b'\Omega$	p	$< A$

wake factors b and b' in terms of a and a', rotational factor x just behind the propeller disk, and the gauge pressure differences Δp_2 and Δp_4. And we can learn the overall efficiency of this propulsive process.

Interference effects are in a sense secondary. But not negligible. In our anthropomorphic haze, we focus only on the airplane. But let us sneak a look at aviation from the point of view of "Molly," an ambient air molecule. She sees the airplane coming—let us say the airplane path is directed just very slightly above and slightly to the side—and thinks: "Well, I'm in for a ride!" Sure enough, first an upwash, then over (or under!) the wing, then a downwash. "Thank goodness I was halfway out along the wing," Molly thinks. "I sure wouldn't want to get mixed up in a *wing-tip vortex*. And the *propeller* is even worse: I get speeded up towards the airplane, chopped at through the propeller, given a corkscrew motion, often banged against the vertical fin, and pushed to the rear until the wake dies out. What a thing!" You get the point: it is an airplane/atmosphere *system*. Each part affects the other. Let us now calculate some effects on the atmosphere.

Thrust and Pressure

The momentum theory looks at thrust two ways: 1) as the time rate of change of airstream momentum and 2) as a net pressure on the actuator disk. If we calculate the thrust both ways, then we can find the relationship between the axial interference factor a at the propeller and the ultimate wake interference factor b. In addition, by focusing on the ratio of useful propulsive work to the total kinetic energy given to the airstream, we can compute the (ideal) efficiency of this process.

Thrust #1. Mass flux (mass passing through some portal in unit time), for a fluid of density ρ passing normally through area A with speed V, is $\rho A V$. (Speed V at the propeller, by definition of the axial interference factor, is actually $V[1+a]$.) The slip-stream air ultimately gains speed increment bV, so

$$\begin{aligned} T &= \text{mass flux} \times \text{speed gained} \\ &= \rho A V (1+a) \times bV \\ &= \rho A V^2 (1+a) b \end{aligned} \qquad (6.10)$$

Thrust #2. Consider thrust as the actuator disk area A times the difference in pressures between the rear (position 4) and front (position 2) surfaces. To do this, use Bernoulli's principle twice, once before and once behind the propeller:

$$p_1 + \tfrac{1}{2}\rho V_1^2 = p_2 + \tfrac{1}{2}\rho V_1^2(1+a)^2 \qquad (6.11)$$

$$p_4 + \tfrac{1}{2}\rho V_1^2(1+a)^2 = p_1 + \tfrac{1}{2}\rho V_1^2(1+b)^2 \qquad (6.12)$$

Solving Eq. (6.11) for p_2, Eq. (6.12) for p_4, and subtracting gives

$$p_4 - p_2 = \rho V^2 b(1 + b/2) \qquad (6.13)$$

so

$$T = A(p_4 - p_2) = \rho A V^2 b(1 + b/2) \qquad (6.14)$$

Equating the final forms of Eqs. (6.10) and (6.14) gives $b = 2a$. Our momentum theory thrust relation is, therefore,

$$T^{MT} = 2\rho A V^2 a(1 + a) \qquad (6.15)$$

The differential form, for the increment of thrust on the disk annulus between stations r and $r + dr$ (which we shall need later on) is

$$dT^{MT} = 2\rho 2\pi r \, dr V^2 (1 + a) a$$
$$= 4\pi r \rho V^2 (1 + a) a \, dr \qquad (6.16)$$

The great bulk of those axial speed gains takes place within one-half the diameter of the propeller. Viscosity is at work throughout this process, slowing the slip-stream to its ultimate dissipation.

We can use Eq. (6.11) to calculate the pressure deficit at position 2. The result is that (positive amount of underpressure) $\Delta p_2 = qa(2 + a)$ where q is the ambient dynamic pressure $\frac{1}{2}\rho V^2$. Similarly, overpressure at position 4 is $\Delta p_4 = qa(2 + 3a)$, always somewhat larger than Δp_2.

Ideal Efficiency

To get a handle on the (ideal) efficiency of this propulsive process, let us look at the ratio of useful to total power:

$$\eta_i = \frac{TV}{P} \qquad (6.17)$$

Because air speed at the disk is $V(1 + a)$, the denominator is

$P = \Delta KE$ per time $= \frac{1}{2} \times$ mass flux $\times$ difference of squared speeds
$$= \frac{1}{2}\rho A V(1 + a)(V_5^2 - V_1^2)$$
$$= \frac{1}{2}\rho A V^3(1 + a)(4a + 4a^2)$$
$$= 2\rho A V^3 a(1 + a)^2 \qquad (6.18)$$

Using our earlier expression for thrust, Eq. (6.15), in the numerator of Eq. (6.17),

$$\eta_i = \frac{2\rho A V^3 (1 + a) a}{2\rho A V^3 (1 + a)^2 a} = \frac{1}{1 + a} \qquad (6.19)$$

Because thrust increases with axial interference factor a, Fig. 6.8 shows that momentum theory ideal efficiency is higher when the disk is more lightly loaded, when thrust is lower.

Efficiency and Power

Having dealt with ideal propeller efficiency, we need to develop one further result for future reference. A relationship between ideal efficiency and propeller power coefficient will be needed in construction of a "general propeller chart" for constant-speed propellers. We know that total power, Eq. (6.18), is

$$P = TV(1+a) = 2\rho A V^2 a(1+a)^2$$
$$= \rho n^3 d^5 C_p \tag{6.20}$$

We also have, from Eq. (6.19),

$$a = \frac{1-\eta_i}{\eta_i} \tag{6.21}$$

Therefore, using $A = \pi d^2/4$, we find

$$C_P = \frac{\pi V^3 (1-\eta_i)}{2n^3 d^3 \eta_i^3}$$
$$= \frac{\pi J^3 (1-\eta_i)}{2\eta_i^3} \tag{6.22}$$

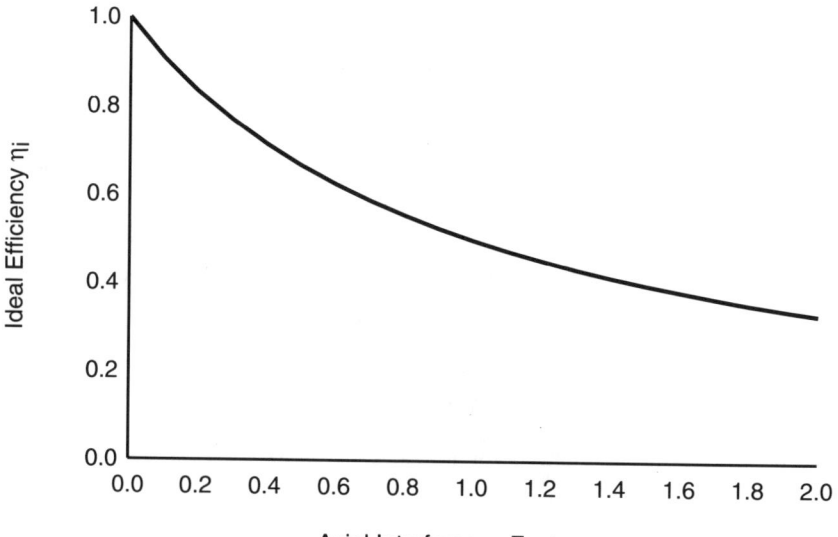

Figure 6.8 Ideal efficiency decreasing with increasing axial interference factor a.

which can be rearranged to give us

$$\frac{\eta_i}{(1-\eta_i)^{1/3}} = \left(\frac{\pi}{2}\right)^{1/3} \frac{J}{C_P^{1/3}} \qquad (6.23)$$

The quantity $J/C_P^{1/3}$ on the right-hand side (RHS) of Eq. (6.23) is interesting in that it does not depend on propeller circular speed n. Only on power P and air speed V (and of course on air density and propeller diameter). As long as you know power and air speed, Eq. (6.23), within its "ideal" limitations, is available for use.

Figure 6.9 is a graphical representation of Eq. (6.23) with approximate "bounding" realistic curves. Ideal efficiency η_i is a theoretical upper limit to propeller efficiency. Even in free air in favorable cases, actual attainable efficiency is only from 80 to 88% of η_i. That is because of additional losses:

- Rotation of the slip stream (which we shall soon consider);
- Profile drag on the blades (the major loss factor);
- Tip losses and hub drag;
- Multiblade interference; and
- Torque pulses from the small number of engine cylinders.

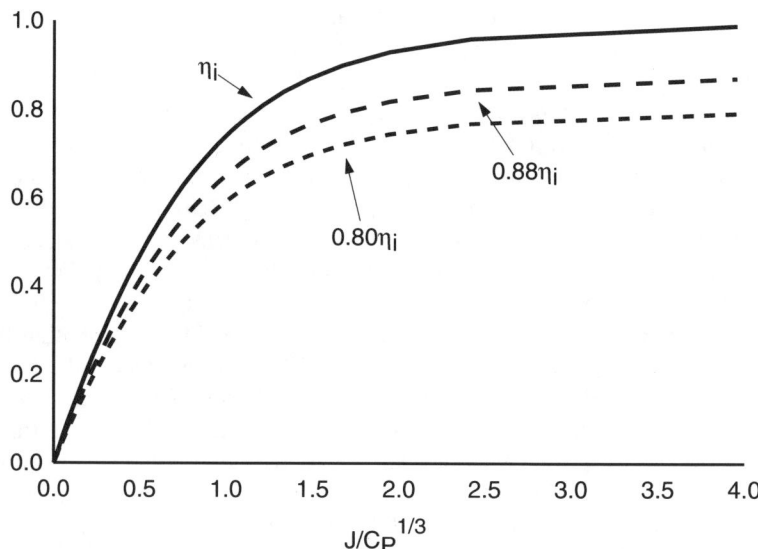

Figure 6.9   Precursor to what will be a general propeller chart.

Slip Stream Rotation and Torque

Pilots know that the propeller slip stream rotates in the same direction as the propeller but not nearly so fast. To calculate some of the details, we focus on one (arbitrary) propeller station r. We do this for two reasons:

1) Two similar objects rotating about a common axis, with the same angular speed ω (units rad/s) but at different stations r, do not have the same angular momentum

$$I\omega = mr^2\omega \tag{6.24}$$

because the moment of inertia I (units slug-ft^2) depends on the square of location r.

2) Propeller sections taken at different stations differ in many respects (breadth, angle, thickness, relative wind speed, and angle). We will need to take these differences into account.

So, we are at station r, asking for the incremental torque dQ absorbed by the disk between r and $r + dr$. The flux is $\rho \times$ area $\times$ speed $= \rho \times 2\pi r \, dr \times V(1+a)$. The propeller is rotating at $\Omega = 2\pi n$ rad/s and our slip stream, in the final wake, is rotating at $\omega_5 = \omega_4 = 2a'\Omega$ (to be discussed). The impressed change of angular momentum per unit mass is $r^2\omega_4 = r^2 2a'\Omega$. So we have

$$\begin{aligned} dQ^{MT} &= \rho \times 2\pi r \, dr \times V(1+a) \times r^2 2a'\Omega \\ &= 4\pi r^3 \rho V(1+a)\Omega a' \, dr \end{aligned} \tag{6.25}$$

The rationale behind the doubling of the at-disk rotational interference factor a', to become $2a'$ just behind the propeller (and also $2a'$ in the ultimate wake), is subtle. The airstream rotation in the vicinity of the propeller is partly due to the bound vortices on the propeller blade and partly due to the trailing vortices. The vortices trailing from the propeller trailing edge wrap up, as do wing vortices, into a single large vortex trailing off the tip. There is also a vortex from the propeller blade root trailing down the propeller shaft (see Fig. 6.10). The bound-to-the-blade vortex line causes equal and opposite angular speeds at positions 2 and 4. But at position 2 there must be no net rotation because the propeller vortex system has not yet reached that locale. Hence, the contribution of the trailing vortices from the tip and the root at position 2 must cancel the contribution from the bound vortices; the two effects are equal and opposite. Hence, the trailing vortex contribution at position 4 (or at position 5, for that matter, ignoring viscosity) must be twice the size of that due to either the bound or trailing vortices alone. At the propeller blade itself, position 3, the interference rotation is due *only* to the trailing vortices and hence is only $a'\Omega$ (see Table 6.5 where the angular speed contributions due to the bound and the trailing vortices are denoted by Bd and Tr, respectively).

Propeller Thrust 161

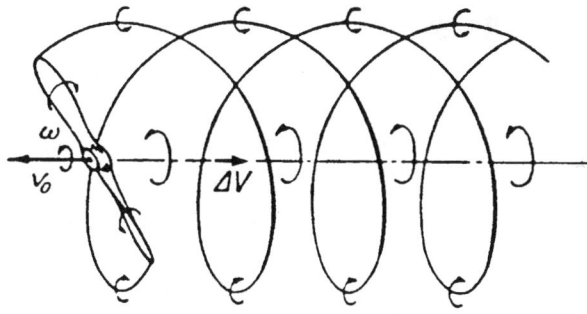

Figure 6.10 Bound and trailing propeller blade vortices.

Table 6.5 presents a tabular picture of bound and trailing vortices' contribution to the angular speed at various locations in the slip-stream. At position 1, as at position 2, there is no rotation. This table describes why $\omega_4 = \omega_5 = 2a'\Omega$. So now we know that, in Table 6.4, $x = b' = 2a'$.

What has the momentum theory done for us? It has given us thrust T in terms of an unknown axial interference factor a. Ideal efficiency η_i is similarly scantily clad. And the torque added to the slip stream annulus at r is in terms of an unknown rotational interference factor a'. A more accurate calculation of ideal efficiency, incorporating the energy loss due to slip stream rotation (we will not go through the details), gives

$$\eta_i = \frac{1-a'}{1+a} \qquad (6.26)$$

Rotational interference factor a' is almost always considerably smaller than axial interference factor a. So where are we? We are waiting for the blade element theory.

Table 6.5 Bound and trailing vortices' contributions to angular speed at various slip stream locations

Position #	2	3	4	5
Final factor	0	a' (by definition)	$2a'$	$2a'$
Bound contribution	$-$Bd	0	Bd	Bd
Trailing contribution	Tr	Tr	Tr	Tr

Blade Element Theory

Physical Picture

The blade element theory, simple unadulterated version, uses the straightforward aerodynamic force reasoning we are all used to:

$$\text{Force} = \text{dynamic pressure} \times \text{area acted on} \times \text{coefficient} \quad (6.27)$$

Figure 6.11 shows the resulting force picture. The total incremental aerodynamic force dF has lift and drag components dL and dD aligned parallel and perpendicular to the resultant airflow of speed W. But the components of dF in the direction of forward motion and in the direction opposing engine torque are dT and dQ/r. Later we will have to add up the increments.

Let us pick at the pieces of Eq. (6.27) with an eye to getting propeller force expressions for increment of thrust dT and increment of torque-producing force dQ/r.

Relative Wind and Dynamic Pressure

The resultant relative wind acting on the propeller segment has speed W given by

$$W^2 = V^2 + (\pi x n d)^2$$
$$= V^2\left(1 + \frac{\pi^2 x^2}{J^2}\right) \quad (6.28)$$

Notice also that the so-called *flow angle*, ϕ, is given by

$$\phi = \beta - \alpha = \tan^{-1}\frac{V}{\pi x n d} = \tan^{-1}\frac{J}{\pi x} \quad (6.29)$$

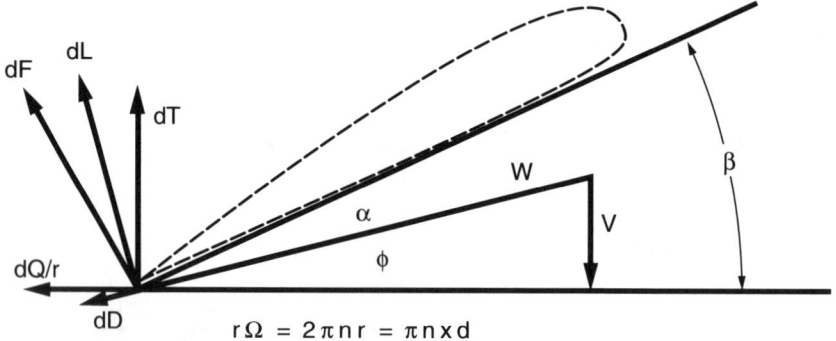

Figure 6.11   Blade element airflows and incremental force components.

so the dynamic pressure is

$$\tfrac{1}{2}\rho W^2 = \tfrac{1}{2}\rho V^2(1+\cot^2\phi) = \frac{\rho V^2}{2\sin^2\phi} \qquad (6.30)$$

Incremental Area

We are taking a little slice of span $dr = R\,dx = (d/2)dx$ of each propeller blade. And there are B blades. Hence, the area acted on is

$$dA = Bc(r)dr = \frac{Bc(x)d}{2}\,dx \qquad (6.31)$$

Aerodynamic Coefficients

There are a couple of problems in using the standard lift-and-drag coefficients for the propeller airfoil: 1) values of those coefficients are hard to come by and 2) increments of lift dL and drag dD do not point in useful directions.

The first problem is by far the most serious. To work with RAF6, airfoils the author uses elaborately interpolated, extrapolated, and curve-fit functions $c_l(\alpha;\tau)$ and $c_d(\alpha;\tau)$ derived from graphs displayed by Weick.[2] There are over 10,000 entries in each table. We will not carry that much apparatus with us for the few sample calculations to follow. Instead, we will use respectable approximations to RAF6 propeller lift-and-drag coefficients by borrowing simpler curve-fit formulas devised by Norris and Bauer.[3] These (as amended by personal communications) are:

$$c_1(\alpha) = 0.09458(\alpha + 4.8)$$

$$c_l(\alpha) = \frac{1.5 + c_1 - \sqrt{(1.5 - c_1)^2 + 0.04}}{2} \qquad (6.32)$$

$$c_d(\alpha) = 0.009 + 0.03(4c_1/3 - 1)^4 \qquad (6.33)$$

The coefficient functions of Norris and Bauer have no dependence on thickness ratio $\tau \equiv t/c$; the preliminary zero lift angle α_0 is taken to be $-4.8\,\text{deg}$, and the lift slope 0.09458 for any blade thickness. Such liberty is not quite realistic, especially in the α_0 matter, but it is the principle that counts; once one sees how the calculations go, one can add refinements. Figure 6.12 is a graph of the Norris and Bauer RAF6 lift-and-drag coefficient functions.

The second problem, orientation of the aerodynamic force components, is taken care of by rotating the coordinate system clockwise by flow angle ϕ (see Fig. 6.11). The effect is to give us coefficients λ_T and λ_P (here we are looking

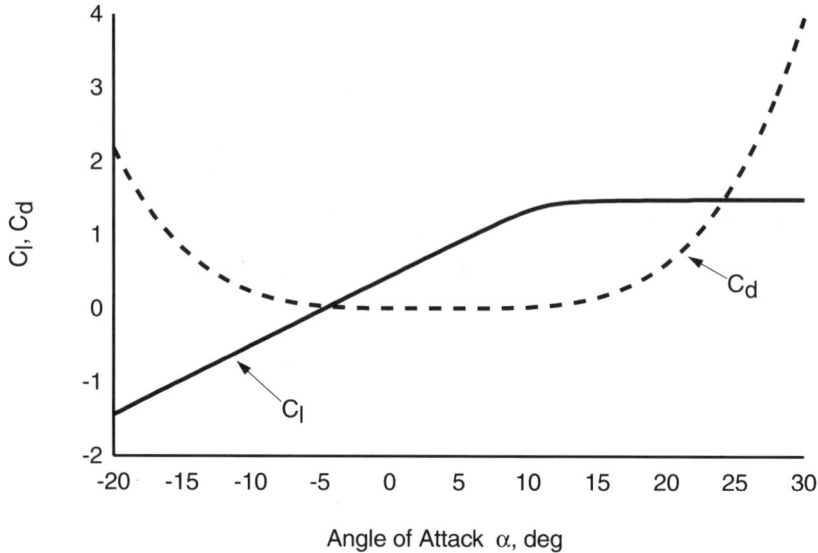

Figure 6.12 RAF6 lift-and-drag coefficient functions of Norris and Bauer.

ahead to our eventual replacement of torque Q by power P), linear combinations of c_l and c_d, for the desired force components:

$$\begin{bmatrix} \lambda_T \\ \lambda_P \end{bmatrix} = \begin{bmatrix} c_l \cos \phi - c_d \sin \phi \\ c_l \sin \phi + c_d \cos \phi \end{bmatrix} \quad (6.34)$$

Blade Element Thrust and Power Coefficient Gradings

Putting together results from the last three subsections, we have our not-quite-final blade element thrust and torque grading formulas:

$$\frac{dT^{BE}}{dr} = \frac{\frac{1}{2} Bc(r) \rho V^2 \lambda_T}{\sin^2 \phi} \quad (6.35)$$

$$\frac{dQ^{BE}}{dr} = \frac{\frac{1}{2} Bc(r) r \rho V^2 \lambda_P}{\sin^2 \phi} \quad (6.36)$$

We are almost home. Our goal, remember, is to get functions $C_T(J)$ and $C_P(J)$. We need to change variables to those coefficients and (numerically) integrate.

Groundwork for that first job has been laid. Equation (6.5) gets us from absorbed torque Q (which always matches, once the propeller has stopped accelerating angularly, engine torque M) to engine power P. Definitions of the propeller thrust and power coefficients, Eqs. (6.7) and (6.8), get us from T, Q to C_T, C_P. And it greases wheels to move from station r to relative station x with

$r = (d/2)x$ and $dr = (d/2)dx$. These changes, with the definition of advance ratio $J \equiv V/nd$, take Eqs. (6.35) and (6.36) to the final blade element thrust and power coefficient gradings formulas:

$$\frac{dC_T^{BE}}{dx} = \frac{Bc}{4d} \frac{J^2(c_l \cos\phi - c_d \sin\phi)}{\sin^2\phi} \qquad (6.37)$$

$$\frac{dC_P^{BE}}{dx} = \frac{\pi x Bc}{4d} \frac{J^2(c_l \sin\phi + c_d \cos\phi)}{\sin^2\phi} \qquad (6.38)$$

It is satisfying to get relatively simple expressions in a relatively complicated problem. A numerical example could be even more satisfying.

Example 6.1 For our simple blade element example we arbitrarily select $J = 0.6$ and $x = 0.8$. We will use the curve-fit geometry of the Cessna 172 McCauley propeller [Eqs. (6.2) and (6.4)], with the Norris and Bauer lift-and-drag coefficients [Eqs. (6.32) and (6.33)].

Step 1. Geometry and Relative Wind
There are $B = 2$ blades of diameter $d = 75$ in. $= 6.25$ ft. The chord $c(0.8) = 4.512$ in. $= 0.3760$ ft. The blade angle is $\beta(0.8) = 16.929$ deg. Flow angle $\phi = \tan^{-1}(J/\pi x) = 13.427$ deg, so angle of attack (AOA) $\alpha = \beta - \phi = 3.502$ deg.

Step 2. Coefficients
Preliminary lift coefficient $c_1(3.502 \text{ deg}) = 0.7852$ gives $c_l(3.502 \text{ deg}) = 0.7715$ and $c_d(3.502 \text{ deg}) = 0.0150$. By Eq. (6.34), then, $\lambda_T = 0.7469$ and $\lambda_P = 0.1937$.

Step 3. Results for $J = 0.6$, $x = 0.8$
The thrust coefficient grading under these circumstances, from Eq. (6.37), is $dC_T^{BE}/dx = 0.1500$; the power coefficient grading, from Eq. (6.38), is $dC_P^{BE}/dx = 0.09777$.

Repeating this calculation for a range of relative stations x gives the graphs of Fig. 6.14 for the thrust and power coefficient gradings for $J = 0.6$.

To get coefficients $C_T(J = 0.6)$ and $C_P(J = 0.6)$ themselves, we must calculate the areas under the curves of Fig. 6.13. The (extended) trapezoidal rule for numeric integration, from Appendix D, is

$$\int_{x_0}^{x_n} f(x)dx \doteq h\left[\frac{f_0}{2} + f_1 + f_2 + \cdots + f_{n-1} + \frac{f_n}{2}\right] \qquad (6.39)$$

where h is the uniform subinterval length and f_i means $f(x_i)$, is accurate enough. To get the final propeller characteristic graphs, in the blade element approximation, all the above must be repeated for a range of J values. Figure 6.14 gives the

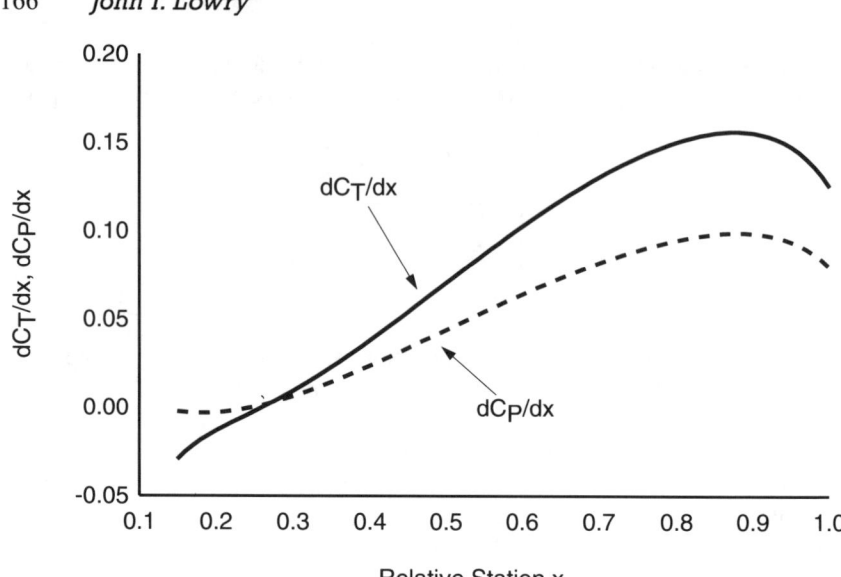

Figure 6.13 Propeller thrust and power coefficient gradings for $J = 0.6$.

final blade element results for this propeller with these assumed lift-and-drag coefficients.

If you compare the blade element results of Fig. 6.14 with the manufacturer's data (Fig. 6.6), you will see our blade element efficiency graph goes too high and ends too soon. Eventually we will see what can be done about those defects. In the meantime, we take up another and more sophisticated version of propeller theory. One final complication before—we promise!—simplification arrives.

Combined Momentum and Blade Element Theory

There are almost as many variations of propeller theory as there are writers on the subject. But almost every such writer includes what most call "the combined momentum theory and blade element theory." But be careful, some call it "the vortex theory." The momentum theory, inconclusive as it was, included the axial (a) and rotational (a') interference factors. And it allowed expressions for differential thrust, Eq. (6.16), and differential torque, Eq. (6.25), in terms of those factors. The blade element theory did not include interference effects, but it was a nice straightforward exposition in terms of propeller geometry and propeller section lift-and-drag coefficients. Equations (6.35) and (6.36) gave the blade element expressions for differential thrust and torque. What could be more natural than to combine them?

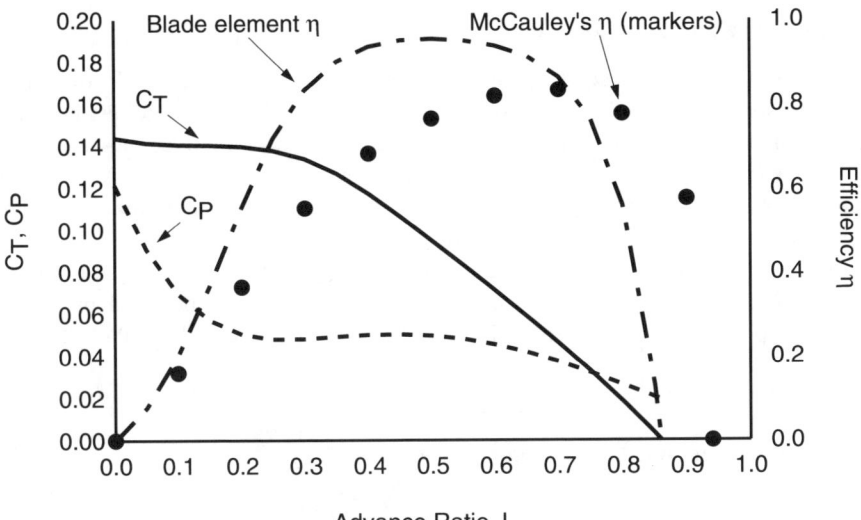

Figure 6.14 RAF6 propeller characteristics, blade element approximation.

Physical Picture

The combined momentum theory/blade element picture, Fig. 6.15, is that of the simple blade element theory (Fig. 6.11) with the addition of interference factors. Those two unknown quantities appear in two new geometric relations:

$$W = V(1 + a)\operatorname{cosec}\phi \tag{6.40}$$
$$W = r\Omega(1 - a')\sec\phi \tag{6.41}$$

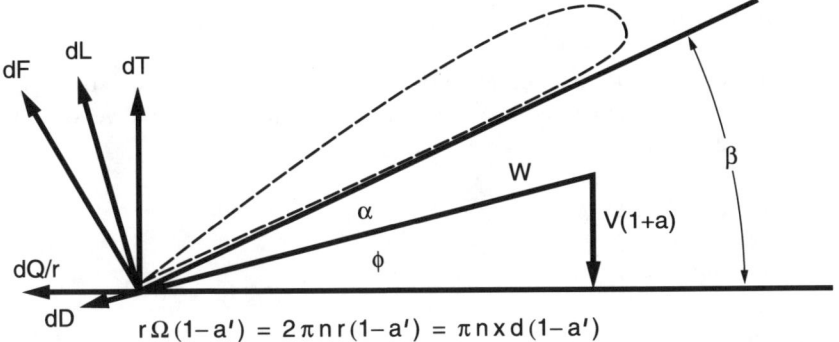

Figure 6.15 Combined momentum and blade element theory airflows and incremental force components.

Mathematical Solution

We cannot hope to succeed by simply dropping the identifying MT from Eqs. (6.16) and (6.25) and the BE from Eqs. (6.35) and (6.36). We had best have as many equations as unknowns: six. The momentum theory involved variables dT, a, dQ, and a'; Eqs. (6.40) and (6.41) bring up two more, ϕ and W. (We can see that AOA $\alpha = \beta - \phi$ so we will not count that as a seventh unknown and relation.) Now, with luck, it is just a matter of algebra.

Appendix E takes the six independent Eqs. (6.16), (6.25), (6.35), (6.36), (6.40), and (6.41), with the use of definition $x \equiv r/R$, $J \equiv V/nd$, and local propeller solidity (proportion of propeller disk annulus actually occupied by blades),

$$\sigma(r) \equiv \frac{Bc(r)}{2\pi r} \tag{6.42}$$

and solves to get our "master propeller equation":

$$J = \frac{\pi x (4F \sin^2 \phi - \lambda_T \sigma)}{(4F \sin \phi \cos \phi + \lambda_P \sigma)} \tag{6.43}$$

where the Prandtl momentum loss factor F, a function of relative blade station x and flow angle ϕ, corrects for both the finite number of blades B and for propeller blade tip losses. It involves this multistep definition:

$$F \equiv \frac{2}{\pi} \cos^{-1} e^{-f}$$
$$f \equiv \frac{B}{2} \frac{(1/x - 1)}{\sin \phi_t} \tag{6.44}$$

where the flow at the blade tip ϕ_t is given by

$$\frac{1}{\sin \phi_t} = \sqrt{1 + \frac{1}{x^2 \tan^2 \phi}} \tag{6.45}$$

For further details on F, see Appendix E.

Example 6.2 To find a solution AOA α (hidden within angle ϕ) for a given propeller geometry and aerodynamics at a given J and given relative station x, follow these steps:

1) Pick a trial AOA α;
2) Compute $\phi = \beta - \alpha$;
3) Compute $F = F(x, \phi)$;
4) Compute $C_l(\alpha)$ and $C_d(\alpha)$;
5) Compute λ_T and λ_P;
6) Compute solidity σ; and
7) Compute the RHS of Eq. (6.43).

If you get the desired given J, you are done; if not, recalculate with a different trial α. Let us work through an example. As with the simple blade element theory example, we use the McCauley 7557 propeller at $J = 0.6$, $x = 0.8$.

Step 1. Choose trial AOA α

For comparison's sake, let us start with trial $\alpha = 3.502$ deg, the solution given by the previous simple blade element theory.

Step 2. Relative wind direction

The blade angle is $\beta(0.8) = 16.929$ deg. Angle $\phi = \beta - \alpha = 13.427$ deg.

Step 3. Compute Prandtl ancillary loss factor f, here equal to 1.0661. Then follow up to find the Prandtl loss factor is $F = 0.7762$.

Step 4. Lift-and-drag coefficients

Preliminary lift coefficient $c_1(3.502 \text{ deg}) = 0.7852$ gives $c_l(3.502 \text{ deg}) = 0.7715$ and $c_d(3.502 \text{ deg}) = 0.0150$.

Step 5. Rotated thrust and power coefficients

By Eq. (6.34) then, $\lambda_T = 0.7469$ and $\lambda_P = 0.1937$. So far we have essentially retraced the simple blade element calculation.

Step 6. Solidity calculation (this will not have to be redone, in spite of iterations, until one moves to a different relative station x)

There are $B = 2$ blades of diameter $d = 75$ in. $= 6.25$ ft. The chord $c(0.8) = 4.512$ in. Blade station $r = 0.8 \times 75/2 = 30$ in. By Eq. (6.42), solidity $\sigma = 0.04787$.

Step 7. Results for trial $\alpha = 3.502$ deg, $x = 0.8$

The RHS of Eq. (6.43) gives $J = 0.46556$, not very close to our target $J = 0.6$. Worth retaining, however, because we will soon want results for a lot of other values of J.

Repeating the above process by hand, or automatically, the correct AOA for this circumstance turns out to be $\alpha = 1.478$ deg. Once $J = 0.6$ is verified, one can use additional formulas in Appendix E to find that axial interference factor a (for this x and J) is 0.1459 and the corresponding rotational interference factor $a' = 0.0102$. This means that *at this station and for this J*, the propeller is drawing in air such that the local axial air speed is about 15% greater than that of the airplane and the immediate slip stream rotation speed (halfway through the propeller) is about 1% less than the propeller angular speed. The thrust and power grading numbers, also from formulas in Appendix E, are $dC_T/dx = 0.1121$ and $dC_P/dx = 0.0826$.

The next step, of course, is to repeat this trial-and-error procedure for a selection of other relative stations from about $x = 0.15$ through $x = 0.95$ at intervals $\Delta x = 0.05$. (Because this is not a square-tipped propeller, blade width and most of the important numbers, certainly the thrust and power gradings, are zero at $x = 1$.) Using a computer spreadsheet program with a Backsolver or SolveFor facility actually makes quite quick work of this process.

Then one numerically integrates the thrust and power gradings—again with the extended trapezoidal (or perhaps extended Simpson's) rule from Appendix D—to find values of C_T and C_P. Treating C as either coefficient, that integration is

$$C(J) = \int_{0.15}^{1} \frac{dC}{dx} dx \qquad (6.46)$$

For this propeller, you would find $C_T(0.6) = 0.0499$ and $C_P(0.6) = 0.0369$. From Eq. (6.9), that makes efficiency $\eta(0.6) = 0.8114$. Figure 6.16 gives this McCauley 7557 two-bladed propeller's characteristics as they come from the combined momentum and blade element theory using these (Norris and Bauer) RAF6 lift-and-drag coefficients.

This is our "last best theory." How good is it? How realistic are the numbers it gives us? We will defer those important questions to briefly discuss a special case: no forward motion of the airplane.

Static Thrust Case ($J = 0$)

The static thrust case, the airplane at the start of its takeoff roll, is an exception. Our previous equations will not do that job because axial interference factor a is undefined when $V = 0$. Trying to make quick and dirty adjustments does not

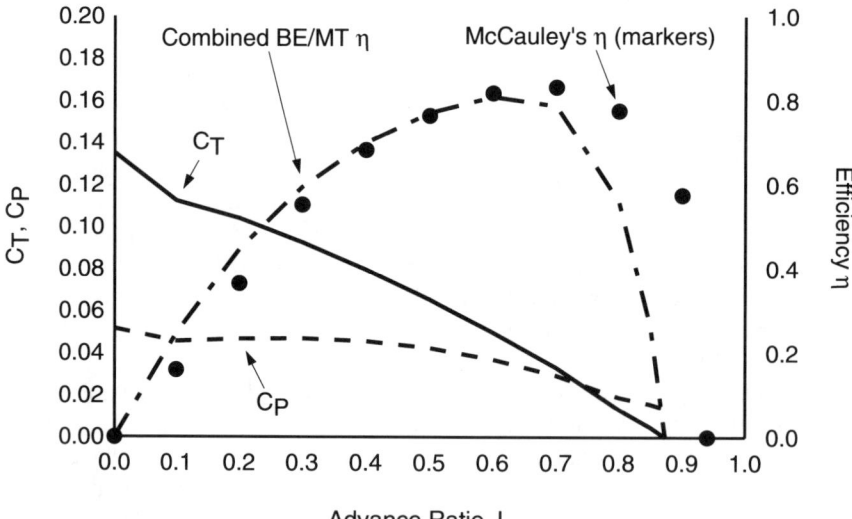

Figure 6.16 Propeller characteristics using combined momentum and blade element theory.

work. The best *theoretical* thing to do is to start the entire derivation over from scratch, different diagrams and all, with $V = 0$. If you do that you will get, instead of our "master" equation, Eq. (6.43), this "static master" equation:

$$4F \sin^2 \phi - \lambda_T \sigma = 0 \qquad (6.47)$$

Equation (6.47) is not original; it can be found (though without the Prandtl loss factor F) in Glauert.[4] In retrospect, one can see that Eq. (6.47) results from imposing the condition $J = 0$ on our full "propeller master equation," Eq. (6.43)—which *is* original so far as we know. But the alternate derivation is needed to get new equations for axial speed v (in lieu of a) and for the thrust and power gradings. In *practical* computational terms, none of this extra consideration and apparatus is needed. The nonstatic equations, using some small but finite J like 0.01, work just fine and give (very nearly) the same static thrust and power grading results as do the static ($J = 0$) solutions.

Errors and Omissions

We now pick up the previous thread in question-and-answer format.

Q: How good are our results?
A: Only fair.

Q: Why so?
A: We have left out some factors and effects.

Q: What can be done about this?
A: More attention to detail.

Following is a brief compendium of where errors lie and of factors we have ignored. These are presented in approximate order of importance and with simple remedies where those are known. Numeric estimates are only rough because they do not take into account individual airplane and propeller characteristics.

1) Slip stream speed greater than true air speed causes increased drag on airframe.
   This lowers effective thrust and, therefore, lowers propeller efficiency about 5%. Multiply calculated values of efficiency by 0.95. For finer work, do a careful analysis of increased drag contributions for the parts of the airframe within the slip stream.
2) Presence of fuselage reduces effective freestream air speed V.
   This places the propeller into operation at a reduced effective advance ratio J, on the average about 8.5% below the nominal value. This means a

calculation for say $J = 1.0$ requires an experimental J of 1.093 (because 91.5% of 1.093 is 1.0). Take the calculated propeller characteristic curves and inflate the abscissa values by about 9.3%, stretching the calculated curves to the right while leaving them anchored at the vertical line $J = 0$. A more accurate slowdown efficiency factor (SDEF) depends on relative station r and can be approximated by a source/sink pair giving streamlines around the fuselage shape. But this finer analysis destroys our ability to consider J to be the single independent variable on which propeller characteristics depend. Having to consider air speed V and propeller speed n separately is, ordinarily and for performance purposes, too much work to be practicable.

3) Uncertainties in lift-and-drag coefficients.
4) Drag of the propeller hub or boss reduces effective thrust.
   Calculate the effect treating the boss as a flat plate ($C_D = 1.11$) or as a spinner ($C_D = 0.35$). But do not forget that effective air speed there is probably less than one-half the nominal aircraft air speed V.
5) Propeller blades twist, under load, so as to *increase* the blade angle.
   If the brake power is less than 100 hp per blade, you can ignore this effect. If bhp/$B > 100$, recalculate the propeller characteristics with all blade angles increased by

$$\Delta\beta \text{ deg} = \frac{\text{bhp}}{100B} - 1 \qquad (6.48)$$

This means increasing all blade angles one-half of a degree if you are calculating for a 300-hp engine direct driving a two-bladed propeller. In case there is a gear box, reduce the effect given by Eq. (6.48) by multiplying it by the ratio of propeller revolutions per minute to engine revolutions per minute. Because larger blade angles mean larger values of J_1 (the advance ratio where C_T and η fall to zero) this effect, as did effect 2, means you should stretch your calculated graphs of propeller characteristics a bit to the right.

6) Torque pulses due to the small number of engine cylinders create an unsteady airflow.
7) Convex lower contours of blade elements near the hub decrease lift coefficient.
   Take the lift coefficient for those elements to be the lift coefficient of the upper portion decreased by three times the ratio of the lower contour thickness to the blade chord:

$$c_l^{\text{Total}} = c_l^{\text{Upper}} - \frac{3 t_{\text{Lower}}}{c} \qquad (6.49)$$

8) Above M 0.85 compressibility effects add noise and diminish efficiency. Many aerodynamics texts include charts for this efficiency correction.
9) Slip stream wake contraction raises the power coefficient a bit.

In typical cases the contraction is about 1% in area. This raises the power coefficient about that same amount and thereby lowers efficiency about 1%.

10) If the propeller shaft is not directly aligned with the airstream, blades are loaded unevenly. There is not much effect on efficiency, however.
11) There are small scale (Reynolds number) effects due to varying air speeds and blade chords.
12) Other speed regimes obviate much of the developed propeller theory. Avoid the fan, windmill, and brake speed regimes.
13) Tip shape has a small effect on propeller efficiency.
    Just as in the case of wing tips, square is worst.
14) In spite of theoretical and experimental evidence that blade elements are independent of one another, this may not be completely true. Because a propeller is essentially a rotating wing of finite aspect ratio, there will be some crossflow. We ignored those small radial airflows through the propeller and in the slip stream.
15) Because propeller blades push back on the air, air resistance bends the blades forward.

Making even casual use of only corrections numbered 1, 2, and 9 modifies the propeller characteristics of Fig. 6.16 to become the more realistic ones of Fig. 6.17. Note the "stretching" to the right and the lowered maximum efficiency.

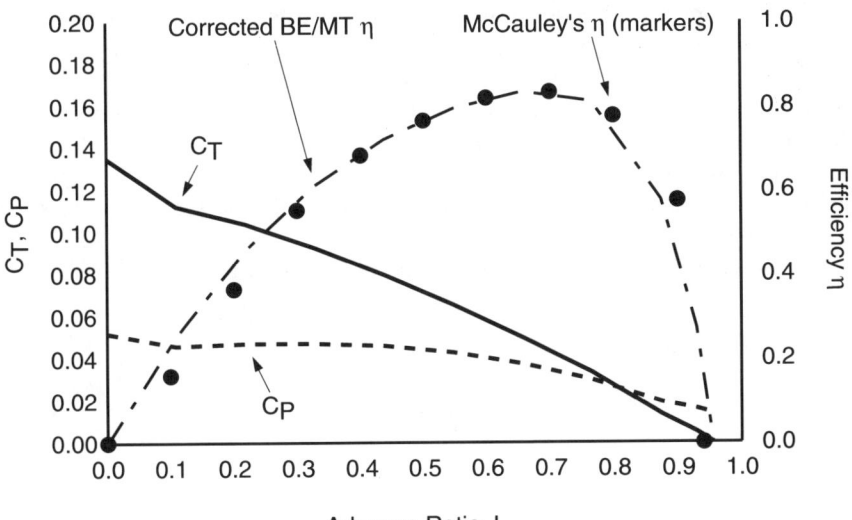

Figure 6.17 Propeller characteristics as in Fig. 6.16, with the addition of corrections numbered 1, 2, 9 (see text).

Our list of possible corrections might well give the reader the impression that one can spend a lifetime studying one propeller. Especially if one bolts copies of that propeller onto different airplanes. Orville Wright was right: there is a lot going on up there. Propeller reality is certainly less sedate than the theory. We can leave the subject at this point, confident that this grounding in propeller culture is sufficient for our airplane performance purposes. We need to know something about propellers and their coefficients, but we will not be using complicated detailed calculational methods. We will do something a lot simpler involving the "propeller polar."

Example 6.3 This introduction to propeller theory may have given you more facility than you realize. As an example take the common experience of having revolutions per minute drop, as speed drops when one starts to climb, at constant throttle, out of level cruise. Here is one way that drop in engine speed with air speed can be explained. We only need three pieces of information:

1) The definition of C_P, Eq. (6.8);
2) The definition of advance ratio $J \equiv V/nd$; and
3) The fact that, in the "working range," graphs of $C_P(J)$ slope down to the right; i.e., $dC_P/dJ < 0$.

At constant throttle (constant manifold pressure), torque is constant. From assumption 1), plus the relation between power and torque, Eq. (6.5), that means there is a constant K such that $C_P = K/n^2$. Now when air speed V drops off, either engine speed n goes down or it does not. If we assume n decreases, we can not tell which way J might go; we are stymied. So let us (provisionally) assume the n increases or stays constant and see what we can figure out. Then, by assumption 2, because V decreases and n increases or stays constant, J decreases. But then, by assumption 3), C_P increases. But then, because $n^2 = K/C_P$, n decreases. So if n increases, then n decreases, which is absurd. So, when V decreases at constant torque, it must be true than n also decreases (just as we know it does).

A more constructive theory,[5] computing differentials with Jacobians, gives

$$\frac{dn}{dV} = \left[d\left(J - \frac{2C_P}{C'_P}\right)\right]^{-1} \qquad (6.50)$$

where C'_P means dC_P/dJ. Again, the fact of negative C'_P ensures dn/dV is positive.

Propeller Polar Diagram

The propeller polar diagram is a graph of C_T/J^2 plotted as a function of C_P/J^2. Why would one care to make such a strange graph? Let us take a look at these new variables.

$$\frac{C_P}{J^2} = \frac{P}{n} \times \frac{1}{d^3 \rho V^2} \qquad (6.51)$$

Equation (6.51) shows that propeller polar variable C_P/J^2 depends on engine torque and, at a given density altitude, on the true air speed. Engine torque is a quite linear function of manifold absolute pressure (MAP). And, for a given density altitude, MAP depends on throttle position. So for a fixed throttle and altitude, C_P/J^2 depends only on air speed. By selecting his air speed, the pilot thereby selects C_P/J^2. Let us take a similar look at the dependent variable:

$$\frac{C_T}{J^2} = \frac{T}{\rho d^2 V^2} \qquad (6.52)$$

The important thing here is that this depends on thrust and air speed and *not* on engine revolutions per minute. That is a good thing because (without a lot of engine and propeller data) we do not have a good way of knowing revolutions per minute. (This is a defect to be remedied in the chapter on cruise and partial-throttle performance.) With an accurate propeller polar diagram, ignorance of revolutions per minute does not hurt much. The propeller polar lets us turn knowledge about throttle position (torque) and air speed into knowledge about thrust. We will use this line of reasoning when we discuss the *bootstrap approach* to computing fixed-pitch airplane performance.

The final interesting and useful fact about the propeller polar diagram or formula is that, over the main operating speeds of the airplane (excepting very low speeds early in the takeoff roll and very high speeds, say diving under power), *the propeller polar is linear.* Figure 6.18, for instance, is the polar for the McCauley 7557 propeller we investigated earlier. This example uses the correction-adjusted combined momentum theory and blade element calculation. Using McCauley's data, the polar is even more nearly linear. If we call the value of advance ratio where efficiency is maximum J_m then Fig. 6.18 uses only Js between 0.50 J_m and 1.17 J_m.

Finding the best straight line through the points of Fig. 6.18 gives us the "linearized" propeller polar (also in Fig. 6.18):

$$\frac{C_T}{J^2} = m\,\frac{C_P}{J^2} + b \qquad (6.53)$$

with (in this particular case) slope $m = 2.086$ and intercept $b = -0.0670$.

There are theoretical reasons why propeller polar diagrams are nearly linear. A competing simplified theory, that of the "representative blade element," says there

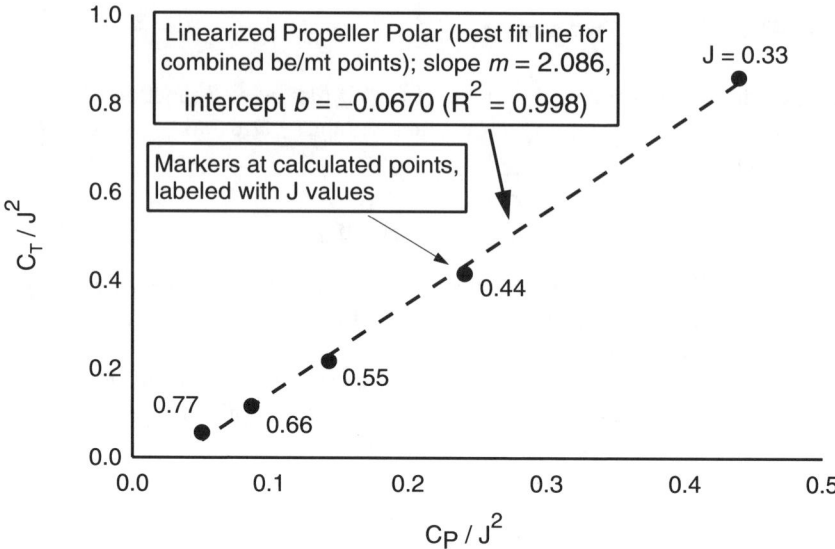

Figure 6.18 Propeller polar diagram from corrected combined momentum and blade element theories calculation.

is some station along the blade (not far from $x = 0.75$) that can "stand in" for the average propeller section. Von Mises[6] uses this representative blade element theory predominantly in his analytic propeller and aircraft performance sections. He includes a "proof" that the propeller polar is approximately linear. One can go farther and tie information about the propeller diameter and average width, together with the blade setting angle and drag coefficient of the representative blade element, to the two numbers (slope and intercept) characterizing the linear propeller polar. We will not do that here, however.

The best argument towards linearity of the propeller polar diagram, to our mind, is that whenever you construct such a diagram (restricted to the "working range") out of believable propeller characteristic charts $C_P(J)$ and $C_T(J)$, you do indeed get quite a straight line. So, for fixed-pitch propellers, we need propeller charts or, for this "reduced" description, propeller polar diagrams. It will turn out that bootstrap approach flight tests obtain the propeller polar experimentally. But what are we going to do about constant-speed propellers?

Constant-Speed Propeller

A proper pair of propeller characteristic graphs $C_P(J)$ and $C_T(J)$—or one of these plus a graph of $\eta(J)$—describes the action of any propeller with a fixed blade angle setting. We usually specify this setting by citing $\beta_{0.75}$, the blade angle at

$x = 0.75$. Often an entire family of related propellers differ only in that regard. But constant-speed propeller blades do not have fixed blade angle settings! Those blades twist in their automated hubs to meet (within limits) the torque and speed demands set by manifold pressure and the propeller speed control. For constant-speed propellers it looks like we will need a whole sheaf of characteristic graphs. If we had them, here is how we might proceed to find propeller efficiency η, the pivotal quantity we need:

- From V, n, and d, calculate J.
- Enter the POH cruise tables with pressure altitude h_p, OAT (giving density altitude h_ρ or its surrogates relative density σ or density ρ), manifold pressure and revolutions per minute, to find shaft power P (or P as a proportion of rated mean sea level power).
- From P, ρ, n, and d, calculate C_P.
- From graphs for C_P vs J for various $\beta_{0.75}$, determine $\beta_{0.75}$ by intersection (with perhaps a little interpolation).
- From the graph giving η vs J for the $\beta_{0.75}$ above (again, perhaps with interpolation), read off η.

The bad news is that you can seldom get your hands on that full set of charts for your particular propeller. The good news is that there is a much easier, though approximate, way that in a sense combines the last two steps above. This alternative has the additional advantage of being general, of not depending on all the details of your particular make and model constant-speed propeller.

General Propeller Charts

The momentum theory, recall, suggested that ideal propeller efficiency η_i is purely a function of $J/C_P^{1/3}$. As usual, reality (including blade drag, mutual interference among blades, and rotation of the slip-stream) is somewhat more complicated. Actual efficiency η is still pretty much a function of $J/C_P^{1/3}$, but the details of just *which* function depends on how big C_P is. And not only C_P but also a "power adjustment factor," X, which depends on the propeller's distribution of blade widths. The first use of this strategy was the general propeller chart pioneered by the Boeing Aircraft Company (BAC)[7] in its production of long-range bombers during World War II. A version of the BAC general propeller chart can be found in Perkins and Hage.[8]

But the difference in scale between say a Bonanza (285 hp) and a bomber or transport sporting four Pratt & Whitney Wasp Majors (2685 hp each) is just too great. The BAC general propeller chart does not give accurate results for general aviation airplanes with constant-speed propellers. What was needed was a new chart, similar to Boeing's but based on data for the smaller general-aviation-sized, constant-speed propeller: a *general aviation* general propeller chart. Data and measurements for a general aviation constant-speed propeller let us construct such a chart (see Fig. 6.19). This chart will play a pivotal role in the constant-speed version of the bootstrap approach. Let us see how it works.

Example 6.4 Our engine is turning over 2400 rpm ($n = 40$ rps) and is putting out 200 hp ($P = 110,000$ ft-lbf/s). We are cruising at density altitude of 5000 ft ($\sigma = 0.8617$, $\rho = 0.002048$ slug/ft^3), at 150 KTAS ($V = 253.2$ ft/s). Propeller diameter is $d = 7$ ft. Accordingly, advance ratio $J = V/nd = 0.9043$; $C_P = P/\rho n^3 d^5 = 0.04993$; $C_P^{1/3} = 0.36824$; $J/C_P^{1/3} = 2.456$. The eight curves of the general aviation general propeller chart are labeled by $C_{PX} \equiv C_P/X$, where X is the power adjustment factor (to be explained below), which depends on the total activity factor (TAF, also to be explained below) of our particular propeller. Assume our blade activity factor (BAF) = 100 and we have two of them; then TAF = 200. It turns out that $X = 0.246$. Hence, $C_{PX} = 0.2030$. We need the phantom graph about halfway between the graphs for $C_{PX} = 0.15$ and for $C_{PX} = 0.25$. Interpolating, we find $\eta = 0.764$ (the Boeing general propeller chart gives $\eta = 0.87$).

Now let us turn, as promised, to describing the propeller BAF and the power adjustment factor. There is actually one additional wrinkle to be ironed out, the SDEF, having to do with obstructed airflow through the propeller. Again, to be explained.

Activity Factor

The tangential air speed of the propeller section at relative station x is proportional to x. Aerodynamic forces vary as the area and as the square of the air speed and hence as $x^2 c(x)\, dx$. Absorbed power goes as force times speed, hence as $x^3 c(x)\, dx$. That is why BAF is defined as

$$\text{BAF} \equiv \frac{100,000}{32} \times \int_{0.2}^{1} x^3 \frac{c(x)}{R}\, dx \tag{6.54}$$

The denominator R (blade radius) is to make BAF dimensionless. The factor in front of the integral sign is to give BAFs reasonable numerical sizes. For commonly shaped blades, BAFs vary from about 70 to 140. Total activity factor, for the entire propeller, is defined as

$$\text{TAF} = \text{number of blades} \times \text{BAF} \tag{6.55}$$

Normally, you do not have a formula for blade width $c(x)$ as a function of relative station x and so evaluate the integral in Eq. (6.54) numerically. Table 6.6 and Eq. (6.55) are the ticket. First you must measure your propeller blade's widths at relative stations $0.20, 0.25, 0.30, \ldots, 0.90, 0.95, 1.00$. If early stations (0.20,

Propeller Thrust 179

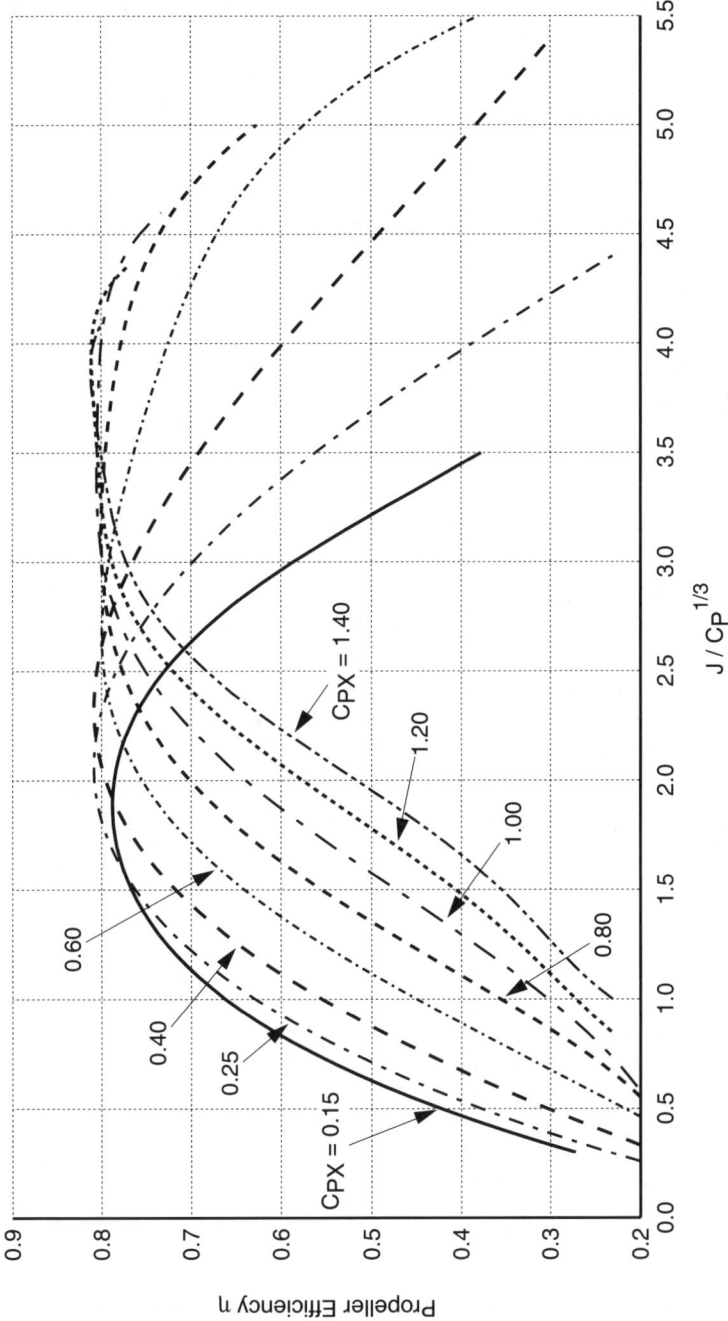

Figure 6.19 The general aviation general propeller chart.

etc.) are hidden by the spinner, just treat them as having zero width. Using the trapezoidal rule from Appendix D,

$$\text{BAF} = \frac{78.125}{R} \times \{f(0.20) + 2[f(0.25) + f(0.30) + \cdots + f(0.95)] + f(1.00)\}$$

(6.56)

Values $f(x)$, as indicated in the BAF data sheet, are products of x^3 and $c(x)$.

Power Adjustment Factor X

Power adjustment factor X for your propeller depends on its TAF according to the (curve-fit) formula (see Fig. 6.20):

$$X = 0.001515 \, \text{TAF} - 0.0880$$

(6.57)

For the McCauley 1C160 propeller on the Cessna 172, for instance, BAF is 87.15, TAF is 174.30, and so $X = 0.1761$. In use, X is divided into C_P to get C_{PX}. It is the value of C_{PX} that tells us which of the eight graphs on the general aviation general propeller chart to use or which two to interpolate between.

Table 6.6 Blade activity factor data sheet

Propeller make & model:		Number of blades:	Propeller diameter:
BAF:		TAF:	Note:
Station x	x^3	Blade width c	$f(x) = x^3 \times c$
0.20	0.0080		
0.25	0.0156		
0.30	0.0270		
0.35	0.0429		
0.40	0.0640		
0.45	0.0911		
0.50	0.1250		
0.55	0.1664		
0.60	0.2160		
0.65	0.2746		
0.70	0.3430		
0.75	0.4219		
0.80	0.5120		
0.85	0.6141		
0.90	0.7290		
0.95	0.8574		
1.00	1.0000		

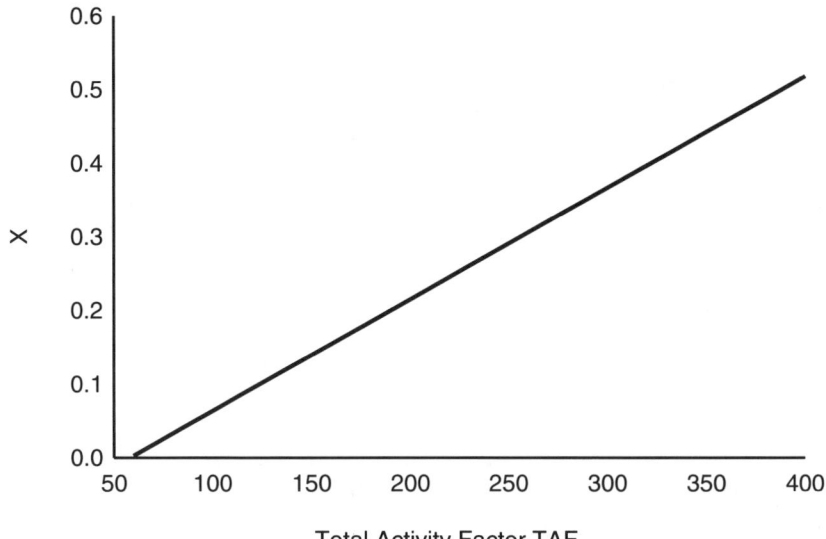

Figure 6.20 Power adjustment factor for constant-speed propellers.

SDEF

As was mentioned, a major propeller correction factor is needed because air just in front of the propeller comes into it considerably slower than the airplane's true air speed. The fuselage (and to some extent the spinner or propeller boss) get in the way and obstruct or deflect the flow. Though the details of this effect depend on the precise fuselage shape and air speed, a reasonably accurate figure can be obtained using only the ratio Z of fuselage diameter to propeller diameter. (The fuselage diameter is to be measured one propeller diameter from the plane of the propeller.) We have adapted graphs made by Diehl[9] from British and American experiments. The SDEF for the tractor propeller is (see Fig. 6.21):

$$\text{SDEF}_T = 1.05263 - 0.00722Z - 0.16462Z^2 - 0.18341Z^3 \qquad (6.58)$$

and, for the pusher propeller,

$$\text{SDEF}_P = 1.05263 - 0.04185Z - 0.01481Z^2 - 0.62001Z^3 \qquad (6.59)$$

The somewhat surprising (larger than unity) leading factors in Eqs. (6.58) and (6.59) are because raw data from which the general aviation general propeller chart was constructed came from a propeller mounted on a small nacelle, not one in free air. Tractor SDEFs from 0.85 to 0.95 are common. A good average seems to be $\text{SDEF}_T = 0.92$, an 8% reduction in efficiency due to airflow obstruction by the fuselage; this figure reduces the nominal maximum "free air" efficiency of 0.85 down to $0.85 \times 0.92 = 0.782$. Because pusher props have more fuselage in their way, their SDEFs are smaller than for tractors.

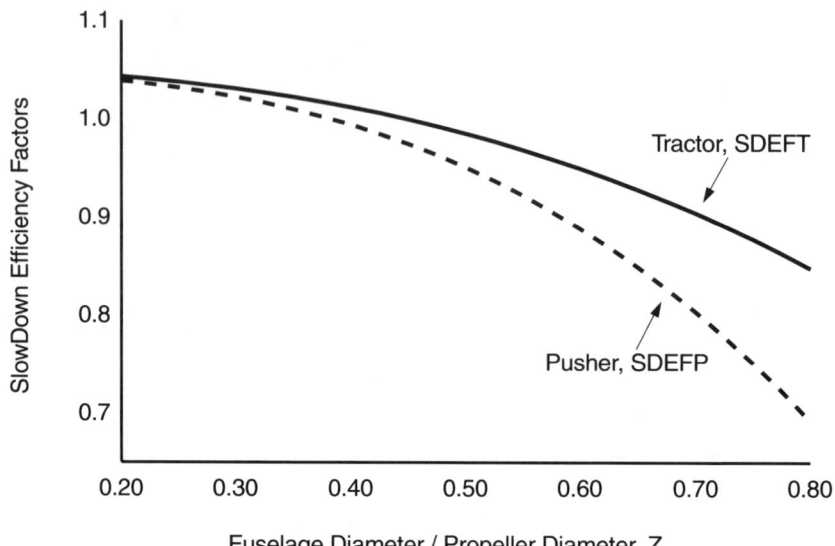

Figure 6.21 SDEF for constant-speed propellers.

The final constant-speed propulsive efficiency approximation is

$$\eta = \text{SDEF}(Z) \times \eta(J/C_P^{1/3}, C_{PX}) \qquad (6.60)$$

Conclusion

As promised, a considerable struggle. We see why even Buckingham[1] wanted to avoid "venturing further into the chaos of screw-propeller theory" (along with Orville Wright). Retreating to the caves of theory to construct finely tuned and well-corrected propeller charts for ordinary performance prediction is usually impractical. For instance, we avoided the question how one might design an *optimum* propeller for a particular propulsion job. The reader interested in that study should consult the papers of Adkins and Liebeck,[10] of Bauer,[11] and of Ribner and Foster.[12] On the other hand, understanding at least a modicum of the theory behind propeller action is a major key to understanding performance of propeller-driven aircraft. Conveniently assuming some constant propeller efficiency—80 or 85% is often cited—can never get that job done. With the linearized propeller polar diagram and the general aviation general propeller chart we have realistic approximations, a middle road that will serve us well.

References

1. Buckingham, E., "On Physically Similar Systems; Illustrations of the Use of Dimensional Equations," *Physical Review*, Vol. 4, No. 4, 1914, pp. 345–376.
2. Weick, F. E., *Aircraft Propeller Design*, McGraw-Hill, New York, 1930.
3. Norris, J., and Bauer, A. B., "Zero-Thrust Glide Testing for Drag and Propulsive Efficiency of Propeller Aircraft," *Journal of Aircraft*, Vol. 30, No. 4, 1993, pp. 505–511.
4. Glauert, H., *The Elements of Aerofoil and Airscrew Theory*, Macmillan, New York, 1943, p. 215.
5. Lowry, J. T., "Why an Engine Slows Down at Onset of Climb," *Journal of Aircraft*, Vol. 33, No. 5, 1996, pp. 1022–1023.
6. Von Mises, R., *The Theory of Flight*, Dover Publications, New York, 1959, pp. 308 ff.
7. Uddenberg, R., "A General Propeller Chart Suitable for a Wide Range of Propellers," Boeing Aircraft Co. Rep. D-4842, 1943.
8. Perkins, C. D., and Hage, R. E., *Airplane Performance Stability and Control*, Wiley, New York, 1949, pp. 149–150.
9. Diehl, W. S., *Engineering Aerodynamics*, Ronald Press, New York, 1936.
10. Adkins, C. N., and Liebeck, R. H., "Design of Optimum Propellers," *Journal of Propulsion and Power*, Vol. 10, No. 5, 1994, pp. 676–682.
11. Bauer, A. B., "A New Look at Optimum Propeller Performance—Going from the Prandtl F Factor to a Vortex-Induced Downwash Analysis," Paper 975560, World Aviation Congress, 1997.
12. Ribner, H. S., and Foster, S. P., "Ideal Efficiency of Propellers: Theodorsen Revisited," *Journal of Aircraft*, Vol. 27, No. 9, 1990, pp. 810–819.

PART II. PRACTICAL AIRPLANE PERFORMANCE

7

Introduction to the Bootstrap Approach

Introduction

The reader may be surprised upon being informed that *the bootstrap approach* (TBA) allows one to closely calculate how an airplane will perform, at any given weight and at any given altitude, through knowledge of only nine aircraft parameters that are not very hard to come by. Despite minor qualifiers to follow, that is a provocative claim! But pilots and flight test engineers who do understand and use TBA get a wealth of accurate quasi-steady (qualifier #1!) flight performance information with very little work.

This chapter introduces TBA by presenting the original full-throttle (or idle throttle, gliding) wings-level version for fixed-pitch, propeller-driven aircraft. Many concrete examples will be given. Later chapters will take up extensions of TBA to maneuvering (turning) flight, partial-throttle operations, and takeoff and landing. Constant-speed propeller airplanes also need nine bootstrap parameters, but numbers for the constant-speed propeller differ slightly from those in the fixed-pitch version. Because there the general aviation general propeller chart must be used, a computer is needed for easy constant-speed aircraft performance calculations.

At this point a TBA initiate is likely to have several questions.

Q1: What are those nine numbers?

A1: With sample values for a Cessna 172, here they are (see Table 7.1).

Q2: For a given airplane, how does one come up with those nine numbers?

A2: Four of them—S, A, M_0, and d—come straight from the airplane's pilots operating handbook or from direct measurement. C comes from analyzing the right-hand side (RHS) of the performance chart in the engine manual[1] (for details see Lowry[2]) or can almost always be simply taken for modern reciprocating aircraft engines as $C = 0.12$. The remaining four are the "hard cases," requiring

Table 7.1 Cessna 172 bootstrap data plate with items attached to appropriate aircraft subsystems

BDP item	Value	Units	Aircraft
Wing reference area, S	174	ft^2	Airframe
Wing aspect ratio, A	7.38		Airframe
Rated MSL[a] torque, M_0	311.2	ft-lbf	Engine
Altitude dropoff parameter, C	0.12		Engine
Propeller diameter, d	6.25	ft	Propeller
Parasite drag coefficient, C_{D0}	0.037		Airframe
Airplane efficiency factor, e	0.72		Airframe
Propeller polar slope, m	1.70		Propeller
Propeller polar intercept, b	−0.0564		Propeller

[a] MSL = mean sea level.

simple climb, glide, and level run performance flight tests occupying a total of about 1 h. A common bootstrap buzz phrase is "Fly the airplane to find out how the airplane flies...."

That does *not* mean to forget theory, install computerized recording instruments, and accumulate flight performance data for a large collection of weights, speeds, and altitudes. Instead, here is what it means and all one has to do.

To get the drag polar parameters C_{D0} and e, one glides repeatedly (usually about half a dozen glides is enough) over the same vertical interval (an indicated 500 ft or so is sufficient) until he or she has maximized the product of air speed and elapsed time. Because that product $V \Delta t$ is the length of the hypotenuse of the flight space triangle and the vertical interval is fixed, maximizing the hypotenuse minimizes the sine of the glide path angle. Hence, it minimizes the glide path angle itself. Figures from that optimum glide—after being processed to find the true geometric vertical interval and corrected with the airplane's air-speed indicator calibration curve (Chapter 2)—will give values of speed for best glide V_{bg} and corresponding best glide angle γ_{bg}. Then formulas in Appendix F, Flight Test for Drag Parameters, let one calculate C_{D0} and e. The glide tests must be flown and the drag polar parameters calculated for each individual flaps/gear configuration of the airplane—or for at least each configuration of performance interest.

The intercept of the linearized propeller polar diagram, b, is obtained from climb tests analogous to the glide tests. One minimizes the product of air speed and elapsed time to find speed for largest climb angle V_x. Having done so, the propeller polar intercept is obtained as

$$b = \frac{SC_{D0}}{2d^2} - \frac{2W^2}{\rho^2 d^2 S \pi e A V_x^4} \qquad (7.1)$$

Finally, to get the slope of the propeller polar, m, one has a choice. One method is to simply make a level full-speed run to find V_M. Then the propeller polar slope is given as

$$m = \frac{2n_0 dW^2}{\Phi(\sigma) P_0 \rho S \pi e A} \left(\frac{1}{V_M^2} + \frac{V_M^2}{V_x^4} \right) \tag{7.2}$$

Alternatively, one can choose to perform repeated climbs to locate speed for best rate of climb V_y. That of course is the speed that minimizes the elapsed time required to cover the vertical interval. Then one gets the propeller polar's slope from

$$m = \frac{2n_0 dW^2}{\Phi(\sigma) P_0 \rho S \pi e A} \left(\frac{3V_y^2}{V_x^4} + \frac{1}{V_y^2} \right) \tag{7.3}$$

When data from the performance flight tests has been massaged, the bootstrap data plate fully filled in, phase I of TBA (see Fig. 7.1) is over. For a given airplane, it is a one-time job. Practical details bearing on these procedures, including on the necessary tools (calibrated air-speed indicator and altimeter, and a stop watch), appear in Lowry[2] and in Appendix B, Bootstrap Approach Inputs and Outputs.

Q3: Precisely what items of information are included in "performance" of the airplane?

A3: Restricting ourselves for the moment to full-throttle or gliding performance with wings level (limitations later to be removed), having knowledge of the airplane's performance means knowing the following V speeds for it at any weight and at any altitude: maximum level V_M, minimum level V_m (seldom achievable, not stall speed V_S, and to be later discussed), speed for best angle of climb V_x, speed for

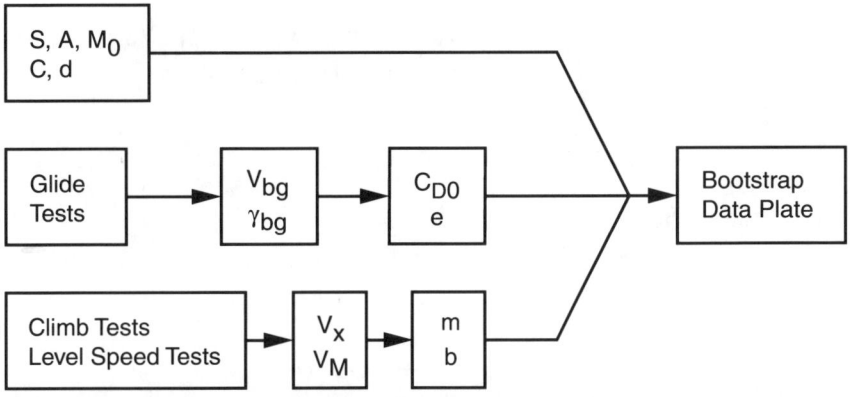

Figure 7.1 The bootstrap approach phase I.

best climb rate V_y, speed for best glide angle V_{bg}, and speed for minimum rate of gliding descent V_{md}. Also knowledge of thrust and drag (parasite and induced) and of rates and angles of climb or descent for the airplane at any weight, altitude, and achievable steady speed. Perhaps a few more details, such as absolute altitude and the speed required there, but the above covers almost all of the ground.

Q4: How do we calculate those performance items?

A4: From simple bootstrap formulas. For an overview, see Fig. 7.2.

Q5: And where do *those* come from?

A5: Bootstrap formulas are derived from the combination of ordinary power-available/power-required analysis with the linearized propeller polar. Deriving them will be our job over the next two sections.

Once basic bootstrap formulas are under our belts, we can take derivatives to see how sensitive the more important performance items (e.g., V_y and perhaps its corresponding maximum rate of climb) are to variations in aircraft gross weight or density altitude. Then we consider what effect winds—headwinds, tailwinds, updrafts, downdrafts—have on the ground-reference V speeds (V_x and V_{bg}) and on their optimal angles of climb and glide. A couple of performance odds and ends, absolute ceilings, and sea level rated propeller efficiency, close out the chapter.

Power-Available/Power-Required (P_{av}/P_{re}) Analysis

Graphical Picture

Power available is thrust times true air speed, $P_{av} \equiv TV$, the useful power output of the engine-propeller subsystem. Power required is drag times true air speed,

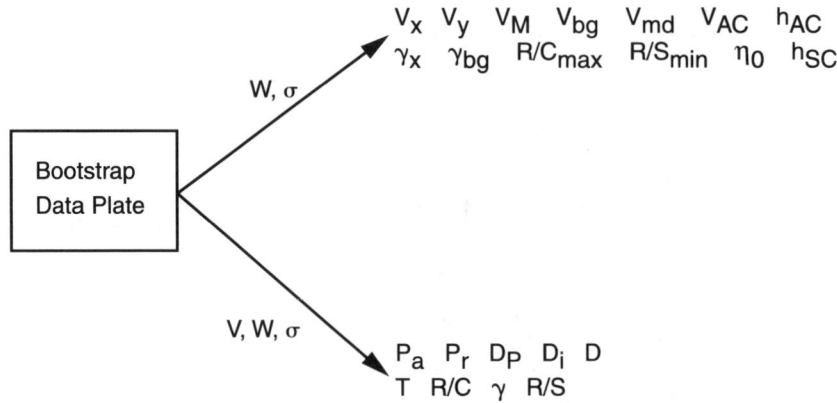

Figure 7.2 The bootstrap approach phase II.

$P_{re} \equiv DV$, the power needed to overcome the total drag force. A nice feature of P_{av}/P_{re} analysis—and the reason for its ubiquity in ground school classes and aviation textbooks—is that graphs of those two quantities, plotted as functions of air speed, show many of the performance numbers a pilot cares about. Figure 7.3 is such a graph for a Cessna 172 at $W = 2400$ lbf and $h_\rho = 7500$ ft. We have supplemented with a third graph, of "excess power" $P_{xs} \equiv P_{av} - P_{re}$, which is needed to demonstrate V_x, speed for best angle of climb.

1) The right-hand intersection of the P_{av} and P_{re} graphs (alternatively, the rightmost zero of the P_{xs} graph) shows where $V = V_M$, maximum level flight speed, under these conditions.
2) The left-hand intersection of P_{av} and P_{re} (or the leftmost zero of P_{xs}) shows where $V = V_m$, minimum level flight speed, would occur if it were not for the fact that (except at quite high altitude) the power-on stall speed is somewhat higher. Because V_m seldom occurs in practice, we only seldom mention it.
3) The maximum value of P_{xs} (alternatively, where P_{av} and P_{re} are farthest apart, or where their slopes are equal, $P'_{av} = P'_{re}$) occurs where $V = V_y$, speed for greatest rate of climb. This V speed is completely independent of headwind or tailwind.
4) The minimum value of P_{re} takes place where $V = V_{md}$, speed for minimum (gliding) descent rate. V_{md} is independent of headwind or tailwind and is always 76% of the speed for best (longest) glide in a calm.

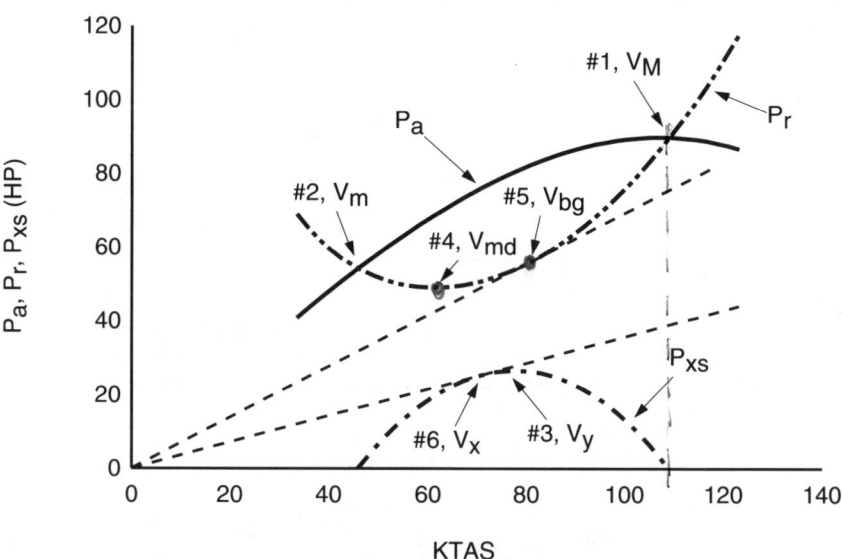

Figure 7.3 Power available, required, and excess. Numerals are keyed to text comments.

5) The minimum value of P_{re}/V (in other words, the minimum value of drag D) takes place where $V = V_{bg}$, speed for best (longest) glide in a no-wind situation.
6) The maximum value of P_{xs}/V (the maximum value of excess thrust $T_{xs} \equiv T - D$) occurs where $V = V_x$, speed for steepest angle of climb (in a no-wind situation).

Table 7.2 recaps these relations by focusing on excess power P_{xs} and power required P_{re}. The next question is "How does one go about getting (for a given airplane at given weight and density altitude) these power-available and power-required curves?"

Traditional Approach to P_{av} and P_{re}

Because power required is the product of drag and true air speed, we need drag as a function of air speed, $D(V)$. In older times "drag polar" meant a plot of total drag coefficient C_D as a function of airplane lift coefficient C_L; in recent times the drag polar appellation has commonly been broadened to mean the formula for total drag force:

$$D = \tfrac{1}{2}\rho V^2 S C_D = \tfrac{1}{2}\rho V^2 S \left(C_{D0} + \frac{C_L^2}{\pi e A} \right) \quad (7.4)$$

It is a fundamental assumption of P_{av}/P_{re} analysis that lift L closely approximates weight W, *even while climbing or descending*. This approximation is good as long as one restricts climb or descent angles to fewer than about 15 deg. Doing so, we have

$$C_L \doteq \frac{2W}{\rho V^2 S} \quad (7.5)$$

and so

$$P_{re} \equiv DV \doteq GV^3 + H/V \quad (7.6)$$

Table 7.2 Steady flight performance V speeds, Cessna 172, 2400#, 7500 ft

V speed	Speed description	Condition	Fig. 7.3 value, KTAS
V_M	Maximum level	$P_{xs} = 0$	109.0
V_m	Minimum level	$P_{xs} = 0$	45.9
V_y	Best rate of climb	$P'_{xs} = 0$	77.7
V_x	Best angle of climb	$(P_{xs}/V)' = T'_{xs} = 0$	70.7
V_{bg}	Best (longest) glide	$(P_{re}/V)' = D' = 0$	80.6
V_{md}	Minimum descent rate	$P'_{re} = 0$	$61.2 = 0.76 \times V_{bg}$

where

$$G \equiv \tfrac{1}{2}\rho S C_{D0} \qquad (7.7)$$

and

$$H \equiv \frac{2W^2}{\rho S \pi e A} \qquad (7.8)$$

We have always known we would need operational variables W and ρ as well as independent variable V, but Eqs. (7.7) and (7.8) tell us that, to graph $P_{re}(V)$, we also need reference wing area S, parasite drag coefficient C_{D0}, airplane efficiency factor e, and wing aspect ratio A. Only the two drag parameters C_{D0} and e pose any problem. A traditional method for getting them has been to do repeated glides at a fairly broad range of air speeds, recording the rate of sink $h^\cdot$ (the time derivative of altitude), and to use the fact that, when gliding ($h^\cdot < 0$),

$$P_{re} = -W h^\cdot \qquad (7.9)$$

There are two ways to see Eq. (7.9). From the power point of view, P_{xs} is what is left over from fighting drag, heating the atmosphere. The only other force to oppose is that of gravity. So the excess power goes into enhancing gravitational potential, mgh, at rate $mgh^\cdot = Wh^\cdot$. When gliding, $P_{av} = 0$ and so $P_{xs} = P_{re}$. Fine. For the force point of view, look back at the force equations for steady flight in Chapter 3; one such equation [Eq. (3.20)] is $T\cos\alpha - D - W\sin\gamma = 0$. In P_{av}/P_{re} analysis, we ignore the slight off-flight-path thrust alignment and get $T_{xs} \equiv T - D = W\sin\gamma$. In this gliding case, moreover, thrust $T = 0$. Multiplying the simplified expression for T_{xs} by air speed V and recognizing that $V\sin\gamma = h^\cdot$ (see Fig. 7.4), we again arrive at Eq. (7.9).

Equations (7.6) and (7.9) then give

$$-W h^\cdot V = G V^4 + H \qquad (7.10)$$

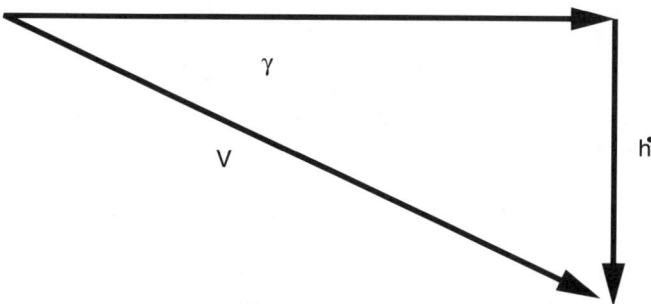

Figure 7.4 Gliding case of $\sin = h^\cdot/V$. Only V is positive.

The next step in the traditional approach to getting the drag polar is to plot the experimentally known left-hand side of Eq. (7.10) against V^4 and determine, by eye or by linear regression, slope G and intercept H. Finally, solve Eqs. (7.7) and (7.8) for C_{D0} and e. So that is where the traditional graph of power required comes from. The bootstrap procedure is quicker because it needs to repeat glides only long enough to find speed and angle for best glide.

The traditional method for finding power available is a different story, requiring much more work and information than does TBA. We can follow the (slightly simplified) procedure of von Mises.[3] He has us start by selecting some arbitrary engine speed n and using that value to enter a curve of full-throttle brake power $P(n)$, thereby finding P. Next, we use the definition of propeller power coefficient C_P [Eq. (6.8)] to find that quantity, then look up advance ratio J on the curve $C_P(J)$ appropriate to the installed propeller. From J, get true air speed $V = ndJ$. Also from J, using the appropriate efficiency curve $\eta(J)$, find η. Finally, calculate P_{av} as ηP and plot one point (V, P_{av}). Repeat for other engine speeds n to get more points.

In other words, the traditional approach to P_{av} requires, one way or another, knowledge of the full (two) propeller charts. Those charts are just what are hard to come by and, even if given, need further correction for the particular cowling and fuselage installation. A murky picture. But the goals of all this, the graphs of $P_{av}(V)$ and of $P_{re}(V)$, are invaluable.

Bootstrap Approach

Basic Derivation

Here is how to get thrust, hence power available, using the linearized propeller polar instead of the full propeller charts. Start with the definitions of power available ($P_{av} \equiv TV$) and propeller thrust coefficient [Eq. (6.7)] to find

$$P_{av} = \rho n^2 d^4 C_T V \tag{7.11}$$

Next, use the linearized propeller polar [Eq. (6.53)] to write

$$P_{av} = \rho n^2 d^4 V(m C_P + b J^2) \tag{7.12}$$

Then use the definition of the propeller power coefficient C_P [Eq. (6.8)] and of $J \equiv V/nd$ to find

$$P_{av} = \frac{mP}{nd} V + \rho b d^2 V^3 \tag{7.13}$$

At first blush this does not help much because we do not know engine speed n or, for that matter, brake power P. But we *do know* the ratio P/n; it is just $2\pi M$ where M is the engine torque. And from what we learned in Chapter 5 about internal combustion reciprocating engines, full-throttle torque M at altitude is just $\Phi(\sigma)M_0$, the power dropoff factor Φ (see Fig. 7.5) times the rated mean sea level (MSL) torque, both of which we can assume are known or knowable. The approved flight manual ordinarily cites, instead of rated torque, MSL-rated power P_0 and rated MSL "red-line" revolutions per minute, $N_0 = 60n_0$. Using these more common specifications, we have our final bootstrap formula for power available:

$$P_{av} \equiv TV = EV + FV^3 \tag{7.14}$$

where

$$E \equiv \frac{m\Phi(\sigma)P_0}{n_0 d} \tag{7.15}$$

and

$$F \equiv \rho d^2 b \tag{7.16}$$

Equations (7.6–7.8) and Eqs. (7.14–7.16) practically carry the entire weight of the bootstrap idea. Our first result will be to simply get an abbreviated form for excess power using the definition of new composite parameter $K \equiv F - G$:

$$P_{xs} \equiv P_{av} - P_{re} = EV + KV^3 - H/V \tag{7.17}$$

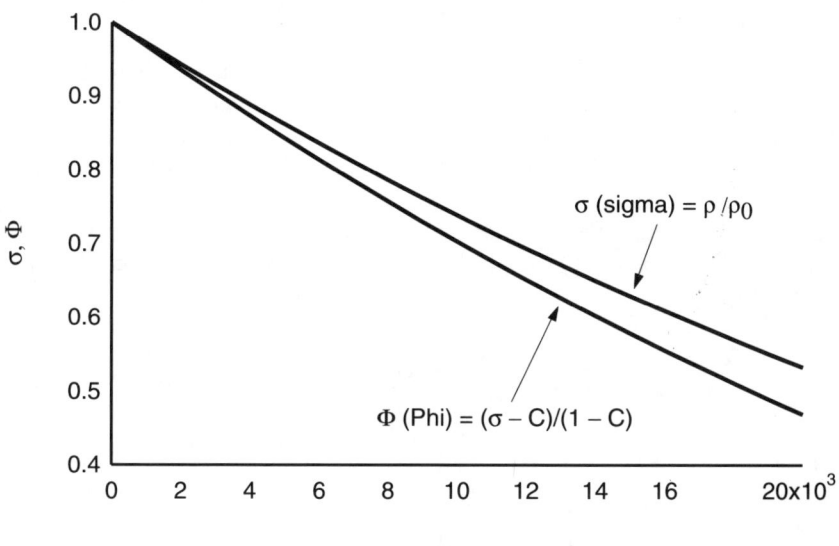

Figure 7.5 Relative density σ and engine altitude dropoff factor Φ for full-throttle torque and power.

The next two subsections use algebra and elementary differential calculus on Eqs. (7.17) and (7.6) to get useful bootstrap performance number results.

Formulas for V Speeds

Using Table 7.2 as a guide to mathematical manipulations, we start with V_M. Set the RHS of Eq. (7.17) equal to zero and multiply through by air speed V. You then have a "biquadratic equation," a quadratic equation in V^2. Solve this with the usual quadratic formula and, using the fact that $K < 0$ forces

$$K = -\sqrt{K^2} \tag{7.18}$$

use the root that gives you

$$V_M = \sqrt{\frac{-E - \sqrt{E^2 + 4KH}}{2K}} = \sqrt{-\frac{Q}{2} + \sqrt{\frac{Q^2}{4} + R}} \tag{7.19}$$

where new composite parameters are defined as $Q \equiv E/K$ and $R \equiv H/K$. The explicit version of the formula for maximum level flight speed is

$$V_M = \sqrt{\frac{-\dfrac{m\Phi P_0}{n_0 d} - \sqrt{\left(\dfrac{m\Phi P_0}{n_0 d}\right)^2 + \dfrac{8W^2(d^2b - SC_{D0}/2)}{S\pi eA}}}{2\rho(d^2b - SC_{D0}/2)}} \tag{7.20}$$

Full-bore Eq. (7.20) will be useful to you in case someone asks how maximum level flight speed depends (everything else being equal) on any particular bootstrap parameter or operating variable (which latter we shall soon consider). Answer—not very simply! It is a relation encountered during development of Eq. (7.20) that allows us to backengineer m via Eq. (7.2).

If you need the formula for *minimum* speed for level flight, that can be obtained from Eqs. (7.19) or (7.20) simply by changing the sign just before the *inner* radical to get

$$V_m = \sqrt{\frac{-E + \sqrt{E^2 + 4KH}}{2K}} = \sqrt{-\frac{Q}{2} - \sqrt{\frac{Q^2}{4} + R}} \tag{7.21}$$

and

$$V_m = \sqrt{\frac{-\dfrac{m\Phi P_0}{n_0 d} + \sqrt{\left(\dfrac{m\Phi P_0}{n_0 d}\right)^2 + \dfrac{8W^2(d^2b - SC_{D0}/2)}{S\pi eA}}}{2\rho(d^2b - SC_{D0}/2)}} \tag{7.22}$$

Next, let us consider V_y, speed for best rate of climb. Table 7.2 says we must take the derivative of Eq. (7.17) with respect to V and set that expression equal to zero:

$$P'_{xs} = E + 3KV^2 + H/V^2 = 0 \tag{7.23}$$

When you multiply through by V^2, you again get a biquadratic equation. Solve it and find

$$V_y = \sqrt{\frac{-E - \sqrt{E^2 - 12KH}}{6K}} = \sqrt{-\frac{Q}{6} + \sqrt{\frac{Q^2}{36} - \frac{R}{3}}} \tag{7.24}$$

An explicit form is

$$V_y = \sqrt{-\frac{m\Phi P_0}{n_0 d} - \sqrt{\left(\frac{m\Phi P_0}{n_0 d}\right)^2 + \frac{24W^2(d^2 b - SC_{D0}/2)}{S\pi eA}}} \Big/ 6\rho(d^2 b - SC_{D0}/2) \tag{7.25}$$

Move on to V_x, speed for best angle of climb. Divide Eq. (7.17) by V, thereby getting an expression for excess thrust T_{xs}, then take the derivative and set equal to zero. Multiply through by $V^3/2$ to get

$$KV^4 + H = 0 \tag{7.26}$$

Solve for V to find the very simple result

$$V_x = \left(\frac{-H}{K}\right)^{1/4} = (-R)^{1/4} \tag{7.27}$$

or, explicitly,

$$V_x = \sqrt{\frac{W}{\rho}} \left(\frac{2}{S\pi eA(SC_{D0}/2 - d^2 b)}\right)^{1/4} \tag{7.28}$$

It is the inversion of this formula that allows us to backengineer b via Eq. (7.1).

These expressions are all for the V speeds in *true* terms, but Eq. (7.28) shows us something interesting about V_x in *calibrated* terms: calibrated air speed $V_{Cx} = \sigma^{1/2} V_x$ does not vary with density altitude. Figure 7.6 gives an example of how all four full-power V speeds depend, in calibrated terms, on density altitude. It also gives stall speed V_S, which preempts minimum level flight speed V_m at low and moderate altitudes.

Composite parameters Q and R were introduced partly to simplify future practical calculations and partly for another reason. Notice that the three full-throttle V speeds—V_M, V_y, and V_x—only depend on the two composite parameters Q and R. This means that there must be some algebraic relation among the three, that they are not all independent. Indeed, after some algebra—one method

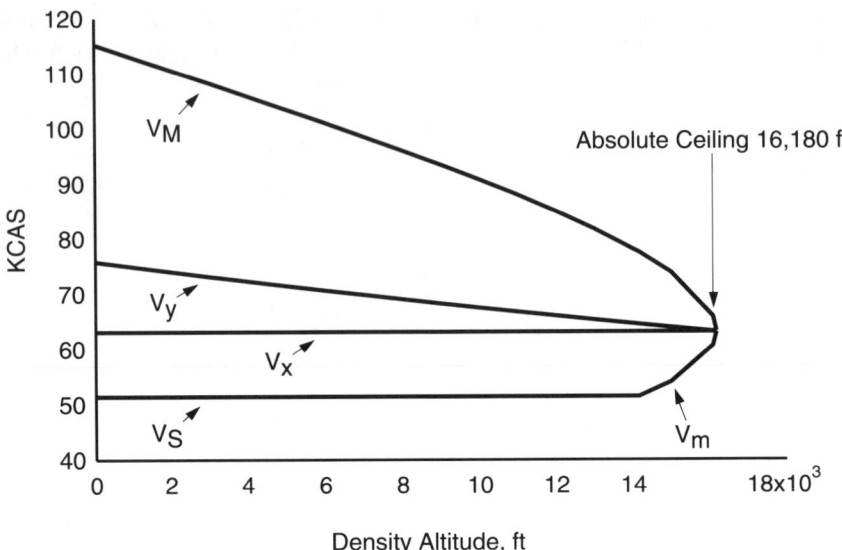

Figure 7.6 How full-throttle V speeds vary with density altitude. Calibrated V_x is constant.

is to look on the three biquadratic Eqs. (7.17), (7.23), and (7.26), from which we obtained the three V speeds, as a system of homogeneous linear equations in the three unknowns K, E, and H and to set the determinant of that system equal to zero—one finds

$$V_x = \left(\frac{3V_y^2 - V_M^2}{V_M^{-2} + V_y^{-2}}\right)^{1/4} \tag{7.29}$$

Another method for getting Eq. (7.29) is to study Eqs. (7.2) and (7.3) for a minute, then solve the obvious relation for V_x. One can solve Eq. (7.29) for either V_M or V_y, but one gets considerably more complicated expressions for those two V speeds. Equation (7.29) is useful as a quick check on the validity of manufacturers' advertised performance figures.

If one makes the same statement as above about full-throttle V speeds V_M, V_m, and V_x—using V_m in lieu of V_y—a considerably simpler relation results:

$$V_x = \sqrt{V_M V_m} \tag{7.30}$$

Speed for best angle of climb, in a calm, is the geometric mean between the minimum and maximum speeds for level flight.

Continuing through the conditions column of Table 7.2, one can similarly find composite and explicit formulas for the two gliding V speeds V_{bg} and V_{md}. But there is a simpler way. When power is off—when $E = F = 0$ (to be discussed)—

V_{md} is simply the speed for best rate of (unpowered) climb and V_{bg} is simply the speed for best angle of (unpowered) climb. With power off, further composite $Q \equiv E/K = 0$ (because $E = 0$) and further composite $R \equiv H/K = -H/G$ (because $F = 0$ and $K \equiv F - G$). Defining a final composite $U \equiv H/G$, one then straightaway has

$$V_{bg} = \left(\frac{H}{G}\right)^{1/4} = U^{1/4} \tag{7.31}$$

explicitly

$$V_{bg} = \sqrt{\frac{2W}{\rho S}} \left(\frac{1}{\pi e A C_{D0}}\right)^{1/4} \tag{7.32}$$

and

$$V_{md} = \left(\frac{H}{3G}\right)^{1/4} = \left(\frac{U}{3}\right)^{1/4} \tag{7.33}$$

or

$$V_{md} = \left(\frac{1}{3}\right)^{1/4} V_{bg} = 0.7598 V_{bg} \tag{7.34}$$

Let us return to the question of $F = 0$ when the engine is off. As defined [see Eq. (7.16)] F does certainly *not* go to zero when you turn the engine off or down. But rather than turning off the engine and pulling up to stop the propeller, it is more convenient to lump drag of the windmilling propeller in with parasite drag. So we pretend $F = 0$ under those low/no power situations. Norris and Bauer[4] have a better, though more elaborate, zero-thrust procedure for getting a consistent drag polar (see Fig. 7.7). This concludes derivations of bootstrap formulas for various V speeds.

Formulas for Thrust, Drag, Rate, and Angle of Climb

Thrust is easy; just divide Eq. (7.14) by V to get

$$T(V) = E + FV^2 \tag{7.35}$$

Notice that propeller efficiency is not mentioned. The linearized propeller polar, with the known tie between (full) throttle and torque, makes it unnecessary to know propeller efficiency. The explicit form is

$$T = \frac{m\Phi(\sigma)P_0}{n_0 d} + \rho d^2 b V^2 \tag{7.36}$$

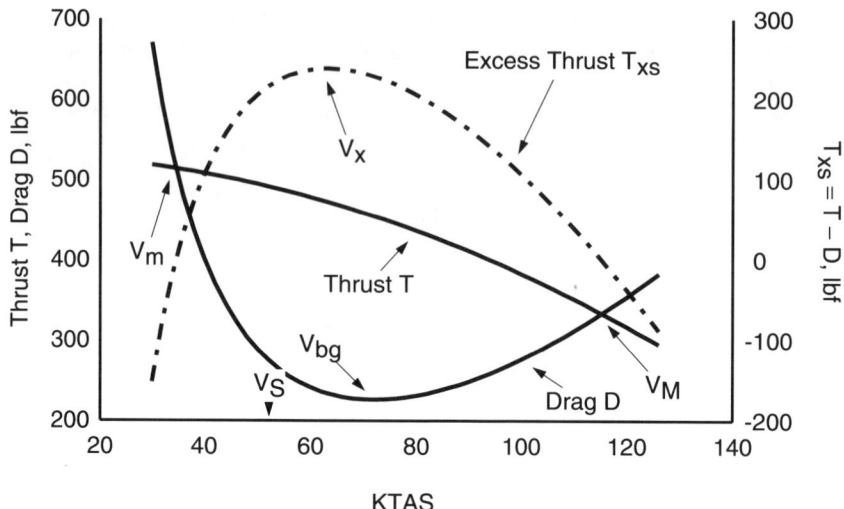

Figure 7.7 Typical bootstrap thrust and drag curves.

Similarly with drag, building off Eq. (7.6) but being more explicit about the division into parasite drag D_P and induced drag (drag due to lift) D_i (see Fig. 7.8):

$$D(V) = D_P(V) + D_i(V) = GV^2 + \frac{H}{V^2} \tag{7.37}$$

The explicit form is

$$D = \frac{1}{2}\rho S C_{D0} V^2 + \frac{2W^2}{\rho S \pi e A} \frac{1}{V^2} \tag{7.38}$$

Rate of climb, as we have discussed, is given by (see Fig. 7.9):

$$h'(V) = \frac{P_{xs}(V)}{W} = \frac{EV + KV^3 - H/V}{W} \tag{7.39}$$

The explicit form is

$$h'(V) = \frac{m\Phi(\sigma)P_0 V}{n_0 dW} + \frac{\rho(d^2 b - SC_{D0}/2)V^3}{W} - \frac{2W}{\rho S \pi e A V} \tag{7.40}$$

One can see from Eqs. (7.25) and (7.40) that there is no possibility of some such a simple rule as "maximum rate of climb varies as weight to the power x." V_y certainly depends on gross weight W, but not according to any power law. And, in the expression for rate of climb, the first two terms on the right are inversely proportional to gross weight W whereas the third term is directly proportional to W. The end result is that, while maximum rate of climb certainly

Introduction to the Bootstrap Approach   201

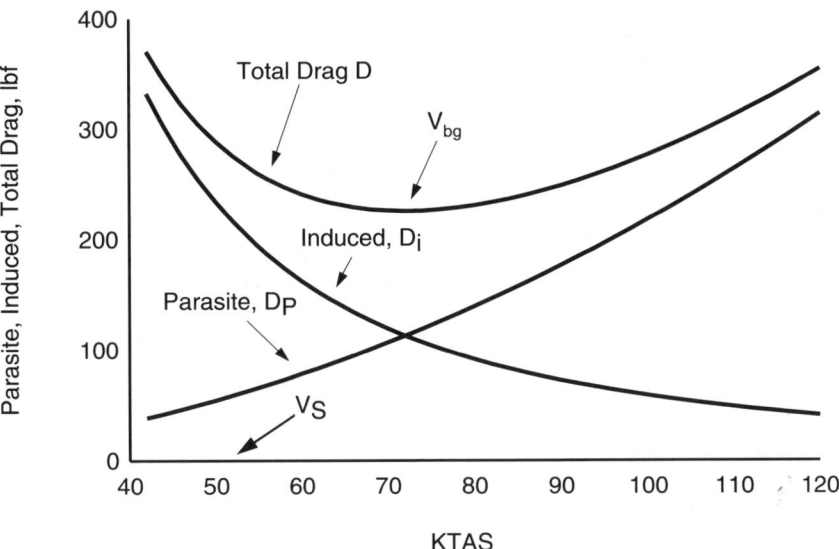

Figure 7.8   Parasite, induced, and total drag curves.

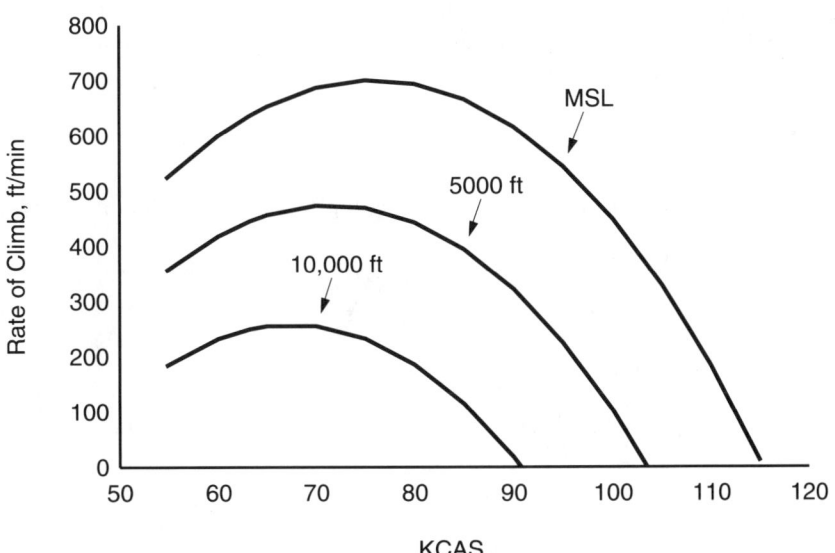

Figure 7.9   Rate of climb plotted against calibrated air speed.

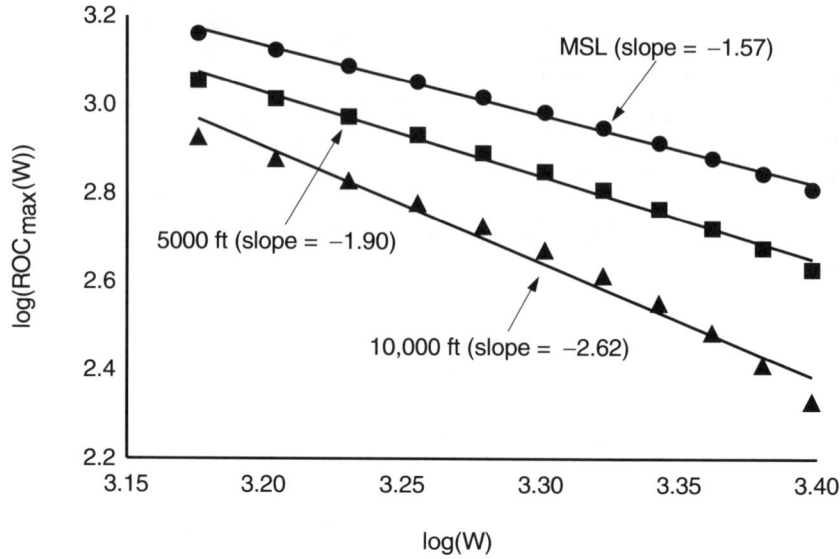

Figure 7.10 Logarithm of maximum rate of climb vs gross weight, three density altitudes.

does depend on gross weight, falling off quite alarmingly with increased weight, it does not depend on weight to any particular power (see Fig. 7.10).

The quick path to angle of climb is to draw a velocity vector diagram (like Fig. 7.4 flipped around the horizontal) and use the fact that $h^\cdot$ is the vertical component of velocity (and that the sine is the ratio of side opposite to the hypotenuse) to get

$$\gamma = \sin^{-1}\left(\frac{h^\cdot}{V}\right) \tag{7.41}$$

Using Eqs. (7.39) and (7.41), the general angle of (full-throttle) climb is (see Fig. 7.11)

$$\gamma(V) = \sin^{-1}\left[\frac{E + KV^2 - H/V^2}{W}\right] \tag{7.42}$$

An explicit version is

$$\gamma(V) = \sin^{-1}\left[\frac{m\Phi(\sigma)P_0}{n_0 dW} - \frac{\rho(SC_{D0}/2 - d^2 b)V^2}{W} - \frac{2W}{\rho S\pi e A V^2}\right] \tag{7.43}$$

Occasionally, you might want to know the size of the *best* angle of climb:

$$\gamma_x = \sin^{-1}\left(\frac{E - 2\sqrt{-KH}}{W}\right) \tag{7.44}$$

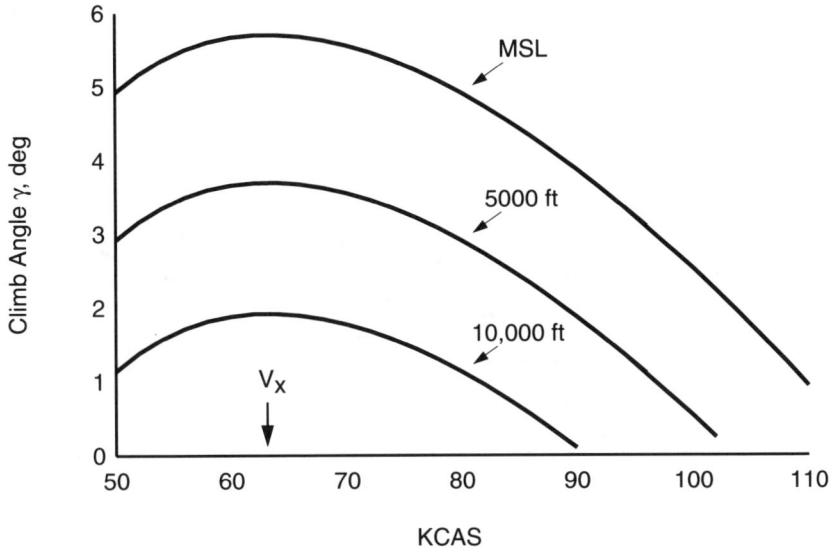

Figure 7.11 Like most general aviation optimal performance parameters, climb angle has a broad maximum.

The explicit version is

$$\gamma_x = \sin^{-1}\left(\frac{\Phi(\sigma)mP_0}{n_0 dW} - 2\sqrt{\frac{(SC_{D0} - 2d^2b)}{S\pi eA}}\right) \quad (7.45)$$

Sometimes it is useful to have specialized gliding versions of some of these formulas. For example, the gliding sink rate is

$$\dot{h}_{gl}(V) = \frac{-GV^3 - H/V}{W} \quad (7.46)$$

The explicit version is

$$\dot{h}_{gl}(V) = -\frac{\rho S C_{D0} V^3}{2W} - \frac{2W}{\rho S \pi e A V} \quad (7.47)$$

Then there is the special case of the *minimum* descent rate (see Fig. 7.12); substituting Eq. (7.33) for V_{md} into Eq. (7.47), one soon gets

$$\dot{h}_{md} = \frac{-4}{3^{3/4}}\sqrt{\frac{W}{\sigma}}(G_0 H_0^3)^{1/4} \quad (7.48)$$

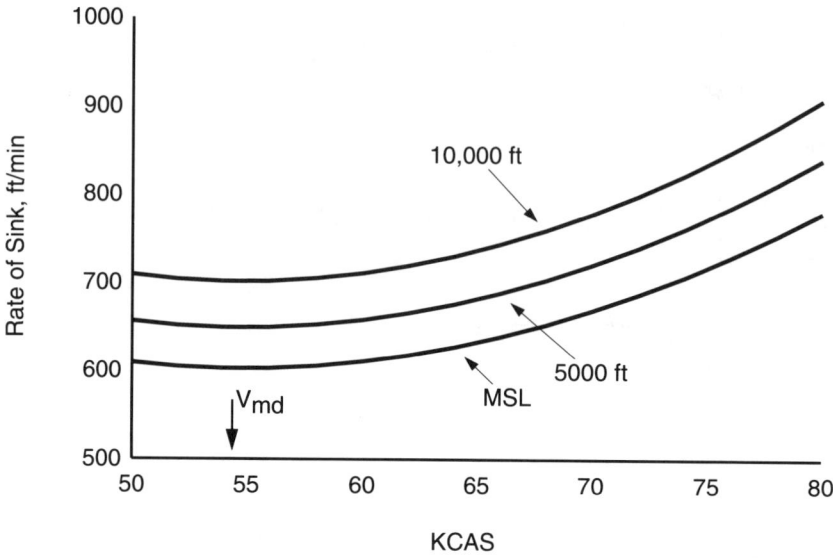

Figure 7.12 Minimum rate of gliding descent is at quite a slow speed.

Where angle of gliding descent is concerned, we have

$$\gamma_{gl}(V) = -\sin^{-1}\left[\frac{GV^2 + H/V^2}{W}\right] \tag{7.49}$$

The explicit form is

$$\gamma_{gl}(V) = -\sin^{-1}\left[\frac{\rho S C_{D0} V^2}{2W} + \frac{2W}{\rho S \pi e A V^2}\right] \tag{7.50}$$

As in the V_x case, the underlying simplicity of V_{bg} (see Fig. 7.13) allows us to get a short version:

$$\gamma_{bg} = -\sin^{-1}\left(\frac{2\sqrt{GH}}{W}\right) \tag{7.51}$$

Explicitly,

$$\gamma_{bg} = -\sin^{-1}\left(2\sqrt{\frac{C_{D0}}{\pi e A}}\right) \tag{7.52}$$

You might notice that the version of Eq. (7.52) used in Appendix F has an inverse tangent in place of this inverse sine. That is because our P_{av}/P_{re} analysis takes weight W equal to lift L while the truth is that $L = W\cos\gamma$; for angles as large as 10 deg there is only a 1.5% relative difference between sine and tangent values.

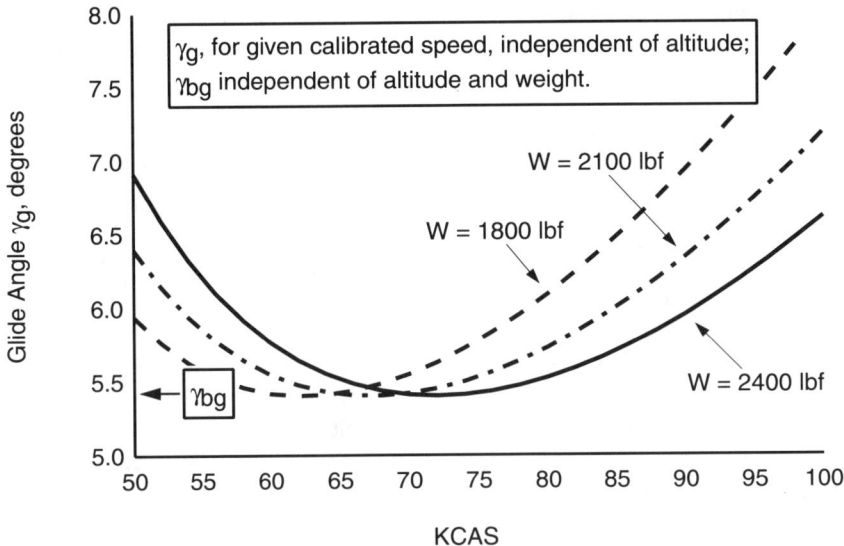

Figure 7.13 You cannot stretch a glide by going slower than V_{bg} (here, about 72 KCAS).

This finishes the main presentation and derivation of bootstrap formulas. And quite a few it was! You may be thinking "the bootstrap approach is only a bunch of formulas." Not *only* formulas, but it is a bunch of them. On the other hand, what else could one expect of a good, clear, quantitative theory? We certainly do not want performance estimates coming from hand waving or rough rules of thumb. The "just add 10% of the..." type rule might be fine for the airplane for which it was initially designed, but when applied to another craft it may be quite far off. TBA formulas are clear and operational and quite accurate. In a quantitative theory, formulas come with the territory.

Next we collect, in one central location, definitions of all of the bootstrap composite parameters with their explicit dependencies on gross weight W and relative air density σ. Using bootstrap composite parameters markedly eases future calculations.

Compendium of Bootstrap Composite Parameters

Almost everything about the airplane's full-throttle, steady-state (nonaccelerating) flight performance depends on the nine BDP items plus two variables weight W and relative atmospheric density σ. But only certain combinations of those nine numbers (called E, F, G, and H) actually occur in TBA formulas for V speeds or for thrust, drag, rate of climb or descent, and angle of climb or descent. In the V speed formulas only certain combinations (called Q, R, U) of those combinations occur.

We remind the reader that if he or she does flight tests, to determine the four harder-to-get BDP parameters, say at 5000 ft and $W = 2200$ lbf, resulting parameters should not depend on those operational choices; BDP parameters only depend on the particular airplane and, for the two drag parameters, on its flaps/gear configuration. Running the flight tests at some other altitude and/or some other weight we would have gotten, within experimental error, the same BDP values.

But performance numbers—rates of climb, V speeds, etc.—obviously do depend on gross weight and density altitude. The easiest way to get performance numbers for a particular case is to employ simple correction factors to the base composite bootstrap parameters and use those corrected figures in the composite version of the appropriate formulas. For our sample Cessna 172, let us take "base" relative air density to be (very reasonably) the MSL value $\sigma_0 = 1.00$. For the base value of the airplane's gross weight, we select maximum gross weight, $W_0 = 2400$ lbf for the sample Cessna 172P. Any relative density and gross weight could have been selected for base values, but these are reasonable and easy to remember.

Now let us review the bootstrap composite parameters and consider their former definitions as being modifications of base bootstrap composites (those with subscript zero). The new composite definitions and their dependence on weight and air density are

$$E = \Phi(\sigma)E_0, \quad \text{with} \quad E_0 = \frac{mP_0}{n_0 d} \tag{7.53}$$

$$F = \sigma F_0, \quad \text{with} \quad F_0 = \rho_0 d^2 b \tag{7.54}$$

$$G = \sigma G_0, \quad \text{with} \quad G_0 = \tfrac{1}{2}\rho_0 S C_{D0} \tag{7.55}$$

$$H = \left(\frac{W}{W_0}\right)^2 \frac{1}{\sigma} H_0, \quad \text{with} \quad H_0 = \frac{2W_0^2}{\rho_0 S \pi e A} \tag{7.56}$$

$$K = \sigma K_0, \quad \text{with} \quad K_0 = F_0 - G_0 \tag{7.57}$$

$$Q = \frac{\Phi(\sigma)}{\sigma} Q_0, \quad \text{with} \quad Q_0 = \frac{E_0}{K_0} \tag{7.58}$$

$$R = \left(\frac{W}{W_0}\right)^2 \frac{1}{\sigma^2} R_0, \quad \text{with} \quad R_0 = \frac{H_0}{K_0} \tag{7.59}$$

$$U = \left(\frac{W}{W_0}\right)^2 \frac{1}{\sigma^2} U_0, \quad \text{with} \quad U_0 = \frac{H_0}{G_0} \tag{7.60}$$

Table 7.3 Bootstrap composite numbers and operational variables, two cases

Variable or composite	Base case, W_0 and MSL	$W/W_0 = 0.75$, 8000 ft
W	2400 lbf $= W_0$	1800 lbf $= W$
σ	1	0.78602
$\Phi(\sigma)$	1	0.75684
E	$531.9 = E_0$	402.6
F	$-0.0052368 = F_0$	-0.0041162
G	$0.0076516 = G_0$	0.0060142
H	$1,668,535 = H_0$	$1,194,062$
K	$-0.0128884 = K_0$	-0.0101305
Q	$-41,270.6 = Q_0$	$-39,738.4$
R	$-129,460,301 = R_0$	$-117,868,335$
U	$218,064,595 = U_0$	$198,538,940$

Every composite has a density altitude correction; only H and the further composites R and U that depend on it have weight corrections.

As a reality check and to provide fodder for numerical examples, at this point we pause to display values of composite bootstrap parameters (Table 7.3) under two different weight/density situations. These values are based on the bootstrap data plate of Table 7.1.

With these concrete figures at our disposal, it is easy to run a few sample performance calculations.

Sample Calculations with Composite Bootstrap Parameters

We now use bootstrap composite parameters to calculate, for the two operational scenarios of Table 7.3, a variety of performance figures. Table 7.4 contains the usual V speeds and attendant optimum rates or angles of climb or descent. Table 7.5 contains various performance figures for the airplane under the two scenarios and at an arbitrarily selected air speed of 75 KCAS. Because we have recently covered a lot of material, the reader should reproduce a few of these results as a check on his or her understanding of it. For ease of checking, we have carried somewhat greater numerical precision than makes strict sense.

Bootstrap Application 1: Variations with Weight or Altitude

The bootstrap approach has provided us with a full set of formulas for predicting the quasi-steady performance of fixed-pitch, propeller-driven airplanes. To recap,

Table 7.4 Sample bootstrap V speeds for two scenarios

Item	Scenario 1, MSL, 2400#	Units	Scenario 2, 8000 ft 1800#	Equation no.
V_M	115.3	KCAS	100.4	19
V_m	34.6	KCAS	29.8	21
V_y	75.8	KCAS	65.9	24
h'_{max}	700.5	ft/min	699.5	24, 39
V_x	63.2	KCAS	54.7	27
γ_x	5.71	deg	5.82	44
V_{bg}	72.0	KCAS	62.4	31
γ_{bg}	−5.40	deg	−5.40	50
V_{md}	54.7	KCAS	47.4	33
h'_{md}	−602.3	ft/min	−588.4	33, 46

those formulas were for six V speeds—V_M, V_m, V_x, V_{bg}, V_y, and V_{md}—as well as for many performance output variables. Among those were optimal figures associated with the last four cited V speeds: γ_x, γ_{bg}, h'_{md}, and h'_{max}. We obtained explicit formulas for the first three of those optimal figures but decided that an explicit formula for h'_{max} would be more cumbersome than a two-step calculation (first calculating V_y, then calculating $h'[V_y]$).

But there is the additional question of how those V speeds or optimal output variables change when one of another of the two operational variables—gross weight W or relative air density σ (or its surrogate h_ρ, density altitude)—change. For two examples: How much does maximum level flight speed V_M decrease if

Table 7.5 Sample bootstrap performance data for two scenarios, $V = 75$ KCAS

Item	Scenario 1, MSL, 2400#	Units	Scenario 2, 8000 ft 1800#	Equation no.
T	448.0	lbf	318.7	35
P_{av}	103.1	hp	82.7	14
D_P	122.6	lbf	122.6	37
D_i	104.1	lbf	58.6	37
D	226.7	lbf	181.2	37
P_{re}	52.2	hp	47.0	6
P_{xs}	50.9	hp	35.7	17
h'	700.2	ft/min	654.3	39
T_{xs}	221.3	lbf	137.5	17/V
γ	5.29	deg	4.38	42

one adds 100 lbf to the airplane's gross weight? or How much does V_M decrease (does it?) if one climbs up 1000 ft? One would expect of our analytic (formula-based) theory that a performance sensitivity analysis can be accomplished, using differential calculus, yielding yet further formulas. Such is the case, but...

Now for the bad news, in three parts. First is the (minor) lack of usable formulas for one of the optimal performance variables. That problem will be addressed through an approximation. Second is the fact that sensitivity or variation formulas are not so important as the original formulas from which they spring. Indeed, one can always check variation simply by calculating twice (once for each value of an operational variable, say first for $W = 2000$ lbf and then again for $W = 2100$ lbf) and subtracting. Third, 10 performance variables times 2 operating variables means a total of 20 cases to consider. Too much!

Next the good news. The required calculations can be done in about four groups; the breadth of work is not so great as first feared. And we do not have to give all of the details. Our sample calculations and several numerical examples should be enough to lead the interested reader to any particular sensitivity result he or she may need.

Variation of V_x, V_{bg}, V_{md} (also V_S), and h'_{md} with W or σ

In this first group, we have a nice simplification. One-half our bootstrap V speeds—V_x, V_{bg}, and V_{md}—as well as stall speed V_S and minimum descent rate h'_{md} are proportional to $(W/\sigma)^{1/2}$. Taking f to stand for any of those five variables, we then have

$$f = \sqrt{\frac{W}{\sigma}} \left[\sqrt{\frac{\sigma}{W}} f \right] \equiv \sqrt{\frac{W}{\sigma}} f_1 \qquad (7.61)$$

where, despite early appearances, f_1 depends on neither gross weight nor altitude. Using Eq. (7.61), it is easy to see that

$$\frac{\partial f}{\partial W} = \frac{1}{2W} f \qquad (7.62)$$

so that, if W increases, then each of the four necessarily positive V speeds also increases; the necessarily negative minimum descent rate decreases (that is, become more negative with increased weight). Similarly, we find that

$$\frac{\partial f}{\partial \sigma} = \frac{-1}{2\sigma} f \qquad (7.63)$$

so that if σ increases, in other words if h_ρ decreases, then each V speed f decreases; minimum descent rate runs contrary and increases (that is, becomes less negative with lower altitude).

While relative air density σ is the natural variable for our V speed expressions, pilots think in terms of density altitude h_ρ. And, in a practical sensitivity analysis, in terms of differentials: how much does such and such a performance number change, under these particular conditions, if altitude increases say 1000 ft. To make the switch of dependent variables, one often needs to make use of

$$\frac{d\sigma}{dh_\rho} \doteq \frac{-\sigma^{0.7651}}{34,174} \qquad (7.64)$$

While Eq. (7.64) may be of use in more complicated circumstances, it is true too that pilots fly according to *calibrated* air speeds. And a practical consequence of Eq. (7.61) is that none of these four V speeds—V_x, V_{bg}, V_{md}, or V_S—considered in calibrated terms, varies one iota with density altitude. If the airplane stalls at 50 KCAS at sea level, it stalls at 50 KCAS at 10,000 ft.

So let us consider a practical example involving only variation in gross weight.

Example 7.1 According to Table 7.4, when our sample Cessna 172 is at 8000 ft weighing 1800 lb, $V_x = 54.7$ KCAS. What would V_x be if, instead, it weighed 100 lb more, 1900 lb? In differential terms, we are thinking

$$dV_x \doteq \frac{\partial V_x}{\partial W} dW = \frac{V_x}{2W} \times 100 = \frac{54.7}{3600} \times 100 = 1.52 \text{ KCAS} \qquad (7.65)$$

So the new V_x, for the higher weight, is about $54.7 + 1.5 = 56.2$ KCAS. An exact calculation to this precision, using Eq. (7.28), also gives $V_x = 56.2$ KCAS. Notice that as far as weight variation is concerned, it makes no difference whether one uses V_x in true or calibrated terms.

Variation of γ_x, γ_{bg} with W or σ

How about variation of the best climb angle or best glide angles themselves? For best glide angle γ_{bg}, matters are as simple as possible: it [Eq. (7.52)] does not vary at all with either weight or altitude:

$$\frac{\partial \gamma_{bg}}{\partial W} = \frac{\partial \gamma_{bg}}{\partial h_\rho} = 0 \qquad (7.66)$$

Having skimmed the cream, it is now necessary to buckle down and actually calculate how best climb angle γ_x varies say with gross weight. It is easier to work with the sine of the angle:

$$\frac{\partial \gamma_x}{\partial W} = \frac{\frac{\partial \sin \gamma_x}{\partial W}}{\frac{d \sin \gamma_x}{d \gamma_x}} = \frac{\frac{\partial \sin \gamma_x}{\partial W}}{\cos \gamma_x} \qquad (7.67)$$

From Eq. (7.44) and realizing that H is proportional to W^2 so that the second term is independent of weight, we have

$$\frac{\partial \sin \gamma_x}{\partial W} = \frac{-E}{W^2} \tag{7.68}$$

and so

$$d\gamma_x = \frac{-E}{W^2 \cos \gamma_x} \, dW \tag{7.69}$$

As expected, more weight means a shallower best climb angle. Let us look at an example.

Example 7.2 Evaluating the derivative at the base point (scenario 2 of Table 7.5), we find that $d\gamma_x = -0.72\,\text{deg}$ and so γ_x for $W = 1900\,\text{lb}$ is about $5.82 - 0.72\,\text{deg} = 5.10\,\text{deg}$. An exact calculation gives $5.14\,\text{deg}$.

What about variation of γ_x with altitude? With one new wrinkle, the calculation is much the same as that just above.

$$\frac{\partial \gamma_x}{\partial h_\rho} = -\frac{\dfrac{\partial \sin \gamma_x}{\partial \sigma} \dfrac{d\sigma}{dh_\rho}}{\dfrac{d \sin \gamma_x}{d\gamma_x}} = \frac{\dfrac{E_0}{W} \dfrac{d\Phi(\sigma)}{d\sigma} \left(\dfrac{-\sigma^{0.7651}}{34{,}174}\right)}{\cos \gamma_x} \tag{7.70}$$

We used Eq. (7.64) to change from density altitude to relative air density. There is still the derivative in the numerator of the right-hand term, but it is straightforward:

$$\frac{d\Phi(\sigma)}{d\sigma} = \frac{1}{1 - C} \tag{7.71}$$

Example 7.3 Gathering numerical pieces from Table 7.3, Eq. (7.70) evaluates to

$$\frac{\dfrac{E_0}{W} \dfrac{d\Phi(\sigma)}{d\sigma} \left(\dfrac{-\sigma^{0.7651}}{34{,}174}\right)}{\cos \gamma_x} \doteq \frac{\left(\dfrac{531.9}{1800}\right)\left(\dfrac{1}{1 - 0.12}\right)\left(\dfrac{-0.7860 2^{0.7651}}{34{,}174}\right)}{0.9948}$$

$$= -8.216 \times 10^{-6} \tag{7.72}$$

If the pilot climbs from his original 8000 to 9000 ft, $dh_\rho = 1000\,\text{ft}$ and so

$$d\gamma_x \doteq -8.216 \times 10^{-6} \times 10^3 = -0.008216\,\text{rad} = -0.47\,\text{deg} \tag{7.73}$$

This approximates the new best climb angle as $5.35\,\text{deg}$. An exact calculation gives $\gamma_x = 5.36\,\text{deg}$.

Variation of V_M, V_m, and V_y with W or σ

To avoid repetition, we will use a couple of algebra/calculus tricks to bring the three remaining V speeds under one umbrella. Referring back to Eqs. (7.19), (7.21), and (7.24), we can consider

$$V_{M/m}^2 = -q \pm \sqrt{q^2 + r} \tag{7.74}$$

where $q \equiv Q/a$ and $r \equiv bR/a$, with $a = b = 2$. Similarly, we can consider

$$V_y^2 = -q + \sqrt{q^2 + r} \tag{7.75}$$

with the difference that for V_y we have $a = 6$ and $b = -2$. Now all three expressions—call any of the three V speeds V_h, for "harder cases"—have been effectively shoehorned into a single form. Add q to both sides, square again (ridding the M/m case of the sign ambiguity!), and simplify to find

$$f(W, \sigma) \equiv V_h^4 + 2qV_h^2 - r = 0 \tag{7.76}$$

Because we fly airplanes according to calibrated air speeds more than true ones, it is convenient to rewrite Eq. (7.76) on that basis. After simplifying, the rewritten form is

$$g(W, \sigma) \equiv aV_{hc}^4 + 2Q_0 \Phi V_{hc}^2 - bW^2 R_0/W_0^2 = 0 \tag{7.77}$$

Next, we take the derivative of $g(W, \sigma)$ with respect to an operational variable, say W, and solve. That first result is

$$\frac{\partial V_{hc}}{\partial W} = \frac{bR_0 W}{2W_0^2 V_{hc}} \frac{1}{(aV_{hc}^2 + \Phi Q_0)} \tag{7.78}$$

Example 7.4 What would be the effect on maximum level flight speed of a 100-lbf gross weight increase for the airplane of scenario 2 (1800 lbf at 8000 ft)? Table 7.4 has $V_M = 100.37$ KCAS $= 169.4$ ft/s calibrated. Because not all terms depend on V_M in the same way, one cannot be as cavalier as before and simply employ knots with the expectation that any needed adjustments will cancel out. The RHS of Eq. (7.78), using other values from Table 7.4, evaluates to -0.00911 ft/s-lbf $= -0.00540$ KCAS/lbf. With a 100-lbf weight increase, the speed increment would be -0.54 KCAS. This gives a new top level speed of 99.83 KCAS. An exact calculation gives 99.80 KCAS. Maximum level flight speed decreases with gross weight, but not markedly.

To assess sensitivity of our "hard case" V speeds to changes in the other operational variable, density altitude, we take the partial derivative of $g(W, \sigma)$

with respect to relative air density σ, convert to density altitude by multiplying by Eq. (7.64), simplify, and find that

$$\frac{\partial V_{hc}}{\partial h_\rho} = \frac{\sigma^{0.7651} V_{hc}}{68{,}348(1 - C)(aV_{hc}^2/Q_0 + \Phi)} \qquad (7.79)$$

For this next example let us take our varied V speed to be V_y. Remember that makes $a = 6$.

Example 7.5 What would be the effect on speed for best climb rate of a 1000-ft climb for the airplane of scenario 2 (1800 lbf at 8000 ft)? Table 7.4 has $V_y = 65.90$ KCAS $= 111.23$ ft/s calibrated. The RHS of Eq. (7.79), using other values from Table 7.4, evaluates to -0.00148 ft/s-ft $= -0.000877$ KCAS/ft. With the 1000-ft density altitude increase, the speed increment would be -0.87 KCAS. This gives a new speed for best climb rate of 65.03 KCAS. An exact calculation gives 65.04 KCAS. In calibrated terms, V_y decreases a bit as altitude increases; in true terms, it increases a little.

Variation of h_{max}' with W or σ

An "exact" calculus approach would not be useful because it would give overly complex results. Recall there is no difficulty in principle: to assess variations we can simply calculate, with known formulas, to find whatever performance figures we need. That was done for Fig. 7.14, which shows how best rate of climb goes for our sample Cessna 172 over its useful vertical range and several weights. If we want variations, differences, then we can subtract. As follows our intuition, maximum rate of climb decreases with added weight or added altitude.

But there is a *nonstrict* approach that will give us a simple, though only approximate, one-step formula for $h_{max}'(W, h_\rho)$ and something of a handle of how best climb rate changes with either weight or density altitude. It uses three approximations: 1) the first few terms of a two-dimensional Taylor series for h_{max}', expanded about some base point $(W_1, h_{\rho 1})$—often taken to be standard weight W_0 and MSL, $h_{\rho 1} = 0$; 2) a binomial expansion to linearize the relation between σ and h_ρ; and 3) ignoring the relatively slight variation of V_y with weight or density altitude. Here is how it works.

A two-dimensional Taylor series expansion of $F(x, y)$ about "point 1," (x_1, y_1), begins as follows:

$$F(x, y) \doteq F(x_1, y_1) + \frac{\partial f}{\partial x}(x - x_1) + \frac{\partial f}{\partial y}(y - y_1) \qquad (7.80)$$

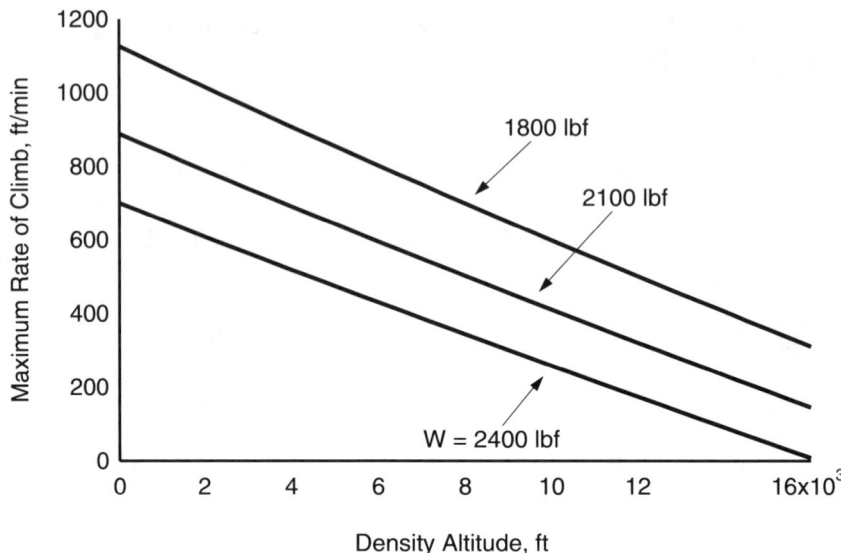

Figure 7.14 Graphs of h'_{max} vs h_ρ for $W = 1800, 2100, 2400\#$.

where the derivatives are to be evaluated at point 1. Translated to our case, this becomes

$$h'_{max}(W, h_\rho) \doteq h'_{max}(W_1, h_{\rho 1}) + \frac{\partial h'_{max}}{\partial W}(W - W_1) + \frac{\partial h'_{max}}{\partial h_\rho}(h_\rho - h_{\rho 1}) \quad (7.81)$$

The first term on the right is simply the rate of climb when $V = V_y(W_1, h_{\rho 1}) \equiv V_{y1}$. For the second term, we need to take the indicated derivative and then evaluate it at point 1. But by our third approximation, above, we do this while considering V_y as (relatively) fixed, at value V_{y1}. Using numeral "1" to denote point 1 throughout, the result is

$$\frac{\partial h'_{max}}{\partial W} \doteq \frac{-\Phi_1 E_0 V_{y1} - \sigma_1 K_0 V_{y1}^3}{W_1^2} - \frac{H_0}{W_0^2 \sigma_1 V_{y1}} \quad (7.82)$$

For the similar derivative with respect to density altitude, we need to change variables by means of

$$\frac{\partial h'_{max}}{\partial h_\rho} = \frac{\partial h'_{max}}{\partial \sigma} \frac{d\sigma}{dh_\rho} \quad (7.83)$$

and then approximately evaluate that last derivative using

$$\sigma = \left(1 - \frac{h_\rho}{145,457}\right)^{4.25635} \doteq 1 - \frac{h_\rho}{34,174} \tag{7.84}$$

which shows us that

$$\frac{d\sigma}{dh_\rho} \doteq \frac{-1}{34,174} \tag{7.85}$$

Putting these various approximate results together, we have

$$h'_{max} \doteq h'_{max}(W_1, h_{\rho 1}) - 60\left[\frac{\Phi_1 E_0 V_{y1} + \sigma_1 K_0 V_{y1}^3}{W_1^2} + \frac{H_0}{W_0^2 \sigma_1 V_{y1}}\right](W - W_1)$$

$$- \frac{1}{569.6}\left[\frac{E_0 V_{y1}}{(1-C)W_1} + \frac{K_0 V_{y1}^3}{W_1} + \frac{W_1 H_0}{W_0^2 \sigma_1^2 V_{y1}}\right](h_\rho - h_{\rho 1}) \tag{7.86}$$

While it does make sense to take starting point 1 at some intermediate point—for our sample Cessna 172 say at $W_1 = 2200$ lbf, $h_{\rho 1} = 5000$ ft—it considerably simplifies matters to take it at MSL and at $W_1 = W_0 = 2400$ lbf. Then Eq. (7.86) is shortened to

$$h'_{max}(W, h_\rho) \doteq h'_{max}(W = W_0, h_\rho = 0) - \frac{60}{W_0^2}\left(E_0 V_{y0} + K_0 V_{y0}^3 + \frac{H_0}{V_{y0}}\right)(W - W_0)$$

$$- \frac{1}{569.6 W_0}\left(\frac{E_0 V_{y0}}{(1-C)} + K_0 V_{y0}^3 + \frac{H_0}{V_{y0}}\right) h_\rho \tag{7.87}$$

where V_{y0} is the speed for best rate of climb, in feet per second, at MSL and at standard weight W_0. Equations (7.86) and (7.87) were adjusted to have units of feet per minute. But after three approximations, it is certainly important to assess how accurate these formulas are.

Example 7.6 Consider our sample Cessna 172 with bootstrap data plate as in Table 7.1. Assume that we want to use a reasonable intermediate base point 1 from which to approximate best rates of climb, the point with $W_1 = 2200$ lbf, $h_{\rho 1} = 5000$ ft. Then Eq. (7.86) obtains. What will our estimate of best rate of climb be for this airplane when it weighs 2300 lbf at 6000 ft? Among the constants in Eq. (7.86), we need $V_y(2200 \text{ lbf}, 5000 \text{ ft}) = V_{y1} = 128.2$ ft/s [from Eq. (7.24)], $h'_{max}(2200 \text{ lbf}, 5000 \text{ ft}) = 583.1$ ft/min [from Eq. (7.39)], and the usual array of composite bootstrap parameters and atmospheric density factors. When all is said and done, Eq. (7.86), for this airplane and this choice of point 1, is simply

$$h'_{\max}(W, h_\rho) \doteq 583.1 - 0.580(W - 2200) - 0.0519(h_\rho - 5000) \quad (7.88)$$

Adding 100 lbf and 1000 ft above the base point, we find $h'_{\max}(2300$ lbf, 6000 ft$) = 473.2$ ft/min. The precise two-step value is 482.1 ft/min, a relative error of less than 2%.

Example 7.7 Let us repeat the above calculation using scenario 1 of Table 7.3 (MSL and standard maximum gross weight) as base point 1. Then Eq. (7.87) obtains. Among the constants in Eq. (7.87), we need $V_y(2400$ lbf, MSL$) = V_{y1} = 128.0$ ft/s and $h'_{\max}(2400$ lbf, MSL$) = 700.5$ ft/min (from Table 7.4). Using the proper base bootstrap composite parameters in Eq. (7.87), we have

$$h'_{\max}(W, h_\rho) \doteq 700.5 - 0.5634(W - 2400) - 0.0464 h_\rho \quad (7.89)$$

This time, subtracting 100 lbf and adding 6000 ft above this new base point, we find $h'_{\max}(2300$ lbf, 6000 ft$) \doteq 478.4$ ft/min. The precise two-step value is 482.1 ft/min, a negligible error and undoubtedly a fluke of our three separate approximations just happening to largely cancel.

This "rough" and completely linearized approximation is nonetheless quite good. If, for instance, in example 7.7 we had asked for $h'_{\max}(2300$ lbf, 1000 ft), still 100 lb light but only 1000 ft above the base point, Eq. (7.89) would have given us 710.4 ft/min; the precise calculated value is 711.4 ft/min. That is close. What one can carry away, for this airplane, is that a weight gain of 100 lb or an altitude gain of 1000 ft, either one, means that best rate of climb will decrease about 50 to 60 ft/min. So, after all, we do have a good handle on variation of best rate of climb with weight and altitude.

Bootstrap Application 2: Absolute Ceiling and Speed There

Service ceiling, usually defined as the highest altitude at which the maximum gross weight airplane can just achieve a 100-ft/min rate of climb, is nothing special. By trial and error, one simply finds the altitude at which the best rate of climb is 100 ft/min. But *absolute ceiling* is something else.

For one thing, unless transported by some high-flying outside agency, you cannot get there. To do so would take an unbounded amount of both time and fuel. But it is still a mildly interesting number to compute, this maximum altitude "where you could stay there if you could get there."

As indicated by Fig. 7.6, at the absolute ceiling all V speeds (and in fact, *all attainable speeds*) coalesce to one. Because $V = V_M = V_y$ there, we know that both excess power $P_{xs} = 0$ and its derivative $P'_{xs} = 0$. (This is just another way to say that the power-available and power-required curves both touch—just!—and are tangent at the pertinent point.) Using the excess power definition, Eq. (7.17), under these conditions gives the dual result

$$V_{AC}^2 = \frac{2H}{E} = \frac{-E}{2K} \qquad (7.90)$$

Subscript AC means absolute ceiling. The second relation in Eq. (7.90) gives the fact that

$$\Phi_{AC} = \frac{2W}{W_0 E_0}\sqrt{-H_0 K_0} = \frac{2W n_0 d}{m P_0}\sqrt{\frac{SC_{D0} - 2d^2 b}{S\pi e A}} \qquad (7.91)$$

and the first relation yields the fact that

$$V_{AC} = \sqrt{\frac{2H}{E}} = \frac{2W}{\sqrt{\sigma_{AC}\Phi_{AC}}}\sqrt{\frac{n_0 d}{\rho_0 S\pi e A m P_0}} \qquad (7.92)$$

One can see that gross weight W plays a simple direct role in calculating Φ_{AC}, the absolute ceiling's power dropoff factor value. Not so simple for V_{AC}, the speed at absolute ceiling, because W appears three places in Eq. (7.92), once openly, twice hidden in σ_{AC} and Φ_{AC}.

To calculate the absolute ceiling density altitude or V_{AC}, one first needs to invert the power dropoff factor definition to find

$$\sigma_{AC} = (1 - C)\Phi_{AC} + C \qquad (7.93)$$

Example 7.8 According to Eq. (7.91), our sample Cessna 172 (of Table 7.3), when stripped down to only 1800 lbf, has absolute ceiling torque/power dropoff factor $\Phi_{AC} = 0.4136$. (Furthermore, $\sigma_{AC} = 0.4839$ and $h_{\rho AC} = 22,805$ ft.) At maximum gross weight $W = 2400$ lbf, on the other hand, $\Phi_{AC} = 0.5515$, $\sigma_{AC} = 0.6053$, and $h_{\rho AC} = 16,184$ ft.

The only problem with these results is that you cannot get high enough to try them directly.

Bootstrap Application 3: Wind Effects on (Mostly) Full-Throttle Climbs

Before we settle down to consider the effects that winds have on climbs—in particular, steepest climbs and others at full throttle—we examine the broader picture. In this work we are not so much interested in the origins or classification of winds as in their performance effects. Still, it is advisable to classify winds, by time scale, as belonging to turbulence, gusts, or steady winds. Pilots try to avoid turbulence or, if they cannot, slow appropriately and tighten their safety belts. Gusts often play havoc with landings or while maneuvering in mountains, but again have little relation to our *quasi-steady* performance problems. So our main concern is with steady winds, whether headwind, tailwind, updraft, downdraft, or crosswind.

Except during landing or takeoff or during ground reference maneuvers (including some instrument flight patterns), we can accommodate steady crosswinds simply by crabbing to an appropriate "wind correction angle." Though crosswinds, headwinds, or tailwinds play a major role in large-scale navigation, for our smaller-scale performance purposes, wind effects impact only those parameters that reference an Earth-based system of coordinates: speed for best angle of climb, angle of climb itself, speed for best glide angle, and glide angle itself. Because glide performance has its own chapter (Chapter 9), here we will focus on climbs. That means effects on speed for best angle of climb V_x and flight path angle γ (whether steepest angle γ_x or path angle at some other speed).

The qualitative situation is clear. Headwinds set you back and steepen your climb angle; tailwinds push you along and make for a shallower angle. See Fig. 7.15 for the pure headwind alternative, which lays out the various speeds and angles. An updraft or downdraft by itself has an obvious performance implication; rate of climb is simply increased (by an updraft) or decreased (by a downdraft) by the speed of the draft; there is no direct horizontal effect. At some point below, we will consider just what happens when headwinds or tailwinds are *accompanied* by vertical air-mass movements. Next we need to ask, "How large are these various wind effects? How much is (say) V_x lowered by a 20-kn headwind? Or by a 300 ft/min updraft? How are best climb angles impacted? What are the quantitative details and explanations?" Before we can answer such questions—graphically and fairly completely, then algebraically

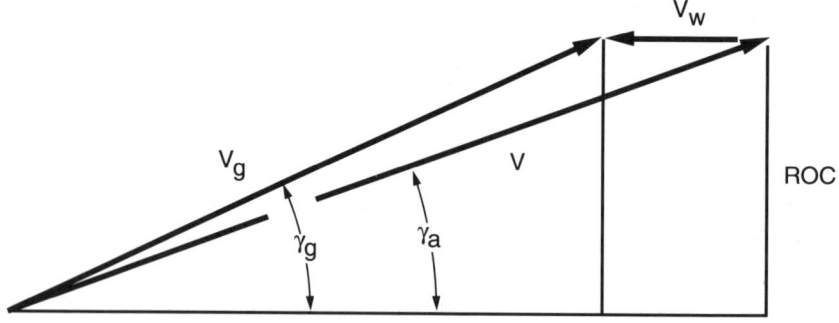

Figure 7.15   Vector and angle relations (side view) for a headwind.

Introduction to the Bootstrap Approach 219

and somewhat spottily—we need some background in the geometry of the problem.

Relative Motion (and an Approximation)

The rule for computing relative motion gets down to this simple vector equation:

$$V_{\oplus p} = V_{\oplus a} + V_{ap} \tag{7.94}$$

Here V stands for velocity, speed together with its direction, $\oplus$ stands for the Earth, p for the airplane, and a for the air. The way to remember this rule is to notice the collapse of the "interior" repeated symbol on the right. That is what counts. In English, the rule reads: velocity of the plane with respect to the (local) Earth (ground velocity) equals velocity of the air with respect to the Earth (wind velocity) plus velocity of the plane with respect to the air (air speed with heading). Of course, Eq. (7.94) works for any three objects. From the physicist's point of view, there is nothing special about the Earth, air, and an airplane. Aviators know better.

Figure 7.15 gave less pedantic abbreviations of the some of the velocities and introduced h, as the rate of climb. Symbol γ is conventionally used for the flight path angle. There are a lot of symbols in Fig. 7.15. Now what might we be after? Perhaps the full-throttle air speed V_a (also simply V in the figure), which, in a given (here) headwind, will give us the largest flight path angle with respect to the Earth, $\gamma_{\oplus p}$, henceforth simply γ_p. It makes sense to call that air speed "V_x in the wind," V_{xw}. But remember, because we fly solely by reference to air speeds, that V_{xw} is an *air speed*. In our graphs and calculations it will be a *true* air speed, but it is easy enough to later convert that to more operational calibrated or even indicated air speed.

So, to find speed for best angle of climb, we want to maximize γ_p with respect to V. To do that, we are going to need an algebraic relationship between those two variables. One such comes from briefly studying Fig. 7.15:

$$h = V_a \sin \gamma_a = V_p \sin \gamma_p \tag{7.95}$$

We need V_p, which, as Fig. 7.15 shows, depends on V_a, V_w and γ_a. Application of the "law of cosines," the Pythagorean theorem for nonright triangles, gives

$$V_p^2 = V_a^2 + V_w^2 - 2V_a V_w \cos \gamma_a \tag{7.96}$$

What saves us in using Eq. (7.96) is that γ_a is a small angle and, therefore, its cosine is quite close to unity. Even cosine 8 deg, probably a larger climb or glide angle we will encounter or achieve, differs from unity by only 1%. This small flight path angle approximation is one we have made consistently and is the one largely responsible for the relative simplicity of TBA and of the P_a/P_r analysis.

With the cosine set equal to 1, the RHS of Eq. (7.96) is a perfect square and we have the simple relationship

$$V_p \doteq V_a - V_w \qquad (7.97)$$

Of course! All we are saying is that, for general aviation aircraft, flight path angles are so shallow—we will eventually be using the gliding version, too, remember—that there is no practical difference between the (slant) air speed and its horizontal component. Figure 7.15 does not support that statement, but that is because we exaggerated the angles so much by using a much larger vertical scale than horizontal scale. None of these airplanes climbs steadily at 30 deg, not even a Bonanza!

Combining Eqs. (7.95) and (7.97), we have

$$\sin \gamma_p = \frac{\dot{h}}{V_p} \doteq \frac{\dot{h}}{V - V_w} \qquad (7.98)$$

This simple but good approximation will be the basis of the following graphic and analytic calculations of V_{xw} when there is a pure headwind or tailwind. When there is also an updraft V_u, or a downdraft V_d, then Eq. (7.94) has another embellishment and becomes

$$\sin \gamma_p = \frac{\dot{h}}{V_p} \doteq \frac{\dot{h} - V_d}{V - V_h} \qquad (7.99)$$

where, just as a positive V_h is to be taken to be a headwind, positive V_d is to be taken to be a downdraft. In a glide, both $\dot{h}$ and γ_p (and its sine) will be negative. But the same relation still holds. It is also true for small angles that

$$\sin \gamma \doteq \tan \gamma \qquad (7.100)$$

Those two trigonometric functions differ by less than 0.001 up until nearly 8 deg.

To investigate how well Eq. (7.94) approximates the "exact" formula, combine Eqs. (7.95) and (7.96) to get

$$\sin \gamma_p(V, V_w) = \frac{\dot{h}(V)}{\sqrt{V^2 + V_w^2 - 2V_w\sqrt{V^2 - \dot{h}^2(V)}}} \qquad (7.101)$$

To examine the validity of the small flight path angle approximation, let us consider an example.

Example 7.9 Consider a sample Cessna 172 airplane with the bootstrap data plate of Table 7.6.

If one uses TBA Eq. (7.39) to evaluate $\dot{h}$ for our sample Cessna at MSL and weighing 2400 lbf, an "exact" trial-and-error calculation using precise Eq. (7.101) gives $V_x = 56.86$ KCAS for a 20-kn headwind, 48.96 KCAS for a 30-kn

Table 7.6 Sample Cessna 172 bootstrap data plate

BDP Item	Value	Units	Aircraft
Wing reference area, S	174	ft^2	Airframe
Wing aspect ratio, A	7.378		Airframe
Rated MSL torque, M_0	311.2	ft-lbf	Engine
Altitude dropoff parameter, C	0.1137		Engine
Propeller diameter, d	6.25	ft	Propeller
Parasite drag coefficient, C_{D0}	0.037		Airframe
Airplane efficiency factor, e	0.72		Airframe
Propeller polar slope, m	1.70		Propeller
Propeller polar intercept, b	−0.0564		Propeller

headwind. Approximate Eq. (7.99) gives, instead, 56.81 KCAS and 48.82 KCAS, respectively. Only negligibly different. So the small flight path approximation is not a problem. We placed exact in quotes above because TBA itself uses the small angle approximation. Hardly anything is exact in aerodynamics or in airplane performance calculations or needs to be.

Graphical Analysis of Headwind/Tailwind Effects

Consider the curve of Fig. 7.16, a plot of rate of climb $h^{\cdot}$ against true air speed. [Figure 7.3 is similar; to be identical, according to Eq. (7.17), only requires division of ordinate P_{xs} by gross weight.] Figure 7.16 is a picture in speed space; speeds in both directions though at much different scales. But speeds, if you multiply them by time intervals, become distances. Doing that then gives a picture in ordinary space, of precisely the same shape, but with this advantage: it is clear that a straight line from the origin to any point on the rate-of-climb curve rises at angle γ from the V axis. (In saying that we have used our slight hedge that actual slant speed and displayed horizontal speed differ by a factor very near unity, the cosine of small angle γ.)

In a no-wind situation, air speed equals ground speed. Now consider three alternative wind situations: calm, a 20-kn headwind, and a 20-kn tailwind. Those winds subtract or add 20 kn to our ground speed. Figure 7.17 gives the picture. The effect of say the headwind is the same as taking the cap-shaped curve of Fig. 7.16 and sliding it V_{hw} to the left. But then *that is* the same (and keeps us in the air-speed picture) as sliding the starting point (origin) of V to the right by amount V_{hw}. So the headwind lowers V_x, raises γ_x, and lowers the rate of climb at V_x. The headwind sets you back so, for best angle, you slow down to let it steepen your angle. The tailwind does the opposite (raises V_x, lowers γ_x, and raises the rate of climb at V_x). The tailwind pushes you along into the barrier ahead so, for best angle, you speed up to let it do as little damage as possible.

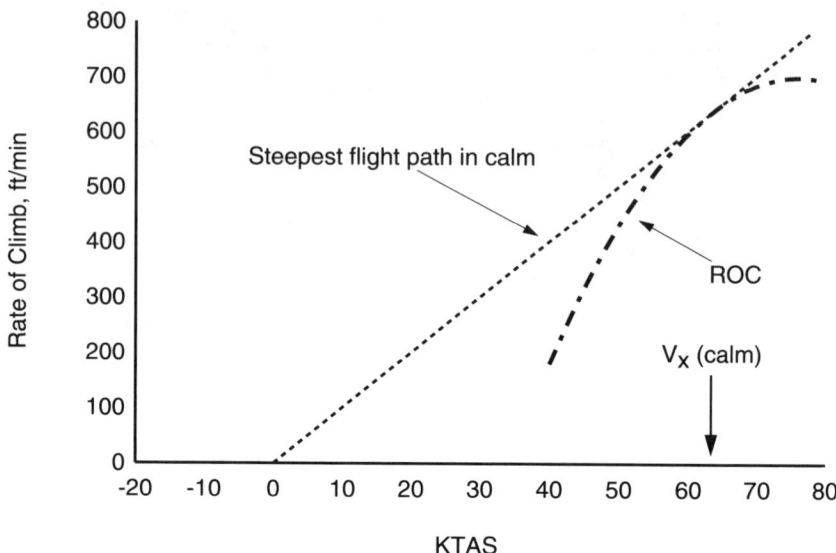

Figure 7.16 Rate of climb vs true air speed, with best-climb flight path in a calm.

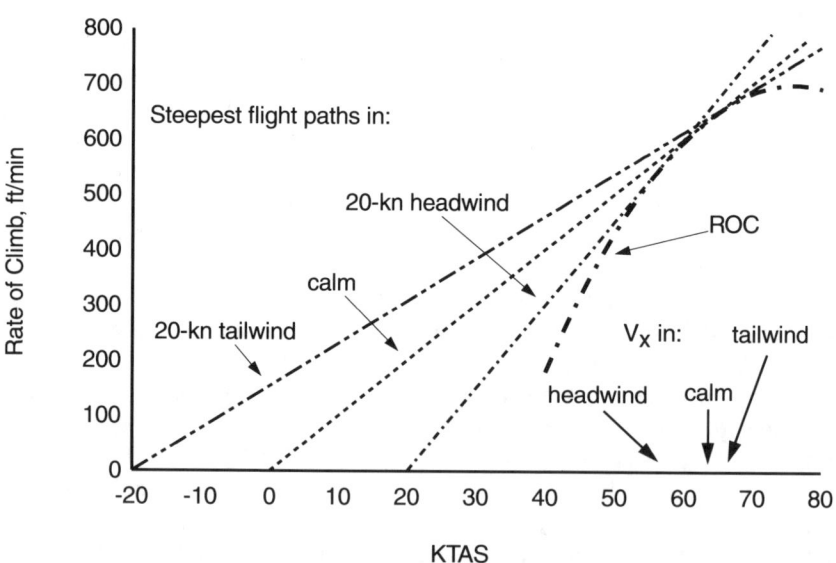

Figure 7.17 Headwind makes for lower ground speed but has no effect on vertical speed. Therefore, angle of climb is greater.

Introduction to the Bootstrap Approach 223

We should also take winds into consideration when trying to achieve long or maximum range. That analysis would use lines tangent to the power-required graph because V_{br} is close to V_{bg}; for details on the cruise situation, see Chapter 10; for the best glide approximation, see Chapter 9. In essence, with a headwind, speed up; with a tailwind, slow down. Wind speeds have to be sizeable, a quarter or more of air speed, to make much difference. Steady winds have no effect on endurance.

Graphical Analysis of Headwinds/Tailwinds Accompanied by Updrafts/Downdrafts

For definiteness, consider a headwind accompanied by a downdraft. Figure 7.18 shows the speed vector relationships in a general (not necessarily optimal) climb under these wind conditions. V_a and $\dot{h}_0$ are air speed and rate of climb in the *calm* atmosphere, with no headwind V_h and no downdraft V_d. V_{gh} is the airplane's ground velocity when there is a headwind but no downdraft. V_{ghd} is the airplane's ground velocity when there is both a headwind and a downdraft.

Under the obvious theory that "what's sauce for the (horizontal) goose is sauce for the (vertical) gander," we can redraw Fig. 7.17 to see the effects on V_x and on γ_x of headwinds or tailwinds with downdrafts or updrafts (see Fig. 7.19). The positive and negative V axes are, as before, where one would relocate the origin of the tangent line in cases of pure headwinds or tailwinds, respectively.

The original (calm) tangent line in Fig. 7.19 hits the rate-of-climb curve at V_{xc}, speed for best angle of climb with calm winds. As the reader can verify by mentally drawing new tangent lines to the rate-of-climb curve, all points below

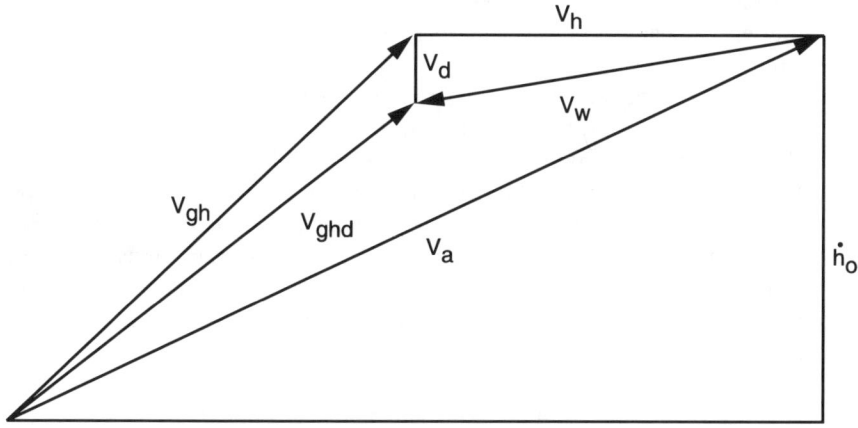

Figure 7.18 Speed vector diagram for airplane climbing into headwind with downdraft.

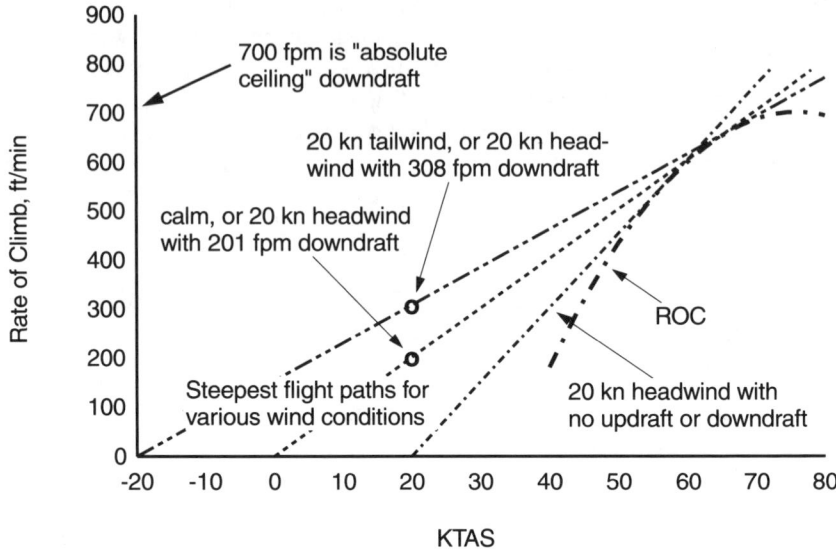

Figure 7.19 Diagram showing effects of combined horizontal/vertical winds on speed for best climb angle and that angle itself.

that calm tangent have $V_x < V_{xc}$ and $\gamma_x > \gamma_{xc}$; all points above the calm tangent (up to the airplane's absolute-ceiling downdraft) have $V_x > V_{xc}$ and $\gamma_x < \gamma_{xc}$. If one draws a scale diagram like Fig. 7.19 (for a given airplane at given weight at given density altitude), one can then read off V_x and γ_x (after compensating for the differing vertical and horizontal scales) for any given headwind/tailwind and downdraft/updraft situation.

We can calculate the numerical details for any given aircraft and wind situation (as in Fig. 7.19) but, as a practical matter, during flight there are too many variables—speeds of the headwind or tailwind and of the downdraft or updraft—one does not know precisely. And even the several known variables—airplane, configuration, weight, and density altitude—influence the numeric details. Not calculations appropriate in the cockpit! Still, by constructing and studying several example graphs, one can get a feeling for the directions and rough sizes of the various effects. The important thing is to not make an adjustment in the wrong direction, such as slowing down when climbing with a downdraft. Table 7.7 gives "rules of thumb" for attaining "best angles" when one has a headwind or tailwind with or without admixture of a downdraft or updraft.

Having exhausted the graphical solution possibilities, it is time to consider some analytic and numeric facts regarding wind effects on performance.

Introduction to the Bootstrap Approach   225

Table 7.7 V_x and γ_x in complex wind situations

Wind situation	Best climb angle speed	Best climb angle
Direct headwind	$V_{xh} < V_x$	$\gamma_{xh} > \gamma_x$
Direct tailwind	$V_{xt} > V_x$	$\gamma_{xt} < \gamma_x$
Headwind with downdraft	$V_{xhd} > V_{xh}$	$\gamma_{xhd} < \gamma_{xh}$
Headwind with updraft	$V_{xhu} < V_{xh}$	$\gamma_{xhu} > \gamma_{xh}$
Tailwind with downdraft	$V_{xtd} > V_{xt}$	$\gamma_{xtd} < \gamma_{xt}$
Tailwind with updraft	$V_{xtu} < V_{xt}$	$\gamma_{xtu} > \gamma_{xt}$

Formula Approach to Wind Effects on Steepest Climbs

We must first take stock. The airplane's flight path angle γ with respect to the air mass is given, in general, by Eq. (7.41) or, making use of TBA, by more explicit Eqs. (7.42) or (7.43). In a no-wind situation, there are precise bootstrap formulas both for the optimal climb angle γ_x [Eqs. (7.44) or (7.45)] and for the speed for best climb angle V_x [Eqs. (7.27) or (7.28)]. When a headwind V_h (to be taken as a tailwind when negative) is present, basic Eq. (7.41) is changed into Eq. (7.98). When both a headwind and a downdraft V_d (to be take as an updraft when negative), that basic formula becomes the even more general Eq. (7.99) for the flight path angle with respect to the Earth, which is

$$\sin \gamma_p \doteq \frac{\dot{h}_c - V_d}{V - V_h} \qquad (7.102)$$

The subscript c is to remind us that the cited rate of climb is for calm wind conditions. Given true air speed V, relative air density σ, and airplane gross weight W, TBA Eq. (7.39) gives us rate of climb (with respect to the air mass) $\dot{h}_c$ in terms of bootstrap composite parameters (see Table 7.8). So if we then also know the wind components V_h and V_d, we have everything we need to compute the airplane's flight path angle γ_p with respect to the ground (see Table 7.9).

Example 7.10   Using the BDP of Table 7.6, which gives composite parameters (at $W = 2200$ lbf and $h_\rho = 4000$ ft or $\sigma = 0.8881$, $\Phi = 0.8737$) of Table 7.7, let us calculate some ground-reference flight path angles when our sample aircraft is climbing full throttle at 80 KCAS = 84.9 KTAS = 143.3 ft/s.

Table 7.9 is fine as far as it goes, but we are even more ambitious. We would also like to know how to modify our cockpit behavior to achieve *best* angle of climb with respect to the Earth under given circumstances. So let us start out to figure that.

Table 7.8 Bootstrap composite numbers and operational variables from BDP of Table 7.6

Variable or composite	Base case, W_0 and MSL	$W/W_0 = 0.917$, 4000 ft
W	2400 lbf = W_0	2200 lbf = W
σ	1	0.8881
$\Phi(\sigma)$	1	0.8737
E	531.9 = E_0	464.7
F	$-0.0052368 = F_0$	-0.0046508
G	$0.0076516 = G_0$	0.0067952
H	$1,668,987 = H_0$	1,579,142
K	$-0.0128884 = K_0$	-0.0114460
Q	$-41,270.6 = Q_0$	$-40,603.4$
R	$-129,495,394 = R_0$	$-137,964,564$
U	$218,123,707 = U_0$	232,389,286

Speed for Best Climb Angle V_{xhd} in Given Headwinds and Tailwinds, Given Updrafts and Downdrafts

To obtain a relation for $V_{xw} \equiv V_{xhd}$, the air speed giving the maximum ground-reference flight path angle γ_{xw}, we substitute Eq. (7.39) into Eq. (7.102), take a derivative with respect to air speed V, set that equal to zero, and simplify. We get an overly complex quintic mess

$$V_h(EV_{xw}^2 + 3KV_{xw}^4 + H) = 2KV_{xw}^5 + 2HV_{xw} + WV_{xw}^2 V_d \quad (7.103)$$

There is no difficulty finding V_{xw} from Eq. (7.103) in principle, but that must be done by trial and error. We no longer have a neat analytic solution.

Table 7.9 Ground-reference flight path angles under various wind conditions

Headwind (+) or tailwind (−), kn	Downdraft (+) or updraft (−), ft/min	Ground-reference γ, deg
0	0	4.0
20	0	5.2
20	200	3.5
20	400	1.7
20	-200	7.0
20	-400	8.7
40	0	7.5
40	200	5.0
40	400	2.5
40	-200	10.1
40	-400	12.7

Introduction to the Bootstrap Approach   227

So we move on to the simpler case of finding V_{xh}, speed for best climb angle when there is a headwind or a tailwind but no palpable updraft or downdraft. With $V_d = 0$, Eq. (7.103) simplifies somewhat to

$$V_h = \frac{2KV_{xh}^5 + 2HV_{xh}}{3KV_{xh}^4 + EV_{xh}^2 + H} \tag{7.104}$$

This has the advantage that the headwind (or, if negative, the tailwind) speed is isolated on the left. Then any trial V_{xh} plugged into the RHS is a solution for *some* headwind or *some* tailwind. And, having calculated, you know which one. In this way, one can quite quickly piece together (see Fig. 7.20) a graphic picture of how speed for best climb angle V_{xh} depends on the strength of headwind or tailwind V_h.

As we have seen both logically and graphically, V_{xh} decreases, from its calm-wind value, when a headwind is present. To obtain a rule of thumb for approximately how much one should slow, when encountering say a small headwind of V_h kn, one can go on a bit further and compute the derivative of Eq. (7.103) with respect to V_h and solve the resulting expression to get

$$\frac{\partial V_{xh}}{\partial V_h} = \frac{EV_{xh}^2 + 12KV_h V_{xh}^3 + 3KV_{xh}^4 + H}{10KV_{xh}^4 + 2H - 2EV_h V_{xh}} \tag{7.105}$$

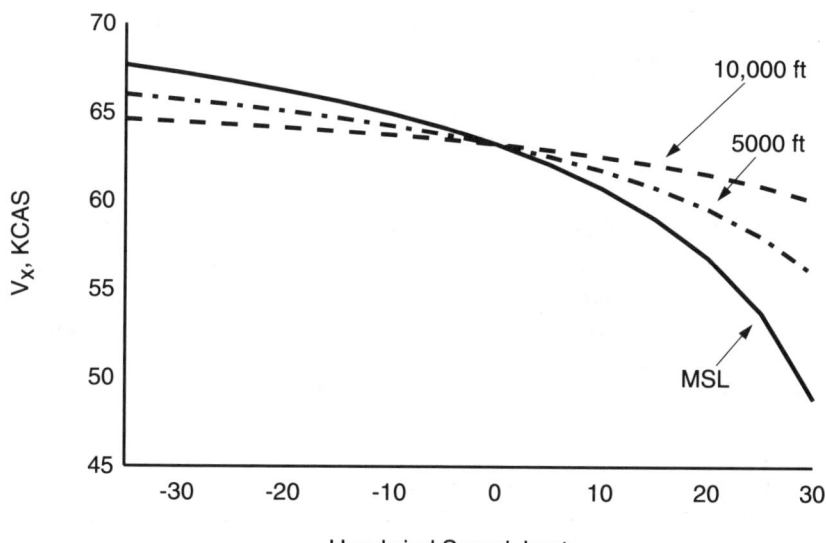

Figure 7.20 V_{xh} vs V_h from -35 kn to $+30$ kn at MSL, 5000 ft, and 10,000 ft.

Again, a mess; but do not despair. Recall that speed for best angle of climb in calm wind, which we can now call V_{x0}, is the fourth root of $(-R) = -H/K$. Using that fact, Eq. (7.105) can be rewritten as

$$\frac{\partial V_{xh}}{\partial V_h} = \frac{1 - 3\left(\frac{V_{xh}}{V_{x0}}\right)^4 - 12\frac{V_h V_{xh}^3}{V_{x0}^4} - \frac{E}{K}\frac{V_{xh}^2}{V_{x0}^4}}{2 + \frac{2E}{K}\frac{V_h V_{xh}}{V_{x0}^4} - 10\left(\frac{V_{xh}}{V_{x0}}\right)^4} \quad (7.106)$$

Now consider the case of relatively light headwinds, taking $V_h = 0$ and $V_{xh} = V_{x0}$. This approximation allows Eq. (7.106) to be rewritten as

$$\frac{\partial V_{xh}}{\partial V_h} \doteq \frac{1}{4} - \frac{E}{8\sqrt{-HK}} \quad (7.107)$$

Referring back to the prior section on absolute ceilings, Eq. (7.91), and doing some minor rewriting with the weight and relative air density dependence of these composite bootstrap parameters, one finds the surprisingly simple result that

$$\frac{\partial V_{xh}}{\partial V_h} \doteq \frac{1}{4}\left(1 - \frac{\Phi}{\Phi_{AC}}\right) \quad (7.108)$$

Because Φ is always greater than Φ_{AC}, the last term on the right is greater than unity, hence $\partial V_{xh}/\partial V_h$ (for light headwinds) is always negative. If one uses the definition of torque/power dropoff factor Φ, a binomial expansion approximation for σ in terms of h_ρ, and treats C as 0.12, one gets the more pilot-friendly but only roughly approximate form

$$\frac{\partial V_{xh}}{\partial V_h} \doteq \frac{h_\rho - h_{\rho AC}}{122,000} \quad (7.109)$$

Example 7.11 According to Eq. (7.91), our sample Cessna 172 (of Table 7.7), when it weighs 2200 lbf, has absolute ceiling torque/power dropoff factor $\Phi_{AC} = 0.5055$. (Furthermore, $\sigma_{AC} = 0.5617$ and $h_{\rho AC} = 18,432$ ft.) At MSL ($\Phi = 1$), we would then have $\partial V_{xh}/\partial V_h \doteq -0.245$ near zero headwind. [Equation (7.109) gives the not-so-close figure -0.151.] So if you are in an optimal climb, in a calm, and you run into a 10-kn headwind, slow down about 2.5 KTAS. At 4000-ft density altitude ($\Phi = 0.8737$), we have, instead, $\partial V_{xh}/V_h \doteq -0.182$ near zero headwind. [Equation (7.109) gives the still not-so-close figure -0.118.] This time, encountering the 10-kn headwind, slow only about 1.8 KTAS. If it is a 10-kn tailwind, speed up about 1.8 KTAS. Figure 7.20 supports this altitude distinction. As one gains altitude, headwinds and tailwinds have increasingly *less* effect on V_{xh}. And the approximation of Eq. (7.109) improves.

In mountain flying, it is often the case that we have to deal with the fuller and more complex situation of a (say) headwind with a (say) downdraft. The other three wind orientations also occur, but let us consider only this one example. Now it is also often the case that the downdraft occurs purely because the air mass is *following the terrain*, so that the total wind vector V_w slopes down at angle θ_t, where subscript t stands for terrain. In this common case we can see a simplification of the full and messy situation presented by Eq. (7.103). That is because the downdraft speed is now simply

$$V_d = V_w \sin \theta_t \qquad (7.110)$$

With that substitution, Eq. (7.103) can be rewritten to look just like the simpler no-downdraft case, Eq. (7.104) with a slight complication to its term featuring bootstrap composite parameter E:

$$V_w = \frac{2KV_{xw}^5 + 2HV_{xw}}{3KV_{xw}^4 + (E - W\sin\theta_t)V_{xw}^2 + H} \qquad (7.111)$$

The only real difference between Eq. (7.104) and Eq. (7.110) is that the latter has $(E - W \sin \theta_t)$ where the former has only E. We got this from the mathematics of the situation, but that added term can be interpreted in terms of forces on the aircraft. E is the bootstrap parameter for static ($V = 0$) thrust and $W \sin \theta_t$ is the weight component pulling back on the airplane when its path is inclined θ_t above the horizontal, a drag-like force opposing thrust. The terrain-following, headwind-with-downdraft problem is precisely like the pure headwind problem, only rotated by angle θ_t. Makes sense.

This concludes our explanation of the effects of winds on climbs and especially on optimal climbs.

Bootstrap Application 4: MSL "Rated" Propeller Efficiency

Without those additional full-throttle climbs we mentioned earlier, recording V and n, propeller efficiency is generally not available to us. But there is an exception. The MSL engine specifications include both power and revolutions per minute, giving us what we need for that one case. Starting at the beginning,

$$\eta = \frac{TV}{P} \qquad (7.112)$$

It is not a good idea to use the manufacturer's claims for maximum level flight speed at MSL—call it V_{M0}—for the V in Eq. (7.112); specifying an airplane's top speed is one place manufacturers tend to be overly optimistic. But it is easy enough (even if you live in a mountainous state where access to sea level is permanently restricted) to use the bootstrap parameters to compute that speed. Let us use η_0 to denote the value of propulsive efficiency corresponding to V_{M0}. Using Eq. (7.14) for $P_{av} = TV$ in Eq. (7.112) then gives us

$$\eta_0 = \frac{mV_{M0}}{n_0 d} + \frac{\rho_0 d^2 b V_{M0}^3}{P_0} = \frac{E_0 V_{M0}}{P_0} + \frac{F_0 V_{M0}^3}{P_0} \quad (7.113)$$

Take for example the airplane typified at sea level in Table 7.3. Equation (7.19), or a glance at Table 7.4, tells us that $V_{M0} = 115.3$ KTAS $= 194.6$ ft/s. Using the base case composite parameters of Table 7.3, Eq. (7.113) then gives $\eta_0 = 0.738$. This is quite a bit lower than the nominal $\eta_0 = 0.80$ or even 0.85 often cited. The more realistic figure is reflected in our maximum level flight speed 115.3 KTAS, whereas the Cessna POH cites (without speed fairings) $V_{M0} = 121$ KTAS.

Conclusions

First we must look more deeply into the validity and accuracy—or, alternatively, into the weak points—of TBA. Because TBA is a combination of P_{av}/P_{re} analysis and the linearized propeller polar, we break our evaluation into those two parts.

Power Picture Assumptions

Power-available and power-required pictures include the following four restrictions and approximations:

1) Only unbanked (bank angle $\phi = 0$) flight is treated. That restriction can be relaxed by rewriting condition $L = W$ as $L \cos \phi = W$ and making corresponding changes to the composite bootstrap parameter for induced drag, H. That will be done in the next chapter on maneuvering performance.
2) Possible offset, due to angle of attack, between thrust direction and flight path, is neglected. Doing so, we neglect a small thrust component aiding lift and (conservatively) slightly underestimate flight path angle γ.
3) Any offset built into the propeller crankshaft orientation relative to the longitudinal axis of the fuselage is neglected. That offset is often a slight droop.
4) The quantity $1 - \cos \gamma$ is treated as a small quantity of second order and that slight difference is ignored. That means treating lift as sensibly equal to weight, $L = W$, where the truth is that (unbanked and with no thrust offsets) $L = W \cos \gamma$.

Linear Propeller Polar Assumptions

The linear propeller polar picture uses three additional assumptions:

5) At given σ, throttle position determines engine torque. This is not new; von Mises[3] mentions that "almost all performance calculations are carried out under this assumption." On the other hand, this assumption is not strictly correct. A graph of P_{br} against engine revolutions per minute at fixed throttle in the operating range—one can get this from the engine manual[1]—has slightly decreasing slope as revolutions per minute increase. Torque for a given point is proportional to the slope of the line from the origin to that point. So if the airplane changes from level to climbing flight, for instance, and the propeller loads down, revolutions per minute decrease and torque increases a little. For the engine in our sample Cessna 172, torque increases only about 4% over a quite large 700-rpm decrease.
6) Full-throttle torque depends on air density via

$$M(\sigma) = \Phi(\sigma) M_0 = \frac{(\sigma - C)}{(1 - C)} M_0 \qquad (7.114)$$

where C is a number close to 0.12, giving the proportion of internal engine losses not responsive to air density. This again is old hat: again see von Mises[3], where he takes $C = 0.15$. Engines have improved. M_0 is the full-throttle sea level torque value. Equation (7.114) lets us get along without detailed engine charts giving $P_{br}(n, \sigma)$.
7) The propeller polar relation between C_P/J^2 and C_T/J^2 is sensibly linear (see Chapter 6); there are two parameters m and b such that

$$\frac{C_T}{J^2} = m \frac{C_P}{J^2} + b \qquad (7.115)$$

Von Mises[3] uses the method of representative blade elements to show that the propeller polar is approximately linear. Graphs made from the full propeller charts support this approximation. So this linearity assumption is not unusual, just the use of it in this P_{av}/P_{re} context. Parameters m and b, incidentally, implicitly *include* such complications as tired engines, dinged propeller blades, and reduced airflow through the propeller due to the fuselage. Recall our early admonition: "Fly the airplane to find out how the airplane flies."

Evaluation

So what is the overall evaluation? We suspect performance numbers plus or minus 3 to 5%. The truth is that not nearly enough well-instrumental trials on a variety of fixed-pitch propeller airplanes have yet been performed. A fertile research field. Accuracy of the various bootstrap numbers will differ in different circumstance and for different airplanes. And with the care or luck of various investigators. While we would never claim consistent accuracy to 1 or 2%, neither would we roll over for those who claim (static ports, and all that) "nothing can be done better than 10%." And there is an additional "social" artifact. Because TBA was devised by the author,[2,5,6] can the reader expect his evaluation to be purely objective? Probably not.

When someone (including you yourself) asks for a calculation, your first response should be "How close do you want it?" Perhaps calculation is not quite as much art as it is science, but knowing what effects to leave out is certainly an important key to *efficient* calculation. "The world is so full of a number of things, I'm sure we should all be ..." well, *totally confused*, if we do not discard many of them.

Say, for example, you are about to climb over a 12,000-ft ridge in your 25-year-old "Classic Clunkerbird." Say further that one of this airplane's nice features is a *perfect* POH performance section 5, all the data you could ever want and all of it infinitely accurate—for an airplane just out of the factory. You are currently at 8000 ft indicated altitude, at 2000 lbf gross weight; the POH says that V_x—speed for best angle of climb—under these circumstances is 65 KCAS. So you pull back and retrim to 65 kn on the air-speed indicator. You have achieved, for now, best angle of climb.

Or have you? Here are some of the effects—some stochastic (chance) and some not—you have ignored:

- The factory-fresh "Clunkerbird" engine was rated at 150 hp at 2600 rpm at MSL. Your airplane, however, is due for an overhaul and only gets an honest 140 hp at sea level.
- It is a warm afternoon and the indicated 8000 ft is actually 9500-ft density altitude. And density altitude is the one that matters.
- The weight-and-balance record suggests your current gross weight is 2000 lbf, but it is out of date. Your actual weight is 2050 lbf.
- Your 65 KIAS is actually 68 KCAS according to your "perfect" air-speed indicator calibration curve.
- It was calm when you left Cody, Wyoming, but up here there is a 15-kn headwind.
- Plus—and this is important—you just happen to have entered a thermal rising at 5 ft/s.
- The POH specifications do not account for those bugs plastered onto the leading edge. Or for those nicks in your propeller blade.

Not a pretty picture. Though the rising thermal helps. How far is your chosen 65 KIAS from the *true* V_x? Three or four kn; say about 5%. The good news is that general aviation performance curves have such shallow minima and maxima that your 5% speed error translates into only about a 2% performance deficit. See Fig. 7.11 for an instance. Weight is an exception; do not be cavalier about weight.

The point is that if random and uncontrollable reality factors dish out 5% errors then it does not make a whole lot of sense to use a theory much more precise than 5%. At least not one you have to pay for. Do what you can to eliminate *systematic* errors, of course. But those who use overly precise but still inaccurate theories or calculations are kidding themselves. The most accurate (and precise) job we can do today in this age of hyperactive computer programs requires mounds of input data worked on and moved forward a hundredth of a second at a stage. And results in reams, if not mountains, of output. Boeing and Martin Marietta can afford to pay for that process. They may even—for supersonic transports or fighters—need it. Most cannot. Most do not.

Advantages of TBA

Pilots (and their bosses, charter/cargo fleet or flying school operators) generally rely on the airplane's pilots operating handbook for performance data. And that performance data (with the possible exception of maximum speed for level flight) is not far from the truth. Performance flight tests are conducted in the U.S. under Federal Aviation Administration regulated guidelines and attested as truthful by the airplane manufacturer. But, even so, there are several problems:

- The airplane engine may be far from factory-fresh, need overhaul, and offer reduced power output.
- The airframe may be dirty or dented, offering increased drag resistance.
- Modifications to engine or propeller or airframe may have been made without the effects of those changes on performance having been properly accounted for.
- The POH performance section is extremely spotty.

This last is the major problem. Here are a few common POH deficits: 1) most V speeds not cited for various gross weights and density altitudes; 2) rate of climb and best glide information for only one gross weight; 3) no maneuvering flight information except for banked stall speeds; 4) only a generalized air-speed indicator calibration curve; 5) very limited takeoff and landing roll data over the possible range of wind, slope, and runway surface conditions; and 6) level cruise performance information ignoring low speeds, some (those for best range and for best endurance) with safety implications. At this point not all of these problems have yet been addressed by TBA, but they all will be by the time we are through. For safe flight operations the airplane's performance envelope must be consulted and adhered to by the pilot. But there is no way the pilot can do that with the performance envelope known so incompletely.

A major advantage of TBA for manufacturers of small airplanes is that design changes—say a different engine—only require, for new performance predictions, new BDP items for the engine. The three subsystems are (relatively) independent. The hedge is this: after you swap out an engine or a propeller, the new combination must again be capable of achieving MSL-rated revolutions per minute at full throttle. And no more than that. Otherwise the MSL rated torque will not be correct and will have to be reevaluated. For instance, an ultralight propeller attached to our sample Cessna 172's Lycoming engine would overspeed; that is not allowed. Putting our sample McCauley 7557 propeller on a little two-cycle Rotax engine would not do the job either. But apart from the need for matching engine and propeller, the subsystems are independent.

The saving graces of TBA are its speed and simplicity. It takes only a day or so to run preliminary flight tests, compute BDP parameters, compute performance numbers, and spot check their validity with further flight tests. And that is the ultimate evaluation of this or of any airplane performance computation technique: Does it work?

To this point, we have stressed the full-throttle, wings-level portion of TBA. Later chapters will consider glides in much greater detail, coordinated turns (level, climbing, and descending), partial-throttle operations, and most aspects of takeoff and landing.

References

1. Textron Lycoming, *Operator's Manual Series O-320, IO-320, AIO-320 & LIO-320*, 2nd ed., Textron Lycoming, Williamsport, PA, 1973.
2. Lowry, J. T., *Computing Airplane Performance with The Bootstrap Approach: A Field Guide*, Flight Physics, Billings, MT, 1995.
3. Von Mises, R., *Theory of Flight*, Dover Publications, New York, 1959.
4. Norris, J., and Bauer, A. B., "Zero-Thrust Glide Testing for Drag and Propulsive Efficiency of Propeller Aircraft," *Journal of Aircraft*, Vol. 30, No. 4, 1993, pp. 505–511.
5. Lowry, J. T., "The Bootstrap Approach to Predicting Airplane Flight Performance," *Journal of Aviation/Aerospace Education and Research*, Vol. 6, No. 1, 1995, pp. 25–33.
6. Lowry, J. T., "Analytic V Speeds from Linearized Propeller Polar," *Journal of Aircraft*, Vol. 33, No. 1, 1996, pp. 233–235.

8

Maneuvering Performance

Introduction

To this point we have kept the airplane moving (either gliding or at full throttle) in a single vertical plane. We will maintain that engine state bifurcation—power setting parameter Π either 0 or 1—but now we let the airplane turn. In the bootstrap approach, this requires but a single simple substitution:

$$L \doteq W \rightarrow L \doteq \frac{W}{\cos \phi} \tag{8.1}$$

where ϕ is the angle of bank. Among the bootstrap first-line parameters, this devolves into a single change:

$$H \equiv \frac{2W^2}{\rho S \pi e A} \rightarrow H \equiv \frac{2W^2}{\rho S \pi e A \, \cos^2 \phi} \tag{8.2}$$

We will henceforth write H as $H(\phi)$; the unbanked case will be $H(0)$. The aerodynamic source of this modification is simply that, to continue to support the airplane's weight as it banks, total lift must be increased (back stick) so that its projection onto the vertical remains at gross weight W. We are neglecting the facts that, in a glide, a small portion of the airplane's drag helps keep it up and, in a climb, a small portion of the airplane's thrust helps support its weight.

The added induced drag, tagging along on the coattails of the necessarily increased lift, is the source of parameter H's augmentation. See Fig. 8.1 which shows power-available and power-required curves for a Cessna 172 at MSL (weight 2400 lbf) at three different bank angles. At 40 deg bank, as you can see from the reduced overlap of the P_{ar} and P_{re} curves the airplane has diminished rate of climb. In a nutshell, because we are concerned only with quasi-steady

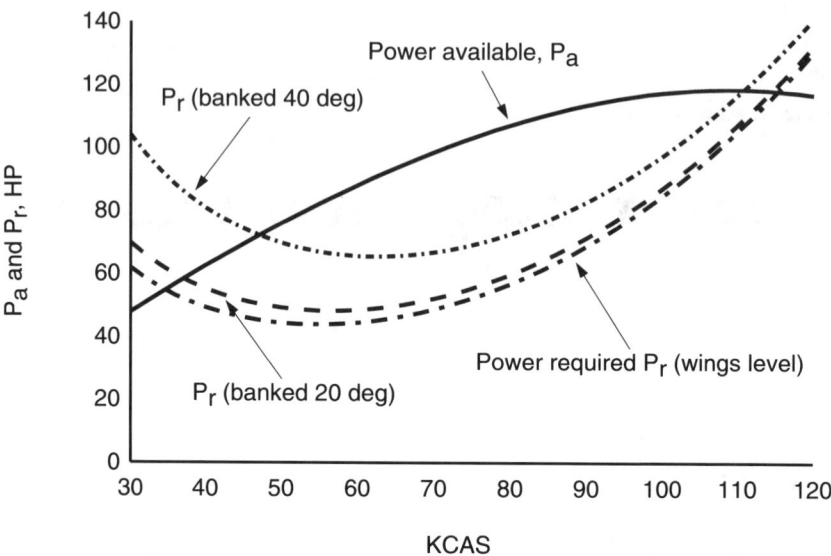

Figure 8.1 Banking the airplane raises the power required curve and rotates it clockwise.

maneuvering—unconcerned with how the airplane ever banked in the first place and how it recovers from that tilt—the two formulas above give us the whole story. In practice, there is quite a bit more to be said.

After a careful review of our dynamic performance assumptions and their expressions in force equations, we begin the maneuvering subject with level turns, both coordinated and uncoordinated. We then discuss structural limitations on general aviation aircraft. The bootstrap approach is then revised, in a thoroughgoing manner, to accommodate maneuvering of any steady sort, including climbing or descending turns. We come to a useful new concept, the "banked absolute ceiling," and look into somewhat strange "crossing" behavior of fixed-pitch values of V_x and V_y. Steady maneuvering charts for those airplanes, providing a wealth of information regarding maneuvering performance, will be displayed; a precise recipe for constructing those charts is given. A graphic example—graphic in both senses—highlights the dangers of high-altitude turns in low-performance aircraft. Particular optimum results for the level case—tightest and fastest turns—are then derived. Those results prove to be quite different than common textbook versions. We will end up with a numerical treatment of maneuvering flight of constant-speed propeller aircraft.

Review of Performance Assumptions

From time to time we need to review the assumptions underlying our performance analyses. This is one such time. They are

- The Earth is flat and the acceleration of gravity g is fixed for any operational altitude at 32.174 ft/s^2.
- The three dynamical inertial terms outlined in Chapter 3 (centrifugal, Coriolis, and rotation deceleration), are all negligible except for the slight diminution of weight from the centrifugal effect, which made the scales indicate a very slightly low weight.
- Buoyancy of the airplane, equal to the weight of air it displaces, is ignored. That small effect also made the scales read a bit light.
- Reynolds number (Re) and Mach number (M) effects, for our general aviation far-subsonic aircraft, can be ignored.
- The airframe's drag polar is accurately quadratic: $C_D = C_{D0} + KC_L^2$.
- Engine torque at given altitude is totally determined by throttle position.
- Aircraft gross weight W, in spite of fuel consumption, is assumed to decrease only negligibly during the course of any single maneuver. This will not be the case, of course, for extended cruises.
- The airplane's engine is always properly leaned for maximum power.
- The propeller polar, for fixed-pitch airplanes, is linear. For constant-speed airplanes, the general aviation general propeller chart is sufficiently accurate.
- In most (but not all) propeller considerations, the effective average speed of air through the propeller is taken to be the same as the "free stream" air speed. i.e., no "slowdown factor." Without this simplifying assumption, propeller coefficients would not depend on only propeller advance ratio J and the analysis would be considerably complicated.
- Flight path angles γ are shallow enough that $\cos \gamma \doteq 1$. Sin γ, however, is generally *not* taken as zero for nonzero γ.
- The airplane's flight is to be considered unaccelerated (except for centripetal acceleration when turning), coordinated (velocity always in the direction the aircraft is pointed), and properly trimmed (no moments).
- Thrust is directly aligned with the flight path angle γ. This ignores any possible thrust offset angle and, moreover, the effect of body angle of attack (AOA) in letting the thrust vector provide a slight lifting effect. Asymmetric propeller loading, so-called "P-factor," is also to be ignored.
- The airframe is rigid (as against elastic) and no rotating or reciprocating parts (engine or control surfaces) need be considered as influencing performance.
- Bank and pitch angles are assumed to commute, that is to give equivalent results no matter in which order they are taken.
- Wing dihedral (or anhedral) angle, if any, is ignored.

- In many cases, no wind. The atmosphere is then stationary with respect to the surface of the Earth. And dry. In a few explicitly stated cases, those assumptions are relaxed. When wind effects are considered, slant air speed and the horizontal component of air speed are usually taken as equal.
- The slight acceleration due to increasingly rarified atmosphere in a constant calibrated air speed climb (or the similar deceleration during descent)—the so-called "kinetic energy" effect—is ignored.
- During takeoff or landing rolls, energy going into spinning up the wheels (a slight deficit on takeoff, a small plus on landing) is ignored. During those maneuvers, in fact, several such small effects will be ignored.
- When making a steady climbing or descending turn, the radius of curvature of the flight path is taken essentially equal to the radius of the cylinder on which it is wound.

It is a long roll we have called. But no more extensive than the list traditionally assumed—explicitly or implicitly—in the vast majority of earlier books on this performance subject. A few such books are specialized exceptions. Our simplifying assumptions should continually be reviewed but, given the vagaries of atmospheres and engines and riggings, they are not crippling. In spite of this manifest imperfection, with careful attention and some repetition useful results can be extracted. We will discuss a few of these points further on.

Flight Controls and Associated Variables

The pilot maneuvers the airplane by manipulating the controls. These, along with the dynamic or kinematic variables each determines, are

- Stick or control column (back and forth). Controls elevator position, which determines wing AOA (thereby lift and induced drag) and air speed V.
- Stick or control column (side to side). Controls ailerons, allowing the airplane to bank (or unbank), leading to (or from) turns. In general, because we treat quasi-steady maneuvers almost exclusively, the roll rates or other detailed motions generated by the ailerons are not of direct interest to us in this work.
- Rudder pedals. Deflects the rudder, causing the airplane to yaw. Proper use ensures coordinated turns. As a concomitant to banking, again not of direct interest to us.
- Throttle. Controls manifold pressure and flow rate of gasoline/air mixture into the engine cylinders and thereby determines the torque developed by the engine. Density altitude is a further important parameter. The propeller transforms this torque into propeller thrust (along with some propeller drag) which, if left to its own devices, will be translated into altitude changes.

- Propeller speed control (constant-speed propeller airplanes only). Controls engine and propeller revolutions per minute, within limits, and thereby engine brake power for given throttle position (manifold pressure reading).

Aircraft Equations of Motion

To maneuver the airplane—change its velocity either in speed or direction or both—the pilot's control inputs must be translated into changes in forces acting on the airframe. The forces that act are of course familiar vectors lift **L**, drag **D**, thrust **T**, and weight **W**. Because jettisoning cargo is seldom an option, we simply take weight **W** as a given parameter always directed downwards towards the center of the Earth. Because we consider only quasi-static maneuvers (the quasi hedge has to do with the slow siphoning away of gasoline and weight, the kinetic energy effect, and the fact that even steady turns are accelerated), the airplane moves from one situation of total force

$$F = L + D + T + W \tag{8.3}$$

to another. Except while turning, $F = 0$; during steady turns, $F = F_c$, the aerodynamically generated centripetal force. It is most convenient to write Eq. (8.3) as three equivalent scalar equations for force components parallel and perpendicular to the instantaneous flight path.

Parallel to the flight path:

$$T \cos(\alpha + \alpha_T) - D - W \sin \gamma = 0 \tag{8.4}$$

Perpendicular to the flight path and in the vertical plane,

$$L \cos \phi - W \cos \gamma + T \sin(\alpha + \alpha_T) = 0 \tag{8.5}$$

Perpendicular to the flight path and in the horizontal plane:

$$L \sin \phi - \frac{W \omega V}{g} = 0 \tag{8.6}$$

In these equations, α is the body AOA, α_T is the angle by which the thrust direction is offset above α, ϕ is the bank angle (not necessarily at all a small angle), γ is the flight path angle (negative when descending), ω is the turning rate (angular velocity) in rad/s. A radian is $180/\pi \doteq 57.3$ deg. V is the true air speed of the airplane in ft/s. The other symbols have their usual meanings: T for thrust, D for drag, L for lift, W for weight, and g for the acceleration of gravity. The attitude or pitch angle θ_i—how much the long axis of the airplane is pointed above the horizon—can easily be calculated from the AOA α and the flight path angle γ: $\theta_i = \gamma + \alpha$.

The three directions represented by Eq. (8.4)–(8.6) are mutually perpendicular. If one computes the sum of the squares of the three, one will indeed find that

$$\mathbf{F} - \mathbf{F}_c = 0 \tag{8.7}$$

where the centripetal force has magnitude

$$F_c = \frac{WV^2}{gR} = WR\omega^2/g = W\omega V/g \tag{8.8}$$

because angular speed $\omega = V/R$, R the radius of curvature of the airplane's path.

We shall find, in the next chapter on glide performance, that setting thrust $T = 0$ means that Eq. (8.4)–(8.6) can often be solved exactly. But our job now is to invoke some of the approximations recently mentioned to make that full set more tractable. Then we have the simpler but still very realistic group

$$T - D - W \sin \gamma = 0 \tag{8.9}$$

$$L \cos \phi - W = 0 \tag{8.10}$$

$$L \sin \phi - \frac{W\omega V}{g} = 0 \tag{8.11}$$

If we had left the $\cos \gamma$ correction to the second term in Eq. (8.10), these would again be consistent, with the total force on the airplane reducing to only the centripetal aerodynamic force. As it is, they are "almost" so.

Coordinated Level Turns

But at this stage it is much more important to understand—approximations or no—the force relations in an approximately level turn. Figure 8.2 gives the picture. It is the horizontal component of the lift vector that provides the centripetal force starting the turn. Aerodynamic side force abaft the airplane's center of gravity and some rudder to counteract adverse yaw keep the turn coordinated.

The force that makes the airplane travel in a circle is the centripetal force, the inwards component of the tilted lift vector, of length $L \sin \phi = WV^2/gR$. If one then substitutes for L in that relation, from Eq. (8.10), cancels weight W, and solves for radius of curvature R, one finds the important turning relations

$$R = \frac{V^2}{g \tan \phi} \tag{8.12}$$

$$\omega = \frac{V}{R} = \frac{g \tan \phi}{V} \tag{8.13}$$

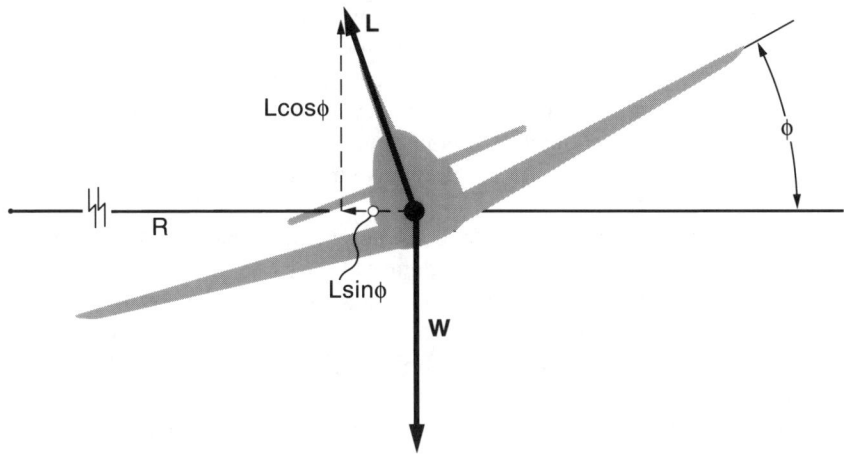

Figure 8.2 The only forces on the airplane, perpendicular to its flight path, are lift **L** and weight **W**.

When the turn is not level, Eq. (8.12) is not quite correct. In those climbing or descending cases, there is an additional factor $\cos\gamma$ in the denominator on the right-hand side (RHS). In addition, Eq. (8.13) needs a slight emendation. While we shall take a close look at these mathematical details in Chapter 9, Glide Performance, for now we only stick a toe into the chilly waters of differential geometry.

A Closer Look at the Flight Path Helix

Let us take a closer look. Our flight path helix can be generated by thinking of the airplane moving at constant speed in a circle while lifting or lowering it, at some other constant vertical speed, perpendicular to the plane of that circle. In the limit of high translation speeds, the path of the moving point would approach a straight line. So "certainly" as we translate at finite speeds the stretched Slinky has a larger radius of curvature, at each point, than does the compressed one. The upshot is that if the flight path radius of curvature is given by $R = V/\omega$ and the planform radius of the cylinder on which the helical flight path is wound is given by r, then closer analysis of the helix gives

$$r = \tfrac{1}{2}R\left(1 + \sqrt{1 - \left(\frac{p}{\pi R}\right)^2}\right) \tag{8.14}$$

where pitch p is the altitude lost in gliding one full "circle." In almost all practical cases, it is good enough to use the approximation that $R = r$.

One further approximation. Say one is making a constant true air-speed climbing turn at full throttle. At lower altitudes, the engine will have more torque and the flight path angle γ will be larger than at high altitudes. So, as the pilot ascends, the spiral path will become somewhat more compressed, with lower pitch p. So the long-term flight path is not quite a helix.

Let us put some numbers into the previous three formulas.

Example 8.1 Take an airplane (it does not matter what kind) traveling at 80 KTAS = 135.0 ft/s, descending at rate of "climb" − 600 ft/min = −10 ft/s, banked 30 deg. Equation (8.12) then gives radius of curvature $R = 981.1$ ft. And Eq. (8.13) implies the yaw rate is 0.1376 rad/s = 7.884 deg/s, which turns into 45.7 s for the airplane to make one complete helical turn, during which it lost $p = 457$ ft in altitude. Plugging these values of R and p into Eq. (8.14), we find that the cylinder onto which the helical flight path is wound has radius $r = 975.5$ ft. As promised, a bit smaller than R.

Turns, Centrifugal Force, and the Federal Aviation Administration

Fitting centrifugal force into a description of how airplanes turn leads to confusion. Mild disorientation is often revealed during "hangar flying" sessions on the subject whether held at the airport or on the Internet. Even official Federal Aviation Administration (FAA) publications line up somewhere between misleading and erroneous in their treatment of centrifugal force in turns. There is a fairly good reason for this confusion: "centrifugal force" has two related but distinct meanings. Before we give them, let us take a look at what the FAA has to say.

Figure 8.3 is taken from *The Pilot's Handbook of Aeronautical Knowledge*.[1] It purports to show the forces (perpendicular to the flight path) acting on an airplane in a level turn. (Kindly overlook the artist's oversight or ignorance in failing to resolve the resultant lift vector correctly into horizontal and vertical components.) The FAA's basic idea here is twofold: 1) that the vertical lift component exactly cancels the airplane's weight, which is correct, and 2) that the horizontal lift component is exactly canceled by a centrifugal force acting on the airplane, which is not. In their discussion, they write "the horizontal component must overcome centrifugal force" in one place and, in another, "The total resultant lift acts opposite to the total resultant load. So long as these opposing forces are equal to each other in magnitude the airplane will maintain a constant rate of turn."

In the FAA's *Flight Training Handbook*,[2] the total resultant load idea is further promulgated (see Fig. 8.4.) In that discussion, after correctly stating that "The

Maneuvering Performance   243

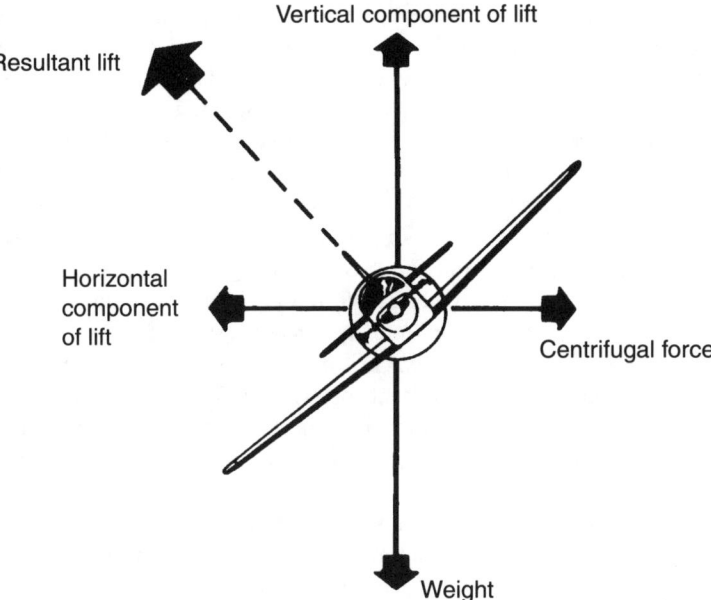

Figure 8.3   The FAA believes that a turning airplane is acted on by a centrifugal force.

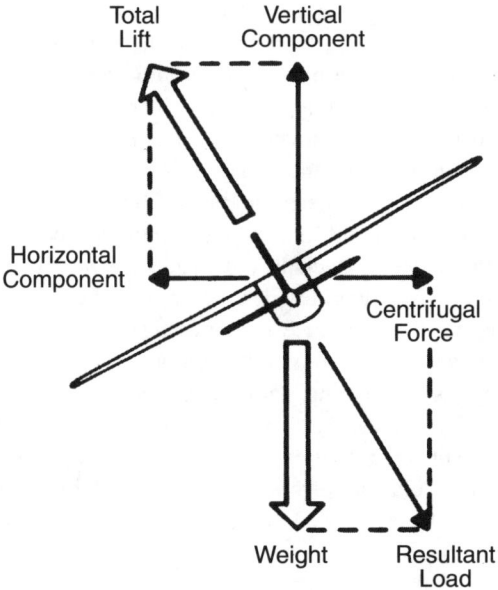

Figure 8.4   The FAA is consistent in believing that a centrifugal force acts on the turning airplane.

horizontal component of lift is the force that pulls the airplane from a straight flight path to make it turn," the FAA goes on to say "Centrifugal force is the 'equal and opposite reaction' of the airplane to the change in direction and acts equal and opposite to the horizontal component of lift." Not so.

Now for the verbal and physical facts. One sense (#1) in which "centrifugal force" is used can be illustrated by the familiar example of a child's hand whirling a rock tied to a string. Newton's third law speaks of "equal and opposite reactions" and, in this case, one of those is the (centripetal—center-seeking) force due to the child's hand (transferred by the string) on the rock. The other is the (centrifugal) force due to the rock (transferred again by the string) on the child's hand. The important point is that, in using sense #1, while the forces do balance, *the two forces act on different objects*. In the turning airplane case, the force causing the airplane to turn is a (centripetal) force caused by the aerodynamic force of the air on the airframe. Its third law partner is the force of the airplane on the air, a (centrifugal) force causing the air to swoosh outwards. If the forces *on the airplane* balanced, it would move in accordance with Newton's first law, in a straight line at constant speed. Because its velocity vector is changing direction, even though not changing length, a turning airplane is accelerating. We grant that this is a particularly simple case of acceleration, but it is acceleration nonetheless.

The second sense (#2), in which "centrifugal force" is used is as one of the so-called "pseudoforces" employed to describe the motion of an object from the point of view of a non-inertial frame of reference. This was discussed earlier in the performance preliminaries chapter (Chapter 3).

Newton's laws of motion are only valid in an inertial frame of reference, one moving uniformly (and therefore not rotating) with respect to "the fixed stars." For the purposes of this discussion let us ignore the rotation and revolution of the Earth and take it to be an approximately inertial frame. For small-scale movements—such as the turning airplane—that approximation is sufficiently valid. While riding on a (noninertial) merry-go-round, the ticket booth undergoes all sorts of gyrations, from our point of view, even though there are no unbalanced forces on the booth. A beetle stuck in the tread of a bike tire similarly sees a violently accelerating outside world which, again, is undergoing no such external forces. The turning airplane, because it is moving not in a straight line and is rotating (yawing), is yet another noninertial frame.

The detailed physics of rotating coordinate systems would lead us too far astray. The simple fact in the uniformly rotating case is that there are two pseudoforces—apparent forces that explain the motions of object seen from non-inertial frames—the outwards-from-the-axis centrifugal force of size $mV^2/R = mR\omega^2$ and the somewhat more complex Coriolis force. These are essentially cancellations of apparent motions of objects, motions that are in fact due to the rotation of the noninertial frame itself. The rotating frame is not subjected to either pseudoforce; instead, those are "forces" on *other objects*, viewed with respect to the noninertial frame.

"Now just a minute!" you might reasonably exclaim at our discarding the reality of the centrifugal force. "When I'm driving my car, with my sunglasses resting on the smooth dashboard, and take a curve to the left, what about the centrifugal (outward) force that sends the glasses sliding to the right?" The explanation is that, in this case, the turning car is providing the frame of reference and the sunglasses are another object whose motion is to be described by an observer fixed in that noninertial frame. The observer sees a pseudoforce acting on the glasses. In an inertial frame (say the Earth) the description is quite different: the glasses did *not* move toward the outside of the curve! As seen by an (inertial) hawk hovering overhead, the glasses moved to the inside, though not as speedily as the car did. Even if the dashboard were completely frictionless, the glasses would merely have continued in a straight line (until they hit the lower right corner of the windshield, of course). A straight line, that is, with reference to *an inertial frame of reference*, which the turning automobile is certainly not. In fact the relative motion of the sunglasses and dashboard is most properly explained by saying that the car was forced (by having the roadway push on the tires, which pushed on the ankle bone, which pushed on...etc.) until the dashboard was pushed toward the inside of the turn and slid beneath the sunglasses. The glasses, if we ignore all friction for a moment, just kept going straight, as does anything when there is no net force on it.

Whether one uses "centrifugal force" in sense #1 or #2, there is no centrifugal force or even centrifugal pseudoforce acting on the turning airplane. A physics professor (and private pilot) at the University of Texas at Austin, Fritz de Wette, has for some years tried to politely disabuse the FAA of their faulty notion of centrifugal force. To no avail.

Coordinated, Slipped, and Skidded Turns

A coordinated turn is one in which the airplane is pointed (relative to the air mass) in the same direction it is moving. A centered inclinometer (or turn-and-bank indicator) ball shows the pilot that he or she is flying a coordinated turn. If you fly a 360 deg circle, coordinated, you must also yaw 360 deg. If you yaw less quickly than your center of mass turns, you are slipping and the ball goes to the low side; if you yaw more quickly than you turn, you are skidding and the ball goes to the high side, to the outside of the turn. When you slip on final approach, to counter a crosswind, you use opposite rudder to prevent the airplane from yawing. When you polish a mirror high on the wall, to take another example, the center of your hand (and the polishing cloth) moves in a circle, but your fingers remain pointing primarily upwards, hardly yawing at all.

To correct a slipping turn, bank less or use rudder to yaw faster. Or speed up. To correct a skidding turn, bank more or use less rudder. Or slow down. A spoonful of rudder, more or less, is the usual prescription. Ultralight airplanes give wonderful demonstrations of turning problems, including a separate secondary aerodynamic effect, "adverse yaw." It seems the smaller the airplane, the harder to fly. But, to a point, the better to learn in.

We can use the centrifugal pseudoforce to analyze the inclinometer's behavior. For simplicity, replace the usual inclinometer—a ball in a fluid-filled arc-shaped race—by a bead constrained to slide (without friction) on a wire circle (see Fig. 8.5). The plane of the wire circle is perpendicular to the long axis of the airplane. In the noninertial frame of the turning airplane, the bead has three forces acting on it: 1) the constraining force of the wire, which we can ignore; 2) weight of the bead, downward; and 3) centrifugal pseudoforce, outward. When the bead has stabilized, has stopped sliding along the wire loop, the components of forces 2) and 3) along the directions tangent to the wire are of equal size but oppositely directed. For simplicity, focusing on accelerations instead of forces, we then have

$$g \sin \theta = (R + r \sin \theta)\omega^2 \cos \theta \qquad (8.15)$$

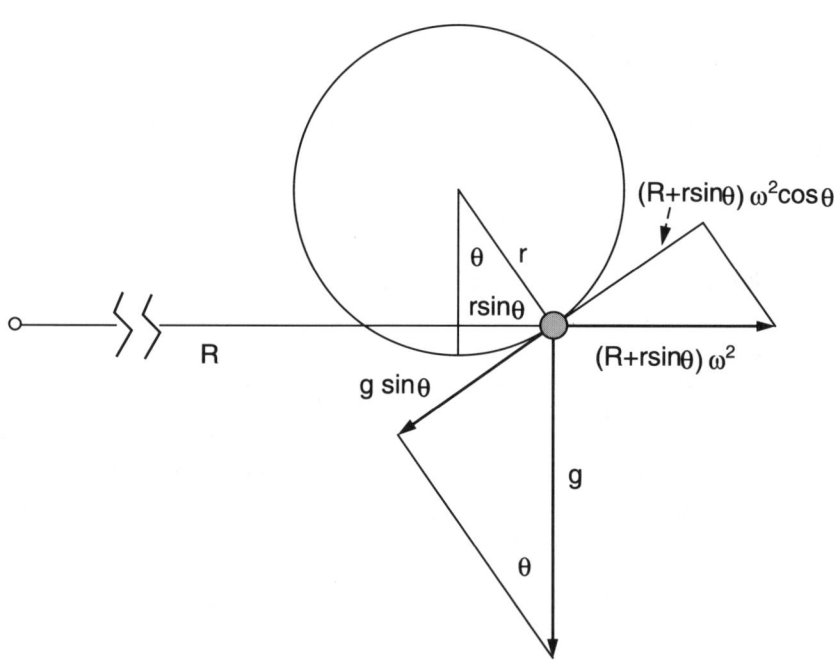

Figure 8.5 Accelerations on the inclinometer ball as seen from the rotating frame of reference of the airplane.

We can ignore $r\sin\theta$, tiny compared to turn radius R. We then have, for the inclinometer, the relation [essentially the same as Eq. (8.12) moved to this new context]:

$$\frac{g}{R} = \frac{\omega^2}{\tan\theta} \tag{8.16}$$

For the airplane in a coordinated turn [via Eq. (8.12) and (8.13)], we know that

$$\frac{g}{R} = \frac{\omega_c^2}{\tan\phi} \tag{8.17}$$

where ω_c is the coordinated turn (therefore, yaw) rate and ϕ is the aircraft bank angle. Saying the inclinometer ball is "centered" means $\theta = \phi$. But, from Eqs. (8.16) and (8.17), this can only be true if $\omega = \omega_c$.

A bit farther down this path, we can see how to use the inclinometer as an ersatz "relative yaw" indicator, an instrument telling us what our actual yaw rate is compared to the yaw rate proper for a coordinated turn (see Figs. 8.6 and 8.7). From Eqs. (8.16) and (8.17) we see that

$$\frac{\omega}{\omega_c} = \sqrt{\frac{\tan\theta}{\tan\phi}} = \sqrt{\frac{\tan(\phi+\epsilon)}{\tan\phi}} \tag{8.18}$$

where ϵ is the angle by which the ball is above being centered. After a little trigonometry,

$$\frac{\omega}{\omega_c} = \sqrt{\frac{\left(1 + \frac{\tan\epsilon}{\tan\phi}\right)}{(1 - \tan\epsilon\tan\phi)}} \tag{8.19}$$

Figure 8.6 gives us a feeling for the workings of our new instrument. Let us calculate a sample point.

Example 8.2 Figure 8.6 is based on an airplane slipping or skidding off its coordinated turn at 100 KTAS (168.8 ft/s) banked 30 deg; the corresponding turn radius is $R = 1534$ ft and the coordinated turn or yaw rate is $\omega_c = 0.110$ rad/s $= 6.31$ deg/s. Assume the airplane is skidding its turn so that the inclinometer is 15 deg to the outside. That makes the actual turn rate, by Eq. (8.19), $\omega = 8.30$ deg/s and the excess yaw rate $\omega - \omega_c = 1.99$ deg/s, a 31.5% relative error.

Using a binomial expansion on Eq. (8.19) somewhat linearizes and certainly simplifies. The result for relative excess yaw rate is

$$\frac{\omega - \omega_c}{\omega_c} \doteq \frac{\tan\epsilon}{\sin 2\phi} \tag{8.20}$$

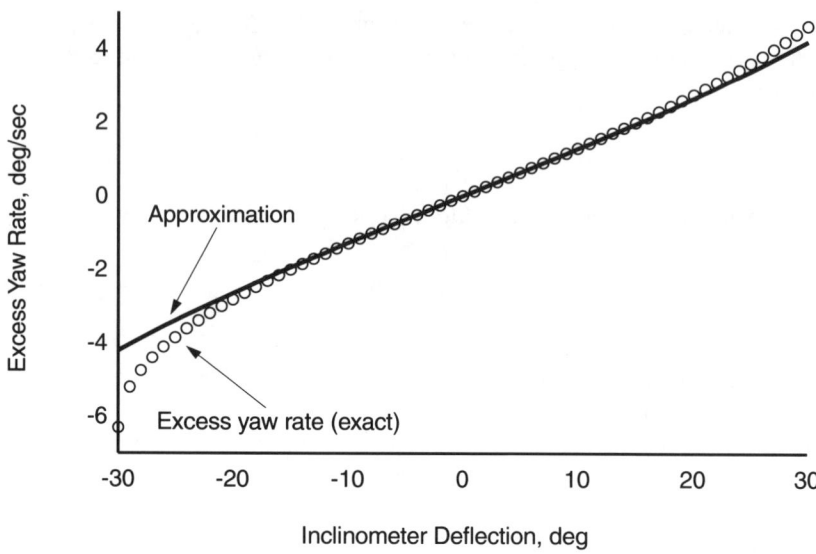

Figure 8.6 The inclinometer makes a passable "excess yaw rate" instrument.

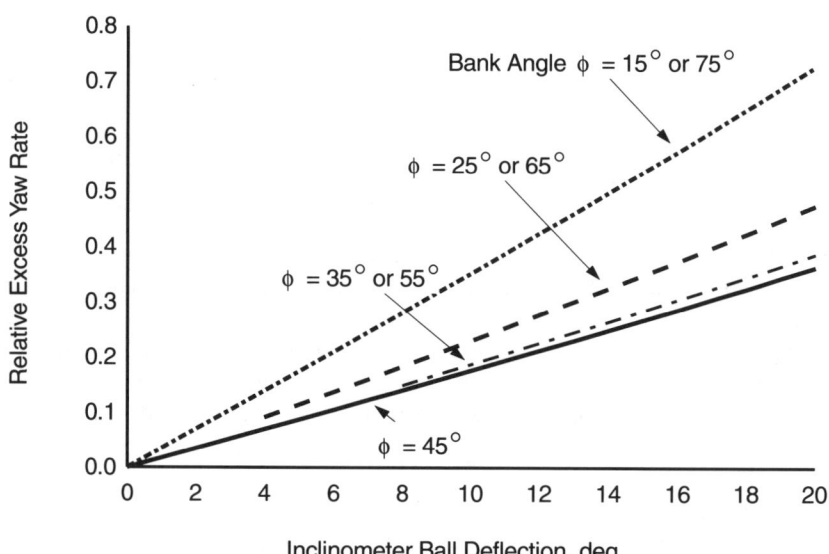

Figure 8.7 The inclinometer gets progressively less sensitive to excess yaw rate as bank angles approach 45 deg; it regains sensitivity beyond that angle.

One can see from Fig. 8.6 that this approximation is a fairly close one. So if we keep in mind that the RHS of Eq. (8.20) is odd in inclinometer (off-coordinated) deflection angle ϵ, and cut the range down to a reasonable 20 deg, we get a chart useful for a much wider range of bank angles (see Fig. 8.7). Because the sine function is symmetric about 90 deg, the RHS of Eq. (8.20), with its double-angle denominator, is symmetric about 45 deg.

Structural Limits in Level Turns

Even though our study is aircraft performance, not design, and almost exclusively steady unaccelerated performance, not aerobatic, we must touch on the V-n diagram if only to enhance our awareness of technical culture. V-n diagrams are plots of load factor n (no relation to propeller revolutions per second) against air speed V with legal or manufacturer-determined structural and ultimate limits shown. Load factor is

$$n \equiv \frac{L}{W} = \frac{1}{\cos \phi} \qquad (8.21)$$

A typical graph looks like Fig. 8.8. Since load factor limits are as much legal and bureaucratic limits as they are technical ones, there are ancillary provisos and much small print. One major restriction is that we are talking only about loads impressed in the lift directions (both upright and inverted). Maneuvers that

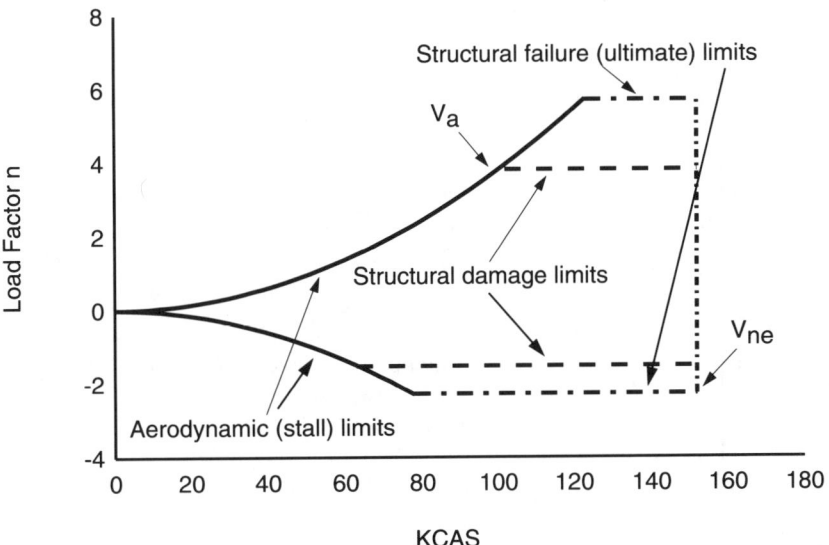

Figure 8.8 V-n diagram for a Cessna 172 at 2400 lbf, flaps up.

include rolling have further and more stringent limits. A second restriction is that, ordinarily, a diagram is drawn for only one gross weight and one flaps configuration; one must create diagrams for the other cases. Figure 8.8, for instance, is only for the Cessna 172 at 2400 lbf with flaps up. V_{ne} is the "never exceed speed," here 152 KCAS. The major limit depicted is the positive structural load limit. In acceleration terms, this is $3.8g$. This limit is related to the maximum level turn bank angle (74.7 deg) the airplane can sustain without damage.

Figure 8.9 narrows focus, in one sense, to only positive limits, starting up from the wings level stall limits ($L = W$, or $n = 1$), but enlarges the V-n conceptual picture by including two weights and two flaps configurations. Maneuvering speeds V_a are those at which the aerodynamic (stall) limit curve meets the structural damage limit line.

A V-n diagram is often loaded down with further information: design cruising speed V_c, design diving speed V_D, and gust load factor lines for those speeds. For details, see Roskam and Lan.[3] The V-n diagram does implicitly include bank angle information, since $n = 1/\cos\phi$, but not in an easy-to-use format. Later in this chapter, we introduce "steady maneuvering charts," in appearance somewhat like "doghouse" plots used to describe fighter aircraft turn performance, which contain *much* more information on an airplane's maneuvering capabilities at a given weight and altitude.

For maneuvering performance information, one cannot rely exclusively or even very much on the V-n diagram. Its basic flaw is that it does not incorporate information about the aircraft power plant—in our case, engine and propeller—

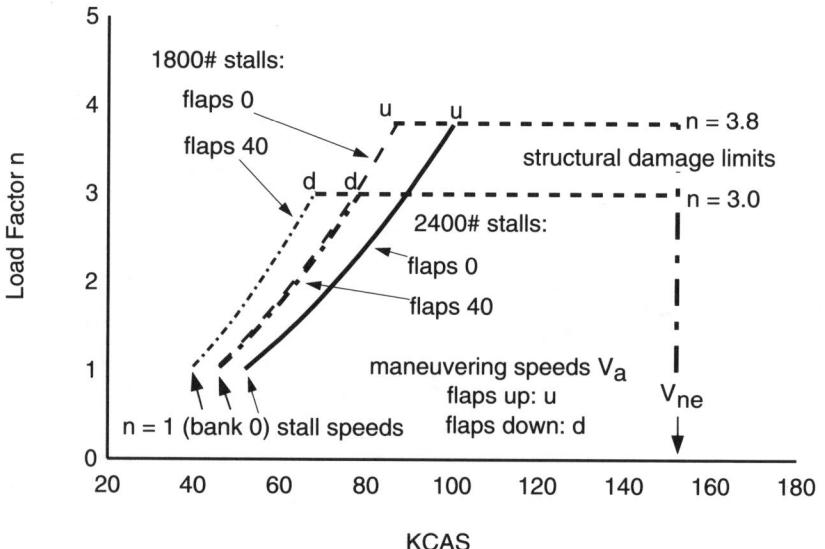

Figure 8.9 Zooming in on the upper right portion of a V-n diagram.

and, in the general aviation case, most of the airplanes' maneuvering capabilities are thrust limited. While it may be true (see any of the following steady maneuvering charts) that the point on the V-n diagram defining maneuvering speed V_a—where banked stall speed meets structural load limit line—provides maximum turn rate and minimum turn radius for *level* flight, general aviation aircraft cannot *maintain* level flight at such extreme bank angles. So that specification is somewhat empty. Because of the large number of variables involved, maneuvering performance is a traditionally difficult subject. So, to the extent possible, using the bootstrap approach, let us simplify it.

Extension of Bootstrap Approach to Steady Full-Throttle Maneuvering Flight

In wings-level, constant-altitude unaccelerated flight to our approximation (no off-axis thrust component), lift L is equal to weight W. When the pilot wants to turn the airplane, he or she banks to some angle ϕ, tilting the lift vector towards the desired direction. To maintain altitude, the pilot also applies sufficient backstick, enlarging the vertical component of lift, to balance weight. Because increased lift means increased induced drag, the pilot must also add power if air speed is to be held constant.

As mentioned, this additional induced drag while turning leads to the only modification of the wings-level bootstrap approach required to encompass quasi-steady-state maneuvering flight:[4]

$$H \equiv H(0) \to H(\phi) \equiv \frac{H(0)}{\cos^2 \phi} \qquad (8.22)$$

The zero in parentheses denotes a value for unbanked, wings-level flight. For most (not all) intents and purposes, banking to angle ϕ is tantamount to increasing gross weight from W to $W/\cos \phi$. Using Eq. (8.22) in the defining relations for bootstrap approach V speeds shows that banking transforms several such speeds precisely as it does the stall speed:

$$V_{S/x/bg/md}(\phi) = \frac{V_{S/x/bg/md}(0)}{\sqrt{\cos \phi}} \qquad (8.23)$$

where the wings-level stall speed is given by

$$V_S(0) = \sqrt{\frac{2W \cos \gamma}{\rho S C_{L\max}}} \doteq \sqrt{\frac{2W}{\rho S C_{L\max}}} \qquad (8.24)$$

For our sample Cessna with flaps up, $C_{L\max} = 1.54$. $V_x(0)$, $V_{bg}(0)$, and $V_{md}(0)$ are given by

$$V_x(0) = \left(\frac{-H(0)}{K}\right)^{1/4} = (-R(0))^{1/4} \tag{8.25}$$

$$V_{bg}(0) = \left(\frac{H(0)}{G}\right)^{1/4} = U(0)^{1/4} \tag{8.26}$$

$$V_{md}(0) = \left(\frac{H(0)}{3G}\right)^{1/4} = \left(\frac{U(0)}{3}\right)^{1/4} \doteq 0.7598\, V_{bg}(0) \tag{8.27}$$

Figure 8.10 shows the effect, at two sample altitudes, that banking has on speeds for best angle and for best rate of climb. V_x and V_y curves cross at the condition for banked absolute ceiling.

Speeds for best rate of climb and minimum or maximum level flight are somewhat more complicated. In the banked case, those three V speeds are:

$$V_{M/m}(\phi) = \sqrt{\frac{-E \mp \sqrt{E^2 + 4KH(\phi)}}{2K}} = \sqrt{-\frac{Q}{2} \pm \sqrt{\frac{Q^2}{4} + R(\phi)}} \tag{8.28}$$

$$V_y(\phi) = \sqrt{\frac{-E - \sqrt{E^2 - 12KH(\phi)}}{6K}} = \sqrt{-\frac{Q}{6} + \sqrt{\frac{Q^2}{36} - \frac{R(\phi)}{3}}} \tag{8.29}$$

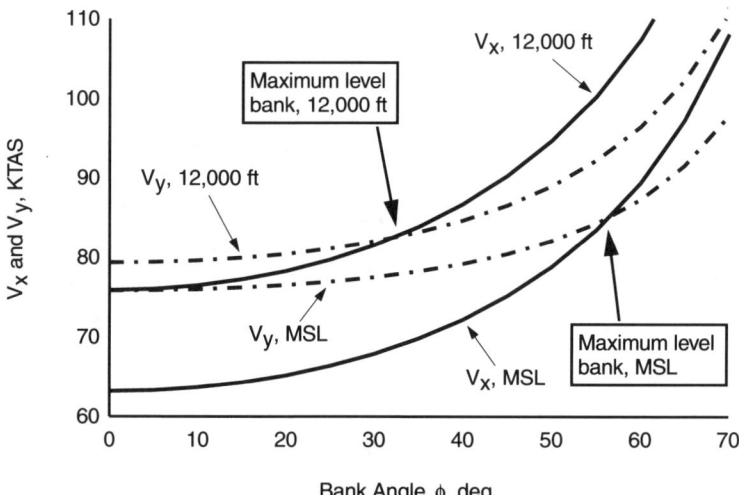

Figure 8.10 Speeds for best rate of climb and best angle of climb cross at the maximum feasible bank angle for level flight at a given altitude.

Maneuvering Performance   253

They now depend, instead of simply on $R = R(0)$ as in our wings-level formulation, on

$$R(\phi) \equiv \frac{R(0)}{\cos^2 \phi} \qquad (8.30)$$

As is shown in Fig. 8.10, V_x is affected more by banking than is V_y. That is because V_x depends on excess thrust and induced drag drops off faster, with increasing air speed, than does the induced power responsible for the speed dependence of V_y.

Figure 8.11 shows the dependence of minimum and maximum level flight speeds and stall speed on bank angle at two different altitudes.

Banked forms for rate and angle of climb are arrived at simply by substituting $H(\phi)$ for $H = H(0)$ in the equations for the wings-level versions of those quantities. The banked versions are given below as Eqs. (8.37) and (8.41). Figures 8.12 and 8.13 show the speed dependence of rate and angle of climb, at mean sea level (MSL), for four specific bank angles. Cutoffs on the left portions of those eight graphs are due to impending stall speeds.

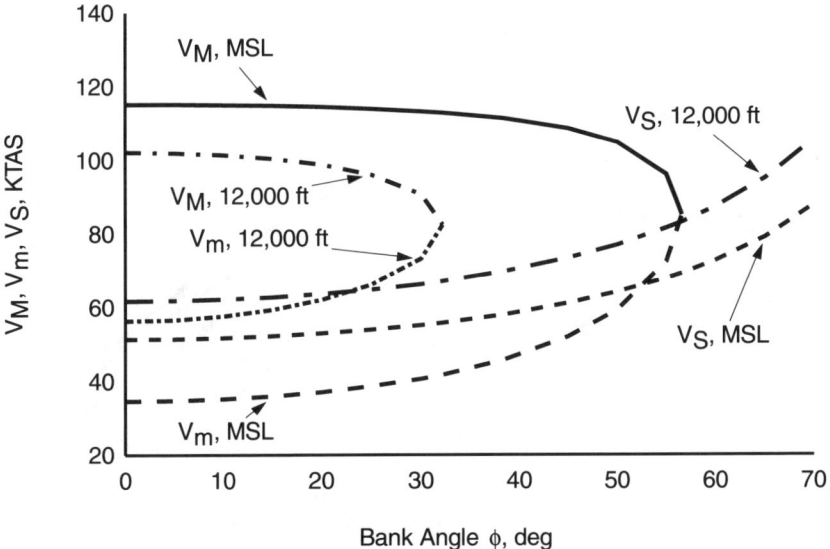

Figure 8.11   Only at relatively high altitudes is minimum speed for level flight higher than stall speed for a considerable range of aircraft bank angles.

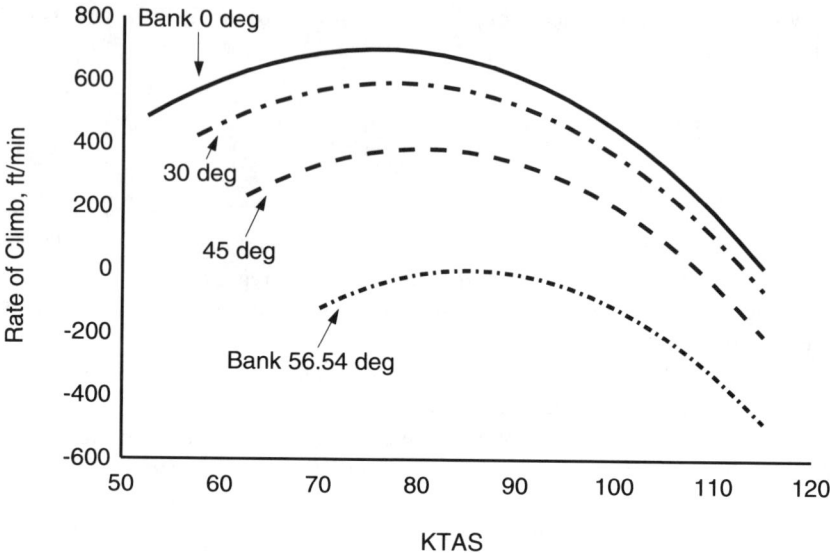

Figure 8.12 Banked 56.54 deg, this airplane can fly level at MSL, but only at 85 KTAS.

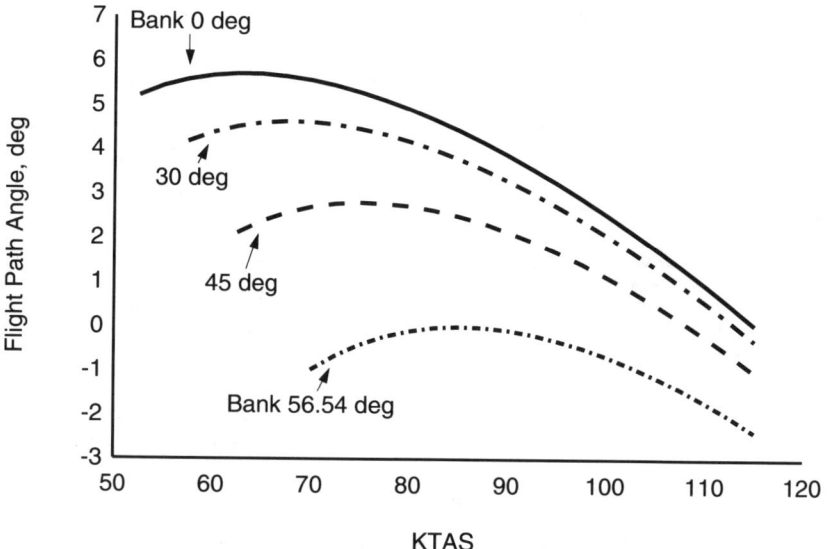

Figure 8.13 Banked 56.54 deg, this aircraft is at its ceiling even though at sea level.

Banked or Unbanked Absolute Ceilings

An airplane's unbanked absolute ceiling, while perhaps a theoretical performance benchmark worthy of record, is not attainable. For the airplane at given weight in given configuration, there are two numbers involved: 1) the ceiling figure, a density altitude, and 2) the air speed the aircraft requires to stay at that altitude (were it dropped there by some higher-flying entity). The airplane's *banked* ceilings, however, are readily achieved. This is, therefore, a much more practical and interesting set of figures, one for each finite bank angle. The airplane's (banked or unbanked) absolute ceiling is characterized analytically by the graphs of power available P_a and power required P_r, using the possibly banked form of composite parameter $H(\phi)$, touching at one point. The expression of that osculation is that both P_{xs} and dP_{xs}/dV are zero. Using

$$P_{xs}(\phi) \equiv P_a - P_r(\phi) = EV + KV^3 - H(\phi)/V \qquad (8.31)$$

to express that dual condition, one can use elementary methods to (somewhat circuitously) derive simple expressions for the required ceiling air speed and altitude for given gross weight and bank angle:

$$V_{AC}(W, \phi) = \sqrt{\frac{2H(\phi)}{E}} \qquad (8.32)$$

$$h_{\rho AC} = 145{,}457(1 - \sigma_{AC}^{0.23494}) \qquad (8.33)$$

The ceiling relative air density is given in terms of the ceiling power dropoff factor

$$\sigma_{AC} = (1 - C)\Phi_{AC} + C \qquad (8.34)$$

which, in turn, is given by

$$\Phi_{AC}(W, \phi) = \frac{2W}{W_B E_B \cos \phi} \sqrt{-K_B H_B(0)} \qquad (8.35)$$

In Eq. (8.35), and many following, former subscripts zero have been changed to subscripts B for "base case"; zeros have been conscripted to denote wings-level-flight.

To get a clear physical picture of what is going on, one can sketch graphs like Figs. 8.12 and 8.13. Figure 8.12 plots rate of climb against air speed for a Cessna 172 (weight 2400 lbf, flaps up, at MSL) for four bank angle values. Unbanked, the airplane has best rate of climb about 700 ft/min at V_y about 76 KCAS. When banked 56.54 deg the airplane will descend unless precisely at 85 KCAS. These are, of course, full-throttle graphs. Each graph ends on the left at the corresponding banked stall speed. Figure 8.13 is very similar but features flight path angle against air speed. Notice that banked values of V_x are less than their V_y counterparts *except* in the case of the largest bank, at which the airplane is at absolute banked ceiling even though at sea level.

Time for calculation of an example.

256   John T. Lowry

Example 8.3   We found in Chapter 7 (Example 7.8) that our sample Cessna 172, wings level and weighing 2400 lbf, had absolute ceiling $h_{\rho AC} = 16,184$ ft. How much lower would this airplane's absolute ceiling be if it were banked 30 deg? And what air speed would be required for it to maintain that ceiling? For concrete data, we recast former Tables 7.1 and 7.3 as Tables 8.1 and 8.2:

To answer these absolute ceiling questions, we can run through Eqs. (8.32–8.35) in reverse order or we can simply make modifications, due to the bank angle, to the previous results in Chapter 7. We choose the latter briefer strategy. Looking at Eq. (8.35) in a relative way gives the useful relation

$$\Phi_{AC}(W, \phi) = \frac{\Phi_{AC}(W, \phi = 0)}{\cos \phi} \qquad (8.36)$$

Table 8.1   Cessna 172 bootstrap data plate with items attached to appropriate aircraft subsystems

Bootstrap data plate item	Value	Units	Aircraft subsystem
Wing reference area, S	174	ft^2	Airframe
Wing aspect ratio, A	7.38		Airframe
Rated MSL torque, M_0	311.2	ft-lbf	Engine
Altitude dropoff parameter, C	0.12		Engine
Propeller diameter, d	6.25	ft	Propeller
Parasite drag coefficient, C_{D0}	0.037		Airframe
Airplane efficiency factor, e	0.72		Airframe
Propeller polar slope, m	1.70		Propeller
Propeller polar intercept, b	-0.0564		Propeller

Table 8.2   Bootstrap composite numbers and operational variables, two cases

Variable or composite	Base case, W_0 and MSL	$W/W_0 = 0.75$, 8000 ft
W	2400 lbf $= W_0$	1800 lbf $= W$
σ	1	0.75602
$\Phi(\sigma)$	1	0.75684
E	$531.9 = E_0$	402.6
F	$-0.0052368 = F_0$	-0.0041162
G	$0.0076516 = G_0$	0.0060142
H	$1,668,535 = H_0$	1,194,062
K	$-0.0128884 = K_0$	-0.0101305
Q	$-41,270.6 = Q_0$	$-39,738.4$
R	$-129,460,301 = R_0$	$-117,868,335$
U	$218,064,595 = U_0$	198,538,940

In our case, this means $\Phi_{AC}(\phi = 30 \text{ deg}) = \Phi_{AC}(\phi = 0 \text{ deg})/\cos 30 \text{ deg} = 0.5515/0.8660 = 0.6368$. That implies [Eq. (8.34)] that $\sigma_{AC}(\phi = 30 \text{ deg}) = 0.6804$, which implies in turn [Eq. (8.33)] that $h_{\rho AC}(\phi = 30 \text{ deg}) = 12{,}582$ ft. The wings-level results included $\sigma_{AC} = 0.6052$ and $h_{\rho} = 16{,}184$ ft.

Banking the airplane at some density altitude h_{ρ} is equivalent to dragging the ceiling down towards that altitude. Bank it enough, and the absolute banked ceiling descends onto the aircraft or even, temporarily, slides beneath it. When our sample Cessna banked 30 deg, its absolute ceiling was reduced from 16,184 to 12,582 ft. If the airplane is at density altitude 12,582 ft, it can certainly bank more than 30 deg, but when it does so its full-throttle best climb angle and rate go negative; the aircraft descends to stabilize at a new and even lower banked ceiling. Each such ceiling of course requires the specific air speed given by Eq. (8.32). To maintain altitude, our wings-level Cessna at absolute ceiling 16,184 ft must fly at 63.2 KCAS; banked 30 deg at absolute banked ceiling 12,582 ft, it must fly at 67.9 KCAS. Both speeds are the pertinent speeds for best angle of climb $V_x(W, \phi)$ which, in the bootstrap approach and expressed in calibrated (more accurately, equivalent) terms, are independent of density altitude.

It is instructive to investigate the range of turning possibilities with this aircraft at its service ceiling ($h'_{max} = 100$ ft/min, at the appropriate V_y). By trial and error on h_{ρ} using

$$\text{ROC}(V) = h^{\cdot}(V) = \frac{P_{xs}(V)}{W} = \frac{EV + KV^3 - H(\phi)/V}{W} \quad (8.37)$$

the maximum gross weight service ceiling is found to be 13,773 ft, with corresponding $V_y = 80.4$ KTAS. The maximum possible bank angle, before one starts down, turns out to be only 24.86 deg, at 82.0 KTAS, with turn radius $R = 1285$ ft. The *minimum* radius turn at this altitude, however (using techniques developed in the third section following), is $R = 1165$ ft at 74.4 KTAS (about 8 kn above the 66.2 KTAS banked stall speed), banked 22.8 deg. This aircraft, if in the straitened circumstance of a narrowing canyon, is about to preclude the possibility of turning towards lowering terrain. Such circumstances cause some general aviation mountain flying accidents and provide a good reason for using steady maneuvering charts. An example of restricted turn capability at high altitude will appear below. Figure 8.14 shows banked absolute ceilings and ceiling air speeds for bank angles up to 60 deg for two aircraft weights.

Let us now consider weight W and density altitude h_{ρ} as given and inquire into the corresponding maximum permissible level-flight bank angle and ceiling speed. First, Eq. (8.33) is inverted to give

$$\sigma_{AC} = \left(1 - \frac{h_{\rho AC}}{145{,}457}\right)^{4.25635} \quad (8.38)$$

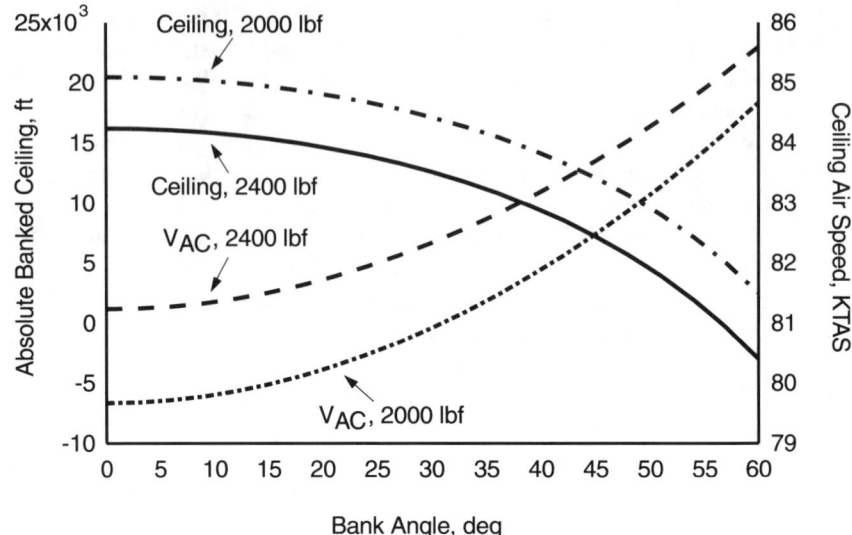

Figure 8.14 The two important banked ceiling numbers, the ceiling density altitude and V_y required to maintain level flight, for two aircraft gross weights.

with Φ_{AC} then given by

$$\Phi(\sigma) = \frac{\sigma - C}{1 - C} \quad (8.39)$$

Using Eq. (8.22) in the osculation condition, Eq. (8.31), then tells us the maximum bank angle for level flight at this altitude is

$$\Phi_{AC} = \cos^{-1}\left[\sqrt{\frac{2 - KH(0)}{E}}\right] \quad (8.40)$$

The corresponding ceiling air speed is given by the same Eq. (8.32) as before. These ceiling manipulations show how easy bootstrap calculations are when one uses composite parameters and how intuitive the "banked absolute ceiling" concept soon becomes.

Steady Maneuvering Charts

A sufficiently complete set of steady maneuvering charts (see Figs. 8.15 through 8.21 for examples) might consist of a dozen or so individual charts for both radius R and turning rate ω outputs: charts perhaps for all the combinations of two gross

Maneuvering Performance 259

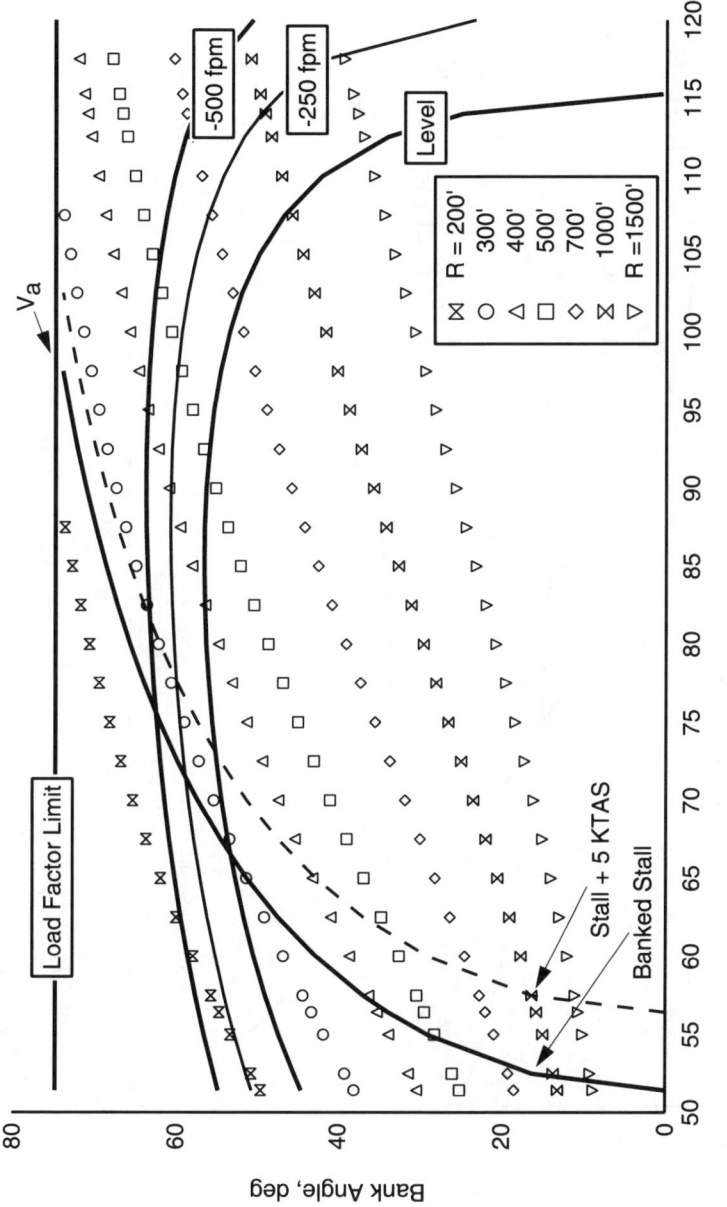

Figure 8.15 Turn radius maneuvering chart for a Cessna 172, MSL, 2400#, flaps up.

260 John T. Lowry

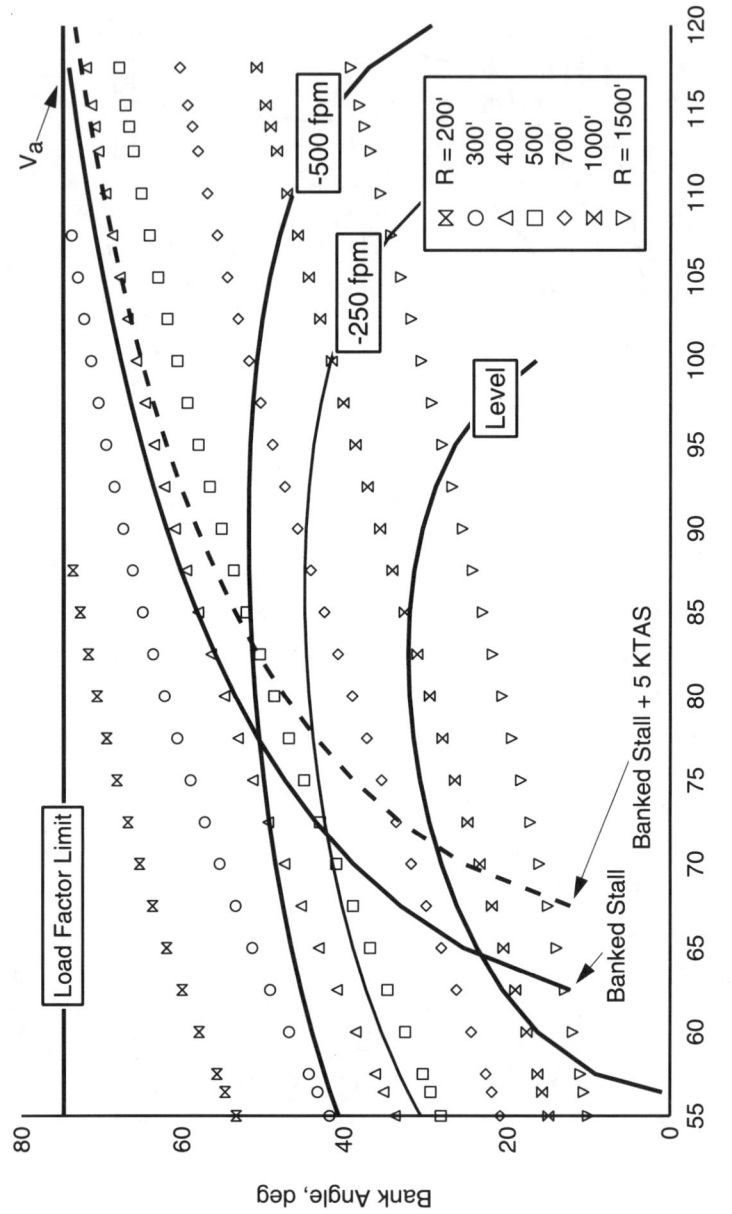

Figure 8.16 Turn radius maneuvering chart for a Cessna 172, 12000 ft, 2400#, flaps up.

Maneuvering Performance 261

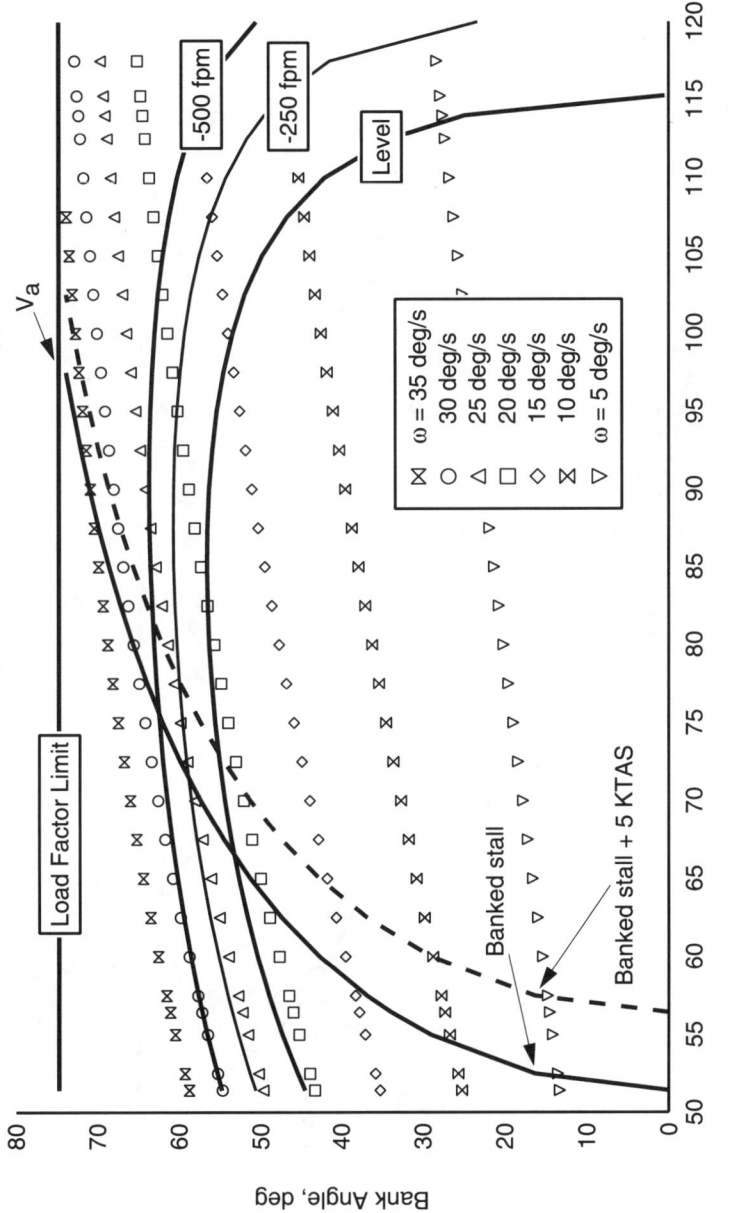

Figure 8.17 Turn rate maneuvering chart for a Cessna 172, MSL, 2400#, flaps up.

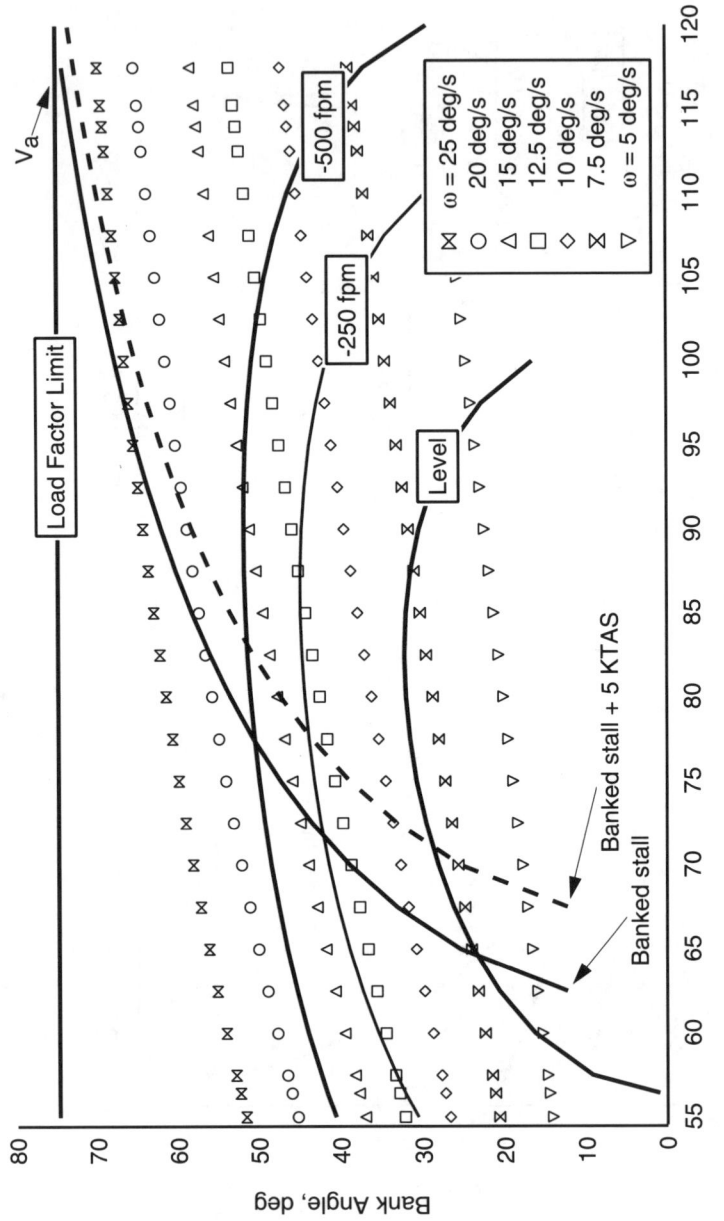

Figure 8.18 Turn rate maneuvering chart for a Cessna 172, 12,000 ft, 2400#, flaps up.

Maneuvering Performance 263

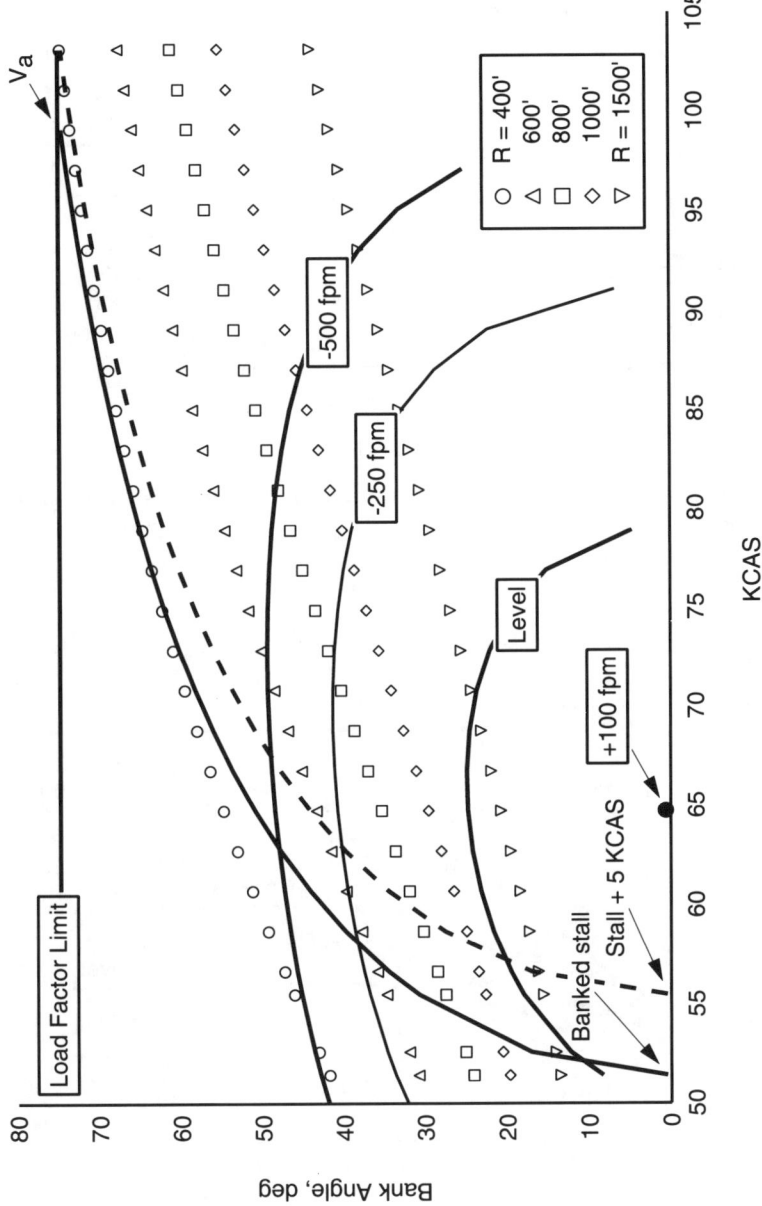

Figure 8.19 Turn radius maneuvering chart for a Cessna 172, 13,773 ft, 2400 #, flaps up.

264   John T. Lowry

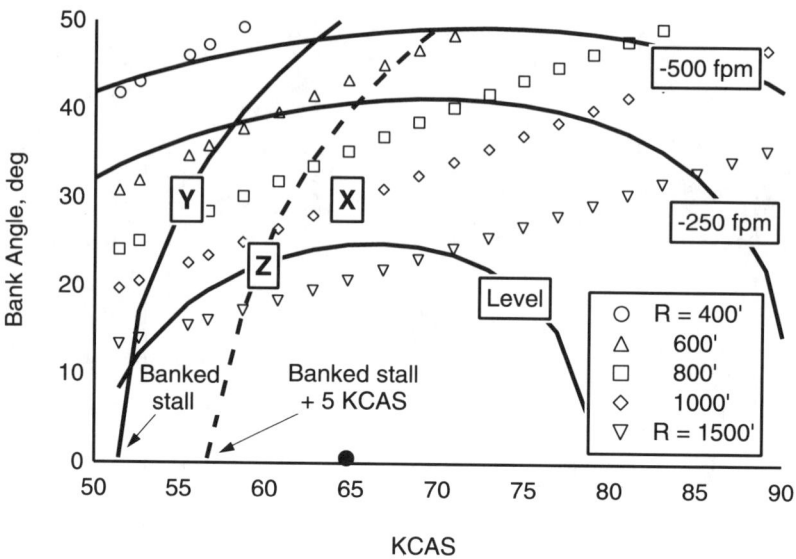

Figure 8.20   Expanded turn radius maneuvering chart for the Cessna 172, 13,773 ft, 2400#, flaps up.

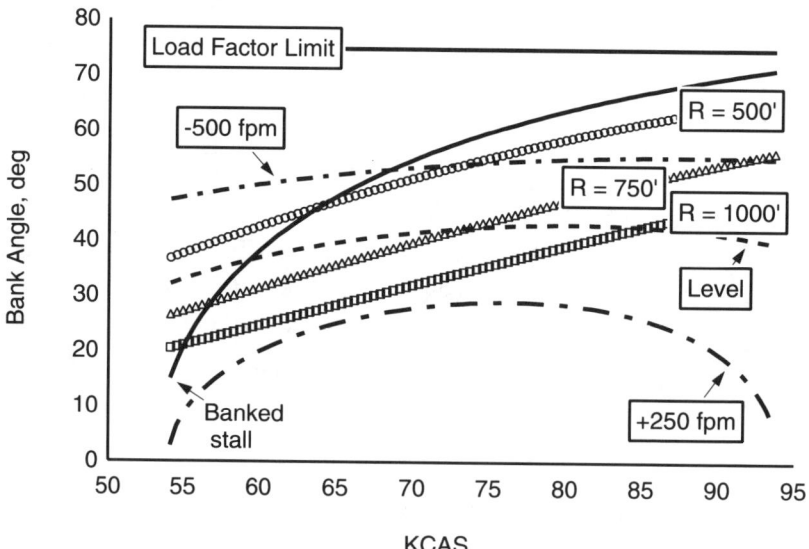

Figure 8.21   Steady maneuvering chart for a Cessna 182, flaps up, 3000 lb, at 10,000 ft.

weights, three density altitudes, and two flaps/gear configurations. A practical minimal set of six charts might span only two weights and three density altitudes.

Because maneuvering possibilities are enhanced by power, at least for low-performance aircraft, these will be full-throttle charts. Because the bootstrap approach is a variant of the venerable power-available/power-required analysis, the small flight path angle is assumed. This means there are small errors (beyond any of the linearized propeller polar itself) in all of the graphs making up the chart. But if one iterates, using

$$\gamma(V) = \sin^{-1}\frac{T_{xs}}{W} = \sin^{-1}\left[\frac{E + KV^2 - H/V^2}{W}\right] \qquad (8.41)$$

selectively for the flight path angle, one finds the errors practically negligible except at extreme descent rates one would never use for steady turns.

The two outputs of interest are given by the standard formulas

$$R = \frac{V^2}{g \tan \phi} \qquad (8.42)$$

$$\omega = \frac{V}{R} = \frac{g \tan \phi}{V} \qquad (8.43)$$

It is clear from Eqs. (8.42) and (8.43) that desirable maneuvering characteristics—small R and large ω—both result from small values of V and large values of ϕ but, as usual in aeronautical work, the devil is in the details. And it is precisely those details that the steady maneuvering charts are designed to elucidate. These charts employ essentially the same independent variables, V and ϕ, as the more common V-n (sometimes V-g) diagrams, but their use is much different. The V-n diagram describes the flight envelope predominantly in terms of quick control inputs, stalls, load factor limits, and the never-exceed speed. It includes negative-g (or inverted flight) possibilities. Quick noncontrol inputs from gusts are often overlaid on the V-n diagram. Our charts, on the contrary, consider only quasi-steady-state maneuvering, steady turns. Though the load factor limit is included, it is hardly ever of serious concern when using a steady maneuvering chart.

Before using either of the steady maneuvering charts, one should answer two questions: 1) How close to the stall is one willing to venture? and 2) How large a descent rate can one live with? If the answers are, respectively, 5 kn and 250 ft/min, then at MSL in our sample aircraft Figs. 8.15 and 8.17 provide a turn radius of about 310 ft and a turning rate of about 25 deg/s, each at about 78 KTAS banked about 60 deg. If the airplane is up at 12,000 ft, however, Figs. 8.16 and 8.18 provide a turn radius of about 560 ft and a turning rate of about 14 deg/s, each at about 77 KTAS and banked about 44 deg.

Notice that if *level* turns were prescribed at 12,000 ft, then the pilot will get minimum turn radius and maximum turning rate at speeds *greater than* the banked stall speed. At higher altitudes, the curves of constant climb or descent rate have emerged from behind the limb of the banked stall curve. Also notice that the standard "maneuvering speed" important in rough air, the speed at which the banked stall curve crosses the load factor limit, is here too large (119 KTAS) to be of significance for steady turns. High-performance aircraft may very well enter a turn with such excess speed (above the stall) that slowing down to enhance turn performance is a viable option. For turning low-performance aircraft, however, the inevitable slowing due to increased induced drag is all they can afford.

To construct steady maneuvering charts for a specific airplane, one must have the nine bootstrap parameters for the desired flaps/gear configuration plus the appropriate positive g limit and C_{Lmax}. The procedure for getting the four harder-to-get parameters (C_{D0}, e, m, and b) was given in Chapter 7. Next a particular weight and density altitude are selected. The following five different kinds of curves can then be drawn.

1) Load factor limit—for the Cessna 172 normal category airplane, the flaps up flight load factor limit is $3.8g$. That means

$$n \equiv \frac{L}{W} = \frac{W \cos\gamma}{W \cos\phi} \doteq \frac{1}{\cos\phi} \leq 3.8 \qquad (8.44)$$

independent of V. The limiting bank is then

$$\phi_{n\,\text{lim}} = \cos^{-1}(1/3.8) = 74.74 \text{ deg} \qquad (8.45)$$

2) The banked (and padded) stall limit—using Eqs. (8.23) and (8.24), allowing for the pad ΔS—we normally suggest about 5 KCAS for this—and inverting,

$$\phi_{S+\Delta S} = \cos^{-1}\left[\frac{2W}{\rho S C_{Lmax}(V_S(\phi) - \Delta S)^2}\right] \qquad (8.46)$$

3) Curves of constant climb or descent rate—using Eq. (8.22) in Eq. (8.37), and solving (using rate of climb $h\dot{}$ in ft/min)

$$\phi_{h\dot{}} = \cos^{-1}\sqrt{\frac{H(0)}{V^2(E + KV^2) - Wh\dot{}V/60}} \qquad (8.47)$$

4) Curves of constant turn radius R—inverting Eq. (8.42)

$$\phi_R = \tan^{-1}\left[\frac{V^2}{gR}\right] \qquad (8.48)$$

5) Curves of constant turning rate ω—inverting Eq. (8.43) (and now using turning rate ω in deg/s)

$$\phi_\omega = \tan^{-1}\left[\frac{\pi V \omega}{180g}\right] \qquad (8.49)$$

Several of these curves (depending on whether one uses true or calibrated air speeds) do not vary with density attitude. Therefore, it does not take a great deal of effort to construct a set of steady maneuver charts for a given airplane.

A Cautionary Tale: Fatal High-Altitude Turn

To illustrate the trouble one can get into by overbanking a thrust-limited airplane at high altitude and the way the steady maneuvering chart can be used in this situation, we relate this (apocryphal) story of Bill and Bob.

Bill and Bob were "goldarn flatlanders" who flew their Cessna 172 out West to see the Rockies and fish and have fun. Their first night in Colorado they camped under the wing at a small airport at a moderate 6000-ft elevation. After a good night's sleep and a leisurely brunch, early the next afternoon they loaded up, topped off, and decided to spend an exploratory hour flying up a broad U-shaped side canyon which, according to the chart, terminated at a 14,000-ft ridge. They did not make it back. The wreckage was found by an elk hunter that fall, rolled up beneath the trees on the side of the canyon at 13,000 ft. It had all *seemed* innocent enough. What had gone wrong?

Figure 8.19 is for the Cessna 172 at maximum gross weight at the service ceiling, calculated to be $h_{SC} = 13,773$ ft for this particular airplane. Under those conditions, the airplane's speed for its 100 ft/min best rate of climb is $V_{ySC} = 80.1$ KTAS. Figure 8.20 enlarges the important portion of Fig. 8.19 and marks several important points on it. There we pick up Bill's Cessna, wings level at full throttle, doggedly climbing slowly but steadily up the canyon. Bob is looking down at the rocks and sparse trees below. Finally he tells Bill the ground is getting closer. By that he means the rocks are getting *uncomfortably* closer. Bill prudently decides it is time to turn around. First he banks left 30 deg, moving the airplane's state (Fig. 8.20) straight up from the barely visible tip of the 100 ft/min curve, at 80 KTAS, to point X. Then Bill notices he is descending a little—not at all stalled, just descending—so he adds a little backstick.

Bill is not used to this high altitude. He also does not realize (lack of performance knowledge, or perhaps preoccupation) that he has gone above his absolute ceiling for 30 deg of bank. By adding angle of attack (AOA) with backstick, induced drag increases and the airplane slows down. On Fig. 8.20, it slides straight left from point X. The turn radius thereby decreases a little, true, but the descent rate gets *worse*. So he adds a bit more backstick. As the steady maneuvering chart makes clear, there is only one place this sad scenario can end: at point Y in a full-power stall.

What should he have done? He should have applied one or more proverbial ounce of prevention:

- Take some qualified mountain flying instruction;
- Buy a higher-performance airplane;
- Circle early to maximum necessary altitude (say 14,000 ft), then fly to the head of the canyon, turn around, and come downhill, towards lowering terrain; or
- Reduce gross weight; leave all excess (nonsurvival) gear in camp.

But once caught up in their high-altitude situation, and not able to afford to lose much altitude—after all, that is what prompted the turnaround—Bill should have banked only about 23 deg and slowed about 6 kn true, going to point Z on Fig. 8.20. The disadvantage of point Z is that the turn radius is fairly large, about 1200 ft. Almost 0.5-statute mile turn diameter. If the canyon had narrowed to less than that, there is essentially nothing Bill and Bob could have done in this airplane except make a precautionary landing onto the smoothest site available. They certainly did not have enough air speed for a fancy wing-over turnaround.

But expecting an ordinary pilot to have the experience and information to bank just so far but no farther and to slow just so much but not slower, is unrealistic. The better moral reads: avoid high altitude maneuvering in low-performance airplanes. Use the bootstrap steady maneuvering chart to learn where and how your airplane can safely maneuver and where it cannot.

Minimum Radius and Maximum Rate Level Turns

As representative of the standard piston-propeller maneuvering theory we take formulas derived by Hale,[5] also cited by Adamson,[6] for level (thrust equals drag) turns. Similar formulas are treated, somewhat more extensively and with fewer approximations, by Miele.[7] Those authors express the variables of interest, R and ω, in terms of the load factor n (standing in for bank angle ϕ) and air speed V. To find optima, they differentiate those expressions with respect to V, set those derivatives equal to zero, and solve. With functions of two variables, one cannot in general be ensured of successfully using that single-variable procedure to find relative minima or maxima. Kaplan[8] has a clear and simple counterexample. Because their resulting expressions are complicated, Hale and Adamson also make warranted discards of small terms. After that work and those approximations, results are to be taken as only indicative because their calculated air speeds for optimal turns are considerably below associated banked stall speeds. That is also often the case when using our alternative calculation procedure.

But not always. At higher altitudes, the analytic optima may prevail. We now compute one of them, speed for minimum-radius level turn $V_{\min R}$ and display the others, which are derived similarly. From Eq. (8.41), because the turn is level ($\gamma = 0$), we have

$$E + KV^2 - H(\phi)/V^2 = 0 \tag{8.50}$$

Use the facts that $H(\phi) \equiv H(0)/\cos^2 \phi$ and that $1/\cos^2 \phi = \sec^2 \phi = 1 + \tan^2 \phi$, we can get an expression for $\tan \phi$ in terms of air speed V:

$$\tan \phi = \sqrt{\frac{KV^4 + EV^2}{H(0)} - 1} \tag{8.51}$$

Then we can use the standard turning relation, Eq. (8.42), to write

$$R = \frac{V^2}{g \tan \phi} = \frac{V^2}{g\sqrt{\frac{KV^4 + EV^2}{H(0)} - 1}} \tag{8.52}$$

Now it is just a matter of taking the derivative of the RHS of Eq. (8.52) with respect to V, setting that equal to zero, and solving to find the speed for the optimum (smallest) radius turn:

$$V_{\min R} = \sqrt{\frac{2H(0)}{E}} \tag{8.53}$$

Similar manipulations, starting from the expression for the minimum level flight speed, Eq. (8.28), result in an expression for the proper bank angle for that shortest-radius turn:

$$\phi_{\min R} = \tan^{-1} \sqrt{1 + \frac{4H(0)}{KQ^2}} \tag{8.54}$$

which shows, because the radicand is less than unity, that $\phi_{\min R}$ is always less than 45 deg.

The search for inputs giving maximum turning rate also gives fairly simple results

$$\phi_{\max \omega} = \tan^{-1} \sqrt{-2 - Q\sqrt{\frac{-K}{H(0)}}} \tag{8.55}$$

and

$$V_{\max \omega} = \left[\frac{-H(0)}{K}\right]^{1/4} = V_x(0) \tag{8.56}$$

Table 8.3 Optimum level turns in standard and bootstrap approaches

Value for optimum	Standard approach	Bootstrap approach
V_{minR}	43.7 KTAS	69.8 KTAS
ϕ_{minR}	30.0 deg	28.0 deg
R_{min}	292 ft	813 ft
$V_{max\omega}$	65.5 KTAS	75.9 KTAS
$\phi_{max\omega}$	45.0 deg	31.0 deg
ω_{max}	16.7 deg/s	8.6 deg/s

Subscripts min R and max ω must be interpreted in a provisional sense. To see whether the banked stall speed is higher, and therefore overriding, the appropriate steady maneuvering chart, or an ancillary calculation, must be consulted.

The corresponding approximations due to Hale[5] are

$$\phi_{min R} \doteq 30 \text{ deg} \tag{8.57}$$

independent of weight or altitude, and

$$V_{min R} \doteq \frac{8W^2}{3 \times 550 \eta \rho S \pi e A (\text{hp})} \tag{8.58}$$

His 550 hp is our ordinary (British engineering units) brake power P. For maximum turning rate, Hale uses

$$\phi_{max \omega} \doteq 45 \text{ deg} \tag{8.59}$$

$$V_{max \omega} \doteq \frac{4W^2}{550 \eta \rho S \pi e A (\text{hp})} = \frac{3}{2} V_{min R} \tag{8.60}$$

Example 8.4 Table 8.3 compares those formulas of Hale with ours for the sample Cessna 172, flaps up, at 12,000 ft at maximum gross weight $W = 2400$ lbf. Both standard approach optimum speeds are less than the actual banked stall speeds and so are not operationally feasible. For this relatively high altitude case, bootstrap approach results, on the other hand, are greater than the actual banked stall speeds and are therefore realistic.

Turn Performance of Constant-Speed Propeller Airplanes

Professional performance test pilots tell the author that, of the two types of propeller airplanes, it is the fixed-pitch ones that are the more problematical, that constant-speed propeller airplanes take considerably less flight test and calculation time. Not with the bootstrap approach, where there are formulas for almost everything about the fixed-pitch type but where one has to use the general aviation general propeller chart for higher-powered airplanes with constant-speed propellers. Nonetheless, we now run through a case using the constant-speed Cessna 182.

Example 8.5 Table 8.4 is a sample bootstrap data plate for the airplane. There will be some differences in the details, but the basic aerodynamics approach is the same as for fixed-pitch airplanes.

Put our sample Cessna 182 initially at density altitude 10,000 ft, weight 3100 lbf, cruising at 2300 rpm with 19 in. MAP. The Cessna Model R182 POH cruise performance table for that altitude, entered with this engine speed and manifold pressure, says we can expect 63% BHP, 148 KTAS, and 11.4 gph fuel consumption rate. We do not care about those last two numbers because 1) this is going to be a short maneuver and 2) our airplane is draggier than the POH archetype so we will use our BDP data for drag.

Not many Cessna 182s go on bombing runs, but let us bank this one 30 deg and push over to a moderate 150 KCAS. We leave the throttle and propeller speed control unchanged. What will the descent rate and turn radius be under these circumstances?

The turn radius is not much of a problem. We do need true air speed in British engineering units. Because 10,000 ft corresponds to $\sigma = 0.7385$, we see 150 KCAS is equivalent to 174.5 KTAS or 294.6 ft/s. Equation (8.12) then straightway [actually, it would be more accurate to first find the flight path angle and correct Eq. (8.12) or (8.42) with a $\cos \gamma$ in the denominator] gives us $R = 4672$ ft.

Table 8.4 Sample bootstrap data plate (Cessna 182)

Bootstrap data plate item	Value	Units	Aircraft subsystem
Wing reference area, S	174	ft^2	Airframe
Wing aspect ratio, A	7.448		Airframe
Rated MSL power, $P_B(\Pi = 1)$	235	hp	Engine
Altitude dropoff factor, C	0.12		Engine
Propeller diameter, d	6.83	ft	Propeller
Parasite drag coefficient, C_{D0}	0.02874		Airframe
Airplane efficiency factor, e	0.72		Airframe
Total activity factor	195.9		Propeller
Fuselage/propeller diameter, Z	0.688		Propeller

The descent rate is given by

$$\dot{h} = \frac{P_{xs}}{W} = \frac{\eta P - DV}{W} \tag{8.61}$$

Drag D will come from the usual bootstrap formulas, using $H(\phi) = H(30\ \text{deg})$, but propeller efficiency η is quite a bit more problematical. It sends us back to the section on constant-speed airplanes and to the propeller thrust chapter (Chapter 6) for the general aviation general propeller chart. There are several steps.

1) Power adjustment factor X comes from the propeller total activity factor (TAF). For TAF = 195.9, $X = 0.2088$.
2) To get propeller power coefficient C_P we use power $P = 0.63 \times 235 \times 550 = 81{,}427$ ft-lbf/s, air density $\rho = 0.001755$ slugs/ft^3, engine speed $n = 2300/60 = 38.33$ rps, and the propeller diameter, 6.83 ft. We find $C_P = 0.0554$ and that $C_P/X = 0.2653$. This last is one of two numbers with which we use the general aviation general propeller chart.
3) The other is $J/C_P^{1/3}$. Advance ratio $J = V/nd = 1.125$, so $J/C_P^{1/3} = 2.951$.
4) With these last two, we use the Chart to find that our provisional propulsive efficiency is approximately 0.720.
5) There is more. The slowdown efficiency factor (SDEF) for this tractor airplane, corresponding to its $Z = 0.688$, is 0.910. So the final estimate of propulsive efficiency is $\eta = 0.91 \times 0.720 = 0.655$.
6) Drag, from the usual bootstrap formula—not forgetting the factor $\cos^2 \phi = 0.75$ in the denominator of the induced drag term—turns out to be 438.2 lbf.
7) From Eq. (8.61), finally, the descent rate is $\dot{h} = -24.44$ ft/s = -1466 fpm. The flight path angle is $\gamma = -4.76$ deg and the turn rate is $\omega = 0.0631$ rad/s = 3.61 deg/s. Not terribly dramatic. Less a bombing run than just a proper response to air traffic control's request to "expedite your descent."

To economically produce steady maneuvering charts for constant-speed propeller airplanes, one needs a computer version of the general aviation general propeller chart. There is one and Fig. 8.21 is an example of a steady maneuvering chart taken from it. But there still remains the fact that these airplanes have an extra "degree of freedom" in their propulsion controls—the propeller speed control—which must be taken into account. That is essentially why, in the bootstrap approach, fixed-pitch airplanes are simpler. But predicting maneuvering performance for either type is quite manageable.

Infamous Downwind Turn

A downwind turn is simply one in which the airplane starts off moving into the wind and then turns so as to move with the wind. For simplicity, let us say it begins *directly* into the wind then turns 180 deg to end moving *directly* downwind. No problem with that maneuver; it is done all the time. The important part of our headline is the word "infamous." That epithet comes about because some beginning pilots—and even a few experienced ones—believe such a turn is fraught with danger. Their (incorrect) reasoning is that, in a downwind turn, the airplane will, unless the pilot takes special corrective action, lose air speed. It will not. But, as in so many controversies, it is instructive to look for reasons why such an air speed loss might *seem* to occur.

Here are the main points bearing on the downwind turn issue:

- Airplanes do tend to slow down, unless one adds throttle, in downwind turns. But that is simply because airplanes tend to slow down in *all* turns, whether to downwind, upwind, or crosswind. Or in a calm. That deceleration is strictly from the added induced drag which, in turn, is due to the increased AOA necessary to maintain the vertical lift component at the airplane's weight. And perhaps a little from the drag of control surfaces deflections. In addition, when banked, stall speed has increased. But in any case, no difference between turning from upwind to downwind or the other way.
- This "killer turn"—that is another common description—is dreaded primarily at low altitudes, when maneuvering close to the ground usually in preparation for landing. It is true that after a turn to downwind, your *ground* speed has increased by twice the wind speed. The problem is a perceptual one: the unwary pilot may translate that undoubted ground speed increase into thinking that the *air* speed has also increased—without checking the airspeed indicator to verify that it has not—and consequently throttle back. Unwarranted.
- If you happen to fly into slackening wind during your turn to downwind—bona fide *wind shear*—then you certainly may have a problem keeping the airplane in the air. Your air speed might drop dangerously... and fast! But again, not a different problem than you would encounter under the same circumstances turning to upwind. If you happened to run into a vacuum, same thing. Practitioners of statistical mechanics say it *could* happen!

Moving to the positive side, what is the correct reason there is no actual difference between downwind and upwind turns? The author has read several not-quite-correct, or at least muddy, explanations. One is "When airborne, you are carried *with* a moving air mass independently of your movement *through* it." Because, in heavier-than-air craft we cannot move with an air mass *without* moving through it, that rationalization is unconvincing. The correct explanation for the phantasmagoric character of the downwind turn "problem" is *All forces acting on the airplane except for gravity, which does not change, are aerodynamic forces, forces due to the airplane's relative movement through the air.* If—considering the two turns, upwind vs downwind—initial air speeds are the same and subsequent forces on the airframe are the same, then the final air speeds are the same. We admit that "mathematical" arguments devoid of numbers leave some readers unconvinced.

Where Did the Extra Kinetic Energy Come From?

Another conundrum, one related to the infamous downwind turn, is this: relative to the air mass, an airplane having just made a (constant air speed and level) downwind turn has not changed its kinetic energy. But relative to the Earth, because it is now moving faster, it certainly *has* increased its kinetic energy! How can that be?

The forces acting on the airplane during its turn, as seen from either the Earth frame of reference or the air mass frame of reference, were identical. All of this is indeed the case. As often happens in questions of this kind, the difficulty is resolved, miraculously, when one performs a detailed calculation.

Which, for the turning airplane, we will not do here. That is because it would take a couple of pages of onerous vector calculus. But we will give the kernel of the idea. And a simplified one-dimensional calculation that correctly models the turning airplane. The basic idea is that, going back to the aircraft performance preliminaries chapter, the change in a body's kinetic energy, if accelerated by force **F** between positions $\mathbf{r}_1$ and $\mathbf{r}_2$, no potential energy change taking place, is

$$\int_{r_1}^{r_2} \mathbf{F} \cdot d\mathbf{r} = m \int_{r_1}^{r_2} \frac{d\mathbf{v}}{dt} \cdot d\mathbf{r}$$
$$= m \int_{v_1}^{v_2} \mathbf{v} \cdot d\mathbf{v}$$
$$= \tfrac{1}{2} m v_2^2 - \tfrac{1}{2} m v_1^2$$
$$= KE_2 - KE_1 \qquad (8.62)$$

With respect to the Earth-based frame and the air-mass-based frame the forces are indeed equal, *but the displacements are not equal*. As the downwind-turning airplane speeds up relative to the Earth, its displacements (say, per second) grow and, consequently, its kinetic energy increases. One cannot simply focus on the forces and neglect the displacements. Any more than one can, when treating internal combustion engines, concentrate on gasoline to the exclusion of air. It takes both. Now let us put some numbers, or at least concrete variables, to a simpler case, one-dimensional and impulsive.

Consider Jennifer, a transcontinental jet flight attendant of weight W, mass $m = W/g$ slugs. She is walking down the aisle, at speed s relative to the airplane, toward the rear of the cabin, to deliver a copy of *Esquire* to a passenger. The airplane is traveling with ground speed V. Relative to the ground, then, Jennifer has ground speed $(V - s)$. And she has initial kinetic energy

$$KE_1 = \tfrac{1}{2}m(V-s)^2 = \tfrac{1}{2}m(V^2 - 2Vs + s^2) \tag{8.63}$$

As she gets near the passenger and hands him the magazine, she pushes on the carpet with her foot, with average force $\langle \mathbf{F} \rangle$ in the airplane's direction of motion, over time interval Δt, and reverses her velocity. Ignore the fact that she also pivots on that foot so she will be able to see where she is going as she retraces her steps. After her dramatic turnabout, Jennifer has (oppositely directed) speed s with respect to the airplane and, relative to it, precisely the same kinetic energy she had before. But relative to the ground, with her new ground speed $(V + s)$, Jennifer has final kinetic energy

$$KE_2 = \tfrac{1}{2}m(V+s)^2 = \tfrac{1}{2}m(V^2 + 2Vs + s^2) \tag{8.64}$$

Therefore, relative to the ground, Jennifer has gained kinetic energy

$$KE = KE_2 - KE_1 = 2mVs \tag{8.65}$$

Not inconsiderable! If she weighs 100 lbf and walks at 4 mph in an airplane going 500 kn, this is an energy gain of 30,777 ft-lbf = 39.56 btu, about the energy given off by burning 40 kitchen matches (in the requisite air). Not recommended in flight.

So Jennifer gained no kinetic energy with respect to the airplane but gained quite a bit of kinetic energy with respect to the Earth. Let us see how we can make sense of this dichotomy using part of Eq. (8.62). We approximate it as

$$\Delta KE = \int \mathbf{F} \cdot d\mathbf{r} = \int \mathbf{F} \cdot \mathbf{v}\, dt \doteq \langle \mathbf{F} \cdot \mathbf{v} \rangle \Delta t = \langle \mathbf{v} \rangle \cdot \Delta \mathbf{p} \tag{8.66}$$

where $\Delta \mathbf{p}$ is Jennifer's momentum change, of size (in either frame) $\Delta p = 2ms$ and directed toward the nose of the airplane.

Taking the time interval Δt, during which she turned around, symmetric about time $t = 0$, her velocity with respect to the airplane is an odd function. Taking the direction of motion of the airplane as positive, her speed relative to the airplane went from $-s$ to $+s$. So her average velocity, relative to the airplane, was zero. Hence her ΔKE, relative to the airplane, was also zero.

With respect to the Earth, the situation was much different. She went from ground speed $(V - s)$ to speed $(V + s)$, so that averages out at V, the speed of the airplane. And her change in kinetic energy, relative to the Earth, and according to Eq. (8.66), is indeed $2mVs$.

To get further insight to the problem—by drawing graphs—one can take force $\mathbf{F}(t)$ during that interval as some Gaussian blip centered at $t = 0$, an even function. Alternatively, one can get back to the downwind turning airplane, using the full array of vector notation and the various sine and cosine functions relating to circular motion, to calculate the details. One will find that a variable downwind turn over time t gives the airplane additional kinetic energy, with respect to the Earth, of amount

$$\Delta KE(t) = mVV_W(1 - \cos \omega t) \qquad (8.67)$$

where m is the mass of the airplane, V its air speed, V_w the wind speed, and ω the rate of turn. If time interval τ is the time for a complete circle, Eq. (8.67) gives the realistic result that $KE(t = 0)$, no turn at all, equals, $KE(t = \tau)$, a full circle; both are zero. And that $KE(t = \tau/2)$, a half-circle to straight downwind, is $2mVV_W$.

Conclusions

The linearized propeller polar or bootstrap approach is easily generalized to encompass maneuvering flight. It gives simple analytic results for the main variables of interest: banked (or unbanked) ceiling specifications; turn radius; turning rate; and specification of conditions needed for optimum level turns. Our specifications for optimum short and quick level turns were more realistic than those suggested by the standard approach and at considerable variance from them. Steady maneuvering charts for a range of weight and altitude situations are fairly easily constructed. Those provide a graphic display of operational limitations and suggest optimum safe turning speeds and bank angles for any given situation. That visual aid is particularly useful for moderately powered trainers, airplanes that are quite thrust limited at higher altitudes. General aviation aircraft manufacturers should consider including steady maneuvering charts in their pilot operating handbooks.

References

1. Illman, P. E., *The Pilot's Handbook of Aeronautical Knowledge*, Tab Books, Blue Ridge Summit, PA, 1991, p. 45.
2. Federal Aviation Administration, *Flight Training Handbook*, Doubleday, New York, no date, pp. 283–285.
3. Roskam, J., and Lan, C.-T. E., *Airplane Aerodynamics and Performance*, DARcorporation, Lawrence, KS, 1997, pp. 593–602.
4. Lowry, J. T., "Maneuvering Flight Performance Using the Linearized Propeller Polar," *Journal of Aircraft*, Vol. 34, No. 6, 1997, pp. 764–770.
5. Hale, F. J., *Aircraft Performance, Selection, and Design*, Wiley, New York, 1984.
6. Adamson, J. C., *Aircraft Performance*, U.S. Military Academy, 1991.
7. Miele, A., *Flight Mechanics-I: Theory of Flight Paths*, Addison-Wesley, Reading, MA, 1962.
8. Kaplan, W., *Advanced Calculus*, Addison-Wesley, Reading, MA, 1952.

9

Glide Performance

Introduction

Glide performance—at least the wings level variety—was treated briefly in Chapter 7, Introduction to the Bootstrap Approach. But this foray will penetrate much deeper. Setting thrust $T = 0$ so simplifies the equations of motion for the airplane that we will be able to get "exact" solutions for flight path angle, rate of descent, and pertinent V speeds. Or at least *more nearly* exact, solutions not taking refuge in the small flight path angle approximation (in which $\cos \gamma \doteq 1$ and lift $L \doteq$ weight W). And we will be able to see differences (both numerically and analytically) between "exact" flight mechanics solutions and the earlier "small γ" ones. For our general aviation aircraft, even most aircraft, differences are normally slight.

Also, having recently covered turning maneuvers, we are now able to treat *banked* glides. Wind effects are usually of greater practical importance for glides than for powered flight. We will consider not only direct headwinds or tailwinds but also updrafts or downdrafts or combinations (headwinds with downdrafts, etc.).

We shall cover some special maneuvers involving gliding: the terminal velocity dive, zooms to gain altitude at the expense of air speed (or dives, antizooms, for the opposite interchange), and the gliding course reversal (return-to-airport maneuver) one *may* want to consider should his or her engine quit soon after takeoff.

We shall skip one topic that often appears in discussions of glide performance—velocity *hodographs*. Those are graphical representations of aircraft glide speeds and angles and the way those are related to the airplane's drag polar. Hodographs are often touted as a good way to go beyond the small flight path approximation. But we have at our disposal (quadratic) drag polars, and aerodynamic force relations, which let us go to beyond that and with more flexibility.

In this computer age, quite a few such older graphical procedures have given way to analytic or numerical methods.

Now for a semimajor caveat: the nagging problem of the "windmilling" propeller. In powered aircraft glides—unless the propeller has seized or the pilot has gone to some pains to stop its rotation—the propeller will be turning at different speeds depending on the steepness of the glide. In our treatment we shall simply lump drag from the windmilling propeller into the airplane (parasite) drag as a whole. That cannot be entirely correct because steeper glide paths make the propeller rotate faster and, presumably, extract more energy from the airstream. More drag, but not necessarily proportional to the square of the air speed. If the propeller is *not* windmilling, for one of the above reasons, we will have a bonus, lowered drag. The author has not found a satisfactory theoretical solution to the windmilling propeller conundrum; each published approach he consulted has differing techniques and quite widely differing numerical estimates. Experiments substantiating a good consistent theory are sorely needed. The linearized propeller polar is of little use in this regime. As advance ratio J approaches the point where thrust coefficient C_T goes to zero, the airplane's polar begins to bend away from its straight line approximation. The bootstrap approach is a performance theory for powered airplanes in the normal flight regime, which this is not.

On the bright side, relative simplicity of glide performance equations means that we can fairly easily calculate a wide range of gliding maneuvers. But not all of them. There are three main forks in the road:

- "Exact" analytic treatment versus the small flight path angle approximation;
- Banked glides versus wings-level glides; and
- Glides with wind effects versus those in calm air.

These make for the eight cases, only five of which we shall consider, detailed in Table 9.1. The double quotes on "exact" theory are to say that, while that theory does not use the small flight path angle approximation, it does have other rough edges such as the windmilling propeller. We will bypass cases 5 and 7

Table 9.1 Glide situations most of which will be treated below

Case #	Theory	Banked/unbanked	Wind/calm	Treated?
1	"Exact"	Banked	Calm	Yes
2	"Exact"	Wings level	Calm	Yes
3	Small γ	Banked	Calm	Yes
4	Small γ	Wings level	Calm	Yes
5	"Exact"	Banked	Wind	No
6	"Exact"	Wings level	Wind	No
7	Small γ	Banked	Wind	No
8	Small γ	Wings level	Wind	Yes

because of bookkeeping complexity; it is not practical to consider either rapid speed changes during a turn or to incorporate details of initial and final headings relative to the wind direction. Nonetheless for a few very necessary calculations—say for litigation support purposes—those cases could certainly be done in some average sense. We skip case 6 because "exact" and wind are contradictory; wind conditions are neither steady enough nor known well enough to pretend to exactitude. Case 8 will have to substitute. Let us get started with our main result while we are still fresh.

Banked or Unbanked Glides in "Exact" Theory, No Wind (Cases 1 and 2)

Gliding Equations of Motion (Exact)

Copied over from Chapter 8, Maneuvering Performance, our full (possibly powered) general aviation force equations are, first parallel to the flight path

$$T \cos(\alpha + \alpha_T) - D - W \sin\gamma = 0 \tag{9.1}$$

then perpendicular to the flight path and in the vertical plane,

$$L \cos\phi - W \cos\gamma + T \sin(\alpha + \alpha_T) = 0 \tag{9.2}$$

and finally perpendicular to the flight path and in the horizontal plane,

$$L \sin\phi - \frac{W\omega V}{g} = 0 \tag{9.3}$$

Simply setting thrust T to zero, and recalling that now flight path angle γ is negative, we have the corresponding "exact" gliding force equations:

$$D + W \sin\gamma = 0 \tag{9.4}$$

$$L \cos\phi - W \cos\gamma = 0 \tag{9.5}$$

$$L \sin\phi - \frac{W\omega V}{g} = 0 \tag{9.6}$$

Aerodynamic forces L and D come to us through their corresponding coefficients, functions of angle of attack (AOA) α, and the usual quadratic drag polar:

$$L = \tfrac{1}{2}\rho V^2 S C_L \tag{9.7}$$

$$D = \tfrac{1}{2}\rho V^2 S C_D \tag{9.8}$$

$$C_D = C_{D0} + \frac{C_L^2}{\pi e A} \tag{9.9}$$

We almost always assume that aerodynamic torques on the airplane have been trimmed away. Therefore, we can ignore torque on the airframe and its coefficient C_M.

Formulas for Flight Path Angle γ, Rate of Descent $\dot{h}$, and Turn Radius R, Case 1 (Exact)

Put yourself in the cockpit as or alongside the pilot. In unpowered flight, the only aircraft parameters that matter are W, C_{D0} (which depends on flaps setting), e, A, and S. Then there is environmental variable ρ. Finally, manipulation of flight controls leads to the pilot's choices for air speed V and bank angle ϕ. Flight path angle γ must be a function of (at most) those variables. Once we get a formula for γ, an expression for rate of descent $\dot{h}$ is immediate and one for turn radius, while not immediate, is forthcoming. But first we need a brief tutorial on gliding along a helical flight path.

More Facts About Helical Flight Paths

If your airplane is in a constant-banked glide at steady speed V (ignoring the fact that a steady *indicated* air speed will mean a very slightly decreasing *true* air speed), its flight path will be a helix wound on some cylinder of radius r. The radius of curvature of the flight path, often called ρ in calculus texts, is somewhat greater than r. If ω is the yaw rate, it turns out that $V \ne r\omega$ and, moreover, that $V \ne \rho\omega$. The correct relation is:

$$V = \sqrt{r\rho}\,\omega \qquad (9.10)$$

so that the correct radius R, in the usual relation $V = R\omega$, is the geometric mean of the two most promising candidates.

A helix is usually described in terms of cylinder radius r and its pitch p, the distance (down, in our case) one moves along it in one revolution. Pitch, yaw rate (angular speed), and descent rate (neglecting signs for the moment) are related by

$$\dot{h} = \frac{p\omega}{2\pi} \qquad (9.11)$$

The relation between cylinder radius r and flight path radius of curvature ρ (in more ordinary parlance, simply R) can be expressed (see Chapter 8) as

$$r = \tfrac{1}{2}\rho\left(1 + \sqrt{1 - \left(\frac{p}{\pi\rho}\right)^2}\right) \qquad (9.12)$$

Glide Performance   281

If, for instance, both p and ρ are 500 ft, $r = 487$ ft, somewhat less. The relation between flight path angle γ and the helix geometric parameters is

$$\sin \gamma = \frac{p}{2\pi r \sqrt{1 + \left(\frac{p}{2\pi r}\right)^2}} \tag{9.13}$$

But in spite of these annoying complications (which we usually ignore), the usual relation between flight path, air speed, and descent rate—$h' = V \sin \gamma$—holds exactly. You can see this by unwrapping the flight path from around the cylinder and flattening it out.

Getting back to our first required result, the flight path angle γ (negative for glides) must arrange itself according to Eqs. (9.4) and (9.5) augmented by Eqs. (9.7–9.9). Dividing Eq. (9.4) by Eq. (9.5) gives

$$\tan \gamma = \frac{-D}{L \cos \phi} \tag{9.14}$$

Using Eqs, (9.7–9.9) then gives

$$\tan \gamma = \frac{-C_D}{C_L \cos \phi} = \frac{-1}{\cos \phi} \left(\frac{C_{D0}}{C_L} + \frac{C_L}{\pi e A} \right) \tag{9.15}$$

The middle term of Eq. (9.15), in the unbanked case, is the simple exact expression commonly serving as backup for the statement that "best glide angle γ_{bg} results when the airplane's AOA makes ratio C_D/C_L a minimum." At this point, however, we are interested in finding a *general* exact expression for *banked* glides. Solving Eq. (9.7) for C_L and using Eq. (9.5), gives us

$$C_L = \frac{2W \cos \gamma}{\rho V^2 S \cos \phi} \tag{9.16}$$

which, when placed in Eq. (9.15), gives

$$\tan \gamma = \frac{-C_{D0} \rho V^2 S}{2W \cos \gamma} - \frac{2W \cos \gamma}{\rho V^2 S \pi e A \cos^2 \phi} \tag{9.17}$$

Multiplying through by $\cos \gamma$ and using two standard trigonometric identities, we get a quadratic equation in $\sin \gamma$:

$$b \sin^2 \gamma - \sin \gamma - (a + b) = 0 \tag{9.18}$$

where we defined auxiliary variables

$$a \equiv \frac{C_{D0} \rho V^2 S}{2W} \tag{9.19}$$

$$b \equiv \frac{2W}{\rho V^2 S \pi e A \cos^2 \phi} \tag{9.20}$$

It is useful to note that product ab is *not* a function of air speed V. The solution of Eq. (9.18) is

$$\sin \gamma = \frac{1 - \sqrt{1 + 4ab + 4b^2}}{2b}$$

$$= \frac{1}{2b} - \sqrt{1 + \left(\frac{1}{2b}\right)^2 (1 + 4ab)} \qquad (9.21)$$

A fully fledged form of the solution is

$$\sin \gamma = \frac{\rho V^2 S \pi e A \cos^2 \phi}{4W} - \sqrt{1 + \left(\frac{\rho V^2 S \pi e A \cos^2 \phi}{4W}\right)^2 \left(1 + \frac{4 C_{D0}}{\pi e A \cos^2 \phi}\right)} \qquad (9.22)$$

The (negative) rate of climb, our second "exact" result, can then be ascertained from the (unwrapped) velocity vector triangle as

$$\dot{h} = V \sin \gamma \qquad (9.23)$$

where $\sin \gamma$ is given by Eq. (9.22).

"Exact" considerations also modify the usual level turn formula [Eq. (8.12) in the chapter on Maneuvering Performance (Chapter 8)] for turn radius as a function of air speed and bank angle. Here is how to find the correct nonlevel turn formula. Return to the last two gliding force equations, Eqs. (9.5) and (9.6), split each around its minus sign, and divide. After canceling weights W on the right, one has

$$\tan \phi = \frac{\omega V}{g \cos \gamma} \qquad (9.24)$$

On the right of Eq. (9.24), multiply by $1 = V/V$, then replace ω/V with $1/R$. (As discussed above, $R^2 = r\rho$, but this stratagem prevents us from having to make a decision between those last two.) Solving for this "radius,"

$$R = \frac{V^2}{g \tan \phi \cos \gamma} \qquad (9.25)$$

Now on to a special glide—the longest possible in calm air—in this "exact" formalism.

Formulas for Speed and Angle for Best Glide, V_{bg} and γ_{bg} (Exact)

Because γ is least when $\sin \gamma$ is, we find γ_{bg} and V_{bg} by taking the derivative of Eq. (9.21) with respect to an auxiliary variable u, given by

$$u \equiv \frac{1}{2b} \qquad (9.26)$$

and setting that derivative equal to zero. After some algebra, the results are

$$V_{bg}(\phi) = \sqrt{\frac{2W}{\rho S}} \times \left[\frac{1}{4C_{D0}^2 + C_{D0}\pi eA \cos^2 \phi}\right]^{1/4} \quad (9.27)$$

$$\sin \gamma_{bg}(\phi) = -\sqrt{\frac{4C_{D0}}{\pi eA \cos^2 \phi + 4C_{D0}}} \quad (9.28)$$

Equation (9.27) shows that speed for best glide increases with either gross weight, altitude, or bank angle. Equation (9.28) shows that calm wind best glide angle does not change with gross weight or with altitude but does decrease (is larger in absolute value) with bank angle. Conversion of Eq. (9.28)—compare it with Eq. (7.52), which used the small flight path angle approximation—to a statement about the tangent of the best glide angle gives the even briefer formula:

$$\tan \gamma_{bg}(\phi) = \frac{-1}{\cos \phi}\sqrt{\frac{4C_{D0}}{\pi eA}} \quad (9.29)$$

Equation (9.29) also has the advantage of showing that $\cos \phi \tan \gamma_{bg}$ is rigorously invariant under banking:

$$\frac{\tan \gamma_{bg}(\phi_2)}{\tan \gamma_{bg}(\phi_1)} = \frac{\cos \phi_1}{\cos \phi_2} \quad (9.30)$$

for any two bank angles ϕ_1 and ϕ_2 with the same airplane in the same flaps/gear configuration.

Looking back at Eq. (9.4), we can see that, because drag $D = -W \sin \gamma$, minimum drag $D_{\min}$ occurs at precisely the condition for best glide, an exact result. It is sometimes said (e.g., Hurt[1]) that "Because of the particular manner in which parasite and induced drags vary with speed (parasite drag directly as the speed squared; induced drag inversely as the speed squared) the minimum total drag occurs when the induced and parasite drags are equal." In fact, the condition for that equality at minimum drag is much more liberal. Consider any drag function D split into parasite component D_P and induced component D_i, which depends on any reasonable function $f(V)$—function and derivative not vanishing—as follows:

$$D(V) = D_p + D_i = rf(V) + \frac{s}{f(V)} \quad (9.31)$$

Using primes now to denote derivatives with respect to V, and taking them:

$$D'(V) = rf'(V) - \frac{sf'(V)}{f^2(V)} \quad (9.32)$$

When D is minimum, $D' = 0$. Multiplying the right-hand side of Eq. (9.32) by f/f', one sees immediately that $D_P = D_i$. In the usual aerodynamic case, of course $f(V) = V^2$. Because the parasite and induced components are equal at V_{bg}, there $C_D = 2C_{D0}$. Then

$$D_{\min} = D_{bg} = \tfrac{1}{2}\rho V_{bg}^2 S(2C_{D0}) = -W \sin \gamma_{bg} \qquad (9.33)$$

and it is only a small step to

$$L_{bg} = L_{\max} = W \cos \gamma_{bg} = \sqrt{W^2 - D_{bg}^2} \qquad (9.34)$$

On the coefficient side, completing the circle, we have

$$C_{Lbg} = \sqrt{C_{D0}\pi e A} \qquad (9.35)$$

This wraps up most of what we want to know about glide performance variables at speed for best glide in calm air, wings level. Except for speed for minimum descent rate, yet to be discussed, Tables 9.2 and 9.3 give a numerical summary of our findings.

Example 9.1 Using the airframe bootstrap data plate (BDP) of Table 9.2, Table 9.3 gives precise values, source relations, and graphic references for the several best glide quantities just described when the airplane is gliding, flaps up, down through 4000 ft. Table 9.4 gives results from applying Eqs. (9.27) and (9.28) to our sample Cessna 172 ($W = 2400$ lbf, $h_\rho = 4000$ ft, $\sigma = 0.8881$) at bank angles 0 and 45 deg. As flaps come down in 10-deg increments, parasite drag coefficient C_{D0} goes up (see Table 4.10) from 0.037 to 0.039, then 0.045, then 0.055, and finally (flaps 40) to 0.066 approximately.

Table 9.2 Cessna 172 bootstrap gliding data plate (flaps up)

BDP item	Value	Units
Wing reference area, S	174	ft^2
Wing aspect ratio, A	7.38	
Parasite drag coefficient, C_{D0}	0.037	
Airplane efficiency factor, e	0.72	

Table 9.3 "Exact" best glide, Cessna 172, 4000 ft, 2400 lbf, flaps up

Glide quantity	Value	Equation no.	Figure no.
V_{bg}	71.84 KCAS	(9.27)	9.1, 9.2
γ_{bg}	−5.38 deg	(9.28), (9.29)	9.1
C_{Lbg}	0.7859	(9.35)	—
C_{Dbg}	0.074	—	—
L_{bg}	2389.4 lbf	(9.34)	9.4
D_{bg}	225.0 lbf	(9.33)	9.4

Table 9.4 Best glide ouputs for Cessna 172, 4000 ft, 2400 lbf, wings level and banked 45 deg

Degrees, δ_f	Degrees, γ_{bg} (0 deg)	KCAS, V_{bg} (0 deg)	Degrees, γ_{bg} (45 deg)	KCAS, V_{bg} (45 deg)
0	−5.38	71.8	−7.58	85.2
10	−5.52	70.9	−7.78	84.1
20	−5.93	68.4	−8.35	81.1
30	−6.55	65.0	−9.22	77.0
40	−7.17	62.0	−10.08	73.5

It is instructive to see that lift here differs from the airplane's weight by only 10.6 lbf, only 0.4%; that bodes well for the small flight path angle approximation to be discussed below.

Figures 9.1 to 9.4 provide additional reality checks, showing data for the Cessna 172 at 4000 ft. On the low speed end, these graphs all terminate at the stall speed. Figure 9.1 is one way to see that this airplane's speed for best glide V_{bg} is about 72 KCAS. Notice that, at that speed, the fuselage is pitched about 2 deg above the horizon. Figure 9.2 presents the *glide ratio* concept, the number of feet the airplane goes forward for each foot it descends. Glide ratio is a more intuitive measure of glide prowess than is the bare glide angle itself. Figure 9.3, a graph of descent rate, has its minimum (at 54.1 KCAS) only very slightly above the stall

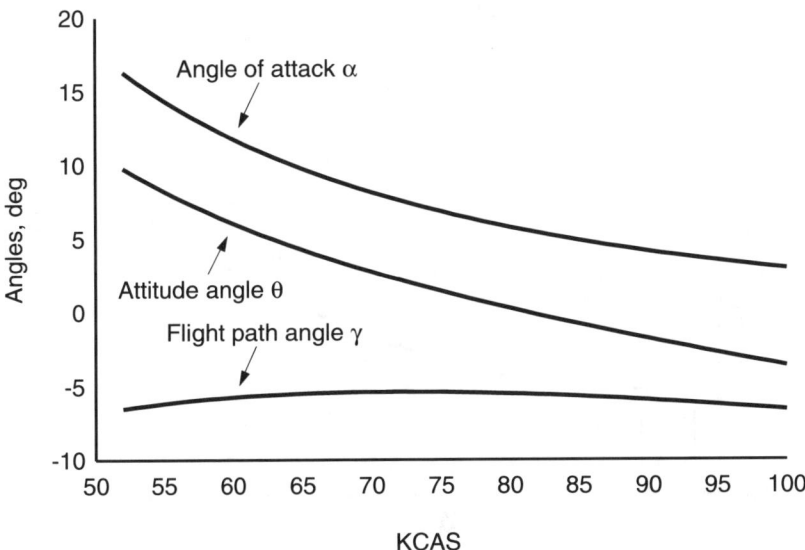

Figure 9.1 Unbanked glide path angle, (AOA), and attitude for a Cessna 172 at 4000 ft, flaps up, 2400 lbf.

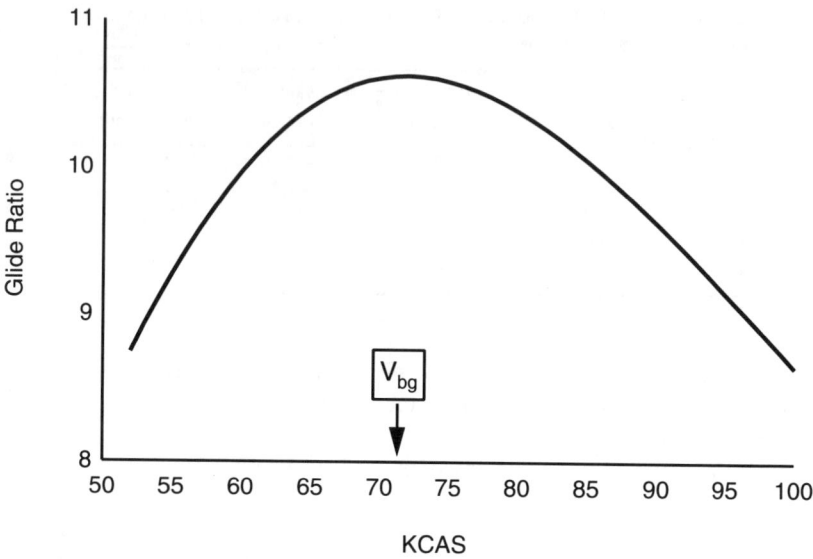

Figure 9.2 Glide ratio, as a function of air speed, for the same airplane as above.

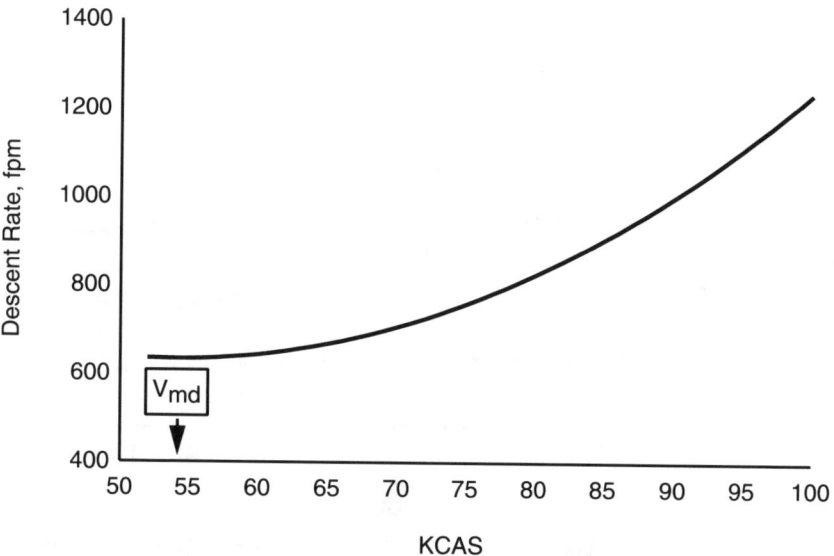

Figure 9.3 Gliding descent rate, as a function of air speed, for the same airplane.

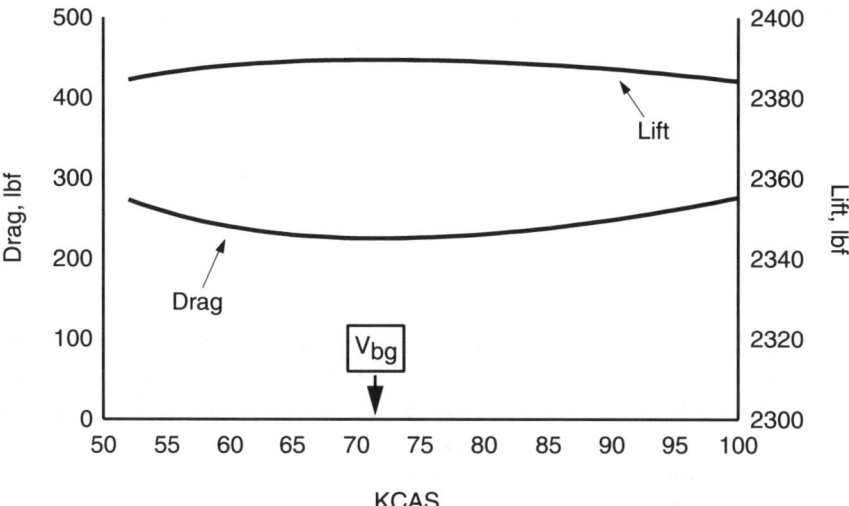

Figure 9.4 Lift and drag forces, as functions of air speed, for the same gliding airplane.

speed (53 KCAS) at which the curve stops. Speed for minimum descent rate V_{md}, for this airplane in this circumstance, while not precisely at the stall speed (53 KCAS), differs from that by only a knot or two. Figure 9.4 gives lift L and drag D numbers; as one can see, when gliding with wings level, lift very nearly equals weight W. Recall that $L^2 + D^2 = W^2$.

Formula for Speed for Minimum Descent Rate V_{md} (Exact)

If we go through the same manipulations for V_{md} as we did in finding an analytic expression for V_{bg}, we get into a real mess. A shortened version follows. Using the same auxiliary variables a, b, and u as above, the job is to find the value of u (proportional to V^2) which makes

$$\frac{dh'}{du} = \frac{d(u^{3/2} - \sqrt{u + u^3 K})}{du} = 0 \tag{9.36}$$

where K is yet another auxiliary variable defined as

$$K \equiv 1 + 4ab = 1 + \frac{4C_{D0}}{\pi e A \; \cos^2 \phi} \tag{9.37}$$

Taking the derivative ends up giving the following biquadratic for auxiliary variable u^2 which is proportional to V_{md}^4:

$$9K(K-1)u^4 + (6K-9)u^2 + 1 = 0 \tag{9.38}$$

Solving Eq. (9.38) for u^2, then rearranging and taking fourth roots, gives the following quite cumbersome (but "exact"!) relation for the speed for minimum descent rate:

$$V_{md}(\phi) = \sqrt{\frac{4W\sqrt{\frac{3 - 2K + \sqrt{9 - 8K}}{6K(K-1)}}}{\rho S \pi e A \cos^2 \phi}} \qquad (9.39)$$

Example 9.2 Using our sample BDP glide parameters in the wings-level subcase ($\phi = 0$), one will find that $ab = 0.0022165$, $K = 1.008866$, $u = 6.0219$ and, finally, $V_{md} = 97.10$ ft/s $= 57.5$ KTAS $= 54.2$ KCAS. This latter value is supported by Fig. 9.3. Notice that neither V_{md} nor V_{bg}, when expressed as calibrated air speeds, varies with altitude. And that each goes up as the square root of the airplane's gross weight. While Eq. (9.39), the "exact" result for V_{md}, is complicated, rest assured that when we get to the small flight path angle approximation we shall come up with a very simple relation for V_{md} in terms of the much less complex V_{bg}. That simpler result can also be obtained directly from Eq. (9.39) by judiciously treating $K - 1 \doteq 0$, $K \doteq 1$, and $1/(K-1) \doteq W^2/(4GH[\phi])$ while manipulating its RHS.

To find the minimum wings-level gliding descent rate itself, one can use Eqs. (9.22) and (9.23) with $V = V_{md}$. Numerical evaluation with our sample parameters gives $\gamma(V_{md}) = -6.245$ deg and $-\dot{h}_{min} = 10.56$ ft/s $= 633.8$ ft/min, values corroborated by Figs. 9.1 and 9.3.

"Instant" Drag Polar Determination from Best Glide Information (Exact)

A common, if not the *most* common, method for determining drag parameters C_{D0} and e (for aircraft for which glide testing is practicable) is to conduct wings-level glide tests at several air speeds V. From experimental rate-of-sink data, one first plots graphs of $\dot{h} V$ as a function of V^4 (this uses a small flight path angle approximation), then uses regression analysis to obtain the best straight line through the experimental points. C_{D0} comes from the fit slope and e from the fit intercept. Here is another method, simpler and more accurate. It is surprising this procedure managed to remain hidden until 1991.

Equations (9.27) and (9.28) show that, under given density, gross weight, and bank angle conditions, V_{bg} and γ_{bg} are determined by C_{D0} and e. We can invert those relations to get formulas determining C_{D0} and e. Multiply the radicand of Eq. (9.28) by C_{D0}/C_{D0} to put that long denominator equal to the long denominator in Eq. (9.27). Square Eq. (9.27). Then solve each of those offshoots for

$$\sqrt{C_{D0}\pi eA \ \cos^2\phi + 4C_{D0}^2} = \frac{2W}{\rho SV_{bg}^2} = \frac{-2C_{D0}}{\sin\gamma_{bg}} \quad (9.40)$$

The second and third terms in Eq. (9.40) immediately give

$$C_{D0} = \frac{-W \ \sin\gamma_{bg}}{\rho SV_{bg}^2} \quad (9.41)$$

On the practical flight test side, to use those formulas one must still run several test glides to experimentally determine V_{bg} and γ_{bg}. Gliding over a fixed vertical space interval in time intervals Δt, V_{bg} is the speed for which product $V\Delta t$ is maximum.

Towards e, simply solve Eq. (9.29) for the airplane efficiency factor and plug Eq. (9.40) into that result. This gives a halfway form, practical for calculation,

$$e = \frac{4C_{D0}}{\pi A \ \tan^2\gamma_{bg} \cos^2\phi} \quad (9.42)$$

Putting Eq. (9.41) into Eq. (9.40) gives the form to best pair with Eq. (9.41):

$$e = \frac{-4W \ \sin\gamma_{bg}}{\rho SV_{bg}^2 \tan^2\gamma_{bg} \cos^2\phi \pi A} \quad (9.43)$$

In practical determinations of the drag polar parameters C_{DO} and e, one will, most likely, perform glide tests with wings level, $\phi = 0$.

Equation (9.41) presents us with a second invariant under banking, $\sin\gamma_{bg}(\phi)/V_{bg}^2(\phi)$:

$$\frac{\sin\gamma_{bg}(\phi_2)}{\sin\gamma_{bg}(\phi_1)} = \frac{V_{bg}^2(\phi_2)}{V_{bg}^2(\phi_1)} \quad (9.44)$$

When combined with the invariant relationship of Eq. (9.30), this allows one to directly obtain best glide speed and angle for any bank angle if one knows those best glide specifications for any other bank angle.

Banked or Unbanked Glides, Small Path Angle Approximation, No Wind (Cases 3 and 4)

Almost any practical work describing complicated systems is replete with approximations. Ours is no exception. Approximations are needed to simplify problems, to make them solvable or at least tractable. One may simply ignore "negligible" variables (e.g., aeroelasticity, angular momentum of rotating machinery, or solar radiation pressure on the airframe), treat slightly varying variables as constants (e.g., neglect changes in gravitational acceleration g with altitude or latitude), or linearize complex "slightly" nonlinear phenomena (e.g., the propeller polar diagram). Sometimes approximations are required for purely technical mathematical reasons. (An example, below, will be Bridges's infinite series solution to the effect of headwind or tailwind on best glide speed; in practice, one eventually truncates the series.) Approximations take many forms.

The "small flight path approximation" used in flight mechanics is of the linearization or small parameter sort. It is unusual in that $\cos \gamma$ is (usually) approximated as unity but $\sin \gamma$ is left as is. The ostensible rationale behind that procedure is that the series expansions of these trigonometric functions (angles γ of course in radians) are

$$\cos \gamma = 1 - \frac{\gamma^2}{2!} + \frac{\gamma^4}{4!} - \frac{\gamma^6}{6!} + \cdots \tag{9.45}$$

$$\sin \gamma = \gamma - \frac{\gamma^3}{3!} + \frac{\gamma^5}{5!} - \frac{\gamma^7}{7!} + \cdots \tag{9.46}$$

and we need keep only leading terms. But in the case of the sine function we do not just keep the leading γ term! We keep the whole thing. The ultimate and actual rationale is that the small flight path angle approximation works well; such approximate solutions (for $\gamma \leq 15$ deg) differ only slightly from the fuller and more accurate solutions one can obtain through more complicated trial-and-error or iterative techniques.

Now that we have well in hand the "exact" solutions to some of our major gliding flight mechanics problems, for comparison purposes in this chapter we take refuge in the small flight path approximation—or trot it out as more succinct—in two different ways. Two ways partly as a tribute to pedagogy, to build variety. First we will make the small γ approximation to the underlying force equations. Solving those will give our first approximations. But then we will turn around, somewhat retracing our steps, and use the small γ approximation directly on our former "exact" *solutions*, via binomial expansion. In this case both approximation approaches give identical results.

Approximating the Force Equations (Small γ)

Taking Eqs. (9.4–9.6) and replacing $\cos \gamma$ by unity in the rewrite of Eq. (9.5), we have

$$D + W \sin \gamma = 0 \tag{9.47}$$

$$L \cos \phi - W = 0 \tag{9.48}$$

$$L \sin \phi - \frac{W \omega V}{g} = 0 \tag{9.49}$$

Notice Eq. (9.48), that we now have lift $L = W/\cos\phi$. (In the unbanked case, it is even simpler, $L = W$.) Just as in the "exact" case, let us go after $\sin\gamma$. It is almost direct. Using Eq. (9.47) with Eq. (9.48) and switching to coefficients, we have

$$\sin\gamma = \frac{-D}{W} = \frac{-D}{L\cos\phi} = \frac{-C}{C_L\cos\phi} \qquad (9.50)$$

Now substitute from the coefficient relations Eq. (9.7–9.9), nothing has changed there, and simplify to find our first small γ approximate result:

$$\sin\gamma = \frac{-C_{D0}\rho V^2 S}{2W} - \frac{2W}{\rho V^2 S\pi e A \cos^2\phi} \qquad (9.51)$$

Equation (9.51) is not only much simpler than its "exact" counterpart, Eq. (9.22), but it looks much different. But we can show that in fact a simple binomial expansion of Eq. (9.22), keeping only lowest order terms, will indeed result in Eq. (9.51).

Approximating the "Exact" Solutions for γ, $h^{\cdot}$, and R (Small γ)

It simplifies the algebra to make use of abbreviations a and b, Eqs. (9.19) and (9.20), and begin with the first form of Eq. (9.21). On the radical portion, use the first two terms of the binomial expansion formula

$$(1+x)^n = 1 + nx + \frac{n(n-1)}{2!}x^2 + \frac{n(n-1)(n-2)}{3!}x^3 + \cdots \qquad (9.52)$$

for $n = \frac{1}{2}$. Equation (9.21) turns into

$$\sin\gamma = \frac{1 - (1 + 2ab + 2b^2)}{2b} \qquad (9.53)$$

When the RHS of Eq. (9.53) is simplified and substitutions from Eqs. (9.19) and (9.20) are made for a and b, one has

$$\sin\gamma = \frac{-C_{D0}\rho V^2 S}{2W} - \frac{2W}{\rho V^2 S\pi e A \cos^2\phi} \qquad (9.54)$$

This is precisely Eq. (9.51), obtained by approximating the underlying "exact" equations. Look back in the bootstrap approach chapter and find the formula for $\sin\gamma$ there, Eq. (7.43). Cross out terms having to do with the engine and the propeller. Put in the weight-amplification factor $1/\cos^2\phi$ due to banking. You then have Eq. (9.54). The bootstrap approach consistently uses the small γ approximation. We shall make much use of that fact in the two subsections following.

Proceeding now to the other two needed results, there is no change in the formula for $h^{\cdot}$. Not so for turn radius R. The small γ formula for R, whether derived from Eqs. (9.48) and (9.49) or approximated straight from the exact solution, Eq. (9.25), is now clearly

$$R = \frac{V^2}{g \tan \phi} \tag{9.55}$$

This is the same as for level turns and is of course a better approximation for nearly level turns.

Approximations for V_{bg} and γ_{bg} (Small γ)

Of the four important bootstrap composite parameters E, F, G, and $H(\phi)$, because we are here considering gliding, only G and $H(\phi)$ matter. Refreshing our memories, those are

$$G \equiv \tfrac{1}{2}\rho S C_{D0} \tag{9.56}$$

$$H(\phi) \equiv \frac{2W^2}{\rho S \pi e A \, \cos^2 \phi} \tag{9.57}$$

A simple expression involving flight path angle γ, rewriting Eq. (9.54), is then

$$D = -W \sin \gamma = GV^2 + \frac{H}{V^2} \tag{9.58}$$

To find V_{bg}, we can set

$$\frac{dD}{dV^2} = 0 \tag{9.59}$$

and solve for V_{bg}. This of course gives us, though it now includes the possibility of banked glides, essentially the result we found in the bootstrap approach chapter:

$$V_{bg} = \left(\frac{H(\phi)}{G}\right)^{1/4} = \sqrt{\frac{2W}{\rho S} \left[\frac{1}{C_{D0} \pi e A \, \cos^2 \phi}\right]^{1/4}} \tag{9.60}$$

This is close to but not identical to the "exact" expression for the same quantity, Eq. (9.27). Putting Eq. (9.60) into Eq. (9.58), and rearranging,

$$\sin \gamma_{bg} = \frac{-\sqrt{4GH(\phi)}}{W} = \frac{-1}{\cos \phi} \sqrt{\frac{4C_{D0}}{\pi e A}} \tag{9.61}$$

The corresponding "exact" expression has a tangent function in lieu of this sine. Because the small path angle approximation consists of treating $\cos \gamma \doteq 1$, and because $\tan = \sin/\cos$, this difference makes sense. And if you decide to derive "small γ instant drag polar" formulas, for C_{D0} and e, from Eqs. (9.60) and (9.61), you will not quite get the same exact drag polar formulas of Eqs. (9.41) and

(9.43). You will get Eq. (9.41) all right, but your analog to Eq. (9.43) will have sines instead of tangents, in the denominator.

Approximations for V_{md} and $h\dot{}_{md}$ (Small γ)

Entirely similar manipulations work for the case of the minimum descent rate. Asking for $h\dot{}$ a maximum (smallest in the negative direction) means that $GV^3 + H/V$ will be a minimum, or

$$3GV^2 - H(\phi)/V^2 = 0 \tag{9.62}$$

which has solution

$$V_{md} = \left(\frac{H(\phi)}{3G}\right)^{1/4} = \left(\frac{1}{3}\right)^{1/4} V_{bg} \doteq 0.7598 V_{bg} \tag{9.63}$$

But why not continue on to get an expression for the minimum descent rate itself? That can be expressed as

$$h\dot{}_{md} = V_{md} \sin \gamma(V_{md}) = \frac{-GV_{md}^3 - H(\phi)/V_{md}}{W} \tag{9.64}$$

Substituting Eq. (9.63) into Eq. (9.64) gives us a new result:

$$h\dot{}_{md} = \frac{-4}{3W}(3GH^3(\phi))^{1/4} \tag{9.65}$$

Because bootstrap composite H is proportional to W^2, $h\dot{}_{md}$ is proportional to $W^{1/2}$. As are so many aircraft performance parameters.

Example 9.3 Just how good *is* the small flight path angle approximation? It is not sufficient to simply focus on $\cos \gamma$ itself, citing such true but not completely pertinent facts such as $0.99 = \cos 8.1$ deg, $0.98 = \cos 11.5$ deg, or even $0.95 = \cos 18.2$ deg. That is because the important performance numbers might go as squares or cubes or some other amplifying function of $\cos \gamma$ or of some other function of γ. Table 9.5 presents alternative numbers for the exact and the small γ approximation. It uses best glide and minimum descent figures for its example simply because we have obtained simple formulas for those. The figures in Table 9.5 are for the Cessna 172, flaps up, 2400 lbf, at mean sea level (MSL). With the single exception of the banked minimum descent rates, there is no practical difference between "exact" figures and small flight path approximation figures. Equal turn radius values, 649.2 ft, are not a coincidence; even though exact and small γ speeds and path angles differ, cancellations make the best glide turn radii always precisely the same.

Table 9.5 makes the small flight path angle approximation pretty convincing. But what would happen, you might ask, if we took a heavier, less glider-like,

Table 9.5 Glide, turn, descent data for Cessna 172, MSL, flaps up, 2400 lbf

	Wings level		Banked 45 deg	
Parameter	Exact	Small γ	Exact	Small γ
V_{bg}, KCAS	71.84	72.00	85.25	85.62
γ_{bg}, deg	−5.38	−5.40	−7.59	−7.65
R_{bg}, ft	NA	NA	649.2	649.2
V_{md}, KCAS	54.22	54.71	63.87	65.06
h'_{md} ft/min	−597.2	−602.5	−1683.0	−1013.2

Table 9.6 Glide, turn, descent data for F104G, MSL, flaps up, 18000 lbf

	Wings level		Banked 45 deg	
Parameter	Exact	Small γ	Exact	Small γ
V_{bg}, KCAS	299.65	300.73	355.10	357.63
γ_{bg}, deg	−6.84	−6.89	−9.63	−9.77
R_{bg}, ft	NA	NA	11324.6	11324.6
V_{md}, KCAS	225.13	228.5	263.51	271.74
h'_{md} ft/min	−3161.5	−3207.4	−8688.9	−5394.1

aircraft? Table 9.6 presents the same kind of data for an F104G, a jet fighter, also at sea level. Its pertinent data are: reference wing area $S = 196$ ft^2, aspect ratio $A = 2.47$, $C_{D0} = 0.018$, $e = 0.644$, gross weight $W = 18{,}000$ lbf. As Table 9.5 shows, the small approximation still holds up (with the exception of the minimum descent rates) quite well. Again, the two identical turn radius values, 11324.6 ft, are not coincidental.

Now we return to "exact" theory to treat a gliding maneuver we hope you never have occasion to consider, the return-to-airport "impossible turn."

Engine Out (Return to Airport) Maneuver (Exact)

If your engine quits soon after takeoff, you may suddenly become interested in the possibility of turning around to return to the airport. What air speed and bank angle should you adopt to maximize angle turned $\Delta\theta$ for each foot of altitude lost Δh?

Because both $h^{\cdot}$ and ω are time rates, combining Eqs. (9.5), (9.6), and (9.23) gives

$$\frac{\Delta\theta}{\Delta h}(V, \phi) = \frac{\omega}{h^{\cdot}} = \frac{g \tan \phi}{V^2 \tan \gamma(V, \phi)} \qquad (9.66)$$

We can solve one-half our problem immediately (and save ourselves from multivariable calculus) by recognizing that the strong inverse speed dependence in Eq. (9.66) can mean only one thing: we should execute the turnaround maneuver as close to the stall speed—stall speed for whatever bank angle we choose—as is possible and prudent. Stall speed at bank angle ϕ with flight path angle γ is given by combining Eqs. (9.5) and (9.7) for the special case of maximum lift coefficient $C_{L\max}$:

$$V_S^2(\phi) = \frac{2W \cos\gamma}{\rho S \cos\phi C_{L\max}} \tag{9.67}$$

Inserting Eq. (9.67) into Eq. (9.66), we get

$$\frac{\Delta\theta}{\Delta h}(V_S(\phi), \phi) = \frac{g\rho S C_{L\max} \sin\phi}{2W \sin\gamma} \tag{9.68}$$

Having specialized our turnaround to the banked stall speed where $C_L = C_{L\max}$, complication is avoided by rewriting our earlier Eq. (9.15) and implicitly defining auxiliary variable k:

$$\tan\gamma = \frac{-1}{\cos\phi}\left(\frac{C_{D0}}{C_{L\max}} + \frac{C_{L\max}}{\pi e A}\right) = \frac{-k}{\cos\phi} \tag{9.69}$$

Using a trigonometric identity to change the tangent function in Eq. (9.69) into the sine function required by Eq. (9.68) and making that substitution, we arrive at

$$\frac{\Delta\theta}{\Delta h}(V_S(\phi), \phi) = \frac{-g\rho S C_{L\max} \sin\phi\sqrt{\cos^2\phi + k^2}}{2Wk} \tag{9.70}$$

The best turnaround bank angle then occurs when

$$\frac{d}{d\phi}(\sin\phi\sqrt{\cos^2\phi + k^2}) = 0 \tag{9.71}$$

Straightforward calculation finds that the best turnaround bank angle is

$$\phi_{bta}(k) = \cos^{-1}\left(\frac{\sqrt{2}}{2}\sqrt{1-k^2}\right) \tag{9.72}$$

Because k^2 will always be positive for any airplane in any configuration, and normally fairly small (for our sample Cessna, at zero flaps, $k = 0.1163$), Eq. (9.72) tells us that the best turnaround bank angle is always slightly over 45 deg. In our sample Cessna case, $\phi_{bta} = 45.39$ deg, only negligibly greater than the 45-deg solution offered by less exact theories (Langewiesche[2]; Rogers[3]) which, explicitly or otherwise, use the small flight path angle approximation.

So, practically speaking, a 45-deg bank it should be. Again speaking practically, one could not afford to fly at the precise banked stall speed even if one knew what it was; for a typical trainer aircraft, a minimum 5-kn pad is in order. For our purposes we will stay with the theoretical optimum. Exact calculation, as sometimes happens, turns out to be simpler than approximation.

Table 9.7 Optimum turnaround for Cessna 172, MSL, 2400 lbf

δ_f	k	ϕ_{bta}, deg	$\Delta h_{min}(180\ \text{deg})$ ft
0	0.1163	45.39	−168.9
10	0.1262	45.46	−162.7
20	0.1347	45.52	−163.0
30	0.1441	45.59	−165.8
40	0.1528	45.67	−169.2

For example, using the optimum of Eq. (9.72) for bank angle ϕ in the numerator of Eq. (9.70) gives the felicitous result that

$$\sin \phi_{bta} \sqrt{\cos^2 \phi_{bta} + k^2} = \frac{(1+k^2)}{2} \qquad (9.73)$$

where k, recall, is the parenthesized group of flap-configuration-dependent aircraft parameters in Eq. (9.69). Using Eq. (9.73) in Eq. (9.70), then taking its reciprocal and multiplying by π radians, we arrive at a simple relation for minimum altitude loss while making a 180-deg turn:

$$\Delta h_{min}(180\ \text{deg}) = \frac{-4\pi k W}{g \rho S C_{L max}(1+k^2)} \qquad (9.74)$$

Factors k and $C_{L max}$ in Eq. (9.74) depend on the airplane's flaps configuration. Minimum relative altitude loss during a turnaround maneuver increases directly with gross weight and inversely with atmospheric density. Next we look into the relative altitude loss effect, for our sample Cessna, of taking different flaps configurations.

Example 9.4 Table 9.7 gives optimum bank angle and altitude loss for each flaps setting for the Cessna 172 at MSL, 2400 lbf. For this aircraft, flaps setting has a negligible practical effect on the efficacy of the return-to-airport, engine-out maneuver. At a time when the pilot's hands (and head) are full, it is nice to be able to ignore at least one aspect of the problem.

Application: Engine-Out (Return-to-Airport) Maneuver

If your engine quits while climbing out after takeoff, should you or should you not try to turn around? To help you answer that question, now follows a calculation that every safety-conscious pilot should repeat, for his particular airplane, several to many times using various values of gross weight; headwind speed; crosswind speed; density altitude; runway length; and perhaps other

Glide Performance   297

variables such as runway surface type, runway slope, and altitude at which the engine stopped. Coupled with at-altitude practice of the suggested maneuver (45-deg banked turn, as close to stall speed as is comfortable), these calculations can be very enlightening.

We will assume, for our version of the problem, a dry level concrete runway, 3000 ft long, situated at a time and place with pressure altitude $h_p = 3000$ ft with runway temperature $10°C = 50°F$. (This makes density altitude $h_\rho = 3110$ ft but, because we will be working in such a restricted range of altitudes, we will simply use $h_\rho = 3300$ ft, with corresponding relative density $\sigma = 0.9069$, for all calculations.) Because we have not treated takeoff yet in detail, we will use Cessna 172 POH takeoff performance data to properly place the airplane during the early part of the exercise. We take the airplane at 2400 lbf with 10-deg flaps (setting for short field takeoff), taking off on runway 36 (to the north). For wind, assume a 12-kn headwind with a 9-kn crosswind from the left; the wind is then 15 kn from 323 deg. The surrounding terrain is assumed level.

We will assume a takeoff and climb out "by the book," with climb out from 50 to 500 ft at the appropriate V_x. We will take it that the engine quits at the moment the airplane reaches 500 ft AGL. According to the Cessna 172 pilots operating handbook,[4] interpreted through the air-speed indicator calibration curve for flaps 10, liftoff speed $V_{LO} = 51$ KIAS $= 56$ KCAS $= 58.8$ KTAS $= 99.3$ ft/s. At 50 ft AGL, speed is $V_{50} = 56$ KIAS $= 59$ KCAS $= 62.0$ KTAS $= 104.6$ ft/s. The POH gives a no-wind distance to liftoff, d_{LO}, as 1140 ft; it gives the no-wind distance to 50 ft, d_{50}, as 2120 ft. We will be making wind adjustments to those figures. Among them is a POH note that one should reduce distances 10% for each 9 kn of headwind. (From more respectable engineering grounds the reduction should be greater; the POH suggestion follows an Federal Aviation Administration mandate that, for takeoff purposes, one should count on only one-half the available headwind and, if taking off with a tailwind, should figure on one and half-times the estimated tailwind speed.)

We need data on the airplane, its BDP (for flaps 10) and composite bootstrap parameters calculated for density altitude 3300 ft and (we ignore any takeoff or climb out fuel burn) gross weight 2400 lbf. Table 9.8 is the BDP; Table 9.9 provides composite parameters.

We shall analyze each segment of this short flight in order. Those segments are between the seven pivotal points described in Table 9.10. Transitions between segments—except for a 4-s pause when the engine quits, for "What's that lack of noise?" and for the answer to sink in—are assumed instantaneous. It is unrealistic to have the turn precisely at the banked stall speed, but we do so to avoid yet another arbitrary parameter. And we ignore the slight additional turn, beyond 180 deg, needed to point towards the end of the runway (minus any wind correction angle, also ignored). In the glide back towards the departure end of the runway, we also neglect to have the pilot retract his 10 deg of flaps. We will include elapsed times (since start of the takeoff roll) because those are needed for some of

Table 9.8 Cessna 172 BDP, Flaps 10

BDP	Value	Units	Aircraft subsystem
Wing reference area, S	174	ft^2	Airframe
Wing aspect ratio, A	7.378		Airframe
Rated MSL torque, M_B	311.2	ft-lbf	Engine
Altitude dropoff factor, C	0.12		Engine
Propeller diameter, d	6.25	ft	Propeller
Parasite drag coefficient, C_{D0}, $\delta_f = 10$ deg	0.039		Airframe
Airplane efficiency factor, e	0.72		Airframe
Propeller polar slope, m	1.70		Propeller
Propeller polar intercept, b	-0.0564		Propeller

Table 9.9 Bootstrap composites and other data for the airplane of Table 9.8

Variable or composite	Value
h_ρ	3300 ft
W	2400 lbf
σ	0.9069
$\Phi(\sigma)$	0.8943
k	0.1262
E	475.7
F	-0.0047495
G	0.0073146
H	1,840,236
K	-0.0120641
Q	$-39,427.8$
R	$-152,537,707$
U	251,582,692

Table 9.10 Return-to-airport maneuver for the airplane of Table 9.8

Point	Description
1	Start of takeoff roll ($V_g = 0$, $V = 12$ KCAS, from headwind)
2	Liftoff ($V_{LO} = 56$ KCAS)
3	At 50 ft ($V_{50} = 59$ KCAS)
4	AT 500 ft ($V = V_x = 63$ KCAS), engine quits
5	Start of turn towards airport [$V = V_S(\phi)$, $\phi = \phi_{bta}$]
6	Completion of 180-deg turn, start of wings-level crabbed glide ($V = V_{bg}$)
7	Aircraft contacts runway (or terrain)

Glide Performance

the calculations and they are instructive in their own rights. At the end of all these annoying little calculations appears a summarizing table.

Segment 1-2, Takeoff Roll

For d_{LO} we need the aforementioned headwind adjustment factor. Because we have a 12-kn headwind, the no-wind distance will be reduced by $10\% \times 12/9 = 13.3\%$, from 1140 ft to 988 ft. The liftoff ground speed will be $(99.3 - 12/0.592468) = 79.0$ ft/s, so the average ground speed during the takeoff roll will be 39.5 ft/s. Elapsed time will therefore be $988/39.5 = 25.0$ s. We are off!

Segment 2-3, Liftoff to 50 ft

Again, let us find an average speed. We lifted off at 99.3 ft/s. By 50 ft, we will have very slightly accelerated to 62 KTAS = 104.6 ft/s. That makes the segment average true air speed 102.0 ft/s. In calm air, our horizontal displacement would be $(2120 - 1140) = 980$ ft and would take $980/102 = 9.6$ s. But, during that time, the 12-kn (20.3-ft/s) headwind component set us back $20.3 \times 9.6 = 194.9$ ft. So we really only progressed over the ground 785.1 ft. But we have cleared that proverbial obstacle.

Segment 3–4, 50 to 500 ft (Engine Failure)

We had decided to climb out, after 50 ft, at the appropriate V_x. How much is that? Because $V_x = -R^{-1/4}$, Table 9.9 tells us $V_x = 111.1$ ft/s. Because the flight path angle is so small, we can also use that same figure for the horizontal component of true air speed. But we need the rate of climb so we know how long it takes to get to 500 ft AGL. For that we use the usual bootstrap formula:

$$h^{\cdot}(V) = \frac{EV + KV^3 - H/V}{W} \qquad (9.75)$$

Substitution from Table 9.9 gives $h^{\cdot}(V_x) = 8.23$ ft/s = 494 ft/min. So to climb 450 ft takes 54.7 s, during which we have progressed over the ground about $111.1 \times 54.7 = 6077$ ft, a nautical mile. Now, the engine quits! Where are we? Taking x directions as straight north of the start of the takeoff roll, y directions to the west, and $h = z$ as AGL altitude (all in feet), we are now at $(x, y, h, t)_4 = (7850, 0, 500, 89.3)$.

Segment 4–5, Getting Ready to Return to Airport

If we were planning to land more or less straight ahead, we would immediately lower the nose to (wings-level) best glide speed, here (from Table 9.7) 70.9 KCAS = 74.5 KTAS = 125.7 ft/s. But there is *something* out there—sharp rocks, an unbroken expanse of thick trees, concrete canyons, freezing cold water,

Table 9.11 Sample Cessna 172 coefficients for various flaps angles

δ_f, deg	C_{D0}	$C_{L\max}$
0	0.037	1.54
10	0.039	1.73
20	0.045	1.84
30	0.055	1.93
40	0.066	2.00

a dozen day care centers all on recess—we do not like. So, today and here, it has got to be the old college try at the old "impossible" turn. Luckily, we have carefully calculated how to optimize our chances.

We need to find the best-turnaround-angle stall speed. But for that we need the bank angle. Using parameter k, from Table 9.9 and Eq. (9.72), that angle is $\phi_{bta} = 45.46$ deg. Using the small flight path angle formula for stall speed

$$V_S(\phi) \doteq \sqrt{\frac{2W}{\rho S \cos \phi C_{L\max}}} \qquad (9.76)$$

and (from Table 9.11) $C_{L\max} = 1.73$, we find that our turnaround air speed is to be $V_s(\phi = 45.5 \text{ deg}) = 102.7$ ft/s $= 57.9$ KCAS. We need to slow a little bit.

Why not zoom? That is, pull up to simultaneously lose air speed and gain altitude. Of course we have only $(111.1 - 102.7) = 8.4$ ft/s $= 5.0$ KTAS $= 4.7$ KCAS excess speed to lose, but it is the principle of the thing. So let us look into the zooming possibility.

Zooming

If one neglects the dissipative effect of drag, zooming merely trades kinetic energy for potential energy:

$$\tfrac{1}{2}m(V_1^2 - V_2^2) = mg(h_2 - h_1) \qquad (9.77)$$

In that case, using our current situation, we would find that we gained (geometric) altitude $(h_2 - h_1) = 27.9$ ft.

But let us, at least approximately, take drag into account. To do that, we need to assume some time interval Δt (which amounts to a flight path, whether sharp and steep or slow and shallow) over which the zoom (xor dive, antizoom) maneuver takes place. It is often said that when the engine quits it takes about 4 s to figure out that it has quit, so we will take that time interval, $\Delta t = 4$ s.

Glide Performance

Energy-dissipation rate is power required, $P_r = DV$, and elementary considerations lead one to the following general formula for the pure gliding (thrust $T = 0$) case:

$$h_2 - h_1 \doteq \frac{V_1^2 - V_2^2}{2g} - \frac{((DV)_1 + (DV)_2)\Delta t}{2gm} \tag{9.78}$$

(When there is still some thrust, $T > 0$, P_r needs only to be replaced by the negative of excess power, $-P_{xs}$.) In our sample calculation, the rightmost (dissipative) term, using composite parameters from Table 9.9, amounts to 43.7 ft. Not good. Our slow zoom has *cost* us altitude. Now of course if we were quicker witted and did the zoom in 1 s, the dissipative term would only be 11 ft, resulting in a small net altitude gain of 17 ft.

So, on second thought, forget the zoom. Just add 4 s to the clock, with $4 \times (111.1 - 20.2) = 364$ ft forward progress, while we gather our wits and roll into a near 45-deg bank.

Segment 5–6, Optimal 180-deg Turn

We are banked 45.5 deg at true air speed 102.7 ft/s. To find the altitude loss, we adjust the proper entry in Table 9.7, dividing by our now not-standard relative air density, 0.9069. We find that $\Delta h_{min}(180 \text{ deg}) = -179.4$ ft. But what about our horizontal "progress" during this turn?

The usual turn radius formula, $R = V^2/g \tan \phi$, tells us our turn radius, *in the air*, was 322 ft. Our slant air travel distance was then the square root of $[(\pi R)^2 + (\Delta h)^2] = 1027$ ft, which took 10.0 s. During that time, the 12-kn (20.2 ft/s) headwind blew us back 202 ft and the 9-kn (15.2-ft/s) crosswind blew us to the east $15.2 \times 10.0 = 152$ ft.

We *instantaneously* roll back wings level and assume both a proper crab angle to head for the departure end of the runway and trim to what? In calm air, we know $V_{bg} = 70.9$ KCAS $= 74.4$ KTAS $= 125.6$ ft/s. But with tailwind of a little over 12 kn, we need to slow by about one-fourth that amount, to 70 KTAS $= 118.1$ ft/s.

Segment 6–7, Glide Toward Runway

First things first: Where are we? If you have kept track, we are at (8012, −492, 321) ft after 103.3 s. Because the departure end of the runway is at (3000, 0, and 0 ft), our nose is pointed 5.6 deg too far to the right. Because we are going to be set to the left by the (now) right crosswind, that is nothing to worry about. Our slant distance from the runway, by Pythagorean theorem, is 5036 ft.

What about our glide angle? The best angle with a tailwind (to be discussed at length below) is a little shallower than the best in calm air (which is, incidentally,

−5.55 deg). The standard bootstrap formula for glide angle, for our true air speed 118.1 ft/s, gives $\gamma = -5.59$; not much difference. In calm air, this would bring us down from our 321 ft AGL altitude in 3279 ft in about 27.8 s. But, during that time, the say 12-kn tailwind will push us along another $27.8 \times 20.2 = 562$ ft. So over the ground we will make a total of 3841 ft. But, still 1195 ft shy of the runway. We did not quite make it! Think positive. Perhaps it is pretty level out there, only waist-high sagebrush and no ditches.

Return-to-Airport Summary

If we had been a couple of hundred pounds lighter, or had had another 5 kn of initial headwind, or had come off a 5000-ft runway, we would have made it. So it is certainly *not* an "impossible" turn; just an *unlikely* one. We did not quite make it back even though we were doing everything "just right." Lightly powered airplanes, with their shallow climb angles, tend to get too far from the runway to be able to make it back.

This is off the strict performance subject, but nothing takes the place of good judgment. Whether or not it is wise to attempt a return-to-airport maneuver depends on many factors: the airplane, the pilot's skill, aircraft weight, density altitude, wind, the runway, and the terrain both close to the airport environment and farther out. The time to think about your response to an engine failure soon after takeoff is before the throttle goes forward.

Consider instituting a new before-takeoff check list item that would lead to a definite piece of advice to yourself, which might run as follows: "Today, under these conditions, I will perform the return-to-airport maneuver, turning left, if and only if the engine goes out when I am between 3600 and 4000 ft (500 to 900 AGL). If I do, I will bank 45 deg and go to 63 KCAS, leaving myself a slight buffer above banked stall speed. If I am not within that range of elevations, I will lower the nose to 71 KCAS and look for the best spot within about 40 deg of my heading."

Table 9.12 is a summary of calculations for the various return-to-airport maneuver segments.

Table 9.12 Sample return-to-airport scenario

Point	Position (x, y, h, t)	KTAS	Note
1	(0, 0, 0, 0s)	12	Start of takeoff roll (12-kn headwind)
2	(988, 0, 0, 25.0s)	58.8	Liftoff ($V_{LO} = 56$ KCAS)
3	(1773, 0, 50, 34.6s)	62	50 ft ($V_{50} = 59$ KCAS)
4	(7850, 0, 500, 89.3s)	65.8	500 ft ($V_x = 63$ KCAS), engine quits
5	(8214, 0, 500, 93.3s)	60.8	Start course reversal at $V_S(\phi)$, $\phi = \phi_{bta}$
6	(8012, −492, 321, 103.3s)	70	180 deg turn complete, start of glide
7	(4195, −117, 0, 131.1s)	0	Aircraft contacts terrain short of runway

As you see, the return-to-airport scenario (refer to Figs. 9.5 and 9.6) played itself out in just a little over 2 min. Most small airplanes will not be able to make it back to a fairly short runway unless there is considerable headwind. But perhaps the airport environment is better for a forced landing than is the terrain farther out.

One cannot take every contingency into account on every takeoff. And one certainly cannot do all these calculations; we are smarter than computers, but we are not as quick. Still, running through a few return-to-airport scenario drills (different winds, different weights, different elevations) is a good idea.

Speed in a Terminal Velocity Dive, V_T

After our complicated turnaround example, here is a short and easy one. To dive at the fastest possible speed, one might think that lift, because it is perpendicular to the flight path, is of no consequence. But that would ignore induced drag, "drag due to lift." So, in fact, maximum diving speed must be at $L = C_L = 0$. Then, in our unpowered case, force Eq. (9.5) says $\gamma = -90$ deg, straight downwards. Furthermore, by force Eq. (9.4), drag D = weight W, so

$$D = W = \tfrac{1}{2}\rho V_T^2 S C_{D0} \qquad (9.79)$$

so the terminal speed, in calibrated terms, is

$$V_{TC} = \sqrt{\frac{2W}{\rho_0 S C_{D0}}} \qquad (9.80)$$

For our sample Cessna 172 with flaps up ($W = 2400$ lbf, $S = 174$ ft^2, $C_{D0} = 0.037$), this gives $V_{TC} = 560.1$ ft/s $= 331.8$ KCAS. Such a high speed dive *might* not be particularly dangerous by itself (assuming the windshield stays in place and that drag tubes in the wings are strong enough), but during the ensuing recovery perpendicular wing loads will go from zero to some very high value unless that pull out is delicate and gingerly. In sum, do not even think about it. Terminal true air speed will be higher at higher altitudes, of course, because air resistance is lower.

Horizontal Wind Effects (No Updraft or Downdraft) on Wings-Level Glides (Case 8)

The qualitative picture is clear. Gliding into a headwind, to make best progress over the ground, one should go faster than the no-wind best glide speed V_{bg0}. The headwind of speed V_w is setting you back, relative to your calm-wind progress over the Earth, by distance V_w each second; to minimize this effect, you speed up.

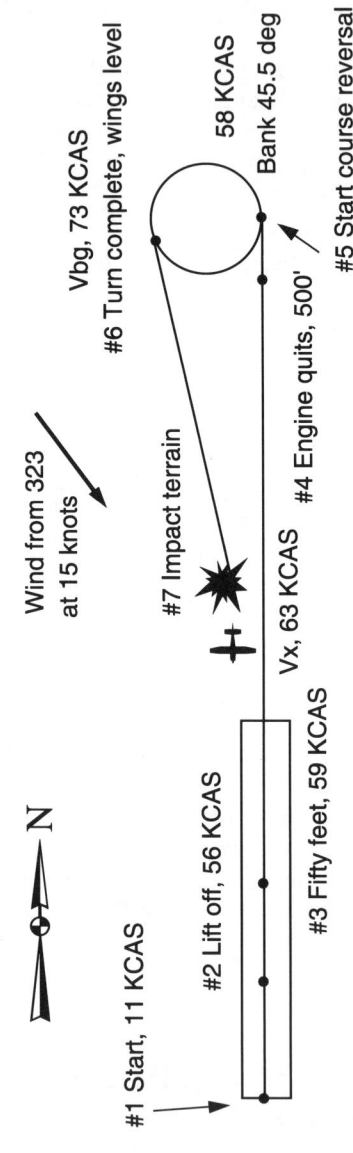

Figure 9.5 Plan view of the return-to-airport maneuver.

Glide Performance 305

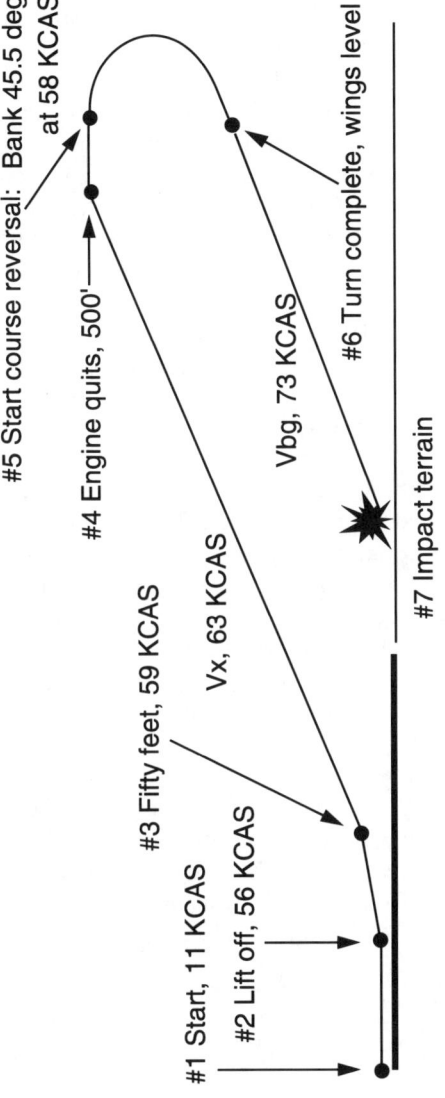

Figure 9.6 Elevation view of the return-to-airport maneuver.

Gliding with a tailwind, on the other hand, you should slow down. The question is: Speed up or slow down by how much?

First we will write down the basic vector equation—and the corresponding vector diagram—having to do with the airplane's motion through the air mass and the air mass's motion over the Earth. Then comes a crucial approximation, that flight path angles are shallow enough that the horizontal component of aircraft speed is tantamount to that speed itself. That is merely another iteration of the small flight path angle approximation and is why this discussion occurs where it does. Next we will show what a graphical solution to the problem would look like. Following that comes the corresponding bootstrap equation, a fifth-degree polynomial unsolvable in closed form. But not unsolvable if one relaxes enough to allow a solution in the form of an infinite series; we present such a solution due to Bridges.[5] Finally, we discuss the dependence of the solution on density altitude.

The rule for computing relative motion gets down to this simple vector equation (first mentioned in Chapter 3 on Aircraft Performance Preliminaries):

$$\mathbf{V}_{\oplus p} = \mathbf{V}_{\oplus a} + \mathbf{V}_{ap} \quad (9.81)$$

Here $\mathbf{V}$ stands for velocity, speed together with its direction, $\oplus$ stands for the Earth, p for the airplane, and a for the air mass. The way to remember this rule is to notice the collapse of the "interior" repeated symbol on the right. That is what counts. In English, the rule reads: velocity of the plane with respect to the Earth (ground velocity) equals velocity of the air mass with respect to the Earth (wind velocity) plus velocity of the plane with respect to the air (air speed with heading). Of course Eq. (9.81) works for any three objects. From the physicist's point of view, there is nothing special about the Earth, the air, and airplanes. Aviators know better.

Figure 9.7 shows you what Eq. (9.81)—after abbreviating ground speed $V_{\oplus p}$ to V_g, wind speed $V_{\oplus a}$ to V_w, and air speed V_{ap} to V—looks like as a vector diagram. Gliding into a headwind, the flight path angle one sees from the ground is steeper than the angle one would see if moving with the air mass. Unless there is a downdraft or updraft, descent rate $h^{\cdot}$ is unchanged.

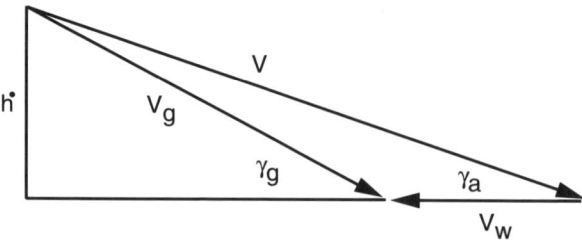

Figure 9.7 Vector diagram, from the Earth point of view, gliding into a headwind.

Glide Performance

There are lots of symbols in Fig. 9.7. Now what are we after? The glide speed V_{bgw} will give us the largest flight path angle (smallest negative angle) with respect to the Earth, $\gamma_{\oplus p}$, henceforth simply γ_g. But remember, because we fly solely by reference to air speeds, that V_{bgw} is an *air speed*. In our graphs and calculations it will be a *true* air speed, but it is easy enough to later convert that to more operational calibrated, or even indicated, air speed.

So we want to maximize γ_g with respect to V. To do that, we are going to need an algebraic relationship between those two variables. One such comes from briefly studying Fig. 9.7:

$$\dot{h} = V \sin \gamma_a = V_g \sin \gamma_g \tag{9.82}$$

We need V_g, which as Fig. 9.7 shows depends on V, V_w, and γ_a. Application of the "law of cosines," the Pythagorean theorem for nonright triangles, gives

$$V_g^2 = V^2 + V_w^2 - 2VV_w \cos \gamma_a \tag{9.83}$$

What saves us, in using Eq. (9.83), is that γ_a is a small angle and therefore its cosine is quite close to unity. Even cosine 8 deg, probably a larger climb or glide angle we will encounter or achieve, only differs from unity by 1%. This small flight path angle approximation again. With the cosine set equal to unity, the RHS of Eq. (9.83) is a perfect square and we have the simple relationship:

$$V_g \doteq V - V_w \tag{9.84}$$

Of course! All we are saying is that, for general aviation aircraft, flight path angles are so shallow that there is no practical difference between the (slant) air speed and its horizontal component.

Figure 9.7 does not support that statement, but that is because we exaggerated the angles so much. General aviation airplanes will not have such steep best glides. Flamed-out turbojets, however, are another matter!

Combining Eqs. (9.84) and (9.82), we have

$$\sin \gamma_g = \frac{\dot{h}}{V_g} \doteq \frac{\dot{h}}{V - V_w} \tag{9.85}$$

This simple but good approximation will be the basis of the following graphic and analytic calculations of V_{bgw}. It is also true, for small angles, that

$$\sin \gamma \doteq \tan \gamma \tag{9.86}$$

Those two trigonometric functions differ by less than 0.001 up until nearly 8 deg.

"Exact" Bootstrap Formula for sin (γ_g)

To investigate how well Eq. (9.85) approximates the "exact" formula, combine Eqs. (9.82) and (9.83) to get the correct expression:

$$\sin \gamma_g(V, V_w) = \frac{h^{\cdot}(V)}{\sqrt{V^2 + V_w^2 - 2V_w\sqrt{V^2 - h^{\cdot 2}(V)}}} \quad (9.87)$$

Equation (9.87) works equally well for climbs or glides. Subscript g stands for ground. Let us consider an example.

Example 9.5 Our job is to support the validity of approximate Eq. (9.85) compared to exact Eq. (9.87) with a sample calculation of each. We consider the unfortunate engine-out airplane of Tables 9.8 and 9.9, the Cessna 172 with flaps 10 deg, weighing 2400 lb, at h_ρ = 3300 ft, wings level. To get off the flat portion of the rate-of-descent curve, put that airplane gliding at 95.2 KCAS = 100 KTAS = 168.8 ft/s true. And have it gliding into a sizeable 25-kn or 42.2 ft/s headwind.

From the gliding version of Eq. (9.75),

$$h^{\cdot} = \frac{P_{xs}}{W} = \frac{-P_r}{W} = \frac{-GV^3 - H/V}{W} \quad (9.88)$$

we soon find that the airplane's rate of "climb" is -19.20 ft/s = -1152.1 ft/min. Evaluation of Eq. (9.85) tells us that flight path angle γ_g is -8.72 deg. On the other hand, the exact form, Eq. (9.87), gives us $\gamma_g = -8.70$ deg. Negligible difference.

The approximation of small flight path angles is again vindicated. Even into a 50-kn headwind, the approximation is probably sufficient. It gives $\gamma_g = -13.15$ deg; the "exact" form for the 50-kn headwind gives $\gamma_g = -12.98$ deg. Nonetheless, exact is in double quotes because the bootstrap approach uses a small angle approximation itself in Eq. (9.88), the bootstrap expression for the gliding rate of climb. Hardly anything is truly exact in aerodynamics or in airplane performance calculations.

Graphical Approach

From performance glide tests, or otherwise, assume you have a graph of (negative) gliding rate of climb $h^{\cdot}$, as a function of air speed V, for a given weight and density altitude (or relative air density): $h^{\cdot}(V; W, \sigma)$. When you are gliding, $P_a = 0$ and so $h^{\cdot} = -P_r/W$ and the peak of $-P_r/W$ is at V_{md}. If you start from the origin and lay off various lines hitting the $h^{\cdot}$ curve, the one of greatest (least negative) slope is the only such line tangent to the $h^{\cdot}$ curve. If you go shallower, you miss the $h^{\cdot}$ curve altogether. See Fig. 9.8 where for a 21.2-kn

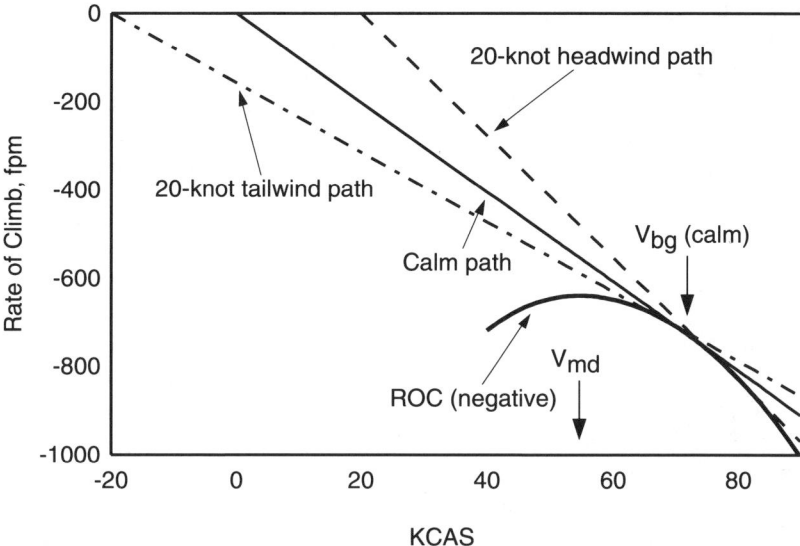

Figure 9.8 Graphical construction for finding speed and angle for best glide, V_{bg} and γ_{bg}, from a rate-of-climb curve.

tailwind, V_{bg} is 69.5 KCAS; for calm wind, $V_{bg} = 72.0$ KCAS; and for a 21.2-kn headwind, $V_{bg} = 75.2$ KCAS. The somewhat oddly sized wind speeds are due to our desire to plot *calibrated* air speeds along the horizontal axis. At 4000-ft density altitude, 20 KCAS = 21.2 KTAS.

Using the approximate "fact" that the slant length from the origin to the curve is the same as the horizontal distance V, Eq. (9.85) tells us that the point of tangency is at the speed that is V_{bg0}, speed for best glide in a calm. Furthermore, starting another tangent line from point $(V_W, 0)$ along the V axis, Eq. (9.85) tells us that *its* point of tangency is at the speed for best glide into a headwind of speed V_w, i.e., at V_{bgw}. If a tailwind is involved instead, one simply moves the starting point of the tangent line, from zero, the other direction.

So that is all there is to the purely graphical approach to finding $V_{bgw}(W, \sigma)$:

- Draw a graph of $h^{\cdot}(V; W, \sigma)$;
- Lay off a tangent from $(V_w, 0)$ to that graph; and
- Go straight up from the point of tangency, to the V axis, and read off $V_{bgw}(W, \sigma)$.

As mentioned, simply treat tailwinds as negative headwinds. If you take starting points for your tangent lines farther to the left, you can see that speed for best glide angle, V_{bgw}, approaches speed for minimum descent rate, V_{md}.

What about the other way, as V_w gets larger? Eventually our approximation—small path angles—breaks down. As V_w approaches aircraft air speed $V = V_a$,

Eq. (9.85) blows up. In practical terms, in a 60-kn wind you have got bigger problems than worrying about how steep your glide path is.

Speed for best glide V_{bg} (refer to Fig. 9.8) is just the unpowered version of V_x, speed for best angle of climb. Indeed, when gliding, V_{bg} is your speed for greatest positive (least negative) flight path angle. In similar fashion, V_{md} is just the unpowered version of V_y. A rate of sink is just a rate of climb that happens to be negative.

Formula Approach

Our guiding analytical light is simply the well-known Eq. (9.88) for excess power and rate of descent. There, G and H are "composite" bootstrap parameters that depend on the four airframe BDP parameters and the two operating variables W and σ. So once we have tied down the BDP, we are in perfect position to use elementary differential calculus to come up with an expression V_{bgw}. Solving that expression will turn out to be something else again.

Computing the derivative of the RHS of Eq. (9.85) using Eq. (9.88) and then setting the resulting expression equal to zero, one gets

$$V_w = \frac{-2GV_{bgw}^5 + 2HV_{bgw}}{-3GV_{bgw}^4 + H} \qquad (9.89)$$

While this general fifth-degree equation can not be solved algebraically in closed form, it can easily be solved numerically. Just plug values for V into the RHS of Eq. (9.89) until you get the desired value of headwind or tailwind. That is how we got the solutions for Fig. 9.8. For a more comprehensive look, at three widely-separated altitudes, see Fig. 9.9, where we used Quattro Pro's SolveFor facility to get the numbers.

While Eq. (9.89) is convenient for seeing what wind speed V_w corresponds to which true air speed of the airplane, there is also something to be learned by rewriting the equation in standard polynomial form and simplifying it a bit by using the bootstrap composite parameters fact that $H/G \equiv U$:

$$V_{bgw}^5 - \frac{3}{2}V_w V_{bgw}^4 - UV_{bgw} + \frac{1}{2}UV_w = 0 \qquad (9.90)$$

There is an even better way, pointed out to the author by its discoverer, Professor Philip Bridges.[5] Working on curve fits to a numerical iteration scheme, he discovered an infinite series solution to Eq. (9.90),

$$\left(\frac{V_{bgw}}{V_{bg0}}\right) = 1 + \left(\frac{V_w}{V_{bg0}}\right) \sum_{n=0}^{\infty} a_n \left(\frac{V_w}{V_{bg0}}\right)^n \qquad (9.91)$$

where the first five values of the constants a_n are: $a_0 = 1/4$, $a_1 = 7/32$, $a_2 = 5/32$, $a_3 = 167/2048$, and $a_4 = 1/64$. That is a sufficient degree of approximation and gives the following result for the optimum glide speed:

$$V_{bgw} \doteq V_{bg0} + \frac{V_w}{4}\left(1 + \frac{7}{8}\left(\frac{V_w}{V_{bg0}}\right) + \frac{5}{8}\left(\frac{V_w}{V_{bg0}}\right)^2 + \frac{167}{512}\left(\frac{V_w}{V_{bg0}}\right)^3 \right.$$
$$\left. + \frac{1}{16}\left(\frac{V_w}{V_{bg0}}\right)^4 + \cdots \right) \quad (9.92)$$

Figure 9.9 shows how V_{bg} varies with direct headwind or tailwind at various altitudes. The spread in V_{bg} values is in a sense an artifact due to winds being ground speeds and our V speeds being expressed as calibrated air speeds. As true air speeds, there would be even less spread between the three V_{bg} graphs.

A commonly cited rule of thumb is to adjust calm-wind values of V_{bg} by adding or subtracting one-half of the speed of the encountered headwind or tailwind. As the curves in Fig. 9.9 show, that rule of thumb greatly overcompensates for light winds. But picking on a sizeable 20-kn headwind (and settling on a moderate altitude of 5000 ft), the graph shows one should add to calm-wind V_{bg} only about 6 kn. In a 20-kn tailwind, it suggests subtracting from calm-wind V_{bg} about 4 kn. Except for quite strong headwinds, the conventional

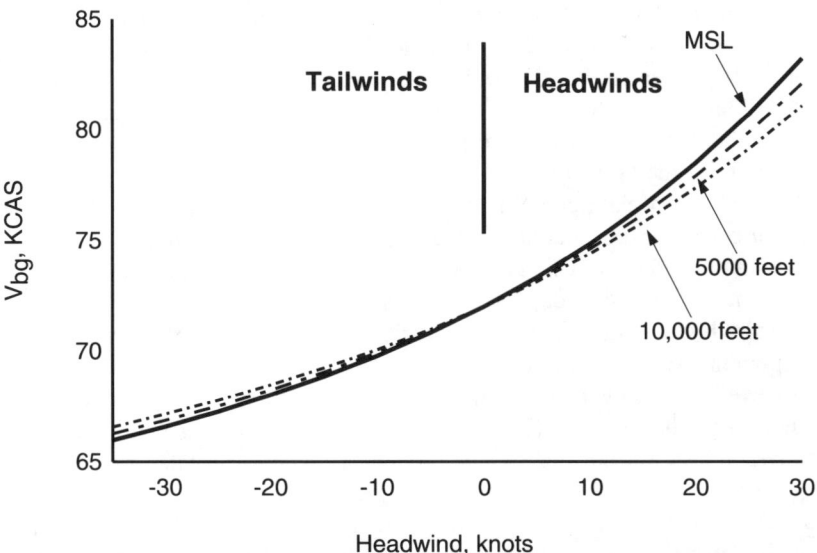

Figure 9.9 Variation of V_{bg} with headwinds or tailwinds at various altitudes.

rule of thumb is about twice what is needed. Adjusting by one-fourth of the encountered headwind or tailwind is a better rule; it is the initial correction factor featured in Bridges's formula, Eq. (9.92).

That revised rule is also supported by calculating the slopes of the graphs at $V_w = 0$. Evaluating the derivative of V with respect to V_w there (by taking the inverse of the derivative of V_w with respect to V) one finds, in the glide case, that the slope there is exactly $\frac{1}{4}$. (In the climb case, the derivative is even smaller and of opposite sign.) But again it is not the small winds one is concerned about.

The corresponding graphs and formulas support a revised rule of thumb for speed for best glide in wind: in a headwind, increase V_{bg}, relative to its calm-wind value, by one-fourth of the speed of the wind. In a tailwind, decrease V_{bg}, relative to its calm-wind value, by one-fourth of the speed of the wind. Only for substantial headwinds (30 kn or more), double the foregoing prescription for V_{bg}.

Weight or Altitude Effects on γ_{bgw} in Direct Headwinds or Tailwinds

In the calm-wind situation, angle for (wings-level) best glide γ_{bg0} does not vary either with airplane gross weight W or relative air density σ. That is a consequence of the well-known fact, seen from Eqs. (9.4) and (9.5), that

$$\tan \gamma = \frac{-D}{L} = \frac{-C_D}{C_L} \qquad (9.93)$$

Because the drag-and-lift coefficients are purely functions of AOA α, the minimum glide angle $-\gamma$ occurs at whatever value of α makes C_D/C_L a minimum. For the Cessna 172 of Table 9.2 (flaps up) we found C_{Lbg} was about 0.786. Referring back to our earlier formulas for the Cessna 172 curves $C_L(\alpha; \delta_f)$, in Table 4.8, best glide implies AOA is $\alpha_{bg} = 7.63$ deg. Because $\gamma_{bg} = -5.38$ deg, this puts the airplane's best glide attitude slightly nose up, 2.25 deg above the horizontal.

So, *in a calm*, weight and altitude have no effect on best glide angle. On the *speed* for best glide, yes, but not on the angle itself. But what about when there *is* a headwind or tailwind? One can see, without calculation, that then weight and altitude *will* affect best glide angle. Here is how.

Consider Fig. 9.8, rate of climb vs air speed, with those few tangent lines (each corresponding to a chosen value of headwind or tailwind) drawn in. Now, in your mind's eye, add many more such tangent lines, those for large tailwinds up through large headwinds. (Even though tangents for large headwinds are "unphysical" in that our approximations break down.) The rate-of-climb graph uniquely determines the set of all such tangent lines.

But that last statement can be turned around! The set of all those tangent lines uniquely determines the rate-of-climb graph. In "mathspeak," the rate-of-climb graph is the unique *envelope* of that set of tangents. This means that a *different* rate-of-climb graph, say one for a different weight or for a different altitude, would have a different set of tangents. But having a different set of tangents means that the set of best glide angles γ_{bgw} will be different. For instance, the tangent line which starts up at $V_w = 20$ kn will be different; its angle with the V axis will be different, hence the corresponding best glide angle γ_{bgw} in a 20-kn headwind (at the same altitude) will be different. So best glide angle, in a nonzero headwind or tailwind, *does* vary with gross weight or with altitude.

But how? To answer that we do need a little calculation, or at least a refined graphical construction. Let us start with additional details on the rate-of-climb graph $h^{\cdot}(V; W, \sigma)$, with parameters W and σ written out explicitly. For instance the graph of Fig. 9.8 is $h^{\cdot}(V; W = 2400, \sigma = 0.8881)$. Going back to our definitions of bootstrap composite parameters G and H, the general formula for poweroff rate-of-climb can be written as

$$h^{\cdot}(V; W, \sigma) = \frac{-1}{W}(GV^3 + H/V) = -\frac{G_0 V^3}{(W/\sigma)} - \frac{(W/\sigma)H_0}{W_0^2 V} \qquad (9.94)$$

What one sees is that poweroff rate-of-climb, for a given airplane in a given flaps/gear configuration (given G_0 and H_0), depends only on air speed and on the *ratio* (W/σ). As do so many important light aircraft performance numbers. Figure 9.10 is an augmentation of Fig. 9.8 with an additional rate-of-climb graph drawn in, that for $h^{\cdot}(V; W = 1600, \sigma = 0.8881)$. Knowing that it is only the ratio (W/σ) that matters, more parsimoniously we could have said that Fig. 9.8 was for $(W/\sigma) = 2400/0.8881 = 2702$ lbf and that the additional graph in Fig. 9.10 was for $(W/\sigma) = 1600/0.8881 = 1802$ lbf. Also keep in mind that, because density altitude h_ρ increases as relative air density σ decreases, lowering (W/σ)—as was done in the second graph of Fig. 9.10—can correspond to lowering, individually, either W or h_ρ.

Figure 9.10 shows how it is that best glide angle in a calm does *not* vary, for a given airplane, with W/σ. The rate-of-climb graph maintains its tangency to the original calm-wind tangent line; speed for best glide changes (getting larger for larger values of W/σ), but angle for best glide stays the same.

Figure 9.10 also gives us "directions" of weight or altitude effects. If W/σ decreases in a headwind, both speed for best glide V_{bg} and angle for best glide γ_{bg} decrease (the glide deteriorating). If W/σ decreases in a tailwind speed for best glide V_{bg} again decreases, but angle for best glide γ_{bg} increases (the glide improving).

Having disposed of effects of direct headwinds or tailwinds, we are ready for the final wrinkle.

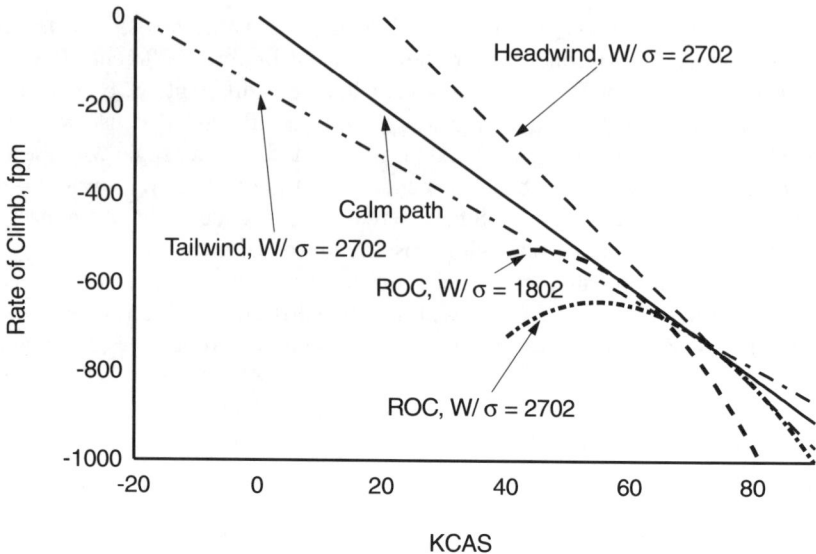

Figure 9.10 Cessna 172 rate-of-climb vs air speed for two values of W/σ.

Downdraft and Updraft Effects on Wings-Level Glides (Small γ, Still Case 8)

When the headwind or tailwind is accompanied by a downdraft or updraft, matters get even more complicated. Let us focus on the downdraft because that is the more problematical case. See the relative motion vector diagram of Fig. 9.11. The downdraft is pushing you into the ground, at so many feet per minute, so you will need to speed up (relative to your direct headwind value of speed for best glide angle, V_{bgh}) to attain the best possible speed, V_{bghd} (speed for best angle of climb with a headwind with a downdraft). Table 9.13 gives the directions the various V speeds move and the angle effects for every possible case. In the case

Table 9.13 Effects of headwinds, tailwinds, updrafts, and downdrafts on best glides

Wind situation	Speed for best glide angle	Best glide angle
Direct headwind	$V_{bgh} > V_{bg}$	$\gamma_{bgh} < \gamma_{bg}$
Direct tailwind	$V_{bgt} < V_{bg}$	$\gamma_{bgt} > \gamma_{bg}$
Headwind with downdraft	$V_{bghd} > V_{bgh}$	$\gamma_{bghd} < \gamma_{bgh}$
Headwind with updraft	$V_{bghu} < V_{bgh}$	$\gamma_{bghu} > \gamma_{bgh}$
Tailwind with downdraft	$V_{bgtd} > V_{bgt}$	$\gamma_{bgtd} < \gamma_{bgt}$
Tailwind with updraft	$V_{bgtu} < V_{bgt}$	$\gamma_{bgtu} > \gamma_{bgt}$

of best glide angles like γ_{bghd}, saying it is greater than γ_{bgh} is meant in the strict algebraic sense, as $-3 > -5$. A greater glide angle is less negative, a shallower angle.

The upshot is that, while it is perfectly possible to get numerical solutions to best glide speeds and angles with headwinds or tailwinds accompanied by downdrafts or updrafts, it is seldom practical to do so. In practice, from the cockpit, speed of an updraft or downdraft can be estimated only very roughly. For safety purposes, more than for performance estimation, Table 9.13 gives the qualitative *directions* in which one should make adjustments if one encounters headwinds or tailwinds accompanied by downdrafts or updrafts. Table 9.13 is strictly for reference and to check answers you get by drawing speed vector diagrams like Fig. 9.11 or analyzing rate-of-climb diagrams such as Fig. 9.10. In the table, updraft and downdrafts are assumed small enough that the no-wind tangent line is not crossed; just as in the V_x climbing case (Chapter 7) that is the dividing line. In addition to this confusion is the possibility of crosswinds, but those are accommodated by crabbing to the appropriate wind correction angle.

Why have we neglected to treat minimum descent rate and speed for it? Because chosen air speed has no *different* effect on descent rate with or without wind. V_{md}, just as does speed for best climb rate V_y, references only the air mass. Granted that a downdraft increases descent rate into the ground, but that downdraft, assumed steady, only adds a constant to the no-wind descent rate. Of course this disclaimer is unresponsive and passive in those cases in which one can *fly out* of the downdraft.

Glide performance is one of few places one can achieve exact (or "exact") solutions to flight mechanics problems. That fact is of interest, but it does not mean that getting those exact solutions is always worth the added effort. The small path angle approximation is quite good enough for most practical purposes, especially for lightly wing-loaded aircraft.

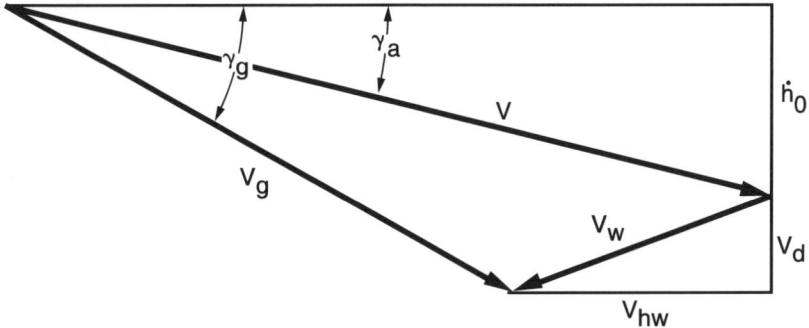

Figure 9.11 Velocity vector diagram, from the Earth point of view, for gliding into a headwind with a downdraft.

Conclusions

From general gliding equations of motion (originally from Newton's laws and the three forces on the airframe: lift, drag, and weight), we obtained exact expressions for flight path angle, rate of descent, turn radius and rate, and gliding V speeds and their associated best glide angle and minimum descent rate for the no-wind case. For comparison, and for shorter formulas, we repeated much of the exact work using the small flight path angle approximation. Then we returned to an exact formulation, for the return-to-airport gliding maneuver, to get formulas for the best turnaround bank angle and for (minimum) altitude lost in that 180-deg turn. Finally, we considered (back in small approximation) wind effects—direct headwinds and tailwinds (including the further compounding of weight and altitude effects) and also headwinds and tailwinds together with updrafts or downdrafts—on over-the-ground best glide air speed and angle.

References

1. Hurt, H. H., *Aerodynamics for Naval Aviators*, U.S. Navy, rev. 1965, p. 94.
2. Langewiesche, W., *Stick and Rudder*, McGraw-Hill, New York, 1944, p. 358.
3. Rogers, D. F., "'Impossible' Turn," *Journal of Aircraft*, Vol. 32, No. 1, 1995, pp. 392–397.
4. Cessna Aircraft Co., *1986 Model 172P Information Manual*, Wichita, KS, 1985.
5. Bridges, P. D., "An Alternative Solution to Optimum Gliding Velocity in a Steady Headwind or Tailwind," *Journal of Aircraft*, Vol. 30, No. 5, 1993, pp. 795–797.

10

Cruise and Partial-Throttle Performance

Introduction

Cruise and partial-throttle operations have traditionally been given short shrift in the aircraft performance literature. Standard textbooks do discuss two important aspects of partial-throttle operation: best range and best endurance. But, almost without exception, they do this for propeller-driven aircraft using an oversimplified theory in which both brake specific fuel consumption rate ($bsfc$) and propeller propulsive efficiency η are taken as constant. These books often include classical Bréguet range and endurance formulas, but derivations of those (e.g., Hale[1] or Perkins and Hage[2]) usually hinge on such assumptions as flight at constant C_L or at constant air speed V; neither are realistic scenarios. Lack of specificity is another problem. Miele,[3] for instance, mentions a general thrust control parameter π for turbojet, turbofan, ramjet, and rocket engines, but none for the propeller airplane. Even if he had, what would be the operational definition of π? (Greek pi stands for power and, though we shall emphasize torque, below we use the capital version of that same letter for our partial-throttle parameter.) Von Mises[4] briefly treats partial-throttle problems using an insupportable approximation in which engine brake moment (torque) is independent of engine speed. If engine speed decreases say because of increased propeller loading under constant throttle—as during transition from level cruise to climb—engine torque *is* closely constant. But not when the revolutions per minute drop is caused by engine throttling.

Practical operational information regarding partial-throttle performance of any given airplane comes from its pilots operating handbook. A cruise performance table typically deals with only one aircraft configuration (clean, flaps, and, if possible, gear up)—a limitation that is understandable—but also with only one gross weight, which is not so easy to accept. For each of several density altitudes (confusingly and unnecessarily split, in the common GAMA-format table, into pressure altitude and outside air temperature), this table presents—for the simpler case of fixed-pitch aircraft—several engine revolutions per minute values with associated true air speeds, per cent ([MSL] mean sea level rated) power, and gallons per hour fuel consumption rate. The range of revolutions per minute and air speeds given is usually shaded towards the higher side. Hardly any POHs give values of V_{be} (speed for best endurance) or V_{br} (speed for best range) even though that knowledge may well be of crucial importance to flight safety. Most POHs have, in addition, range profiles and endurance profiles. Again, these are for one (initial) weight and one configuration. They typically include graphs giving the range (nautical miles) or endurance (hours and minutes) associated with several combinations of cruise air speed V, percent power, and density altitude.

Well, the reader might ask, what more could one want? Fair question.

How about range and endurance figures for other configurations? What if, for instance, a hapless instrument pilot puts his gear down for approach to a socked-in airport, can not retract the gear because of some mechanical or electrical problem, and must divert to another field? Because his 45-min IFR pad was only figured for the clean configuration, gear-down range information now becomes critical. How about speeds for best range and best endurance for a variety of gross weights and at least for several (terrain clearance) altitudes? One can voice similar objections about the brevity of the POH cruise performance table: more weights, and a wider range (at the lower end) of revolutions per minute and air speeds should be represented. And while we are scratching out our wish list, why not include columns for propeller thrust and efficiency? The GAMA-format cruise performance table is better than most of its predecessors, but it is certainly not sufficient. How could it be? When it was designed, there was no adequate partial-throttle performance theory available.

Straying from the straight and level, might it not be useful to include at least some data on the airplane's performance during standard maneuvers such as steady climbing turns or descents? For example:

Question #1. Consider the airplane at 6000 ft, 2400 lbf, flaps up, making a standard rate (3 deg/s) turning 300 ft/min descent at 90 KCAS. What bank angle ϕ and engine speed N does this require?

Question #2. Consider the same airplane and situation (2400 lbf, 6000 ft, flaps up). What revolutions per minute are required for the airplane to descend along a 3-deg glide slope (in calm air) at 90 KCAS?

Once you have understood our extended bootstrap approach theory of partial-throttle operation you will be able to answer *all* of the above query and wish list items and more. That theory will rehabilitate the partial-throttle performance stepchild. And elevate it. If not into Cinderella, at least into respectability. A first fair warning, however: this is a fledgling theory (Lowry[5]), extensively unchecked by experiment, still awaiting the good offices of a competent pilot/investigator. A second fair warning: he or she, and even the reader, may find the theory's logic and algebraic reasoning somewhat tortuous; straightforward examples will help keep you on the path.

We shall begin by quantifying "partial throttle." Next we lay out new concepts and expressions needed for the partial-throttle extension. Then we show how, by collecting a little bit of level cruise flight test data, one can move beyond the propeller polar to get graphs of the airplane's installed propeller power coefficient, thrust coefficient, and propulsive efficiency. Specific endurance and specific range are introduced. A better cruise performance table is constructed, and scaling rules that simplify that work are derived. Two concrete partial-throttle examples, a banked descending turn (question #1 above) and a descent along a glide slope (question #2 above), are laid out in detail. Partial-throttle absolute ceilings are discussed. We will end with a brief discussion of unanswered questions and future opportunities.

The Bootstrap Approach Partial-Throttle Model

Definition of Bootstrap Power Setting Parameter Π

Pulling back to partial throttle, in our formulation, is equivalent to replacing the actual engine by a "virtual" engine of diminished capacity, one with reduced MSL "maximum" torque. Because throttle is tied to torque, partial or reduced throttle means reduced torque. (Another warning, on nomenclature: the words are backwards. "Full throttle" actually means not throttling the engine at all and "reducing throttle" actually means to throttle it more. For historic reasons we stay with the common misconstruction.) From a description of the "derated" engine, our analysis proceeds to use—in a way as yet obscure to the reader—one of several bootstrap formulas to calculate torque needed for a given operational situation, a given air speed associated either with level cruise or with a known rate or angle of climb or descent (with or without turning). Our concrete operational definition of power setting Π is:

$$\Pi \equiv \frac{M}{\Phi(\sigma)M_B} \quad (10.1)$$

M is engine torque at whatever altitude, revolutions per minute, and throttle setting obtains; $\Phi(\sigma)$ is the altitude power dropoff factor; and M_B is base MSL-rated torque at full throttle. The denominator measures maximum possible torque at altitude with full throttle and with fuel–air mixture leaned for maximum power. This definition differs in kind from that of the percent power numbers found in POH cruise tables. In them, for instance, 70% brake horse power means 70% of *MSL-rated full-throttle power at maximum rated revolutions per minute*. In our definition, $\Pi = 1$ always corresponds to *full throttle under ambient conditions*.

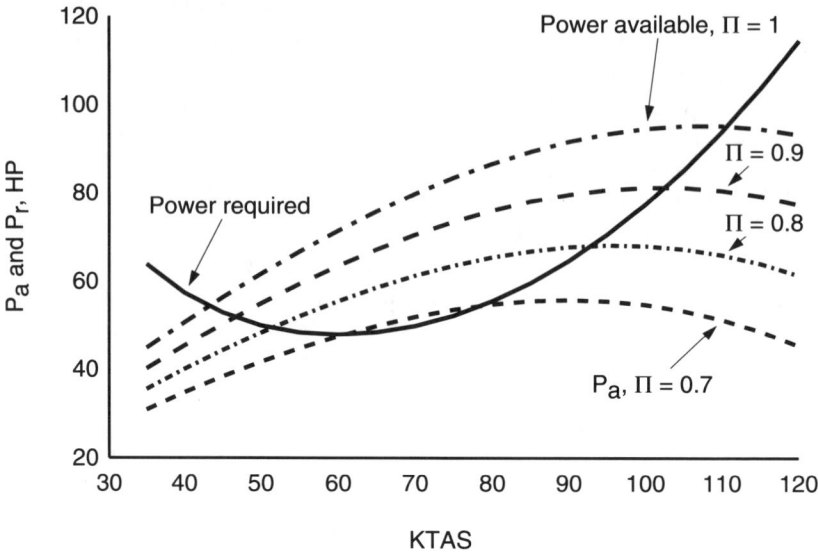

Figure 10.1 Throttling reduces engine torque and lowers the P_a (power-available) curve.

New Wine from (Mostly) Old Bootstrap Bottles

Reviewing basic bootstrap formulas, one sees that engine torque M appears only in composite parameter E. Given the bootstrap data plate (BDP) for the airplane and altitude, power setting $\Pi \leftrightarrow M \leftrightarrow E$; if you have any one you have any other. (Symbol $\leftrightarrow$ means "corresponds to.") We can motivate deceptively simple Eq. (10.1) by plotting power-available/power-required curves for several values of partial throttle parameter Π (see Fig. 10.1), based on the BDP of Table 10.1 and on case 2 of the sample composite bootstrap parameters in Table 10.2. One sees in Fig. 10.1 that 6000 ft is near absolute ceiling for $\Pi = 0.7$.

Table 10.1 Sample bootstrap parameters for Cessna 172

BDP item	Value	Units	Aircraft
Wing reference area, S	174	ft²	Airframe
Wing aspect ratio, A	7.378		Airframe
Rated MSL torque, $M_B(\Pi = 1)$	311.2	ft-lbf	Engine
Altitude dropoff parameter, C	0.12		Engine
Propeller diameter, d	6.25	ft	Propeller
Parasite drag coefficient, C_{D0}	0.037		Airframe
Airplane efficiency factor, e	0.72		Airframe
Propeller polar slope, m	1.70		Propeller
Propeller polar intercept, b	-0.0564		Propeller

Cruise and Partial-Throttle Performance

Table 10.2 Sample composite bootstrap parameters, two full-throttle situations

Variable or composite	Case 1	Case 2
h_p	MSL	6000 ft
W	2400 lbf $= W_B$	2400 lbf $= W_B$
σ	1	0.8359
$\Phi(\sigma)$	1	0.8135
E	$531.9 = E_B$	432.7
F	$-0.0052368 = F_B$	-0.0043772
G	$0.0076516 = G_B$	0.0063956
H	$1,668,535 = H_B$	1,996,192
K	$-0.0128884 = K_B$	-0.0107729
Q	$-41,270.6 = Q_B$	$-40,165.4$
R	$-129,460,301 = R_B$	$-185,297,929$
U	$218,064,595 = U_B$	$312,118,214$

A somewhat more detailed review of bootstrap formulas shows that a little additional knowledge about the airplane and its motion—weight W, air speed V, and either rate of climb $\dot{h}$ or flight path angle γ (including the possibility of those last two variables being zero, level cruise) implies knowledge of torque M and therefore again knowledge of E and of Π. That comes from the basic bootstrap relations surrounding maximum (or, it turns out, minimum) level flight speed, rate of climb, and flight path angle. Let us here repeat those three formulas as Eqs. (10.2), (10.4), and (10.5):

$$V_{M/m} = \sqrt{\frac{-E \mp \sqrt{E^2 + 4KH}}{2K}} = \sqrt{-\frac{Q}{2} \pm \sqrt{\frac{Q^2}{4} + R}} \qquad (10.2)$$

In the partial throttle case, $V_{M/m}$ becomes level cruise speed V_{cr}. Which speed, the high or the low one? It turns out that it does not matter. Either choice for solution for E (or Q) leads to this relation for torque M:

$$M(\Pi, \sigma) = \Phi(\sigma) M_B(\Pi, \sigma = 1) = \sigma K_B \left(\frac{R - V_{cr}^4}{V_{cr}^2} \right) \times \frac{d}{2\pi m} \qquad (10.3)$$

Example 10.1 If one decided, under these case 2 circumstances, to reduce throttle and cruise (still at 6000 ft) at 93 KTAS, Eq. (10.3) would tell us that torque was reduced to $M = 202.80$ ft-lbf. Dividing M by the maximum possible full-throttle torque at that altitude, $\Phi(\sigma) M_B(\Pi = 1, \sigma = 1) = 0.8135 \times 311.2 = 253.16$ ft-lbf, gives $\Pi = 0.80$. In corroboration, the right-hand intersection of the power-available curve for $\Pi = 80\%$ with the power-required curve does occur at $V = 93$ KTAS.

The rate of climb formula is

$$ROC(V) = h'(V) = \frac{P_{xs}(V)}{W} = \frac{EV + KV^3 - H/V}{W} \tag{10.4}$$

and the flight path angle connection is

$$\gamma(V) = \sin^{-1}\frac{T_{xs}(V)}{W} = \sin^{-1}\left[\frac{E + KV^2 - H/V^2}{W}\right] \tag{10.5}$$

Solving both those last two relations for torque gives

$$M(\Pi, \sigma) = \Phi(\sigma)M_B(\Pi, \sigma = 1)$$

$$= \begin{cases} \left[\dfrac{Wh'}{V} + \sigma K_B\left(\dfrac{R - V^4}{V^2}\right)\right]\dfrac{d}{2\pi m} \\ \left[W\sin\gamma + \sigma K_B\left(\dfrac{R - V^4}{V^2}\right)\right]\dfrac{d}{2\pi m} \end{cases} \tag{10.6}$$

What is missing in all of this is information about the engine circular speed n. How does one come by that? In quite a roundabout way. First off, the definition of the power coefficient

$$C_P \equiv \frac{P}{\rho n^3 d^5} \tag{10.7}$$

of propeller advance ratio

$$J \equiv \frac{V}{nd} \tag{10.8}$$

and the relation between power P and torque M

$$P = 2\pi n M \tag{10.9}$$

give us a relation

$$\frac{C_P}{J^2} = \frac{2\pi M}{\rho d^3 V^2} \tag{10.10}$$

for the independent variable in the linearized propeller polar. But because C_P depends only on advance ratio J, so does C_P/J^2. And, assuming invertibility, J depends only on C_P/J^2. So *if* we had a graph or chart giving us knowledge of J as a function of C_P/J^2, we would be able to find J and then, quite easily,

$$n = \frac{V}{dJ} \tag{10.11}$$

That *if* will be the subject of the next subsection. First we discuss a couple of objections, or at least doubts, the reader might have.

Objection #1. Composite bootstrap parameter E and engine torque M have been associated so far only with full-throttle operations. Values of M smaller than base-rated torque M_B have certainly been found at full throttle, but those have been smaller only because of altitude dropoff factor $\Phi(\sigma)$. How can we understand, intuitively, this changed convention?

Answer #1. The basic idea behind this bootstrap extension is that an engine at partial throttle may be thought of as a smaller engine at full throttle. If someone took a hacksaw and cut an inch off your throttle control plunger and then screwed the knob back on over the cut, you would suddenly have an airplane of diminished performance. But a neophyte pilot would have no easy way of distinguishing between this sabotaged partial-throttle situation and a "virtual" engine characterized by reduced base MSL full-throttle torque $M_B(\Pi < 1)$. The reader should keep in mind that we are implying the possibility of *many different* virtual engines, each with its own derated MSL torque value $M_B(\Pi < 1)$; which particular virtual engine we use depends on assumed air speed, altitude, and flight path. We also assume any such virtual engine has the same power dropoff parameter C as does the real one. That assumption is supported by the fact that C is quite close to 0.12 for a wide variety of aviation internal combustion gasoline engines of various displacements and various numbers of cylinders.

Objection #2. Another possible objection to our virtual derated engine procedure might be linearized propeller polar parameters m and b were calculated using data from full-throttle flight tests. How can one be sure the same values of those parameters will be obtained in the partial-throttle case?

Answer #2. One answer to this objection comes from reviewing formulas from which m and b were originally calculated [Lowry,[6] Eq. (18) there for b and either Eqs. (19) or (20) for m]. There one sees that the relation for b does not involve engine torque or any variable that depends on engine torque. And while either relation for m does indeed involve base engine torque M_B, either could just as well be looked upon as a relation for the product mM_B. If one had originally inadvertently (say) overestimated M_B because one's old tired engine needs an overhaul, one would have calculated as a consequence a larger and precisely compensating (though not correct) value for m. Because that same overestimate of M_B continues to plague the partial-throttle development, the inadvertent compensation in m continues to do its job. A shorter answer to this objection comes from remembering that the correct propeller polar is a relation only having to do with the propeller. One of the most useful features of the bootstrap approach is its separation of the airplane into three essentially independent parts: airframe, engine, and propeller. Utility of dimensionless coefficients and other quantities in the propeller polar hinges on the fact that their correct values do not depend on the particular engine used to determine them.

Beyond the Propeller Polar

At this point, both for definiteness and for use in sample calculations, we describe one prescription from which one may determine $J(C_P/J^2)$. There are others. And, in fact, we recommend that if one is implementing the bootstrap approach on an airplane, he or she keeps a running record of cockpit readings taken during a wide variety of steady climbs, descents, and level cruises. Each of those can be arranged to give an additional $J(C_P/J^2)$ datapoint.

Our procedure is as follows. Take the airplane to a given density altitude at a known gross weight (these may be calculated after the flight) and a given flaps/gear configuration. Fly level and stabilized at various (indicated) air speeds and record the associated engine speeds. The AGARD manual suggests taking 3 to 10 min to let the aircraft stabilize at each level air speed and suggests moving down from the highest speed instead of up from the lowest. Using the airplane's air-speed indicator calibration curve, correct indicated air speeds to calibrated. As an example, consider our sample Cessna 172 with flaps up at a density altitude of 3000 ft, gross weight 2300 lbf (Table 10.3, case 3). Data as in Table 10.4 (fabricated from propeller charts for the McCauley 7557 propeller on this airplane and an assumed installed "slowdown factor" of 8.5% as suggested by Norris and Bauer[7]) are collected (first and third columns) and lightly massaged (second and fourth columns) to get advance ratio J (fifth column).

To conveniently use these data to construct $J(C_P/J^2)$, use Eqs. (10.3) and (10.10) and the relation between composite parameter R and speed for best angle of climb in calm air

$$V_x = (-R)^{1/4} \tag{10.12}$$

eliminating torque M to piece together

$$\frac{C_P}{J^2} = \left[1 + \left(\frac{V_x}{V_{cr}}\right)^4\right] \times \frac{(-K_B)}{\rho_0 d^2 m} \tag{10.13}$$

In this way, we can readily find C_P/J^2 for given level flight speed V_{cr}.

Table 10.3 Sample composite bootstrap parameters, two full-throttle situations

Variable or composite	Case 1	Case 3
h_ρ	MSL	3000 ft
W	2400 lbf $= W_B$	2300 lbf
σ	1	0.9151
$\Phi(\sigma)$	1	0.9035
E	531.9 $= E_B$	480.6
F	$-0.0052368 = F_B$	-0.0047923
G	$0.0076516 = G_B$	0.0070021
H	$1{,}668{,}535 = H_B$	$1{,}674{,}526$
K	$-0.0128884 = K_B$	-0.0117944
Q	$-41{,}270.6 = Q_B$	$-40{,}748.5$
R	$-129{,}460{,}301 = R_B$	$-141{,}976{,}614$
U	$218{,}064{,}595 = U_B$	$239{,}146{,}256$

Table 10.4 Data from simulated level cruise flight test

KCAS	KTAS	N	n	J
50	52.27	2021	33.68	0.42
55	57.49	1959	32.65	0.48
60	62.72	1946	32.43	0.52
65	67.95	1967	32.78	0.56
70	73.17	2013	33.54	0.59
75	78.40	2076	34.60	0.61
80	83.63	2153	35.88	0.63
85	88.85	2239	37.32	0.64
90	94.08	2332	38.87	0.65
95	99.31	2431	40.52	0.66
100	104.54	2534	42.24	0.67
105	109.76	2640	44.01	0.67

"Just a minute!" we hear the reader protesting as he or she reviews Eq. (10.13). "Do you mean to tell me that V_x is independent of throttle position?" Precisely. The best climb angle itself, of course, very *much* depends on engine torque, but with identical gross weight, altitude, and bank angle, any throttle setting gives identical values for V_x. Composite parameter E, the only first-line composite parameter involving torque, plays no role in the definition of V_x.

Now, another point about this pivotal performance speed. Recall that we mentioned, on the question of whether to use the form of Eq. (10.2) for maximum level flight speed or for minimum, that it did not matter. Solution Q, and then M is the same for either. Being able to ignore that distinction implies a relationship between V_M, V_m, and R and it is satisfyingly simple:

$$V_x = \sqrt{V_M V_m} \qquad (10.14)$$

V_x is the geometric mean between bootstrap minimum and maximum speeds for level flight.

With these raw and immediately derived performance flight test data in hand, Eq. (10.13) with standard bootstrap information on the pertinent value of V_x (bootstrap composite parameter $R = -141{,}976{,}614$ corresponds to $V_x = 109.2$ ft/s $= 64.7$ KTAS) then provides corresponding values of C_P/J^2. One gets a graph such as Fig. 10.2; that is the major result obtained from the supplementary level cruise flight test. Additional valuable information, propeller power and thrust coefficient graphs, and a propulsive efficiency graph, can be gained from it. To be able to calculate with the graph, it is convenient (though

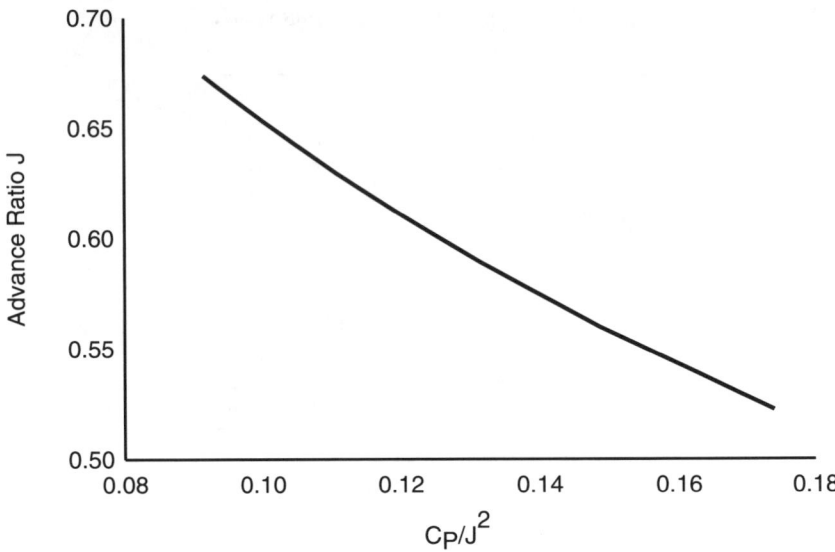

Figure 10.2 Propeller advance ratio J as a function of C_P/J^2.

devoid of any deeper meaning) to fit a curve to it. In our sample case, one such fit is given by

$$J = [6.9145 - 5.9501 e^{-C_P/J^2}]^{-1} \qquad (10.15)$$

In fact, the range of J values given by the data in Table 10.4 is somewhat too small to cover some sample calculations below. That is where, in practice, reconsideration of early full-throttle climbing data or even an additional flight test employing partial-throttle descents timed over some vertical interval would come in handy. Instead of belaboring experimental exigencies, we will simply use the curve fit of Eq. (10.15) as though it subsumed all of our requirements. In several later sections, we shall demonstrate the use of this somewhat unusual propeller chart [Fig. 10.2 or Eq. (10.15)] and our theory with sample calculations. First, however, one further data input item must be tied down.

Brake Specific Fuel Consumption Rate

Unlike extension of the full-throttle, wings-level theory to include maneuvering (which required only a slight change in the induced drag term), extension to partial-throttle operations requires additional data on both engine and propeller. To calculate fuel usage, we must know bsfc rate. For our sample Cessna 172 with its Lycoming O-320-D2J engine, as for almost all general aviation aircraft, the appropriate engine manual is readily available and includes the necessary graph, from which we obtain

$$c = \begin{cases} 0.45, & \text{hp} \leq 122 \\ 0.51, & \text{hp} > 122 \end{cases} \quad (10.16)$$

The units of c are lbm/h/hp. Aviation gasoline density is approximately 6 lbm/gal (U.S.). Rather than a piecewise constant *bsfc*, one may find that the engine is better modeled by a continuous graph, probably with a minimum near 70% rated full-throttle power. Keep in mind that Eq. (10.16) is only an example.

Calculation of GAMA-Format Cruise Performance Table

Such a table is for a given type of airplane in given flaps/gear configuration, wings level, at given gross weight. The table is entered first with pressure altitude and outside air temperature, then with any of several values of engine speed N in revolutions per minute. Dependent data consist of percent rated MSL brake power, true air speed V in KTAS, and fuel flow rate in gallons per hour. Table 10.5 shows POH cruise performance entries for one density altitude for the Cessna 172 airplane without speed fairings. Because fuel flow comes directly from brake power, it is convenient to cut entries for each density altitude back to triples (N, V, P). Though any one of those three might be considered a second independent variable, that role usually falls to N. It will turn out, in our alternative bootstrap cruise table later on, that N is *not* a good choice for independent variable.

In theoretical developments we use torque M, rather than power P, because torque is most closely tied—through manifold absolute pressure, and brake mean effective pressure—to throttle position. And partial-throttle operation is our subject. So we will consider the slightly transformed cruise performance table, which has, for each of several density altitudes, entries (V, M, N). To calculate two of those from the third, two functions of single variables are required. The

Table 10.5 Cessna 172 POH cruise performance at 6000 ft

RPM	%bhp	KTAS	gph
2600	77	117	8.6
2500	69	111	7.8
2400	63	105	7.0
2300	57	99	6.4
2200	52	93	5.9
2100	47	86	5.5

first, of form $M(V)$, is Eq. (10.3). The second, of form $N(V)$ [only for our particular curve fit, Eq. (10.15)] is

$$N = \frac{60V}{d}[6.9145 - 5.9501 \, e^{-C_P/J^2}] \qquad (10.17)$$

where C_P/J^2 comes from Eq. (10.13). Let us consider an example.

Example 10.2 Let us calculate cruise performance details for our sample Cessna 172, flaps up, 2400 lbf, at 6000 ft, for cruise air speed $V = 95$ KTAS $= 160.3$ ft/s. From Table 10.2 (case 2), the altitude factors are $\sigma = 0.8359$ and $\Phi = 0.8135$. $V_x = (-R)^{1/4} = 185{,}297{,}929^{1/4} = 116.7$ ft/s. From Eq. (10.3), torque $M = 207.5$ ft-lbf. From Eq. (10.13), $C_P/J^2 = 0.1046$. From Eq. (10.17), $N = 2393$ rpm. With circular speed, we can then calculate engine power $P = 2\pi n M = 52{,}007$ ft-lbf/s $= 94.56$ hp. Because the rated MSL power is 160 hp, the engine is developing 59.1% of full power.

At this power level, from Eq. (10.16), bsfc $= 0.45$ lbm/h/hp; fuel consumption rate is therefore 7.09 gph:

$$\text{gph} = \frac{c \times \text{hp}}{6} \qquad (10.18)$$

To compare our sample Cessna 172 (BDP as in Table 10.1) with the archetype from the Cessna POH (Table 10.5), in Table 10.6 we use two lines interpolated from the POH table flanking the results of our calculation. The bootstrap airplane is neither as clean nor as efficient.

Forcing POH and bootstrap revolutions per minute to match (rows 2 and 3 in Table 10.6), the POH airplane is almost 10 KTAS faster. Forcing air speeds to match (rows 3 and 4 in Table 10.6), the bootstrap airplane requires faster engine speed and has considerably higher fuel consumption.

Propeller Charts $C_P(J)$, $C_T(J)$, $\eta(J)$

Propeller blade elements, under most steady conditions, move in two mutually perpendicular directions, tangentially and longitudinally. We are here ignoring the possibility of differentially loaded propeller blades, as in a steep climb or descent, leading to P-factor; that can be adjusted for in retrospect. The usual two

Table 10.6 Comparison of bootstrap and POH entries

Source	rpm	%bhp	TAS	gph
Interpolated from POH	2393	62.6	104.6	6.96
Bootstrap calculation	2393	59.1	95.0	7.09
Interpolated from POH	2233	53.6	95.0	6.06

directions of motion require two independent scalar propeller relations. The linearized propeller polar

$$\frac{C_T}{J^2} \doteq m\frac{C_P}{J^2} + b \tag{10.19}$$

is one of these but, because only one, can never do the full job. That is why the supplementary level cruise flight test (or some other test recording engine speed) is necessary. Considering what we gain from it, performing that brief test is a very small price to pay. With that additional information, we can go far beyond the limitations of the linearized propeller polar and construct a full set of (installed!) propeller charts for the airplane. This development is of extreme importance and utility. And quite easy to do.

There are many types of propeller charts. From any independent pair of them, one can construct all of the others. We will construct the usual graphs of coefficients $C_P(J)$ and $C_T(J)$ and supplement that pair with a graph of propulsive efficiency $\eta(J)$. These graphs will be for those values of advance ratio J encountered in flight tests.

The power coefficient, because we have pairs $(C_P/J^2, J)$, is immediate by multiplying abscissas by J^2. Next we can use our linearized propeller polar, Eq. (10.19), to calculate matching values of C_T/J^2. From those one finds $C_T(J)$. Then we get matching values of propulsive efficiency η from

$$\eta(J) = J \times \frac{C_T/J^2}{C_P/J^2} \tag{10.20}$$

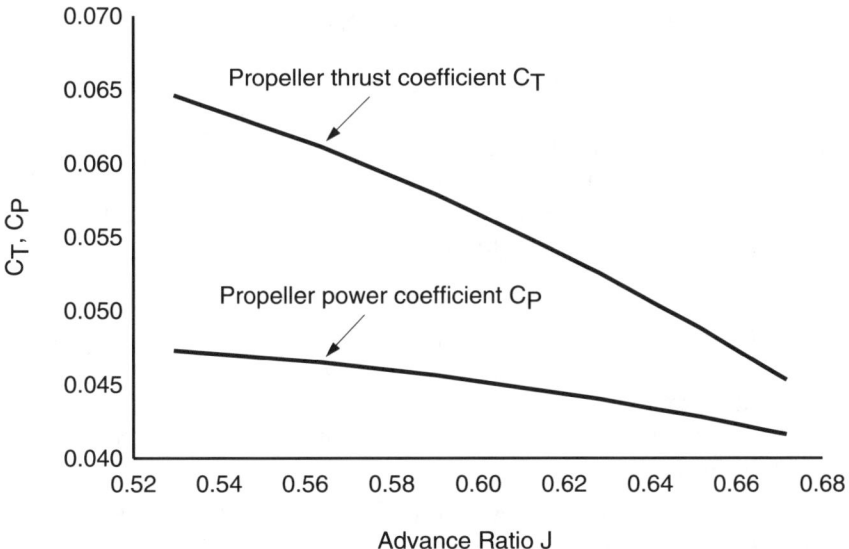

Figure 10.3 Propeller power and thrust coefficients as functions of advance ratio J.

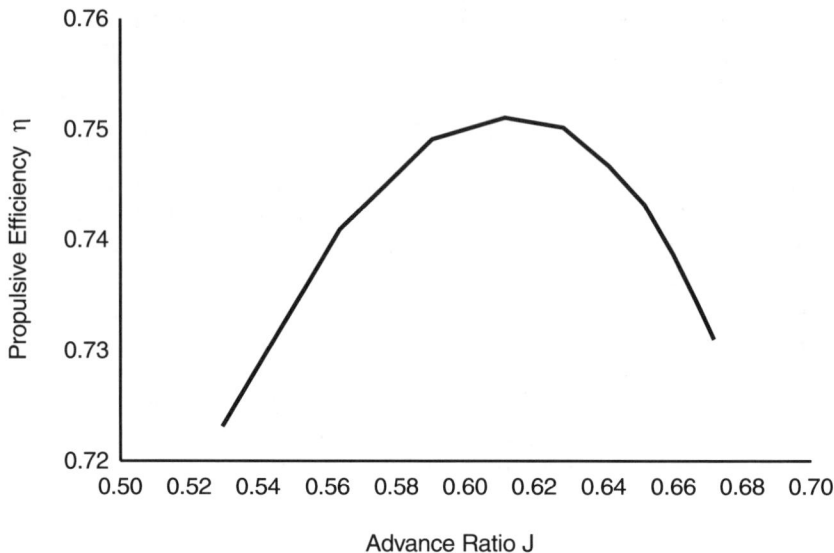

Figure 10.4 Bootstrap partial-throttle procedures give realistic values for installed propeller propulsive efficiency.

That is all there is to it. Figures 10.3 and 10.4 show the relevant graphs for the experimentally accessible range of J values.

Specific Endurance, Range, and V_{be}, V_{br}

How long an airplane can stay aloft (endurance) and how far it can go without landing to refuel (range) depend on how much fuel it carries and the rate at which that fuel is consumed. Range and endurance values depend on fuel load, gross weight of the airplane, altitude at which it flies (assumed in level flight), and how fast it travels. In the range case, headwinds or tailwinds make a difference. As the airplane travels, its weight decreases according to

$$\dot{W} = -cP \qquad (10.21)$$

As gross weight W decreases, less lift is needed to keep the airplane up. Slower speed, or a smaller angle of attack, will suffice. For the same speed, induced drag goes down. Because less thrust is required, engine power can be reduced. It is a slowly varying yet dynamic situation. The practical operational questions pilots have in this regard are

1) For any given circumstance, what are the speeds for best (longest) endurance, V_{be}, and best (longest) range, V_{br}?

2) If we fly continually at (ever-changing) V_{be}, what will maximum endurance E (not the similarly named bootstrap composite parameter) be? If we fly, instead, at ever-changing V_{br}, what maximum range R (again not the bootstrap parameter) can we expect?

Considering the situation at a given point in time, V_{be} is the speed that maximizes *specific endurance*, given by

$$\text{specific endurance} = \frac{1}{cP} \qquad (10.22)$$

with units h/lbm of fuel. To actually find endurance E one must integrate Eq. (10.22) from the initial to the final aircraft weight (von Mises[8]). Using an unrealistic assumption of constant propeller efficiency and constant bsfc, that integration can be performed analytically. In our realistic nonsimplified case, the integration must be done numerically.

Similarly, V_{br} is the speed that maximizes *specific range*, given by

$$\text{specific range (no headwind)} = \frac{V}{cP} \qquad (10.23)$$

with units n mile/lbm of fuel. To find actual range R, in the realistic case, one would numerically integrate Eq. (10.23) from the initial to the final (empty of fuel) weight (von Mises[8]). Flying into a headwind of speed V_{hw}, the item V_{brw} (the extra subscript w means a headwind or tailwind is involved) maximizes as

$$\text{specific range (headwind)} = \frac{V_g}{cP} = \frac{V - V_{hw}}{cP} \qquad (10.24)$$

In Eq. (10.24), as is usual, tailwinds are accommodated by treating them as negative headwinds. V_g is speed made good over the ground.

At given W and σ (and in a given headwind or tailwind, in the range case), one finds these new V speeds essentially by trial and error. Referring to the cruise performance table, which we shall soon augment, one needs to take a trial air speed, find the corresponding torque, then C_P/J^2, and then engine speed N. Then power P, bsfc c, and finally the specific endurance or range figure as given in one of the last three equations. When this has been done for a range of air speeds, V_{be} is the speed that maximizes specific endurance and V_{br} or V_{brw} the one maximizing specific range.

Because $J(C_P/J^2)$ is given as numerical data (though we have fit a curve to our sample data), this is essentially a numerical process. We shall go through an example after we construct an augmented cruise performance table.

Bootstrap Cruise Performance Table

The combination of Tables 10.7 and 10.8 is what we have in mind for an adequate and serviceable cruise performance table. Brief bootstrap calculations and the use of density altitudes allow construction of a cruise performance table with considerably more information, though with fewer columns, than the typical GAMA-format cruise table. This newer specimen illustrates several points.

Table 10.7 Bootstrap expanded cruise performance table for Cessna 172, 2400 lbf, 6000 ft, flaps up, part 1 of 2

KCAS	KTAS	N	$\%P$	gph
100	109.4	2662	78.9	10.7
95	103.9	2556	70.5	8.5
90	98.4	2454	63.1	7.6
85	93.0	2359	56.9	6.8
80	87.5	2272	51.7	6.2
75	82.0	2196	47.5	5.7
73 V_{br}	79.8	2169	46.2	5.5
70	76.6	2134	44.5	5.3
65	71.1	2093	42.8	5.1
62 V_{be}	67.8	2081	42.5	5.1
60	65.6	2079	42.7	5.1
55	60.2	2104	44.6	5.4

Table 10.8 Bootstrap expanded cruise performance table for Cessna 172, 2400 lbf, 6000 ft, flaps up, part 2 of 2

KCAS	KTAS	$\%\Pi$	T or D	η
100	109.4	98.4	283.5	0.735
95	103.9	91.6	298.1	0.739
90	98.4	85.4	311.9	0.744
85	93.0	80.0	324.9	0.748
80	87.5	75.5	337.2	0.751
75	82.0	71.8	348.8	0.751
73 V_{br}	79.8	70.6	353.2	0.750
70	76.6	69.2	359.6	0.747
65	71.1	67.9	369.7	0.735
62 V_{be}	67.8	67.8	375.3	0.722
60	65.6	68.1	379.0	0.711
55	60.2	70.4	387.6	0.671

Cruise and Partial-Throttle Performance 333

1) Near the bottom of the table, $V(N)$ is double valued; $N = 2100$, for instance, occurs for a speed near 55 KCAS and also for a speed near 66 KCAS. *Therefore, N is not an appropriate independent variable for a wide-span cruise performance table.* This defect does not show in current POH cruise tables because those are prematurely cut off, on the lower end, at relatively high air speeds. (Considering that the POH Cessna is about 10 KTAS faster than our sample airplane, the POH cruise table only goes down to settings for approximately our 78 KTAS.) Air speed is the appropriate independent variable because that choice makes the other performance items single valued. We chose calibrated (equivalent) air speed partly because it is closest to the operational indicated air speed and partly for a reason having to do with cruise table scaling to be discussed below.

2) KTAS is included for two reasons. As in the POH cruise tables, it is needed for navigation. But also for trial-and-error calculation of speed for best range (with or without a headwind or tailwind).

3) Power setting Π is of course a new feature. It is calculated either as $M_B(\Pi < 1)/M_B(\Pi = 1)$, which would be $(207.5/0.8135)/311.2 = 82.0\%$ for the recent 95 KTAS example, or as $M(\Pi, \sigma)/\Phi(\sigma)M_B(\Pi = 1) = 207.5/(0.8135 \times 311.2) = 82.0\%$.

4) Thrust is calculated from the usual bootstrap relation:

$$T(V) = EV + FV^2 \qquad (10.25)$$

Because the craft is at partial-throttle, $E_B(\Pi < 1) = \Pi E_B(\Pi = 1)$ is the correct base composite factor to use in calculating E. Because this is level flight, using Eq. (10.26) for drag avoids having to even consider throttle setting.

$$D(V) = D_P(V) + D_i(V) = GV^2 + \frac{H}{V^2} \qquad (10.26)$$

5) Propeller efficiency η is another new feature. Because our extended flight test data were fabricated, efficiency calculated from thrust, air speed, and power gives results a couple of percents higher than does Table 10.7. Accurate experimental data would give more consistent results.

6) It seems to us imperative that a good cruise performance table include V_{br} and V_{be}. Let us first deal with the endurance side. In Table 10.7, V_{be} was found as the speed for which consumption rate gallons per hour is a minimum (alternatively, where $1/cP$ is maximum) (see Fig. 10.5). Specific endurance peaks for this airplane under these circumstances at about 62 KCAS. The graph dislocation on the lower right is from bsfc c switching from its low to its high value. The same is true of the two graphs following.

In the usual simplified theory, in which both propeller efficiency η and specific fuel consumption c are taken to be constant, V_{be} is speed for

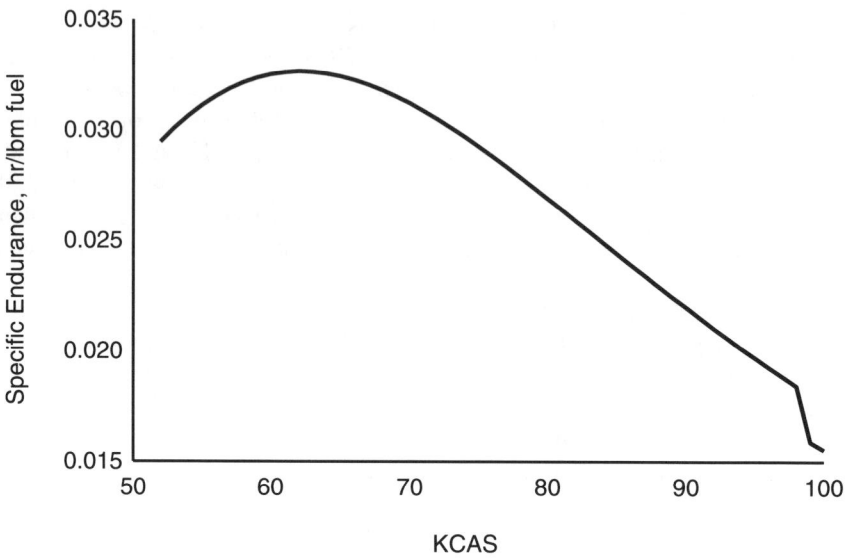

Figure 10.5 Specific endurance as a function of air speed.

maximum $C_L^{3/2}/C_D$ and speed for minimum descent rate

$$V_{md} = \left(\frac{H}{3G}\right)^{1/4} = \left(\frac{U}{3}\right)^{1/4} \doteq 0.7598 V_{bg} \qquad (10.27)$$

For our sample Cessna at 6000 ft, 2400 lbf, $V_{md} = 54.7$ KCAS. But that is more than 7 KCAS below our (correct!) best endurance speed $V_{be} = 62$ KCAS. The discrepancy is due to the moderately sharp decrease in propeller efficiency at lower values of J.

7) Now for V_{br}, speed for best range. First, in calm air. This is the speed for V/cP, or $V/$gph, is maximum (see Fig. 10.6). Speed for best range in calm air, for this airplane under these circumstances, is at 73 KCAS. A temporary column off to the side gives the nod, for this circumstance, to $V_{br} = 73$ KCAS $= 79.8$ KTAS. In the simplified theory in which efficiency η and specific fuel consumption c are constant, V_{br} is speed for best glide

$$V_{bg} = \left(\frac{H}{G}\right)^{1/4} = U^{1/4} \qquad (10.28)$$

For our sample Cessna at 6000 ft, 2400 lbf, $V_{bg} = 72.0$ KCAS, only 1-kn below our $V_{br} = 73$ KCAS. Our calculations improve realism in that η varies with air speed and, closely following the engine manual for the Cessna 172's Lycoming O-320-D2J engine, c is taken [Eq. (10.16)] to be only piecewise constant. Pilots should be able to base crucial range or endurance decisions on fuller data, as given by Table 10.7.

Cruise and Partial-Throttle Performance   335

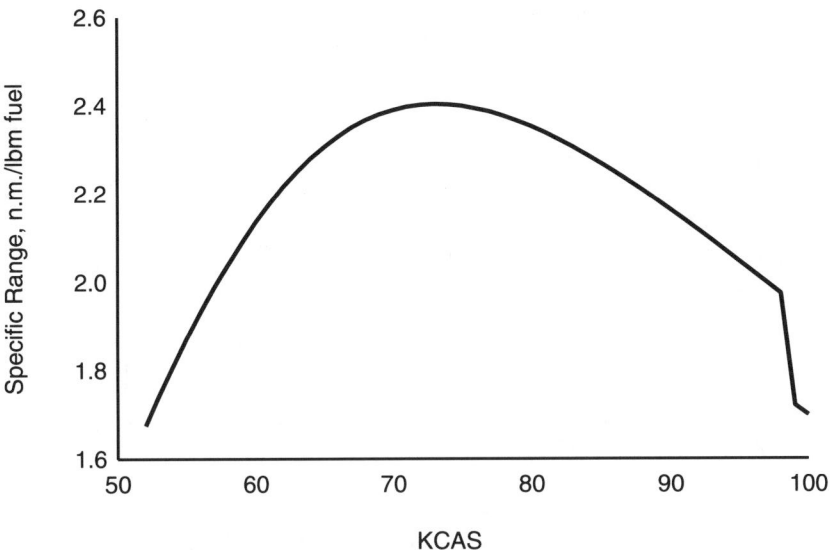

Figure 10.6   Specific range in calm air as a function of air speed.

8) With a headwind or tailwind, finding V_{brw} is somewhat cumbersome. But let us do a sample calculation for say a 20-kn headwind.

Example 10.3   See Fig. 10.7. With headwind $V_{hw} = 20$ kn, best range is achieved when $(KTAS - 20)/cP$ is maximized. The upper curve is the calm-wind specific range, as in Fig. 10.6. The lower curve is V_{hw}/cP, 20 times the graph in Fig. 10.6. The middle curve, the difference between the other two, is the operative one here. Notice, just as in the Pa/Pr curves, that maximum distance between the first two curves mentioned, where their difference peaks (for the Pa/Pr curves, at V_y), is where the two curves have the same slope. The same relation holds here: V_{brw} is about 78 KCAS. Notice that with a headwind the maximum value of specific range is lower than for a calm. That only makes sense, as does the fact that maximum range itself, the integral over weights, will be lower.

Going off to the side of Table 10.7, calculate a new column $(KTAS - 20)/gph$. For the selection of air speeds featured, maximum of that new column is near 80 KCAS. If one uses a finer grid, the maximum is closer to $V_{brw} = 78$ KCAS. So for best range, one needs to fly somewhat faster when bucking a 20-kn headwind. The procedure is straightforward enough, but finding V_{brw} for a variety of gross weights, altitudes, headwinds, and tailwinds entails lots of number crunching. But it is nice to have at least a few sample points in a fuel pinch far from a safe haven.

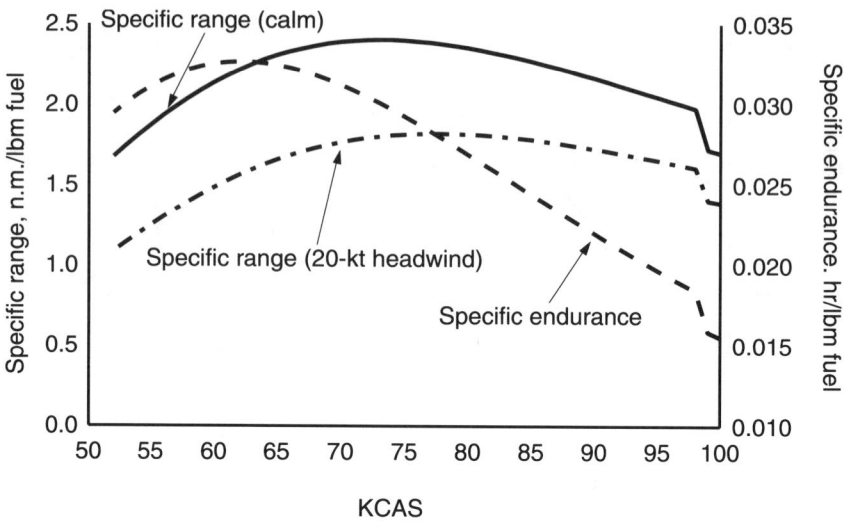

Figure 10.7 Figuring specific range in a 20-kn headwind.

Cruise Performance Scaling Rules

The sample level flight data in Table 10.3 gave us the pivotal propeller coefficient relation $J(C_P/J^2)$. But if we had been particularly interested in the airplane's cruise performance at that particular altitude and weight (3000 ft, 2300 lbf) we could have simply used that tabular data, with the $V_{M/m}$ relation Eq. (10.3), to find, for each featured cruising speed, $M_B(\Pi < 1)$ and then M, P, or $\%P$, and then gallon per hour fuel consumption. We had the germ of the cruise table for $W = 2300$ lbf and $h_\rho = 3000$ ft in our hand. Because information in Table 10.3 was all that was needed to predict cruise performance at other weights and altitudes, one suspects there are rules for deriving cruise performance data for any W and σ from cruise performance data for any *single* W and σ.

Indeed there are. The easiest way to see that and decipher the cruise performance connection or scaling rules, is to look again at Eqs. (10.12) and (10.13) with the definitions of various composite bootstrap parameters. Then Eq. (10.13) can be rewritten as

$$\frac{C_P}{J^2} = \left[\frac{H_B}{W_B^2 K_B}\left(\frac{W}{\sigma V^2}\right)^2 + 1\right] \times \frac{(-K_B)}{\rho_0 d^2 m} \qquad (10.29)$$

Equation (10.29) shows us that two cruise situations related by having identical values of

$$\frac{V\sqrt{\sigma}}{\sqrt{W}} = \frac{V_C}{\sqrt{W}} \qquad (10.30)$$

are two situations characterized by identical values of C_P/J^2. Therefore, by identical values of J and of identical values for C_P, C_T, and η, items that depend only on J. And not so incidentally, in level flight, by identical values of C_L and C_D.

Focusing on identity of J values gives the scaling rule for engine speed N. Our hypothesis is

$$\frac{V_1\sqrt{\sigma_1}}{\sqrt{W_1}} = \frac{V_2\sqrt{\sigma_2}}{\sqrt{W_2}} \tag{10.31}$$

and, because $J_1 = J_2$,

$$N_2 = \frac{V_2}{V_1} N_1 = \sqrt{\frac{W_2}{W_1}} \sqrt{\frac{\sigma_1}{\sigma_2}} N_1 \tag{10.32}$$

As situation 1, consider a fixed-pitch airplane in a given configuration at MSL ($\sigma_1 = 1$), $W_1 = 2400$ lb, requiring 2500 rpm to fly level at 100 KTAS = 100 KCAS. If that airplane goes to 8000 ft ($\sigma_2 = 0.7860$) weighing 1800 lbf, at air speed given by

$$\frac{V_C}{\sqrt{1800}} = \frac{100}{\sqrt{2400}} \tag{10.33}$$

or $V_C = 86.6$ KCAS, then from Eq. (10.32) its engine must be turning 2442 rpm. Similar machinations with definitions of propeller power and thrust coefficients C_P and C_T, with the engine speed scaling rule just derived, give Table 10.9. For abbreviation we used $w_{12} \equiv W_1/W_2$, $\sigma_{12} \equiv \sigma_1/\sigma_2$, and so forth.

Again, the importance of these scaling rules is that they let one use known cruise performance data for a fixed-pitch propeller airplane, data good for only one weight and one altitude, to calculate cruise performance for that airplane at any other weight and any other altitude.

Table 10.9 Bootstrap cruise performance scaling rules (for $V_{c2}/\sqrt{w_2} = V_{c1}/\sqrt{w_1}$)

Variable	Scaling rule
Engine speed	$N_2 = \sqrt{(w_{21}\sigma_{12})}N_1$
Engine power	$P_2 = w_{21}^{3/2}\sqrt{\sigma_{12}}P_1$
Engine torque	$M_2 = w_{21}M_1$
Power setting	$\Pi_2 = w_{21}\Phi_{12}\Pi_1$
Thrust	$T_2 = w_{21}T_1$
Drag	$D_2 = w_{21}D_1$
$X \in \{J, C_P, C_T, \eta, C_L, C_D\}$	$X_2 = X_1$

Partial-Throttle Operation at Given Speed, Turn Rate, and Climb or Descent Rate

The best way to demonstrate the calculation wrinkles demanded by climbing or descending turns (including wings-level climbs and descents as a special case) is to go through an example (previous question #1).

Example 10.4 Consider our sample Cessna 172 at 6000 ft, 2400 lbf, flaps up (case 2 of Table 10.2), making a standard rate (3 deg/s) turning 300 ft/min descent at 90 KCAS. Here are steps and formulas required to compute bank angle ϕ, engine speed N, and further performance items of interest.

1) Use the standard turn relation

$$\phi = \tan^{-1}\left(\frac{\omega V}{g}\right) \quad (10.34)$$

Three deg/s is 0.05236 rad/s and $V = 90$ KCAS $= 98.44$ KTAS $= 166.15$ ft/s. Hence, bank angle $\phi = 15.13$ deg. This will modify bootstrap composite H for additional induced drag.

2) A useful variation of the top portion of Eq. (10.6), directly using composite parameter values as in Table 10.2 and the general banking prescription [Eq. (10.39)], is

$$M(V, \dot{h}, \phi; W, \sigma) = \frac{d}{2\pi m}\left[\frac{\dot{h}W}{V} - KV^2 + \frac{H(0)}{\cos^2\phi V^2}\right] \quad (10.35)$$

Substituting the proper BDP and composite items (case 2 of Table 10.2) and bank angle gives $M = 177.15$ ft-lbf. (Recalling altitude power dropoff factor Φ, at 6000 ft, is 0.8135, and that MSL base torque is 311.2 ft-lbf, power setting is $\Pi = M/\Phi M_B = 70\%$.)

3) Use Eq. (10.10) to find that $C_P/J^2 = 0.0831$. From Eq. (10.17), $N = 2295$ rpm. The desired descending turn, maintaining 90 KCAS, results from banking 15 deg and throttling back to about 2300 rpm.

4) From the linearized propeller polar [Eq (10.19)], $C_T/J^2 = 0.0849$ and so propeller efficiency, during the early stages of this partial-throttle descent, is $\eta = 0.6950 \times 0.0849/0.0831 = 71.0\%$.

In summary, calculation input data (to the pilot, given conditions and goals) consists of 1) BDP for $\delta_f = 0$; 2) $W = 2400$ lbf; 3) $h_\rho = 6000$ ft; 4) $V = 90$ KCAS; 5) $\dot{h} = -300$ ft/min; and 6) $\omega = 3$ deg/s. The calculated outputs (to the pilot, required control inputs) are 1) $\phi = 15.1$ deg; 2) $N = 2295$ rpm; and 3) whatever elevator manipulation is required to maintain 90 KCAS.

Partial-Throttle Descent at Given Speed Along Given Approach Path

Again, an example (previous question #2) is called for.

Example 10.5 Consider the sample Cessna 172 at 2400 lbf at 6000 ft with flaps up. What revolutions per minute are (initially) required for the airplane to descend along a 3-deg glide slope (in calm air) at 90 KCAS?

The glide slope is -0.0524 rad, with sine -0.0523. The formula similar to Eq. (10.35), but featuring flight path angle, is

$$M(V, \gamma, \phi; W, \sigma) = \frac{d}{2\pi m}\left[\frac{W \sin \gamma}{V} - KV^2 + \frac{H(0)}{\cos^2 \phi V^2}\right] \quad (10.36)$$

In our example, bank angle $\phi = 0$. $M = 142.88$ ft-lbf (hence, $\Pi = 142.88/(0.8135 \times 311.2) = 56.4\%$). $C_P/J^2 = 0.0670$, by Eq. (10.10), and $N = 2153$ rpm by Eq. (10.17). To descend along the 3-deg glide slope, at 90 KCAS, the pilot should throttle back to about 2150 rpm.

In summary, calculation input data (to the pilot, given conditions and goals) consist of 1) BDP for $\delta_f = 0$; 2) $W = 2400$ lbf; 3) $h_\rho = 6000$ ft; 4) $V = 90$ KCAS; 5) wings level; and 6) $\gamma = -3$ deg. The calculated outputs (to the pilot, required control inputs) are 1) $N = 2153$ rpm and 2) whatever elevator manipulation is required to maintain 90 KCAS.

A continuing question. Assuming that indicated air speed and physical throttle position are maintained down to 2000 ft, and that the latter implies constancy of Π, what will the revolutions per minute be at this lower altitude? At 2000 ft, $\sigma = 0.9428$ and altitude dropoff factor $\Phi = 0.9350$. True air speed has decreased to 92.7 KTAS = 156.4 ft/s. Torque M will be increased by the ratio of dropoff factors, $0.9350/0.8135 = 1.15$, to 164.31 ft-lbf. C_P/J^2 is also increased by that factor because air speed appearing in the denominator of Eq. (10.10) is effectively calibrated (equivalent). Then $C_P/J^2 = 0.0771$. By Eq. (10.17), $N = 2110$ rpm. Engine speed has dropped by only 10 rpm/1000 ft; difficult to notice even in very calm air.

Partial-Throttle Absolute Ceilings

Because it is well known that an airplane cannot possibly achieve its absolute ceiling in finite time with finite fuel unless deposited there by some higher-flying entity, calculation of absolute ceilings and speeds there appears pointless, at most a comparative benchmark. But, in fact, getting to absolute ceiling is impossible only when one is "confined" to wings-level, full-throttle flight. Banking or throttling back certainly do not give the airplane a upwards boost; those control manipulations work the other way around, dragging the absolute ceiling down. So "absolute" ceiling, in relative terms, is attainable by flying high, full throttle and wings level and then banking or throttling. In so reducing the ceiling, some interesting facts emerge. Moving from the full-throttle wings-level ceiling relations

$$\Phi_{AC}(W, \Pi = 1, \phi = 0) = \frac{2W}{W_B E_B}\sqrt{-H_B(0)K_B} \tag{10.37}$$

$$V_{AC}(W, \Pi = 1, \phi = 0) = \frac{W}{W_B}\sqrt{\frac{2H_B(0)}{\sigma_{AC}\Phi_{AC}E_B}} \tag{10.38}$$

to the possibly banked, possibly partial-throttle case is easy. Use the general prescription for banking

$$H \equiv H(0) \rightarrow H(\phi) \equiv \frac{H(0)}{\cos^2 \phi} \tag{10.39}$$

and the similar prescription for partial throttle

$$E_B(\Pi = 1) \rightarrow E_B(\Pi < 1) = \Pi E_B(\Pi = 1) \tag{10.40}$$

with the composite formula for V_x, Eq. (10.12), to get

$$\Phi_{AC}(W, \Pi, \phi) = \frac{1}{\Pi \cos \phi}\Phi_{AC}(W, \Pi = 1, \phi = 0) \tag{10.41}$$

$$V_{CAC}(W, \Pi, \phi) = \sqrt{\frac{W}{W_B \cos \phi}}V_{cxB} \tag{10.42}$$

The parenthesized zero, as in Eqs. (10.37–10.39), denotes a value for unbanked (wings-level) flight. For most intents and purposes, banking to angle ϕ is tantamount to increasing gross weight from W to $W/\cos\phi$. If, for instance, one goes from full to three-fourths throttle ($\Pi = 0.75$), ceiling dropoff factor Φ increases by one-third and ceiling density altitude is considerably diminished, say from 16,000 ft ($\Phi = 0.5556$) to 8,582 ft ($\Phi = 0.5556 \times 4/3 = 0.7409$) and similarly, if one banks 41.4 deg. If the airplane is currently above the new banked or throttled ceiling, the airplane descends to it.

We have chosen to express the air speed required at the ceiling, Eq. (10.42), in calibrated terms to simplify matters. V_{cxB} is the calibrated speed for best angle of climb under base conditions. For a given airplane in a given configuration, it is just a number, independent of altitude. The interesting aspect of Eq. (10.42) is that it shows calibrated air speed at absolute ceiling, while it does depend on weight and bank angle, is independent of ceiling altitude and power setting Π.

Throttling back certainly lowers the ceiling, but calibrated air speed needed to maintain level flight at that lower ceiling is precisely the same as before.

Example 10.6 For a full numerical example, consider the airplane described in Table 10.10, case 4 ($W = 2400$ lbf, altitude 12,000 ft). What if, wings level at 12,000 ft, the pilot throttles back to $\Pi = 0.75$. What will the "absolute" ceiling, and the calibrated air speed to maintain it, then become?

From Eq. (10.41), using composite parameter values from Table 10.2, $\Phi_{AC}(2400, 1, 0) = 0.5514$. Because the pilot then goes to three-quarters throttle, the new $\Phi_{AC}(2400, 0.75, 0) = 0.7352$. Inverting the power dropoff formula to get

$$\sigma = (1 - C)\Phi + C \tag{10.43}$$

one finds that this corresponds to $\sigma_{AC}(2400, 0.75, 0) = 0.7670$, which by

$$h_\rho = 145{,}457(1 - \sigma^{0.23494}) \tag{10.44}$$

gives $h_{\rho AC} = 8790$ ft. Absolute ceiling air speeds are always V_x and, because this does not vary (in calibrated terms) when wings are kept level at constant weight, it is convenient to compute at the original 12,000 ft altitude: $V_x = (-R)^{1/4} = 269{,}435{,}170^{1/4} = 128.12$ ft/s $= 75.91$ KTAS. Using the value for relative air density at that altitude, $\sigma = 0.6932$, $V_{cx} = 63.2$ KCAS. Having pulled back to $\Pi = 0.75$, the airplane, if kept at optimum air speed 63.2 KCAS, descends to 8790 ft.

The importance of V_x, and especially of V_{cx}, is a red thread running through the bootstrap approach to propeller aircraft performance. Numerous earlier investigators' schemes for simplified performance calculations (e.g., Kerber[9] or Diehl[10])

Table 10.10 Sample composite bootstrap parameters, two full-throttle situations

Variable or composite	Case 1	Case 4
h_ρ	MSL	12,000 ft
W	2400 lbf $= W_B$	2400 lbf $= W_B$
σ	1	0.6932
$\Phi(\sigma)$	1	0.6513
E	531.9 $= E_B$	346.5
F	$-0.0052368 = F_B$	-0.0036300
G	$0.0076516 = G_B$	0.0053039
H	$1{,}668{,}535 = H_B$	2,407,100
K	$-0.0128884 = K_B$	-0.0089339
Q	$-41{,}270.6 = Q_B$	$-38{,}779.5$
R	$-129{,}460{,}301 = R_B$	$-269{,}435{,}170$
U	$218{,}064{,}595 = U_B$	453,840,065

used other dimensionless speed ratios. More often than not, base speed was chosen as maximum level flight speed (either at sea level or at altitude) or as level flight speed for maximum propeller efficiency. Very often those were taken to be the same. The treatment by Perkins and Hage,[11] which uses V_{bg} as the base speed, is exceptional. V_{bg} is close to V_x in concept. In fact V_{bg} is speed for best angle when gliding. The two speeds differ because V_x has an extra term in its denominator relating to propeller drag. In our work V_x, because it depends only on drag characteristics of the airframe and of the propeller (and in fact only on the ratio between induced and parasite components), plays the natural base role.

Conclusions

Three specific issues pertaining to this theory require further thought or experiment:

1) Π is defined under conditions of maximum power leaning while cruise performance usually assumes a "recommended lean" mixture. Our sample Cessna 172P POH quantifies recommended lean as 50°F rich of peak EGT. Because the engine manual shows that setting providing slightly over 99% of best power, there is little to no problem in our case. In other cases some modest adjustment might be needed.
2) Does C, the BDP altitude dropoff parameter, have the same value for the virtual derated engine as for the actual one? For practical purposes, it probably does. Because that factor varies little among actual engines of widely varying displacement and power, the two are likely very close. In addition, no performance item depends critically on the precise value of C.
3) A more important question is "To what extend does power setting variable Π correspond to spatial position of the throttle control plunger?" In other words, if one pulls back the throttle control 2 in. at 10,000 ft, thereby throttling back to say $\Pi = 0.55$, then leaves the throttle control alone, will it still be $\Pi = 0.55$ at 2,000 ft? Because of temperature expansion in the throttle and carburetor linkages and the vagaries of volumetric efficiency, we doubt that Π will be precisely maintained constant. In a sample problem, for lack of sufficient reason otherwise, we assumed constancy of Π, but we do not know how correct or incorrect that assumption is. A few simple experiments involving partial-throttle ceilings (because in those cases, air-speed dependency is so easy to deal with) would settle the issue at least for a single airplane.

Supportive experiments are called for generally. There are simplifying bootstrap assumptions concerning torque and throttle, small flight path angles, and the quadratic drag polar, plus a new set of assumptions connected to this partial-throttle extension. Any lengthy sorites of this type, no matter how well or convincingly argued, always requires experimental verification. Experiment, when all is said and done, is the final arbiter of the value of a theory. While small emendations to this one may be needed, it does provide a realistic method for calculating detailed fixed-pitch, partial-throttle performance.

References

1. Hale, F. J., *Introduction to Aircraft Performance, Selection, and Design*, Wiley, New York, 1984, pp. 123–132.
2. Perkins, C. D., and Hage, R. E., *Airplane Performance Stability and Control*, Wiley, New York, 1949, pp. 183–194.
3. Miele, A., *Flight Mechanics I: Theory of Flight Paths*, Addison-Wesley, Reading, MA, 1962, pp. 50, 110.
4. Von Mises, R., *Theory of Flight*, Dover, New York, pp. 406–407.
5. Lowry, J. T., "Fixed-Pitch Propeller/Piston Aircraft Operations at Partial Throttle," *Journal of Propulsion and Power*, Vol. 15, No. 3, 1999.
6. Lowry, J. T., "Analytic V Speeds from Linearized Propeller Polar," *Journal of Aircraft*, Vol. 33, No. 1, 1996, pp. 233–235.
7. Norris, J., and Bauer, A. B., "Zero-Thrust Glide Testing for Drag and Propulsive Efficiency of Propeller Aircraft," *Journal of Aircraft*, Vol. 30, No. 4, 1993, pp. 505–511.
8. Von Mises, R., op. cit., p. 461.
9. Kerber, L. V., "Airplane Performance," *Aerodynamic Theory, Vol. V*, edited by W.F. Durand, Dover, New York, 1935.
10. Diehl, W. S., *Engineering Aerodynamics*, Ronald Press, New York, 1936.
11. Perkins, C. D., and Hage, R. E., op. cit., pp. 168–171.

11

Takeoff Performance

Introduction

"Takeoff is optional." But not for us. For without it there is little to interest the aviator. Analyzing takeoff puts us into the new realm of dynamics, the study of motions of objects under forces. In our previous flight mechanics, forces and moments on the airframe were balanced; yes, the airplane moved, but only (quasi-)statically, at constant velocity. The exception was steady turning flight. Then the airframe was accelerated, true, but in only a quite simple fashion, moving (with respect to the air) in a circular or helical path. During takeoff, the airplane will (should!) move in a sequence of straight lines, but with new and varying forces acting on it—runway reaction, rolling friction, possibly contaminant drag.

So our job will be to delineate those forces, quantify them, and see their effects on the airplane's speed and position during the three phases of takeoff—from a standing start to initiation of rotation, a brief rotation to liftoff, and a climb out to an altitude of 50 ft AGL. Under only slightly idealized conditions, we will be able to piece together more or less "exact" solutions to the takeoff problem: mainly, how much distance (and, to some extent, time) is traversed during each takeoff phase. We will obtain and study various useful approximations to our lengthier analytical solutions. We will see what effect takeoff headwinds or tailwinds have. Some additional algebraic adjustment will result in a perturbation approach to calculating distance to rotation and liftoff. We will study the question: When should we takeoff uphill into a headwind and when downhill with a tailwind? We will come up with a single simple (though approximate) number—a takeoff power parameter (TOPP)—for relatively quick and dirty predictions of minimum takeoff distances.

Takeoff Phases

The typical small general aviation aircraft pilots operating handbook recognizes only two takeoff phases: 1) from start of roll through liftoff and 2) from liftoff to an AGL altitude of 50 ft (assuming flat terrain). We will split their first phase into two: up until start of rotation and from then until liftoff. That lets us account for variation in the airplane's attitude and changing forces on the airframe during rotation. So our picture assumes the three phases depicted in Figure 11.1 with the kinds of performance numbers we will be calculating. Our work will be specialized to airplanes with tricycle landing gear; taildraggers have a further brief initial phase during which the tail wheel is still in contact with the ground.

Cessna 172 Sample Takeoff Numbers

For an initial reality check (and subsequent comparison with calculated estimates), let us see what the Cessna 172 pilots operating handbook (POH) short field takeoff performance page gives for their airplane at mean sea level (MSL) (standard conditions) weighing 2400 lbf, flaps 10, full throttle, on a level dry concrete runway with no wind. They cite $d_{LO} = 892$ ft, $V_{LO} = 55.8$ KCAS (having used the POH air-speed calibration table), $d_{50} = 1628$ ft, and $V_{50} = 59.3$ KCAS. We typically rotate at their recommended liftoff speed.

Liftoff speed V_{LO} is slightly greater than stall speed. Much higher than that and we would be burning up runway and rubber for no good reason; much lower and the airplane would not fly. For flaps 10, stall speed V_S for this airplane is given a few pages earlier in the POH as 49 KCAS (for most forward center of gravity). So (POH value) $V_{LO} = 1.14\, V_S$. Liftoff speeds commonly are 10 to 20% greater than corresponding stall speeds. We will call that proportionate excess x; so for these POH values $x = 0.14$. We normally select $x = 0.15$.

Where did V_{50}, here 56 KIAS = 59.3 KCAS, come from? Speed for best climb angle V_x for this weight and flaps setting is 62.7 KCAS. A Cessna report[1] goes into this minor question in major detail. It mentions that this "obstacle clearance

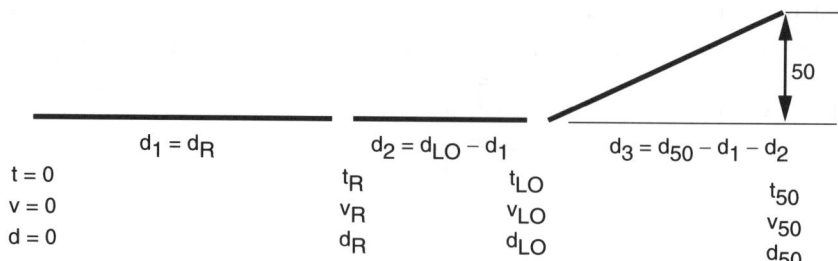

Figure 11.1 The three phases of the takeoff maneuver.

speed" is certainly not V_x (as we have seen it is not) and that "test pilots normally select a speed close to 1.2 V_S (which would be, here, 58.8 KCAS) for obstacle clearance." With 0.5-kn leeway, that might well have been the Cessna POH selection. Thinking ahead, one will be accelerating to climb out at either V_y (the usual choice) or at V_x (our preference). V_{50} can be looked upon as a way station toward one of those speeds. We shall normally take the POH book value in our calculations for comparison with our computed d_{50} estimates. We want to make as few changes as possible.

Cessna 182 Sample Takeoff Numbers

Running through similar numbers for the Cessna 182, MSL standard day, 3100 lbf (maximum gross weight for this model of this airplane), flaps 20, cowl flaps open, full throttle, 2400 rpm (giving rated power 235 bhp), level dry concrete runway with no wind, the 182 POH gives $d_{LO} = 820$ ft, $V_{LO} = 57.5$ KCAS, $d_{50} = 1570$ ft, and $V_{50} = 62.0$ KCAS.

The forward c.g. stall speed for flaps 20 is 52 KCAS, so the liftoff excess over stall is only $x = 0.11$. $V_{50} = 1.19 V_S$, essentially the nominal 1.2 figure mentioned in the Cessna report.

The two Cessna POHs say nothing about *times* to liftoff or to 50 ft. Those are not normally of any interest, though one way to calculate takeoff numbers with wind involves calculating time to liftoff. Both POHs give some guidance on takeoff roll adjustments for headwinds or tailwinds and for taking off on short dry grass instead of concrete. Their tables include many density altitudes (in the usual manner, unnecessarily split between pressure altitude and OAT) and three representative gross weights.

For future reference, Table 11.1 collects these results for the two Cessnas. These are "base case" figures at MSL, no wind, for takeoff from a dry level concrete runway.

Table 11.1 Important takeoff numbers (POH) for two Cessnas

Variable	Cesna 172	Cessna 182
Short field takeoff flaps, deg	10	20
Gross weight, lbf	2400	3100
Stall speed, V_S, KCAS	49	52
Distance to liftoff, d_{LO}, ft	892	820
Speed at liftoff, V_{LO}, KCAS	55.8	57.5
$x \equiv (V_{LO}/V_S - 1)$	0.14	0.11
Distance to 50 ft, d_{50}, ft	1628	1570
Speed at AGL 50 ft, V_{50}, KCAS	59.3	62
Ratio V_{50}/V_S	1.21	1.19

Takeoff Forces

"If you know the forces (and initial conditions), you can figure out the motion," said Newton. Airframe forces directed along the runway are thrust, drag, rolling friction, and sometimes (on sloped runways) a weight component. Runway reaction ("normal" forces) and lift are also operative. Those affect the takeoff roll, but only indirectly through interaction on drag (a little) and on frictional retarding force (a lot). We need to understand how those work. So let us begin (Table 11.2) with a catalog of all major forces on the airframe during each takeoff phase. Then we will comment on each type force. The various forces at work during takeoff, at first glance, seem many and varied. In fact they are quite easy to keep track of.

Thrust, T

For fixed-pitch airplanes, we have a good theory, with two minor caveats, for thrust. First, at rest (static), and very early in the run, propeller blades are partly stalled. This is mirrored by the Cessna 172 POH citation that full throttle static revolutions per minute are about 2360 ± 60 rpm instead of the 2700 rpm rated in flight at MSL. The second caveat is that the linearized propeller polar is a poorer

Table 11.2 Takeoff forces

Force type and component	Phase 1, up to rotation	Phase 2, during rotation	Phase 3, liftoff to 50 ft
Thrust, T	Yes, ignore static thrust deficit	Yes	Yes
Parasite drag, D_p	Yes	Yes	Yes
Induced drag, D_i	Yes, in ground effect	Yes, in ground effect	Yet, mostly out of ground effect
Lift, L	Yes, in ground effect	Yes, in ground effect	Yes, mostly out of ground effect
Weight, parallel to motion	$-W \sin\theta$	$-W \sin\theta$	$-W \sin\gamma$
Weight, perpendicular to motion	Countered by runway reaction and lift	Countered by runway reaction and lift	$-W \cos\gamma$
Runway reaction, N	Yes	Yes	No
Rolling friction, F_f	Yes	Yes	No
Tire/wheel runup	Small, ignore	No, or very slight	No
Net force parallel to motion	$T - D - F_f - W \sin\theta$	$T - D - F_f - W \sin\theta$	$T - D - W \sin\gamma$
Net force perpendicular to motion	$L - W \cos\theta - N$	$L - W \cos\theta - N$	$L - W \cos\gamma$

approximation to actual dynamic propeller behavior at very low air speeds. But because these effects are only small to moderate and transitory we shall ignore them. Then in terms of our bootstrap parameters, we have, for fixed-pitch (FP) airplanes,

$$T_{FP}(V) = E + FV^2 \qquad (11.1)$$

where of course parameters E and F both depend (somewhat differently) on relative air density σ.

For our base case Cessna 172, $E = 531.9$ lbf and $F = -0.005237$ slug/ft. So, being somewhat optimistic (the partially stalled propeller), initial takeoff thrust is 531.9 lbf. At rotation, with $V_R = 55.8$ KTAS $= 94.2$ ft/s, thrust has diminished to 485.4 lbf.

For constant-speed propeller airplanes, we recommend use of the general aviation general propeller chart (GAGPC). But there is a problem. In the static $V = 0$ situation, $J = \eta = 0$ and we cannot use the GAGPC. We need an alternative method for finding constant-speed propeller static thrust. There are many such (e.g., von Mises,[2] McCormick,[3] Crawford,[4] McCormick,[5] and Diehl[6]), most of which do not agree with one another. Of course one could go back to Chapter 6, Propeller Thrust, and do a full-blown propeller analysis for the static situation, but that would be calculational overkill. We will hedge by using an average of the methods of Diehl and McCormick.

To use Diehl's method, one needs his "static thrust coefficient" K_{TO} to put into

$$T_S = \frac{K_{TO}\,\text{hp}}{\text{rpm}\,d} \qquad (11.2)$$

where T_S is the static thrust estimate in pounds force and d is the propeller diameter in feet. K_{TO} comes either from a graph, Figure 11.2, or the corresponding curve fit formula, Eq. (11.3). The corresponding formula is a standard sigmoid:

$$K_{TO} = 22{,}893 + \frac{79{,}614}{(1 + e^{(\beta_{0.75} - 18.69)/4.296})} \qquad (11.3)$$

where the three-quarter radius blade angle $\beta_{0.75}$ is in degrees. For the Cessna 182 propeller, the low pitch stop is 15.8 deg at the $r = 30$ in. station. Because the propeller radius is $R = 41$ in., that is close enough. Putting in the Cessna 182 numbers gives $K_{TO} = 75{,}600$. Because hp $= 235$, rpm $= 2400$, and $d = 6.83$, we find $T_S = 1084$ lbf. This is just over twice the corresponding Cessna 172 figure.

McCormick's method, much newer, only requires entering a graph, Figure 11.3, or using a simple formula, Eq. (11.4), with the propeller's *disk loading* figure, hp/(disk area). McCormick's method, besides being much more current (propellers have changed considerably since 1932), also has the advantage of using readily available input data. The static thrust values, however, may be optimistic. For the Cessna 182 at MSL, his graph indicates about 6.4 hp/ft^2.

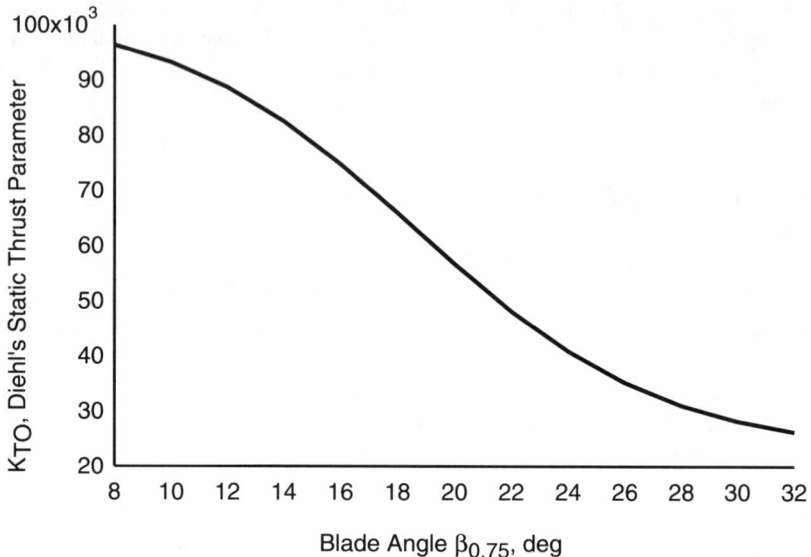

Figure 11.2 Diehl's graph is one way to estimate static thrust for a constant-speed airplane.

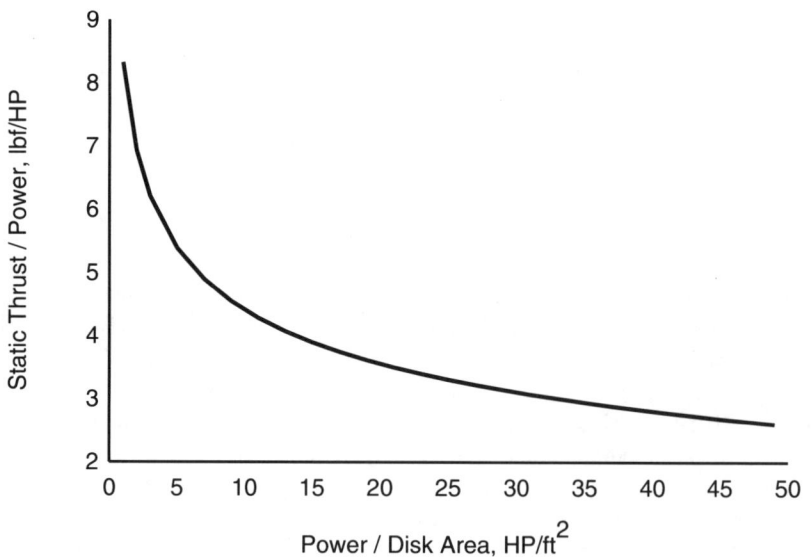

Figure 11.3 McCormick's graph is another way to determine constant-speed static thrust.

A curve-fit formula for McCormick's graph is

$$\frac{T_S}{\text{hp}} = \sqrt{-03.6426 + \frac{72.976}{\sqrt{\text{hp}/DA}}} \quad (11.4)$$

where DA, the propeller disk area, is 36.67 ft² for the Cessna 182 propeller of diameter 82 in. The formula [Eq. (11.4)] then gives $T_s/\text{hp} = 5.02$ and, consequently, $T_S = 1180$ lbf. Only 8% variation between the estimates of Diehl and McCormick in this generalized treatment is not bad.

When we need it, we shall use the two methods' average value, $\langle T_S \rangle = 1132$ lbf.

In the "reasonable" air-speed case (20 KTAS or greater), all the data needed to use the GAGPC *will* be available. With full throttle and constant revolutions per minute, air speeds V during the takeoff run will translate directly to values of advance ratio J; thrust values, after some work, will be forthcoming. In fact, for most constant-speed propeller aircraft, $T(V)$ fairly closely follows a cap-shaped quadratic pattern as in Eq. (11.1). One of our later approximation schemes will only need to sample $T(V)$ at two places, once at the beginning of the takeoff roll and again at rotation speed.

Drag, D

There are three varieties of drag to consider: parasite, induced (now with the added wrinkle of ground effect), and "contaminant drag." That latter will take some work. We start at the top.

Parasite Drag
There is nothing out of the ordinary with parasite drag; it remains

$$D_P = GV^2 = \tfrac{1}{2}\rho V^2 S C_{D0} \quad (11.5)$$

Induced Drag
D_i is quite a different matter. While this "drag due to lift" and its modification due to ground effect was treated in Chapter 4, Aerodynamic Force, let us here run through the main points and find some associated Cessna 172 numbers. In coefficient form,

$$D_i^I = \tfrac{1}{2}\rho V^2 S C_{Di}^I \quad (11.6)$$

where superscript I means in ground effect and where

$$C_{Di}^I \equiv (1-\sigma)C_{Di}^O \quad (11.7)$$

Subscript O means out of ground effect, given here with additional subscript GR meaning ground run,

$$C_{Di}^O = \frac{C_{LGR}^2}{\pi e A} \tag{11.8}$$

Factor σ, no relation to relative air density (distinctive small Greek letters are in short supply), is approximated by

$$\sigma \doteq e^{-4.22(h/B)^{0.768}} \tag{11.9}$$

where h is the (average) height of the wing (of span B) above the ground. For the Cessna 172, $h/B \doteq 0.16$, making $\sigma \doteq 0.356$. Body angle of attack (AOA), during the ground roll is about 5 deg, so that (Table 4.8 in the Aerodynamic Force chapter) $C_{LGR} \doteq 0.082 \times 5 + 0.34 = 0.75$. Out of ground effect, the induced drag coefficient $C_{LGR}^2/\pi e A = 0.0337$. In ground effect, this will be moderated by factor $(1 - \sigma)$, to $C_{Di}^I = 0.0217$, only 64% of the out-of-ground-effect value. $C_{DP} = C_{D0}$, recall, is 0.039 for flaps 10. So induced drag, up until rotation, is only about one-half of parasite drag. These figures will vary somewhat with aircraft loading and the state of the nose gear strut, but small changes will not much matter. You can get an accurate ground roll AOA by loading the airplane as though for takeoff on a level hard surface and measuring heights of the leading and trailing edges, h_L and h_T, and the average wing chord c, and computing

$$\alpha_{GR} = \sin^{-1}\left[\frac{(h_L - h_T)}{c}\right] \tag{11.10}$$

Assuming, as is most likely, you are dealing with an airplane *other* than the Cessna 172, legwork collecting the necessary numbers is in order. Even if you have not built up practical aerodynamic characteristics for the airplane (as outlined in the Aerodynamic Force chapter), you may be able to find your wing in Abbott and von Doenhoff[7] or in some similar reference. Even so, zero lift AOA α_0 is a problem. But you will not be far off, when it includes your airplane, by using Table 11.3 to get α_0 for $\delta_f = 0$ and then correcting for takeoff flaps δ_f using

$$\Delta\alpha_0 \equiv \alpha_0(\delta_f) - \alpha_0(\delta_f = 0) = \frac{C_{Lmax}(\delta_f = 0) - C_{Lmax}(\delta_f)}{a_I^A} \tag{11.11}$$

After some routine algebra on Eqs. (4.60) and (4.61) for the out-of- and in-ground-effect lift curve slopes, one finds the relation between them is

$$a_I^A = \frac{a_O^A}{\left(1 - \dfrac{180 a_O^A \sigma}{\pi^2 A}\right)} \tag{11.2}$$

Table 11.3 Important numbers (zero flaps) for common general aviation wings

Airplane	NACA no.	a_O^A	$\alpha_0(\delta_f = 0)$
Beech Bonanza	23012	0.105	−1.2
Cessna 150, 152, 170, 172, 177	2412	0.103	−2.1
Mooney M20K Root	63_2–215	0.112	−1.3
Tip	64_1–412	0.109	−2.8
Average		0.110	−2.1
Piper Archer, Lance, Cherokee, Warrior, Seneca	65_2–415	0.106	−2.8

Contaminant Drag

D_C when there is runway contamination, "foreign" material, is a major new factor. The most common runway contaminants are water, snow, and slush. Tire number, size, and pressure, contaminant depth, and ground speed are the major influences.

Contaminant drag is approximately given (for instance, by various NASA and Federal Aviation Administration [FAA] technical reports[8–11]) by

$$D_C = 1.94 \rho_S d V^2 \left[2 w_M C_{DCM} \sqrt{\frac{\delta_M + d}{w_M}} - \left(\frac{\delta_M + d}{w_M}\right)^2 \right.$$

$$\left. + w_N C_{DCN} \sqrt{\frac{\delta_N + d}{w_N}} - \left(\frac{\delta_N + d}{w_N}\right)^2 \right] \quad (11.13)$$

where

D_C = contaminant drag, in lbf;
ρ_S = contaminant specific density relative to water;
d = contaminant depth, ft;
V = ground speed, ft/s;
w_M = width of unloaded main tire, ft;
w_N = width of unloaded nose tire, ft;
δ_M = deflection of main tire, ft;
δ_N = deflection of nose tire, ft;
$C_{DCM} = 1.4$, if $V \leq V_{PM}$, and $= \max\{1.4 - 0.0180(V - V_{PM}), 0.3\}$, if $V > V_{PM}$;
$C_{DCN} = 2.6$, if $V \leq V_{PN}$, and $= \max\{2.6 - 0.0328(V - V_{PN}), 0.6\}$, if $V > V_{PN}$;
V_{PM} = main tire planing speed $= 15.2 \, (P_M/\rho_S)^{1/2}$, ft/s,
V_{PN} = main tire planing speed $= 15.2 \, (P_N/\rho_S)^{1/2}$, ft/s;
P_M = main tire inflation pressure, psi; and
P_N = nose tire inflation pressure, psi.

Planing speed factor 15.2 may be more familiar as 9, its value when planing speed is in knots.

Because tire pressure (if we ignore contributions from tire sidewall stiffness) supports all of the airplane's stationary weight, there is a relation between deflected tire geometry, tire pressure, and the weight it supports. A circular disk of radius R, if one cuts off a cap of deflection (radial) thickness δ, has an exposed chord of length

$$l = 2\sqrt{\delta(2R - \delta)} \quad (11.14)$$

with inverse relation

$$\delta = R\left(1 - \sqrt{1 - \frac{l^2}{4R^2}}\right) = R - \sqrt{R^2 - l^2/4} \quad (11.15)$$

Example 11.1 Let us take our usual Cessna 172 and ask what the "standard" tire deflection might be expected to be. The single 5.00-5 (radius 6.5 in., width 5.0 in.) nose tire has a recommended pressure of 34 psi; the two 6.00-6 (radius 7.5 in., width 6.0 in.) main tires have recommended pressures of 28 psi. The airplane weight, 2400 lbf, is distributed—normal weight and balance assumed—83% on the mains, 17% on the nose wheel. Call those factors f_{WB}. The length of "flattened" tire, l, is given by

$$l = \frac{f_{WB}W}{NPw} \quad (11.16)$$

where N is the number of such tires sharing the load, tire pressure P is in psi, and both l and tire width w are, here, in inches. For the main tires ($N = 2$), we get $l = 5.93$ in.; for the nose tire, $l = 2.40$ in. Using Eq. (11.14) twice, deflection of the main tire will be $\delta = 0.61$ in. and deflection of the nose tire will be $\delta = 0.11$ in. Those deflections compare well with actual measurements, so perhaps sidewall stiffness does not badly perturb our rough calculation.

Example 11.2 Because we have accumulated this data on Cessna 172 tires, let us continue on to calculate contaminant drag for 1/8 in. standing water. Always citing the main tire figures first, the planing speeds will be 80.4 ft/s (47.6 kn) and 88.6 ft/s (52.5 kn). So one will expect hydroplaning during the latter portion of the takeoff ground run. Taking the POH liftoff speed of 55.8 kn or 94.2 ft/s, just before liftoff the C_{DC} coefficients will be 1.15 and 2.42. Because speeds never will get high enough for this airplane to invoke maximum conditions on the coefficients, those coefficients will vary linearly between onset of hydroplaning and liftoff. To simplify our calculation, we assume an average main tire value $C_{DCM} = 1.30$, $C_{DCN} = 2.55$. Once all of the numbers for this situation have been plugged into Eq. (11.13), we find

$$D_C = 0.0132 V^2 \qquad (11.17)$$

This is 1.64 times the corresponding parasite drag force. Just before liftoff ($V_{LO} = 55.8\,\text{KTAS} = 94.2\,\text{ft/s}$), contaminant drag would amount to 116.7 lbf, quite a sizeable retarding force. Takeoff, on a badly contaminated runway in a light airplane, is a subtle danger for two reasons. While all may seem fine early in the run, the contaminant drag effect builds up as the square of ground speed and may eventually overpower the engine. In addition, one will expect to reach hydroplaning speed, with consequent skidding and possible loss of control, fairly late in the maneuver.

Lift, L

Even though lift is not directed along the runway, it affects the takeoff run by making the airplane progressively "light on its feet." More lift means less force pressing the tires to the runway surface, therefore, less rolling friction. Again, ground effect plays a minor role to enhance lift. Equation (4.63) in the Aerodynamic Force chapter is

$$C_L^I = \frac{C_L^O}{\left(1 - \dfrac{180\sigma a_O^A}{\pi^2 A}\right)} \qquad (11.18)$$

Because the out-of-ground-effect lift coefficient for our sample Cessna 172 was 0.75, we find from Eq. (11.18) that the in-ground effect value is 0.808, about 8% greater. We can use this ground-effect excess to help us compute the airplane's body AOA at liftoff, α_{BLO}. At liftoff, lift equals weight, so

$$C_{LLO} = \frac{2W}{\rho V_{LO}^2 S} \qquad (11.19)$$

For our sample Cessna 172, standard conditions, this is $C_{LLO} = 1.313$. But we can get by with only $1/1.08 = 93\%$ of this out-of-ground-effect value, or $C_L = 1.22$, which requires only $\alpha_{BLO} = 10.7\,\text{deg}$. Getting slightly ahead of ourselves, into the second takeoff phase, this means that a "normal" 3-deg/s rotation rate will have done its job after only $(10.7-5.0)/3 = 1.9\,\text{s}$. Actually (we assume the performance test pilot is trying for a "maximal performance" takeoff) the pilot would likely rotate about twice that fast, bringing the rotation time interval down to 1 s.

Weight, W

On a flat runway, weight is opposed by the combination of lift and runway reaction N. But when the runway slopes up at angle θ, weight component $F_W = W \sin \theta$ pulls back on the airframe. If the runway slopes down, that same weight component helps the airplane along. Using positive angles for upsloping runways and negative for downsloping ones automatically takes care of the sign of this retarding force.

Runway Reaction, N

During the ground run there is no net force perpendicular to the runway surface; N always adjusts itself so that

$$N + L - W \cos \theta = 0 \tag{11.20}$$

Rolling Friction, F_f

We do not care about static friction—assuming that we can eventually get the airplane moving—because overcoming it does not use up any runway. But rolling friction is a major negative player always working against the existing motion (as long as the airplane is moving on the runway) of size

$$F_f = \mu N = \mu (W \cos \theta - L) \tag{11.21}$$

Rolling friction coefficient μ depends on the type surface on which the rubber tires roll. Table 11.4 gives values for the common types and coefficient values.

Tire and Wheel Runup Neglected

As the airplane accelerates down the runway, a very small portion of the propulsive energy is transferred to rotational kinetic energy of the tires and wheels instead of to kinetic energy of the airframe. We shall ignore this effect,

Table 11.4 Approximate rolling friction coefficients for common runway surfaces

Runway surface type	Rolling friction coefficient μ
Dry smooth concrete	0.02
Broken-up dry asphalt	0.03
Hard dirt	0.04
Short grass	0.05
Wet concrete or asphalt	0.05
Tall grass	0.1
Soft field	0.2

which amounts, for our standard Cessna 172 taking off at seal level weighing 2400 lbf, to a diminution of net force of about 3.5 lbf.

Net Takeoff Force (from Start to Rotation)

Putting the previous results together using the definitions of lift and drag coefficients we have, for $V < V_R$ (neglecting the complicating possibility of runway contamination both because it so seldom occurs and because it messes up the pretty analytic picture),

$$F_{\text{Net}}(V) = T - D - F_f - F_W = (E - \mu W \cos\theta - W \sin\theta)$$
$$+ \left(K - \frac{\rho S f_{GE} C_{LGR}^2}{2\pi e A} + \frac{1}{2}\mu \rho S C_{LGR}^I \right) V^2 \qquad (11.22)$$

where the first equality is true of all airplanes but the second only applies to fixed-pitch ones. K is the usual propeller drag plus airframe parasite drag bootstrap composite parameter $F - G$. In spite of its seeming complexity (including the fact that density–altitude dependence is buried in several terms), this last relation is in essence (because the coefficient of the V^2 term will always be negative) simply

$$F_1(V) = a^2 - b^2 V^2 \qquad (11.23)$$

(for convenience in future work, take "constants" a and b to be the positive square roots) and can readily be integrated to find the space length or time interval of this portion of the run. We shall soon do this. But first we look at two graphic examples showing the progressing sizes during the takeoff run of the various force components.

Graphic Interlude: Takeoff Forces Parallel to Aircraft Motion

In the base case (sea level, maximum gross weight, flaps set for short field, bare dry concrete, no slope, no wind), Fig. 11.4, each of thrust, drag, and rolling friction is quadratic in the air speed. Figure 11.4 gives the *magnitudes* of each force component; keep in mind that drag and rolling friction always work in the direction *opposite* to thrust force. Rolling friction starts out at 48 lbf ($0.02W$) and decreases as lift grows. Notice that propeller drag, due to bootstrap parameter b [not the b in Eq. (11.23)], makes thrust drop off about 50 lbf during the run. Drag quadratically increases with speed until, at about 30 KCAS, it surpasses rolling friction as a retardant. Overall, net force on the airframe drops about 25%, about 128 lbf, during the run.

The higher altitude sloped grass takeoff case, Fig. 11.5, looks considerably less promising. At 2000 lbf, now the airplane can lift off at only 50.9 KCAS. But up at 5000 ft, on a (short) grass runway sloped up 2 deg, it will still be at quite a

358 *John T. Lowry*

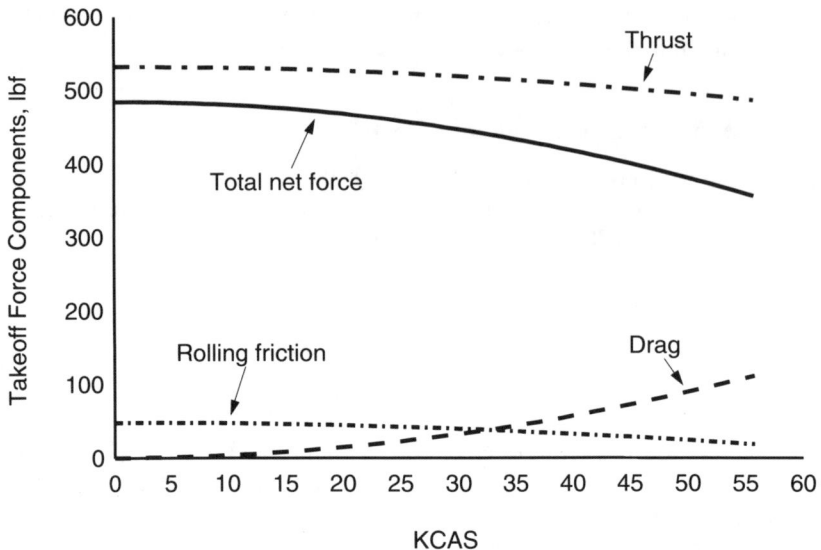

Figure 11.4 Takeoff force components, fixed-pitch base case.

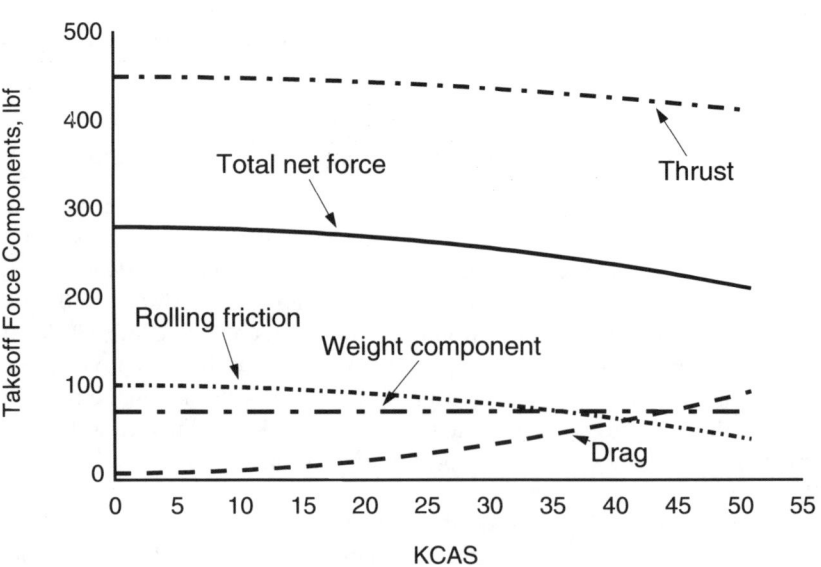

Figure 11.5 Takeoff force components, far from base case.

disadvantage, *vis a vis* takeoff roll up to rotation, when compared with the base case of Fig. 11.4. Keep in mind both graphs use *calibrated* air speeds, so true air speeds in Fig. 11.5 are higher than in Fig. 11.4. And we are not yet ready to talk about the distances covered. On grass, rolling friction, in spite of the weight reduction, has more than doubled over the base case, starting off at 100 lbf. Drag is unchanged. But the further addition of runway upslope means a constant 70 lbf retarding weight component force on the airframe. Net force is considerably depressed. Now of course one does have to ask "Why is the pilot is taking off up hill without a headwind?" Just for drill.

We now have a firm grip on the fairly complicated picture of forces acting during takeoff. Aviation needs got us into this mess, but only physics, supported by some mathematics, will get us out of it. Given the forces appearing in the airplanes' equations of motion, one must essentially integrate to find first speeds, then again to find displacement distances. With forces depending only on speed, there are some easy calculus tricks letting us eliminate time in favor of distance. In the most general case, say on a runway with changing slopes, forces get more complicated, depending on distance covered as well as on speed. One may need to use numerical techniques (e.g., the Runge–Kutta method for numerically solving second-order ordinary differential equations) that allow the investigator to more precisely model the dynamic situation during rotation, those undulating runways, etc. We will not in this book—with the exception of numerical integration using the trapezoidal rule or Simpson's rule—we draw a line just this side of numerical methods. A few indefinite integrals (Appendix C) are all we need. Now let us look into the physics of the takeoff situation.

Physics Interlude: Dynamics When Force Depends on Speed, $F = F(V)$

There are essentially four variables involved in our takeoff dynamics: time t, acceleration $a = \ddot{x} = F/m$, speed $V = \dot{x}$, and displacement $d = x$ (no relation to the excess liftoff speed parameter). We do not need to deal with vector notation because the airplanes are moving in straight lines; plus and minus signs will be enough. With four variables, there are $4 \times 3 = 12$ different relations among them. That is overkill; though any are accessible, we do not need them all (e.g., $x(a)$), displacement as a function of acceleration, is seldom of any use. Despite some arbitrariness in their selection, only about half of the twelve feasible dynamical relations need concern us. Of the other half, one is given. And the other five are accessible if ever they are needed. Table 11.5 presents the survivors.

Starting from Newton's second law,

$$F = ma = m\frac{dV}{dt} = m\frac{d^2x}{dt^2} \tag{11.24}$$

Table 11.5 Important functional relations for takeoff

	(t)	(x)	(V)	(a)
t	—	No $t(x)$	Yes $t(V)$	No $t(a)$
x	Yes $x(t)$	—	Yes $x(V)$	No $x(a)$
V	Yes $V(t)$	Yes $V(x)$	—	No $V(a)$
a	Yes $a(t)$	No $a(x)$	Given $a(V)$	—

the job is to come up with the six needed relations when force $F = F(V)$. The force relation, for our purposes, is one of only three types: 1) with neither contaminant drag nor wind:

$$F_1(V) = a^2 - b^2 V^2 \qquad (11.25)$$

2) with wind but without contaminant drag:

$$F_2(V) = AV^2 + BV + C \qquad (11.26)$$

and 3) with both wind and contaminant drag:

$$F_3(V) = AV^3 + BV^2 + CV + D \qquad (11.27)$$

Because our canvas is relatively narrow, we should be able to push this program through without running out of paint. We shall consider the most common cases in detail. But first, notes on how to acquire the necessary relations.

1) To get $t(V)$, multiply Eq. (11.24) by $dt/F(V)$ and integrate. Using subscript i for the initial situation and f for the final one, you get

$$\Delta t = t_f - t_i = \int_{t_i}^{t_f} 1 \, dt = m \int_{V_i}^{V_f} \frac{dV'}{F(V')} \qquad (11.28)$$

We will consistently refer here to the short list of integrals in Appendix C. Those we need for takeoff are denoted I_1 through I_9. To perform the integration, use I_2 when $F = F_1$, use I_4 when $F = F_2$, and use I_6 when $F = F_3$.

2) To get $V(t)$, invert the result of 1 just above. Because those relations will involve logarithms or inverse hyperbolic tangents, your inversions for $V(t)$ will involve exponentials or the tanh function. In more complicated exponential cases, you may find it useful to employ

$$\exp\left[\sum_j c_j \ln f_j(V)\right] = \prod_j (f_j(V))^{c_j} \qquad (11.29)$$

3) To get acceleration $a(t)$, differentiate $V(t)$.
4) To get $x(t)$, integrate $V(t)$ obtained in 2 above:

$$x_f - x_i = \int_{t_i}^{t_f} V(t') dt' \qquad (11.30)$$

If $V(t)$ involves exponentials, use I_8 or its more general form, I_9. If tanh is involved, use I_3.

5) Finding $x(V)$ is one of the most important cases. If net force is in its simplest form, $F = F_1 = a^2 - b^2 V^2$, rewrite Newton's second law as follows:

$$F_1(V) = m\frac{dV}{dt} = mV\frac{dV}{dx} = \frac{m}{2}\frac{dV^2}{dx} \tag{11.31}$$

Then multiply by dx/F_1 and integrate. The way station is

$$\Delta x = x_f - x_i = \int_{x_i}^{x_f} 1\, dx' = \frac{m}{2}\int_{V_i}^{V_f} \frac{dV'^2}{a^2 - b^2 V'^2} \tag{11.32}$$

Change integration variable from V'^2 to z, then use I_1. If net force is in one of the more complicated forms, F_2 or F_3, stop at the next to last term in Eq. (11.31), then use I_5 or I_7, respectively.

6) To find $V(x)$, invert the (logarithmic) result of 5 just above. You may want to use Eq. (11.29) in the ensuing simplification.

Some of these same techniques will be useful in calculating takeoff numbers for constant-speed propeller airplanes, but (unless you find you can successfully curve fit the GAGPC net force results into one of our three previous forms) the necessary integrations will all need to be done numerically.

Application: Takeoff Phase 1, Base Case, Fixed Pitch, No Wind, No Contamination

We know the net takeoff force acting on our fixed-pitch airplane follows the simple relation of Eq. (11.23), net force $F = F_1$. Taking the sample Cessna 172 at maximum gross weight, MSL, flaps 10, on dry level concrete, no wind or contamination, a little calculation shows that $a^2 = 483.9$ and $b^2 = 0.01448$. Therefore, $a = 22.00$ and $b = 0.1202$. Rotation speed is $V_R = 94.2$ ft/s. The airplane's mass is $m = W/g = 2400/32.174 = 74.6$ slugs. We run roughly parallel to the recent physics interlude and ask several questions.

Questions #1: What is $t(V)$ in general? And what is $t(V_R)$ in particular?

Integral I_2 applies. The $\tanh^{-1}$ form is quickest, but the logarithmic form works fine if one does not have immediate access to hyperbolic functions and their inverses. The result is

$$t(V) = \frac{W}{gab}\tanh^{-1}\left(\frac{bV}{a}\right) \tag{11.33}$$

Plugging in $V = V_R$, we find, in this no-wind situation, that the time interval from the start of the takeoff roll, up to initiation of rotation, is $t(V_R) = 16.05$ s.

Question #2: What is V(t) in general?

Equation (11.33) is easily inverted to

$$V(t) = \frac{a}{b} \tanh\left(\frac{gabt}{W}\right) \tag{11.34}$$

If one inserts $t_R = 16.05$ s, $V_R = 94.2$ ft/s is recovered.

Question #3: What is acceleration a(t) in general?

Taking the derivative with respect to time of Eq. (11.34), we have (because the derivative of the hyperbolic tangent tanh is sech^2, the square of the hyperbolic secant):

$$a(t) = \frac{ga^2}{W} \text{sech}^2\left(\frac{gabt}{W}\right) = \frac{ga^2}{W}\left[1 - \tanh^2\left(\frac{gabt}{W}\right)\right] \tag{11.35}$$

We see, for instance, that indeed, at $t = 0$, we have $a(t=0) = F_1(V=0)/m = a^2/m = 483.9/74.6 = 6.49$ ft/s^2 $= 0.202g$. By the time (16.05 s later) the airplane is up to rotation speed, acceleration has slacked off to 4.77 ft/s$^2 = 0.148g$.

Question #4: What is displacement x(t) in general?

Displacement is the integral of speed and the integral of the hyperbolic tangent is (see I_3) the logarithm of the hyperbolic cosine. Here

$$x(t) = \int_0^t V(t')dt' = \frac{a}{b}\int_0^t \tanh\left(\frac{gabt'}{W}\right)dt' = \frac{W}{gb^2} \ln \cosh\left(\frac{gabt}{W}\right) \tag{11.36}$$

If one puts in $t_R = 16.05$ s, one has a long-winded way of finding that d_R, distance to start of rotation, is 794.2 ft. The next subsection has a better way.

Question #5: What is displacement x(V) in general?

Evaluation of the integral in Eq. (11.32) is, here, via I_1 with $A = a^2$ and $B = -b^2$. The result is

$$x(V) = \frac{m}{2}\int_0^V \frac{dz}{a^2 - b^2 z} = \frac{-W}{2gb^2} \ln\left(1 - \frac{b^2}{a^2}V^2\right) \tag{11.37}$$

Substituting $V = V_R = 94.2$ ft/s gives $x(V_R) \equiv d_R = 794.4$ ft.

Question #6: What is air speed $V(x)$ in general?

In this case, the inversion of result 5, Eq. (11.37), is straightforward. The result is

$$V(x) = \frac{a}{b}\sqrt{1 - e^{-2b^2 x/m}} \qquad (11.38)$$

Substituting $x = d_R = 794.4$ ft, $V_R = 94.2$ ft/s.
We will consider another important special case, including wind effects, below. Though not in such excruciating detail.

"Constant Average Force" Approximation

Back to aviation. Because the initial force on the airframe (no wind) is $F_0 = a^2$ and the force at rotation speed is $F_R = a^2 - b^2 V_R^2$, a more operational way of looking at Eq. (11.37) comes from looking at "net force dropoff parameter" b^2 (no relation to the propeller polar b) as

$$b^2 = \frac{F_0 - F_R}{V_R^2} \qquad (11.39)$$

and then at the formula for distance to rotation as

$$d_R = \frac{W}{2g} \frac{V_R^2}{F_0 - F_R} \ln \frac{F_0}{F_R} \qquad (11.40)$$

This shows that length of takeoff roll increases with gross weight, increases as the square of the rotation speed (though inverse with the size of the net force dropoff parameter b^2), and increases with the ratio of initial to final force. But a much more immediate realization of Eq. (11.37), though approximate, comes about by rewriting Eq. (11.40) using the first term of a little-known approximation to the logarithm function, for positive arguments x, and better for x close to unity,

$$\ln x \doteq 2\left(\frac{x-1}{x+1}\right) + \frac{2}{3}\left(\frac{x-1}{x+1}\right)^3 + \cdots \qquad (11.41)$$

using $x = F_0/F_R$. The bottom line is that

$$d_R \doteq \frac{WV_R^2}{g(F_0 + F_R)} = \frac{WV_R^2}{2g\langle F_{\text{Net}}\rangle} \qquad (11.42)$$

where the angle brackets denote an ordinary average, the arithmetic mean. This approximation is better if the net force on the airframe drops off only relatively little during the course of the ground run. Equation (11.42) is just what one would expect under the constant-acceleration kinematics we are familiar with from high school physics: kinetic energy gained equals force times distance.

How good is the "constant average" force approximation in our Cessna 172 base case? Using $F_0 = 483.9$ lbf, and $F_R = 355.7$ lbf, we find $d_R \doteq 788.4$ ft. Our exact value was 794.4 ft. Negligible difference, in this case, though the slight error is to the nonconservative side. We shall explore the "constant average" force approximation in greater depth a few sections later. First we would like to get the airplane all the way up into the air. And then return to wind effects.

Calculation of Lengths of Takeoff Phases 2 and 3

Of the three takeoff phases—to rotation, to liftoff, and to 50 ft—we have completed "exact" and approximate (constant average force) calculations of only the first. (The exact value was 795 ft; the approximate value was 788 ft.) But in fact there is very little to do in phase two.

Calculating $d_2 = d_{LO} - d_R$ is quite simple. Assume a rotation rate; we have taken it to be 6 deg/s. Assume the airplane stays essentially at V_R during rotation. (If, for larger heavier general aviation aircraft, speed constancy during rotation is not realistic, substitute an average speed, $\frac{1}{2}[(V_R + V_{LO}])$ Calculate the body AOAs for both liftoff and during the ground run; subtract those. One then has

$$d_2 \equiv d_{LO} - d_R \doteq \frac{(\alpha_{LO} - \alpha_R)V_R}{\omega_R} \qquad (11.43)$$

For our sample Cessna 172, base case, while calculating lift we found that $\alpha_{LO} = 10.7$ deg, $\alpha_{GR} = 5.0$ deg, $V_R = 94.2$ ft/s, and we assumed rotation rate $\omega_R = 6$ deg/s. That makes $d_2 = 89.5$ ft. Taking the exact d_R value (794.4 ft), our estimate is that $d_{LO} = 883.9$ ft. The POH makes it 892 ft. We will not ordinarily expect this kind of (less than 1% error) accuracy.

Now for the third phase, $d_3 \equiv d_{50} - d_{LO}$. Our method is to average V_{LO} (55.8 KCAS = 55.8 KTAS = 94.2 ft/s) and $V_{50} = 59.3$ KCAS = 59.3 KTAS = 100.1 ft/s to get $V_3 = 57.6$ KCAS = 57.6 KTAS = 97.2 ft/s. Then we simply find the rate of climb for V_3 from ordinary bootstrap formulas to see how much time it takes for the airplane to gain 50 ft. Then one turns that time interval into a horizontal distance:

$$d_3 \doteq 50\sqrt{\frac{V_3^2}{h_3'^2} - 1} \doteq \frac{50V_3}{h_3'} \qquad (11.44)$$

The rightmost term in Eq. (11.44) proclaims our usual approximation that horizontal and slant distances, for these relatively low-powered aircraft, are essentially equivalent.

How do the numbers come out? For $V = V_3 = 97.2$ ft/s, under these circumstances (do not forget that $\delta_f = 10$ deg), we find that $h_3' = 558.1$ ft/min = 9.30 ft/s. So $d_3 \doteq 520.2$ ft. The Cessna 172 deg POH, on the other hand, implies

$d_3 \equiv d_{50} - d_{LO} = 1628 - 892 = 736$ ft. That makes our estimate of d_3 a whopping 29.3% lower than the book value. We do not know why the large discrepancy exists or where the truth of this matter lies. Perhaps because distance to clear a 50-ft obstacle has corporate legal liability implications, the manufacturer was being overly cautious, as they were in their overly conservative citation of best glide ratio.

But the nicest feature is that we have, at this point, fairly accurate—and not overly complex—ways of estimating distances for each of the three airplane takeoff phases.

Let us add one easy wrinkle. What if the airplane takes off on this same runway but sloped up 2 deg? Using the "constant average" force technique, the effect is quite easy to calculate. A 2 deg upslope lessens net accelerating force on the 2400-lbf airframe by a constant $W \sin \theta = 2400 \times 0.0349 = 83.8$ lbf throughout the ground run, changing $\langle F_{\text{Net}} \rangle$ from level value 419.8 lbf down to sloped value 336.0 lbf. The ratio of net average forces is $419.8/336.0 = 1.25$, so distance to rotation will be 25% longer on the sloped runway, increasing from the original 795 to 994 ft. There are no changes to d_2 and d_3 (unless you now decide to measure the 50-ft height off the sloped runway).

Application: Takeoff Phase 1, Base Case, Fixed Pitch, Headwind, No Contamination

Because the usual takeoff wind is a headwind, we will use that for our examples. (Except for the brief condition that a net tailwind deserves a different, higher, and unknown parasite drag coefficient, a tailwind, for our purposes, is just a negative headwind.) Our treatment will also ignore those normally minor perturbations due to a crosswind component, small effects from control deflections and side force. Taking off into a headwind affects the length of the takeoff roll in two ways: 1) lower ground speed is required for liftoff and 2) the takeoff roll now begins with nonzero drag force.

Simplified constant-force treatments ignore the second effect. Then, because the required rotation ground speed is

$$V_{gR} = V_R - V_{hw} \qquad (11.45)$$

and d_R is proportional to V_{gR}^2, it is easy to see that the effect of the headwind is to shorten d_R by factor

$$\left(1 - \frac{V_{hw}}{V_R}\right)^2 \qquad (11.46)$$

Equation (11.46) must overstate the headwind effect (understate the headwind takeoff roll distance) because of the second effect, the increased drag due to wind.

The official FAA prescription accounts for this by lowering the exponent 2, in Eq. (11.46), to 1.85. They explain that their modification was "developed empirically and is an acceptable method for correction of low wind conditions." As we shall see (Fig. 11.6), application of either Eq. (11.45) or the FAA's version with lowered exponent both give quite good approximations to the exact formulation, which we now obtain.

Distance to Rotation in Wind, d_{Rw}

With a good handle on the forces and on the kinematics, the "exact" solution to the take-off-in-the-wind problem should be fairly near at hand. There are two different, but completely equivalent, ways to proceed:

- Take a coordinate system fixed with respect to the air. One first calculates distance to rotation with respect to the air (from initial air speed $V_i = V_w$ to final air speed $V_f = V_R$, then corrects that (overly large) figure by subtracting the distance the air has moved during the time needed for the takeoff roll to rotation, $V_w \Delta t$. Of course one has to do an extra calculation to find Δt, but we know how to do that.
- Take a coordinate system fixed to the ground. Then one has initial (ground) speed $V_{gi} = 0$ and final ground speed $V_{gf} = V_R - V_w$.

Both frames are inertial. There is no good reason for preferring one over the other. Von Mises[2] choose the first. We choose the second, the ground-based system. Using subscripts g for ground speeds and ground displacements, Newton's second law then reads

$$F(V) = G(V_g) = m \frac{dV_g}{dt} \quad (11.47)$$

where the force expression, in terms of ground speed, using $V = V_g + V_w$ in $F(V)$, is

$$G(V_g) = -b^2 V_g^2 - 2b^2 V_w V_g + (a^2 - b^2 V_w^2) = AV_g^2 + BV_g + C \quad (11.48)$$

That defines convenient constants A, B and C. To get a form suitable for integrating, to find the ground roll up to rotation in the wind, d_{Rw}, one can again insert 1 in the form dx_g/dx_g. But this time, because of the V_g term, there is no advantage to integrating with respect to V_g^2. Multiplying the resulting expression by dx_g, and dividing by $G(V_g)$, one gets

$$d_{Rw} = \int_0^{d_{Rw}} 1 \, dx_g = m \int_0^{V_R - V_w} \frac{V_g \, dV_g}{AV_g^2 + BV_g + C} \quad (11.49)$$

One then uses indefinite integrals I_4 and I_5 to find (as is too often said), "after some algebra," that

$$d_{Rw} = \frac{W}{2gb}\left[\frac{1}{b}\ln\left(\frac{(a-bV_w)(a+bV_w)}{(a-bV_R)(a+bV_R)}\right) + \frac{V_w}{a}\ln\left(\frac{(a-bV_R)(a+bV_w)}{(a+bV_R)(a-bV_w)}\right)\right]$$
(11.50)

One should keep in mind that Eq. (11.50) only applies when the thrust function is quadratic in the air speed and of the particular form $T(V) = E + FV^2$, where constants E and F may depend on gross weight and air density and do not necessarily have to be those eponymous bootstrap parameters. This algebraic form restriction is actually not a very strong constraint even for constant-speed propeller aircraft. For instance, it is the same assumption made by von Mises, in his treatment of the takeoff problem, for all aircraft. Diehl,[6] on the other hand, assumed a linearly decreasing thrust function.

Figure 11.6 shows the effects of headwinds and tailwinds in the Cessna 172 sea level maximum gross weight "base" case. When one assumes identical values of distance to rotation with no wind d_{R0}, there is little distinction among the "exact" and the two simplified (no initial wind drag) treatments. And do not take the spread at large tailwinds in Fig. 11.6, seriously; there is little reason to assume (as we did) that the airplane's drag coefficient at negative air speeds corresponds to that at ordinary positive air speeds.

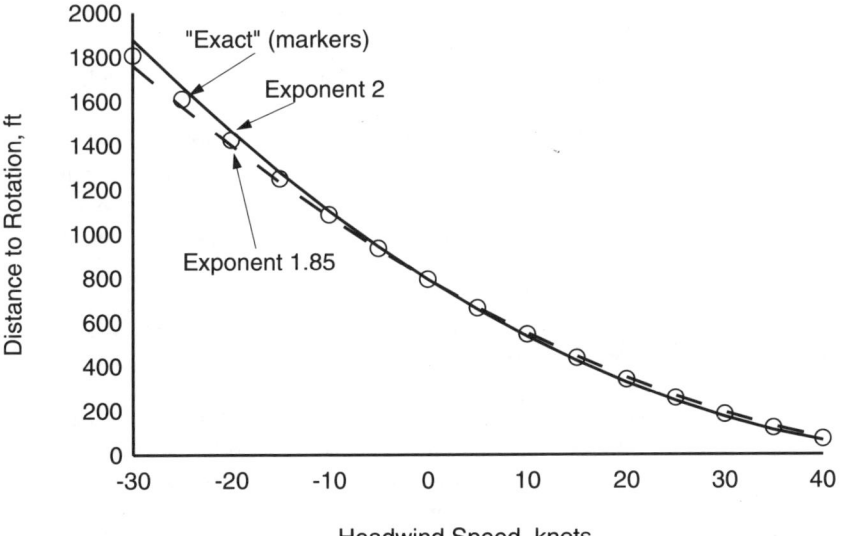

Figure 11.6 Rotation distance vs wind speed, three models.

If one sets $V_w = 0$ in Eq. (11.50) one gets, as makes sense, precisely Eq. (11.37) evaluated for d_{R0}.

Time to Rotation in Wind, t_{Rw}

You may occasionally want to find the length of time it takes the airplane to get to rotation speed V_R with a wind. Start again from Newton's second law for this situation,

$$G(V_g) = m \frac{dV_g}{dt} \tag{11.51}$$

Multiply both sides by $dt/G(V_g)$ and integrate:

$$t_{Rw} = \int_0^{t_{Rw}} 1 \, dt = m \int_0^{V_R - V_w} \frac{dV_g}{G(V_g)} \tag{11.52}$$

Then substitute for G using Eq. (11.48) and evaluate the right-hand integral using I_4. The result is

$$t_{Rw} = \frac{W}{2gab} \ln\left(\frac{(a + bV_R)(a - bV_w)}{(a - bV_R)(a + bV_w)}\right) \tag{11.53}$$

Setting $V_w = 0$ will of course let Eq. (11.53) serve also in the windless case. Notice that, except for a factor $-V_w$, Eq. (11.53) is precisely the rightmost term in Eq. (11.50). Von Mises[2] has several approximations to these "exact" formulas; he uses, for two guiding parameters, initial and final net force values in lieu of our a and b.

The reader conversant with the cases presented will be able, should the need arise, to set up integrals or otherwise find formulas for any of the remaining relations among the four kinematic variables. Now we move to expand the purview of the "constant average force" approximation previously derived.

Kinematics Under Uniform Acceleration

A few pages back we showed how, when the ratio of force at start of the takeoff roll F_0 to force at initiation of rotation F_R is not much different than unity, the exact expression for d_R can be closely approximated by a formula that looks like a holdover from uniform acceleration mechanics. The "constant average" force was taken to be $\frac{1}{2}(F_0 + F_R)$. And moreover, at least for our example, the numbers supported that approximation.

Important Uniform Acceleration Relations

Here we solidify that approximation with brief formulas that will allow the reader to use it under varying circumstances. We take quantities with subscripts 0 to refer to initial times and we explicitly take $t_0 = 0$. Acceleration a is of course a constant for any single problem. The most basic relations are then

$$V(t) = V_0 + at \tag{11.54}$$

and

$$d(t) = d_0 + V_0 t + \tfrac{1}{2} a t^2 \tag{11.55}$$

Because we so often want to relate distance traveled with air speed attained, it is useful to derive, from Eqs. (1.54) and (1.55), two further relations:

$$d(V) = d_0 + \frac{V_0}{a}(V - V_0) + \frac{1}{2}\frac{(V - V_0)^2}{a} \tag{11.56}$$

and

$$V(d) = \sqrt{V_0^2 + 2a(d - d_0)} \tag{11.57}$$

The "70/50" Takeoff Rule, No Wind

The reader has more than likely encountered this rule, which reads "If you do not have at least 70% of your (calibrated) liftoff speed, when half the available runway length is behind you, then abort your takeoff."

This rule follows from the assumption of uniform acceleration during takeoff. There is a hidden assumption of no wind. Under that no-wind condition, here is a derivation of the rule.

Assume the airplane starts from rest ($V_0 = d_0 = 0$) at one end of a runway, of length $L = d_{LO}$ (barely long enough). From Eq. (11.57), $V^2(d) = 2ad$. Taking the situation at liftoff, we know that (barely sufficient) acceleration $a = V_{LO}^2/2L$. Therefore, at midfield, when $d = L/2$, we must have

$$V^2(L/2) = 2\left(\frac{V_{LO}^2}{2L}\right)\left(\frac{L}{2}\right) = \frac{V_{LO}^2}{2} \tag{11.58}$$

So, taking square roots,

$$V(L/2) = \frac{V_{LO}}{\sqrt{2}} \doteq 0.707 V_{LO} \tag{11.59}$$

That is with no wind. And you see, the rule should more accurately be called the "71/50" takeoff rule.

The "70/50" Takeoff Rule With Wind

Here is a way to see that the no-wind rule, Eq. (11.59), cannot work when there *is* wind. Assume your airplane requires $V_{LO} = 60$ KCAS and that you have a 42-kn headwind at MSL. You start your takeoff roll. Having satisfied Eq. (11.59) after rolling 1 ft, your surmise that a 2-ft long runway will be sufficient. Wrong. Here is how a doctored-up derivation, for the correct with-wind rule, goes.

The following complications come from the fact that sufficient takeoff distance refers to the ground, but we fly the airplane according to air speed. Not only that, by *calibrated* air speed, V_C. Assume a headwind of size V_w. The relation giving ground speed is

$$V_g = \frac{V_C}{\sqrt{\sigma}} - V_w \qquad (11.60)$$

For reasons to be revealed, rewrite Eq. (11.60) twice, once for the situation at (barely adequate) liftoff and once for midfield:

$$V_{gLO} = \sqrt{2aL} = \frac{V_{CLO}}{\sqrt{\sigma}} - V_w \qquad (11.61)$$

and

$$V_g(L/2) = \sqrt{2aL/2} = \frac{\sqrt{2aL}}{\sqrt{2}} = \frac{V_C(L/2)}{\sqrt{\sigma}} - V_w \qquad (11.62)$$

Solve both Eqs. (11.61) and (11.62) for $(2aL)^{1/2}$ and set the two longer expressions equal:

$$\frac{V_{CLO}}{\sqrt{\sigma}} - V_w = \frac{\sqrt{2}V_C(L/2)}{\sqrt{\sigma}} - \sqrt{2}V_w \qquad (11.63)$$

Finally, solve for the required minimum calibrated air speed at midfield to find

$$V_C(L/2) = \frac{V_{CLO}}{\sqrt{2}} + \sqrt{\sigma}\left(1 - \frac{1}{\sqrt{2}}\right)V_w = 0.707V_{CLO} + 0.293\sqrt{\sigma}V_w \qquad (11.64)$$

where V_w will be negative for tailwinds. So the more accurate and complete takeoff rule should read "If you do not have at least 71% of your calibrated liftoff speed, plus (or minus) 29% of the 'calibrated' headwind (or tailwind) speed, when half the available runway length is behind you, then abort your takeoff."

If you are going to recite this rule to yourself, do it *before* beginning your takeoff roll. Otherwise you may be past midfield before all of those words can get out.

Practical Perturbation Approach for Distance-to-Rotation Calculations

We have seen how to do minimum takeoff distance calculations in detail. While a few sets of attendant computations may be fun, takeoff conditions are extremely varied even for the same airplane. It is onerous to trek through a hundred slightly different technical combinations.

Especially in view of the fact that such calculations are seldom better than say $\pm 5\%$, we need something quicker even though only approximate. This section and the next provide those more practical approaches.

First, let us review those impinging variables, the influences affecting distances to rotation, liftoff, and 50 ft. The Cessna 172 POH short field takeoff distance performance page lists or implies seven factors: gross weight, relative air density, flaps setting, runway surface type, runway slope, runway contamination, and wind. There are also further seldom-mentioned assumptions in all of those variations:

- The engine is at full throttle and properly leaned for maximum static revolutions per minute. And presumably in like-new condition.
- Tire inflation pressures, we can assume, are correct at 34 psi (nose) and 28 psi (main) or higher. It is extremely instructive to tow your airplane around in a level hangar with a spring scale (100 to 200-lbf calf-weighing variety) with tires at various pressures. You will find that rolling friction coefficient μ changes markedly.
- Though it makes little difference at the low speeds during takeoff, we presume the airframe surface is clean and smooth.

Changes in major variables affect distance to rotation d_R (our main concern) in different ways. Wind is a special case, but we know how to handle it either precisely or approximately. Runway surface type, slope, and contamination—when comparison is made with some standard runway—are felt as additional forces. Added weight lengthens takeoff because there is greater mass to be accelerated, but there are also force effects (greater rolling friction and, on sloped runways, a retarding or accelerating weight component). Flaps setting affects parasite drag, but also rolling friction (through changed lift); in addition, varied flaps allow varied rotation and liftoff speeds. It is a mess. Reality... what a thing! What are we to do?

One perhaps superficially attractive stratagem might be to *linearize* the takeoff problem by computing lots of partial derivatives (of either d_R or $F_{\text{Net}}(V)$ on which it depends) and consider small variations about some standard takeoff configuration and condition. One would then have, taking F as either d_R (in which case integrals would be differentiated using Leibniz's rule) or F_{Net}, and using just two variables (instead of all seven, say only $x = \theta$, runway slope, and $y = W$, gross weight) as examples and stand-ins:

$$F(x, y) \doteq F(x_1, y_1) + \frac{\partial F}{\partial x}(x - x_1) + \frac{\partial F}{\partial y}(y - y_1) \qquad (11.65)$$

where both derivatives are to be evaluated at the "standard" condition (x_1, y_1). But this truncated Taylor series approach seems to be not a good one for at least two reasons. One is that some partial derivatives are messy. We do not want to trade a precise mess for nothing better than an approximate mess. The second, and perhaps more telling, is that some of the underlying parameter variations may not be at all "small." Runway slope (from horizontal) is normally small. But weight variations (from some standard) certainly might not be small; the same can be said of contaminant drag or wind speed.

Derivation of Perturbation Approach

What we propose instead—under the general theory that it is just as well to be hanged for stealing a sheep as for stealing a goat—is to use the "constant average force" approximation, whenever that can be reasonably supported, and consider variations of its few factors. Interspersing a few "exact" calculations helps one build confidence and realism. For review, here is what the "constant average force" approximation looks like when adjusted for possible wind:

$$d_R \doteq \frac{W V_R^2}{2g \langle F_{\text{Net}} \rangle} \left(1 - \frac{V_w}{V_R}\right)^2 \qquad (11.66)$$

$\langle F_{\text{Net}} \rangle$, recall, is just the average of the initial and final net forces on the airframe, $\frac{1}{2}(F(V = 0) + F(V = V_R))$. This approximation is simple, but perhaps not quite as simple as it looks. Weight, for instance, appears there not only explicitly but also buried in V_R and in $\langle F_{\text{Net}} \rangle$. Relative air density σ does not appear explicitly in Eq. (11.66) but it too is buried in V_R (because that is a true air speed) and in $\langle F_{\text{Net}} \rangle$ (as an influence on thrust, drag, and rolling friction). Actually, as we shall see in the next section on the takeoff power parameter, weight and relative air density influence distance to rotation *almost* only through their ratio W/σ. There is a slight hedge, but only very slight. What this means is that one might do well to perform about a dozen exact "base" calculations, using various values of w/σ (where the small-letter version of weight, w, is defined as W/W_0, and W_0 is the airplane's maximum gross weight, 2400 lbf for our sample model of the Cessna 172). Interpolation between flanking values of $d_R(w/\sigma)$ will then take care of those two important variables. If one is interested in the truly *shortest* takeoffs, changing flaps settings is not an option. So we are left with only frank added or subtracted forces, which come from runway surface type, runway slope, or runway contamination.

In its simplest terms, the question before us goes like this: "If the base value of $\langle F_{\text{Net}} \rangle$ is $\langle F_B \rangle$ = 419.8 lbf (for $w/\sigma = 1$) and the corresponding base value of d_R is $d_{RB} = 794.4$ ft (as we found it to be), what will $d_R = d_{R1}$ be if we impose an additional constant incremental force $\Delta F_1 = -50$ lbf (in the retarding direction) on the airframe?[11] We want a simple formula, based on Eq. (11.66), to answer this kind of question straight away.

You can probably see how to do this. Think of the base version of Eq. (11.66) as

$$d_{RB} = \frac{Z}{\langle F_B \rangle} \tag{11.67}$$

which means that we have defined

$$Z \equiv \frac{WV_R^2\left(1 - \frac{V_{hw}}{V_R}\right)^2}{2g} = d_{RB}\langle F_B \rangle \tag{11.68}$$

Now the new perturbed case, in these terms, is

$$d_{R1} = \frac{Z}{\langle F_B \rangle + \Delta F_1} \tag{11.69}$$

Substituting for Z from Eq. (11.68) and rearranging slightly, we then have what we need:

$$d_{R1} = \frac{d_{RB}}{1 + \frac{\Delta F_1}{\langle F_B \rangle}} \tag{11.70}$$

Equation (11.70) shows us how to find the new "perturbed" distance to rotation in terms of the unperturbed base case distance d_{RB}, the base case average net force $\langle F_B \rangle$, and the perturbing force ΔF_1. In a practical aviation setting—preparing for takeoff from some sloped and windy mountain strip—it makes more sense to eliminate $\langle F_B \rangle$ in Eq. (11.70) in favor of Z, using Eqs. (11.67) and (11.68). One then has

$$d_{R1} = \frac{d_{RB}}{1 + \frac{2g d_{RB} \Delta F_1}{WV_R^2}} = \frac{1}{\frac{1}{d_{RB}} + \frac{2g \Delta F_1}{WV_R^2}} \tag{11.71}$$

As makes sense, a negative ΔF_1 gives a smaller denominator, hence a larger distance to rotation d_{R1}. In the sample question we posed above, we find $d_{R1} = 901.8$ ft. This is 107.4 ft longer than d_{RB}. Keep in mind there is *no* approximation, beyond the stretch of the constant average force approximation itself, in Eq. (11.70) or (11.71).

There are alternative forms of Eqs. (11.70) and (11.71) that stress incremental distances $\Delta d_1 \equiv d_1 - d_B$ (we suppress the "rotation" subscript R) instead of ratios d_1/d_B. Those use a useful (refugee from field theory) formula:

$$\frac{1}{1+x} = 1 - \frac{1}{\left(1+\frac{1}{x}\right)} \tag{11.72}$$

Using Eq. (11.72) with obvious substitutions for x turns Eqs. (11.70) and (11.71) into

$$\Delta d_R \equiv d_{R1} - d_{RB} = \frac{-d_{RB}}{1 + \frac{\langle F_B \rangle}{\Delta F_1}} = \frac{-d_{RB}}{1 + \frac{WV_R^2 d_{RB}}{2g\Delta F_1}} \tag{11.73}$$

Putting our sample figures into Eq. (11.73) of course gives us the same 107.4-ft increment as before. Again, there is no further approximation.

But, because $\langle F_B \rangle / \Delta F_1$ is usually much larger than unity, Eq. (11.73) does lead us immediately to an approximation, even quicker and only slightly dirtier:

$$\Delta d_R \equiv d_{R1} - d_{RB} \doteq \frac{-\Delta F_1 d_{RB}}{\langle F_B \rangle} = \frac{-2g\Delta F_1}{WV_R^2} \tag{11.74}$$

This approximation gives $\Delta d_R \doteq 94.6$ ft instead of the "exact" 107.4-ft value given by Eq. (11.73). That might or might not be accurate enough. For relatively smaller incremental forces, of course, the approximation of Eq. (11.74) will be better. Now let us consider some concrete examples of this perturbation approach.

Example 11.3 For a perturbation approach example, let us take an uphill takeoff. Consider our flat dry concrete MSL maximum gross weight base case modified. The pilot needs to take off under the same conditions, on the "same" runway, but now sloping up 2 deg. The added force on the airplane (we ignore the attendant slight diminution of the rolling friction force, only 6 parts in 10,000) is then

$$\Delta F_1 = -W \sin \theta = -2400 \sin 2 \text{ deg} = -83.8 \text{ lbf} \tag{11.75}$$

Because $d_{RB} = 794.4$ ft and $V_R = 94.2$ ft/s, Eq. (11.71) suggests that this uphill takeoff will involve $d_{R1} = 994.4$ ft. The exact value, from modifying a^2 (from 483.9 to 400.1 lbf) and integrating, gives 997.2 ft.

Example 11.4 For another perturbation approach example, consider a takeoff on short grass. Again the base case is modified. This time by substitution of a short grass runway for the original cement. That amounts to substituting $\mu_2 = 0.05$ for the original $\mu_B = 0.02$. Because rolling friction starts large but drops off to near zero by $V = V_R$, our new average force increment is only one-half of the start-of-roll figure:

$$\Delta F_2 = -W\cos\theta(\mu_2 - \mu_B)/2 = -2400(0.05 - 0.02)/2 = -36.0 \text{ lbf} \quad (11.76)$$

This time, Eq. (11.71) suggests $d_{R2} = 869.5$ ft, about 9.5% greater. The Cessna 172 POH suggests we should add, for distance to liftoff on dry grass, a 15% increment. Perhaps their grass is somewhat longer than ours. An exact calculation is a bit more complicated in this case because rolling friction coefficient μ appears not only in a^2 but also in b^2 (because friction force is diminished by lift, which increases with speed). [See Eq. (11.22) for details.] But making that calculation gives an exact distance to rotation on grass of 898.3 ft; this time it is 13% greater than for the base case. Our grass grew a bit taller while we were crunching those last few numbers.

When to Take Off Uphill Into the Wind Rather Than Downhill With the Wind

Derivation of Formula for Break-Even Headwind

Combining the distance-to-rotation perturbation technique, Eq. (11.71), with the approximate wind-effect formula, Eq. (11.46), gives us a way to solve the conundrum implied by this section's header: uphill or down? It turns out that, for a given airplane in a given situation, there is a "break-even headwind" V_{hw}^{BE}. V_{hw}^{BE} is the headwind speed for which it makes no difference, with regard to distance to rotation, whether we take off uphill into the wind or downhill with it. If the actual V_{hw} is greater than V_{hw}^{BE}, we are better off to go uphill; if actual V_{hw} is less than V_{hw}^{BE}, we are better off heading downhill. We take perturbed situation 1 the uphill choice, with angle $\theta > 0$, and situation 2 the downhill choice, with angle $\theta < 0$.

With two perturbed situations at issue, we have to take an extra moment to keep our ideas straight. For instance, there are now two factors Z, with different winds,

$$Z_1 \equiv \frac{WV_R^2\left(1 - \frac{V_{hw}}{V_R}\right)^2}{2g} = Z_0\left(1 - \frac{V_{hw}}{V_R}\right)^2 \quad (11.77)$$

and

$$Z_2 \equiv \frac{WV_R^2\left(1 + \frac{V_{hw}}{V_R}\right)^2}{2g} = Z_0\left(1 + \frac{V_{hw}}{V_R}\right)^2 \quad (11.78)$$

The "break-even" condition on V_{hw}^{BE} is that $d_{R1} = d_{R2}$, where

$$d_{R1} = \frac{Z_1}{\langle F_B \rangle + \Delta F_1} = \frac{Z_0 \left(1 - \frac{V_{hw}^{BE}}{V_R}\right)^2}{\langle F_B \rangle - W \sin \theta} \qquad (11.79)$$

and

$$d_{R2} = \frac{Z_2}{\langle F_B \rangle + \Delta F_2} = \frac{Z_0 \left(1 + \frac{V_{hw}^{BE}}{V_R}\right)^2}{\langle F_B \rangle + W \sin \theta} \qquad (11.80)$$

After some routine algebra, we find that

$$V_{hw}^{BE} = V_R \times \frac{\sqrt{1 + \frac{W \sin \theta}{\langle F_B \rangle}} - \sqrt{1 - \frac{W \sin \theta}{\langle F_B \rangle}}}{\sqrt{1 + \frac{W \sin \theta}{\langle F_B \rangle}} + \sqrt{1 - \frac{W \sin \theta}{\langle F_B \rangle}}} \qquad (11.81)$$

where one will probably want to make use of the fact that

$$\frac{W \sin \theta}{\langle F_B \rangle} = \frac{2g d_{RB} \sin \theta}{V_R^2} \qquad (11.82)$$

The Break-Even Headwind Rule

Therefore, we have the Break-Even Headwind Rule: If the actual wind coming down the sloped runway has greater speed than is given by the right-hand side (RHS) of Eq. (11.81)—or given approximately by the RHS of Eq. (11.83)—takeoff uphill into the wind. Otherwise, take off downhill with the wind.

Approximations to Break-Even Headwind Formula

A pattern as complicated as Eq. (11.81) is hard to interpret. But if one is confronted only with a relatively small slope [or, going back to Eqs. (11.79) and (11.80), a relatively small wind speed], a binomial expansion gives the much simpler result

$$V_{hw}^{BE} \doteq \frac{d_{RB} g \sin \theta}{V_R} \qquad (11.83)$$

An even more practical version, suitable for cockpit pre-take-off checks, comes from converting to wind speed in (ground) knots on the left, to calibrated knots on the right, using the close linearity of the sine function for small angles, and approximating a combination of factors equal to 0.1971 as 1/5:

$$V_{hw}^{BE} \doteq \frac{d_{RB}\theta \deg\left(1 - \dfrac{h_\rho}{70,000}\right)}{5 \text{ KCAS}(V_R)} \text{ kn} \qquad (11.84)$$

Example 11.5 What is the break-even headwind speed for our base case Cessna 172 ($d_{RB} = 794.4$ ft, $V_R = 94.2$ ft/s $= 55.8$ KCAS) on a 2-deg dry concrete runway? The RHS of Eq. (11.82) evaluates to 0.2010. The RHS of Eq. (11.81) is then 9.56 ft/s or 5.67 kn. The cockpit-ready approximate version, Eq. (11.84), gives 5.69 kn. Perhaps that close agreement is only fortuitous. An important fact to keep in mind is that, for a given actual wind and slope, different airplanes may very well need to take off in opposite directions. For more powerful airplanes, break-even headwinds are smaller. One might very well encounter a case in which of three airplanes contemplating takeoff, one should go uphill, another should takeoff downhill, and for the third it might not matter which direction was chosen. Of course terrain clearance and possible wind shear are practical factors that, even though we have ignored them in this calculation, need to be considered.

Is the Optimal Break-Even Headwind Takeoff Good Enough?

An ancillary question may have occurred to the reader: "What does it profit a pilot to take off (in a wind on a slope) in the proper direction but still end up in the trees near the departure end of the strip?" In other words, what guarantee is there that the *best* takeoff direction will be good enough?

It turns out, if you follow our "break-even headwind rule," you can be assured that even your longest takeoffs will only be a little longer than you would have if that same runway were level and with no wind. Only a little longer? To be explained.

There are two parts to the supportive reasoning. The first part is to ensure you that the longest takeoff you will have, for a given airplane on a given slope, is one for which the wind is precisely the break-even headwind V_{hw}^{BE}.

Here is how you can see that fact. The runway has a fixed slope. Assume, for starters, there is no wind. Then of course you take off downhill and your distance to rotation is somewhat less than d_{RB}. Next, let a small tailwind come up. Your takeoff run will still be downhill and will be a little longer. Let the tailwind speed increase. Your takeoff run gets longer and longer up to the break-even wind value, at which point you will get the same length takeoff by changing to an uphill takeoff. Any wind stronger than the break-even value will only reduce your (uphill) takeoff run. So the break-even wind gives you the *longest* distance to liftoff.

But, second, what is the *value* of that longest distance to rotation at the break-even headwind? Now we have to compute. Using the uphill takeoff version, Eq. (11.79), to define distance-to-rotation for the maximal break-even headwind case, we have

$$d_{RBE} = d_{RB} \frac{\left(1 - \frac{V_{hw}^{BE}}{V_R}\right)^2}{\left(1 - \frac{W \sin \theta}{\langle F_B \rangle}\right)} \qquad (11.85)$$

Calling, for brevity,

$$M \equiv \frac{W \sin \theta}{\langle F_B \rangle} \qquad (11.86)$$

and then dividing by d_{RB}, substituting from Eq. (11.80) for V_{hw}^{BE}, and taking square roots, we have

$$\sqrt{\frac{d_{RBE}}{d_{RB}}} = \frac{1 - \left(\frac{\sqrt{1+M} - \sqrt{1-M}}{\sqrt{1+M} + \sqrt{1-M}}\right)}{\sqrt{1-M}} \qquad (11.87)$$

Simplifying and then squaring, one sees that the break-even takeoff will only be a little larger than from the flat calm base case runway under the same circumstances:

$$d_{RBE} = \frac{2d_{RB}}{1 + \sqrt{1 - \frac{W^2 \sin^2 \theta}{\langle F_B \rangle^2}}} \doteq d_{RB}\left(1 + \frac{W^2 \sin^2 \theta}{4\langle F_B \rangle^2}\right) \qquad (11.88)$$

Example 11.6 For our Cessna 172 base case ($W = 2400$ lbf, $\langle F_B \rangle = 419.8$ lbf), Eq. (11.88) gives $d_{RBE} = 1.010 d_{RB}$ for a 2-deg slope. Even for a (seldom encountered) 5-deg slope, the longest distance to rotation is only about 7% longer than for the flat calm base case, $d_{RBE} = 1.071 d_{RB}$.

A "Paperless Office" Version of the Previous Calculation

For small slopes, which means small break-even headwinds V_{hw}^{BE}, a binomial expansion in rewrites of Eqs. (11.79) and (11.80) is in order. Under that approximation, it turns out that $d_{RBE} \doteq d_{RB}$ can be demonstrated without calculation at all and even without much manipulation. Those rewritten relations are

$$\frac{d_{R1}}{d_{RB}} \doteq \frac{\left(1 - \frac{2V_{hw}^{BE}}{V_R}\right)}{\left(1 - \frac{W \sin \theta}{\langle F_B \rangle}\right)} \tag{11.89}$$

and

$$\frac{d_{R2}}{d_{RB}} \doteq \frac{\left(1 + \frac{2V_{hw}^{BE}}{V_R}\right)}{\left(1 + \frac{W \sin \theta}{\langle F_B \rangle}\right)} \tag{11.90}$$

For brevity letting $A = 2V_{hw}^{BE}/V_R$ and $B = W \sin \theta / \langle F_B \rangle$, the break-even headwind distance-to-rotation ratio is now characterized by

$$\frac{1-A}{1-B} = \frac{1+A}{1+B} \tag{11.91}$$

Doing the division a little differently (this is the "not much" manipulation):

$$\frac{1-A}{1+A} = \frac{1-B}{1+B}. \tag{11.92}$$

Because the function on either side of Eq. (11.92) has a single-valued inverse (itself, strangely enough!)—no horizontal line hits its graph in more than one place—this can only mean that $A = B$. But that means that either side of Eq. (11.91) is just unity. But, in turn, that means the break-even headwind gives us back d_{RB}, the distance to rotation with no slope and no wind to this approximation.

By now we have dealt completely and accurately enough with the vagaries of various runways. Now it is time to turn our attention to two of the most important factors influencing a given airplane's distance to rotation d_R and thereby its distance to liftoff d_{LO}: gross weight and density altitude.

Takeoff Power Parameter

This approximation concept piggybacks on the "constant average force" approach. The TOPP, a number attached to each airplane (with an implicit understanding that some chosen standard base case runway and atmosphere is lurking in the background), allows easy adjustments to d_R (thereby to d_{LO}) to account for nonstandard weight or density altitude. The airplane's TOPP can be

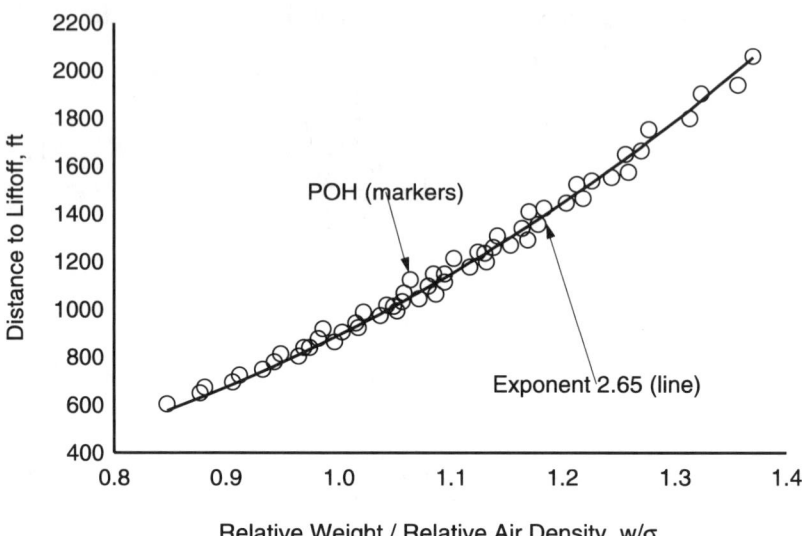

Figure 11.7 Distance to liftoff vs w/σ for our sample Cessna 172.

determined, as will be shown below, either from theory or experiment. Once the airplane's TOPP number is known, the required adjustment is this simple:

$$d_R(w, \sigma) \doteq d_{R0} \times \frac{\text{TOPP}\left(\frac{w}{\sigma}\right)^2}{1 + (\text{TOPP} - 1)\left(\frac{W}{\sigma}\right)} \quad (11.93)$$

where $w \equiv W/W_0$, W_0 any standard weight but normally maximum gross weight. An interesting aspect of Eq. (11.93) is that d_R there depends on weight and air density only through the combination w/σ. There is a very slight fudge in that assertion (to be cooked below), but it provides considerable simplification in calculating and tabulating takeoff performance (see Fig. 11.7, where in curve-fit fact $[w/\sigma]^{2.65}$ does a good job for this particular airplane). We went to relative weight $w \equiv W/W_0$ to make the numbers more indicative.

Relation Between TOPP and Exponent α in $(w/\sigma)^\alpha$

Our starting point is that

$$\frac{d_R}{d_0} \doteq \frac{Tz^2}{1 - (1-T)z} \doteq z^\alpha \quad (11.94)$$

where $T \equiv \text{TOPP}$ and $z \equiv (w/\sigma)$. First, get rid of the z^2, invert, and define β as $2 - \alpha$:

$$\frac{1 - (1 - T)z}{T} \doteq z^{\beta} \qquad (11.95)$$

Now (crux of the pitch), because z varies around unity, substitute $1 + x$ for z. Then do a binomial expansion, simplify, and subtract 1 from each side:

$$\frac{(T - 1)x}{T} \doteq \beta x \qquad (11.96)$$

Divide by x [which is only zero when $z = 1$, in which case Eq. (11.94) holds for other reasons], then recall $\alpha = 2 - \beta$, along with $T = \text{TOPP}$, to find

$$\alpha \doteq 1 + \frac{1}{\text{TOPP}} \qquad (11.97)$$

This approximate relation suggests that, if exponent $\alpha = 2.65$, then $\text{TOPP} \doteq 0.606$. We shall see.

Proper Order of Approximations

Before getting into the algebraic derivations of $d_R(\text{TOPP})$ and the somewhat complicated expression for TOPP itself (in terms of more fundamental quantities), we step back to make sense of the entire takeoff calculation picture. As usual, we ignore the possibility of runway contamination drag, or at least ignore cases with hydroplaning.

We are in good shape in having, at least for fixed-pitch propeller airplanes, a relatively exact theory of the forces involved in takeoff and in having closed-form indefinite integrals for calculating d_R from those forces. We have a similar though less accurate process for dealing with constant-speed propeller aircraft; there takeoff thrust must be calculated numerically (using the GAGPC) and succeeding integration to d_R must also be performed numerically (with trapezoidal or Simpson's rules).

Doable, but messy. With many different takeoff scenarios to deal with, adequate approximation methods are much to be desired. Our major technique involves the "constant average force" assumption, in which forces need be evaluated only at the beginning and end of the run. But end of the run means $V = V_R$, which depends on weight (because lift is involved) and density altitude (because we need the true air speed). There is no "one formula does all" solution to the problem.

Variations from the base case (MSL, $W = W_0$, flat concrete runway, no wind) are due to various values of relative air density σ, relative gross weight w, runway slope θ, runway surface type μ, and wind speed up or down the runway, V_{hw}. In pursuing this more global situation, in which several variables have nonbase values, it is important to impose approximations in some correct order. For instance one could not correct for wind *before* correcting for weight because weight figures, through rotation speed V_R, into the wind correction. While there may certainly be more than one satisfactory order in which approximations are imposed, "just any old order" will not do. Our suggestion, assuming all five corrections are needed, is as follows:

1) Use the TOPP method to compute $d_{R1}(w/\sigma)$ with Eq. (11.93). But at this early stage one will also, for later purposes, need to find the modified average force $\langle F_1(w/\sigma) \rangle$ and modified rotation speed $V_{R1}(w/\sigma)$.
2) Use the perturbation approach to find a ΔF_2, which depends on slope θ and surface type μ. ΔF_2 will involve modified weight $W = wW_0$ and relative density σ. Then

$$d_{R2} = \frac{d_{R1}(w/\sigma)}{1 + \frac{\Delta F_2}{\langle F_1 \rangle}} \quad (11.98)$$

3) Finally, correct for headwind ($V_{hw} > 0$) or tailwind ($V_{hw} < 0$), using

$$d_R = d_{R3} = d_{R2}\left(1 - \frac{V_{hw}}{V_{R1}}\right)^2 \quad (11.99)$$

Or some slightly smaller exponent. That is all there is to it. Now back to TOPP.

Derivation of Expression for d_R (TOPP)

The job is to make explicit the weight and relative air density dependence of each factor and term in the constant-average force expression for distance to rotation:

$$d_R \doteq \frac{WV_R^2}{2g\langle F_{\text{Net}} \rangle} \quad (11.100)$$

where $\langle F_{\text{Net}} \rangle$ is given by Eq. (11.22) and angle brackets mean the average of initial and final values. If we think of Eq. (11.22) as

$$F_{\text{Net}} = A + BV^2 \quad (11.101)$$

then

$$\langle F_{\text{Net}} \rangle = \tfrac{1}{2}(\langle F_{\text{Net}}(V=0) \rangle + \langle F_{\text{Net}}(V=V_R) \rangle) = A + \tfrac{1}{2}BV_R^2 \quad (11.102)$$

For ease of handling, we next dissect each of the "constants" in the average net base-case force expression as

$$A = E - \mu W = A_1 + A_2 \quad (11.103)$$

and

$$B = \left(F - G - \frac{\rho S f_{GE} C_{LGR}^2}{2\pi e A} + \tfrac{1}{2}\mu\rho S C_{LGR}'\right) = B_1 + B_2 + B_3 + B_4 \quad (11.104)$$

E, F, and G are the usual bootstrap composite parameters. We run into the problem we alluded to earlier as soon as we begin to tease apart

$$E = \left(\frac{\sigma - C}{1 - C}\right)\frac{mP_0}{n_0 d} \quad (11.105)$$

The power dropoff factor cannot be expressed as σ times anything independent of σ. We do so anyway, treating its numerator as

$$\sigma - C = \sigma\left(1 - \frac{C}{\sigma}\right) \doteq \sigma\left(1 - \frac{C}{\langle\sigma\rangle}\right) \quad (11.106)$$

where $\langle\sigma\rangle$ is suitable "average" relative air density for takeoffs. For someone usually taking off at density altitudes between 2000 and 4000 ft, for instance, a $\langle\sigma\rangle$ of 0.9151, corresponding to $h_\rho = 3000$ ft, might be appropriate. Even choosing a value for $\langle\sigma\rangle$, which is 5000 ft too low, in density altitude, only causes a 2% error in distance to rotation. These d_R estimates are not extremely accurate; that is why we adopt personal safety factors. In this relatively high desert climate, with large temperature variations between summer and winter, it might make sense to calculate two (slightly different) seasonal values of TOPP. But, again, the inaccuracy is only slight.

Example 11.7 We combine our derivation with a numerical example using those ubiquitous Cessna 172 parameters. We shall take $\langle\sigma\rangle$ as unity, tantamount to taking the power dropoff factor to be σ itself. So then we have

$$A_1 = E = \sigma\frac{mP_0}{n_0 d} = \sigma\frac{1.70 \times (160 \times 550)}{45 \times 6.25} \quad (11.107)$$

where we have substituted in our standard Cessna 172 numbers.
Next, we have an easy case,

$$A_2 = -\mu W = -w\mu W_0 = -w \times (0.02 \times 2400) \quad (11.108)$$

Now we start to work on the B terms.

$$B_1 = F = \sigma \rho_0 d^2 b = \sigma \times (\rho_0 \times 6.25^2 \times (-0.0564)) \tag{11.109}$$

$$B_2 = -G = -\tfrac{1}{2}\sigma\rho_0 S C_{D0} = \sigma \times (-\tfrac{1}{2}\rho_0 \times 174 \times 0.039) \tag{11.110}$$

$$B_3 = \frac{-\sigma\rho_0 S f_{GE} C_{LGE}^2}{2\pi e A} = \sigma\left(\frac{-\rho_0 \times 174 \times 0.64 \times 0.75^2}{2\pi \times 0.72 \times 7.378}\right) \tag{11.111}$$

$$B_4 = \tfrac{1}{2}\sigma\rho_0 \mu S C_{LGR}^l = \sigma \times (\tfrac{1}{2}\rho_0 \times 0.02 \times 174 \times 0.808) \tag{11.112}$$

The last separate entity we have to deal with is

$$V_R^2 = (1+x)^2 V_S^2 = \left(\frac{w}{\sigma}\right)\frac{(1+x)^2 2 W_0}{\rho_0 S C_{L\max}} = \left(\frac{w}{\sigma}\right) \times \left(\frac{1.15^2 \times 2 \times 2400}{\rho_0 \times 174 \times 1.73}\right) \tag{11.113}$$

End of numerical example; time to assemble results. Note that all four B terms have the same σ dependence. A is more complicated; it amounts to

$$A = \sigma A_1' + w A_2' \tag{11.114}$$

where the primes indicate that no w or σ dependence remains. If one rewrites Eq. (11.100) for d_R using this primed-constants pattern, calls $(w/\sigma) = z$, multiplies top and bottom by σ, and divides through to get a term unity in the denominator, one will get

$$d_R = \frac{\alpha z^2}{1 + \beta z} \tag{11.115}$$

where

$$\alpha = \frac{w_0 V_R^{2'}}{2g A_1'} \tag{11.116}$$

and

$$\beta = \frac{(2A_2' + B' V_R^{2'})}{2A_1'} \tag{11.117}$$

At this point, Eq. (11.115), there are still two composite constants. We promised there would be only one. The second can be eliminated by seeing that the $(w/\sigma) = 1$ case, $d_{RB} = d_{R0}$, is given by

$$d_{R0} = \frac{\alpha}{1+\beta} \tag{11.118}$$

so α can be eliminated as $d_0(1+\beta)$, making

$$d_R = d_{R0}\left(\frac{z^2(1+\beta)}{1+\beta z}\right) \tag{11.119}$$

Finally, defining

$$\text{TOPP} \equiv 1 + \beta = 1 + \frac{(2A'_2 + B'V_R^{2'})}{2A'_1} \quad (11.120)$$

and using the implicit definitions given earlier, we have, explicitly,

$$\text{TOPP} = 1 + \frac{n_0 d}{mP_0}\left[-\mu W_0 + \left(\frac{d^2 b}{S} - \frac{C_{D0}}{2} - \frac{f_{GE}C_{LGR}^2}{2\pi eA} + \frac{\mu C'_{LGR}}{2}\right)\left(\frac{(1+x)^2 W_0}{C_{L\max}}\right)\right] \quad (11.121)$$

Using Eq. (11.119), we get our previous Eq. (11.93):

$$d_R(w, \sigma) \doteq d_{R0} \times \frac{\text{TOPP}\left(\frac{w}{\sigma}\right)^2}{1 + (\text{TOPP} - 1)\left(\frac{w}{\sigma}\right)} \quad (11.122)$$

Arithmetic with Eq. (11.107) *et seq.* will convince the reader that, for our constant companion Cessna 172, TOPP $\doteq 0.7895$. (From the backengineered power-law relation, we got TOPP $\doteq 0.606$.) Base distance to rotation d_{R0}, recall, is about 794.4 ft from an exact calculation, 788.4 from the constant average force approximation. For airplanes with more powerful engines (which quite a few Cessna 172s do have now), TOPP is larger. That is why it is called the takeoff *power* parameter.

Example 11.8 Say our sample Cessna takes off on its standard flat paved calm runway, but with weight 2250 lbf at density altitude 5000 ft ($\sigma = 0.8617$). Then $(w/\sigma) = 1.088$. Using TOPP $= 0.7895$ in Eq. (11.122), $d_R = 1.212 d_{R0} = 963$ ft. To get distance to liftoff, still assuming a 1-s rotation, we must recognize that V_R is somewhat higher, by factor $(w/\sigma)^{1/2} = 1.043$, or $V_R = (94.2)(1.043) = 98.3$ ft/s. So $d_{LO} = 963 + 98 = 1061$ ft. The nearest case from the Cessna POH suggests d_{LO} about 1110 ft, about 5% higher.

It is very useful, once you have obtained the TOPP value for your airplane, to construct a cockpit-friendly graph like Fig. 11.8 from following the admittedly tortuous path to an airplane's takeoff power parameter. Two caveats, however: 1) Fig. 11.8 is for a flat dry paved calm runway; further adjustments may need to be made; and 2) no safety factor is attached; after all adjustments are in, we suggest adding 15%.

Backengineering TOPP From Experimental or POH Takeoff Values

One can also estimate TOPP from two good values of d_R for well-separated ratios of relative weight to relative air density. Let us take an example using POH data.

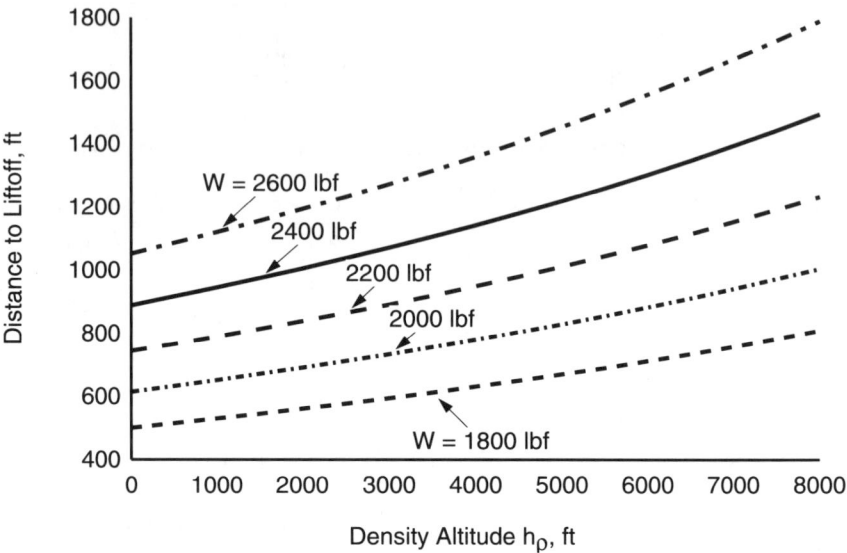

Figure 11.8 TOPP takeoff graph for a Cessna 172, various weights, on a flat dry paved calm runway.

Consider our Cessna 172 in two extreme cases: 1) fairly low weight (2000 lbf) and low density altitude (MSL), with $d_{LO} = 585$ ft; and 2) high weight (2400 lbf) and fairly high density altitude (4000 ft, $\sigma = 0.8881$), with $d_{LO} = 1232$ ft. Because our calculation is for d_R and not for d_{LO}, we must subtract reasonable values of d_2. To do that, we must know the rotation speed during the (say) 1-s of rotation. The general formula is

$$V_R(W, \sigma) = (1+x)V_S(W, \sigma) = (1+x)\sqrt{\frac{2W}{\rho_0 \sigma S C_{L\max}}} \qquad (11.123)$$

Equation (11.123) says the low weight low altitude case has $V_R = 86$ ft/s, hence $d_2 = 86$ ft, and the high weight high altitude case has $V_R = 100$ ft/s, hence $d_2 = 100$ ft. That makes the low low $d_{R1} = 499$ ft, the high high $d_{R2} = 1132$ ft. Store this information for later use.

With data at hand, we proceed to the necessary algebra. Using abbreviations

$$z \equiv \frac{w}{\sigma} \qquad (11.124)$$

and

$$r \equiv \frac{d_{R1} z_2^2}{d_{R2} z_1^2} \qquad (11.125)$$

Table 11.6 Inferred TOPP values for several light aircraft

Aircraft type	TOPP from POH
Piper PA-28 Cadet	0.49
Cessna 172	0.65
Cessna 152	0.65
Cessna 177RG	0.72
Cessna 182RG	0.75
Mooney M20K	0.96
Cessna 414	1.02
Piper Twin Comanche B	1.17
Piper PA-34 Seneca II	1.20
Cessna 208B	1.22

manipulating Eq. (11.122) gives the result that TOPP is (for $\langle \sigma \rangle = 1$):

$$\text{TOPP} = \frac{(1-z_2) - (1-z_1)r}{(z_1 r - z_2)} \qquad (11.126)$$

In this POH case, $z_1 = 0.833$ and $z_2 = 1.126$. Substituting into Eqs. (11.125) and then (11.126), we find $r = 0.8055$ and TOPP $= 0.5726$, not very close at all to our earlier bootstrap-computed value 0.7895 (but close to the backengineered power-law value 0.606). Taking a representative set of a dozen takeoff pairs from the Cessna 172 POH gave an average TOPP of about 0.65; still not close. The problem, the bad news, is that TOPP is a relatively rapidly decreasing function of d_R. The good news, on the other hand, is that d_R is relatively slowly decreasing function of TOPP. Table 11.6 gives TOPP values from representative POH takeoff data for a selection of single engine and light twin aircraft. Always check TOPP predictions against the airplane's POH, or experimental data, before using them for the first few times.

While for purposes of rough comparison it is interesting to compute TOPP values for a variety of aircraft types from their POHs (as in Table 11.6), one should not take those individual values very seriously. Much better to take the actual airplane, in two fairly widely differing density altitude and loading situations, and *measure* distances to rotation. Three takeoffs with closely clustered results in each of the two situations should be sufficient.

Conclusions

Takeoff may be optional, but its theory is difficult. Those additional factors (rolling friction, ground effect, relative motion with respect to the Earth as well as the air, attitude changes, etc.) make predicting takeoff performance more problematic than predicting steady flight performance. Still, we have found it perfectly possible to come up with good estimates.

A next-to-last arrow in your prediction quiver should be that of relative, versus absolute, predictions. Say you have a high-performance airplane, or a taildragger—cases we skirted—and, no matter how carefully you assemble the various effects, your estimate of d_{LO} (for some particular circumstance) comes out say 20% higher than your carefully measured takeoff value. (Note: Suspect your engine.) With that discrepancy temporarily banished to some mental backburner, use that 20% overestimate to "adjust" similar predictions made for other circumstances of density altitude, weight, runway surface type, and so forth. An estimate 20% too high means actual distance was about 17% lower than the estimate. Keep thinking, but do not stop flying! Divide your attention, adjust, and go on. You will not be far off. The best and easiest way to learn to make good aircraft performance predictions is to make them; then compare with experiment.

Finally, a general cautionary note on takeoff calculations. Nowhere along the line did we append safety factors. In practice, we all need personal safety factors. For distance to liftoff, this author adds a safety pad of 15% to his calculations. The FAA mandates that one should use only one-half of the available headwind speed but should use one and one-half times the tailwind speed. One must be safety conscious but also realistic. Attaching an adequately safe, but parsimonious, deck to one's house, for instance, actually only requires about one-third the supporting timber demanded by an official architects' handbook prepared by the lumber industry. The trouble with compounding safety factors, as is done by some government agencies and professional institutes, is that unbelievably conservative rules result. So conservative that they are largely ignored. It is better to create your own personal minima, from experience and calculation, and stick with them.

References

1. Thompson, W. D., "Selection of Flap Settings and Airspeeds for Takeoffs and Landings," Cessna Aircraft Company Rep. No. F-182P-2, 1973.
2. Von Mises, R., *Theory of Flight*, Dover, New York, 1959, p. 353.
3. McCormick, B. W., *Aerodynamics of V/STOL Flight*. Academic Press, New York, 1967.
4. Crawford, D. R., *A Practical Guide to Airplane Performance and Design*, Crawford Aviation, 1981, p. 136.
5. McCormick, B. W., *Aerodynamics, Aeronautics, and Flight Mechanics*, Wiley, New York, 1979, p.98.
6. Diehl, W. S., "The Calculation of Take-Off Run," NACA TR 450, 1932.
7. Abbott, I. H., and von Doenhoff, A. E., *Theory of Wing Sections*, Dover, New York, 1959.
8. NASA TN D-1376.
9. NASA TR R-64.
10. NASA TN D-2056.
11. FAA Rep. 308-3X.
12. FAA, "Flight Test Guide for Certification of Part 23 Airplanes," Advisory Circular AC-23-8A, 1989.

12

Landing Performance

Introduction

"But landing is mandatory." And, if not mandatory, it is at least advisable that we begin by circumscribing the complex landing maneuver, cutting it back to manageable proportions. We take the view that "letting down," getting into the traffic pattern, and all in-air approach portions down to 50 ft AGL, lined up with the runway, merely involve combinations of flight maneuvers we have previously treated. Pilots operating handbooks (POHs) break landing distance d_L into two portions: 1) d_{LA}, in the air, from 50 ft AGL to the runway, and 2) d_{LG}, the ground roll to a full stop. Certification regulations[1] define "landing distance" as the sum of those two portions. We shall not only treat both parts but will need to break them up into somewhat finer pieces (see Fig. 12.1):

- d_{LA1}, from 50 ft to just off the runway, say 2 ft AGL. Initial speed will be some specified V_{50}; flaps will (usually) be well extended.
- d_{LA2}, from just off the runway to touchdown. Speed will bleed off due to aerodynamic drag down to V_{TD} (TD for touchdown); flaps stay extended.
- d_{LG1}, from touchdown through wheel spin-up. We take this as a 1-s interval—the reader is free to vary that—stabilizing the airplane on the runway. Speed decreases first from V_{TD} to some slightly lower value told us from impulsive calculation of wheel spin-up and then from aerodynamic braking, flaps still extended and rolling friction ignored because the airplane barely has weight on the runway. Angle of attack (AOA) still quite large.
- d_{LG2}, rotation down to all wheels on the runway. We take this as a 2-s interval—again user-selectable—during which flaps fully retract. Drag and rolling friction are operating. Speed continues to bleed off. No brakes yet. Angle of attack ends at the same ground-roll figure as for takeoff.

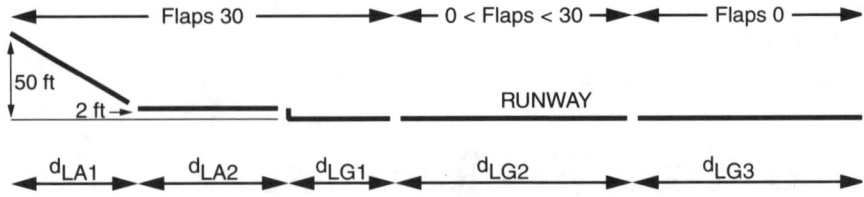

Figure 12.1 We consider two in-air and three on-runway portions of the overall landing maneuver.

- d_{LG3}, from all wheels on the surface to a full stop. Heavy braking on the main wheels, rolling friction on the nose wheel. Air speed decreases to ambient headwind V_{hw} and ground speed decreases to zero.

We shall find that the first in-air portion of the landing maneuver is quite idiosyncratic (pilot technique!) and resists realistic and reproducible analysis. We will "backengineer" it from the desired result—a time-honored but totally dishonest procedure—so as not to leave the reader in the dark as to how to get around that manifest difficulty.

Even the landing ground roll is not all that straightforward. To achieve maximal landing performance—shortest possible landing roll under the circumstances—the pilot lands near stall speed (touchdown speed $V_{TD} = 1.1 V_S$ is often specified; see Roskam and Lam[2]), then executes several fairly intricate flaps and brakes timing and control decisions. We will approximate and sample those myriad variations of pilot skill and experience.

Although effects of aircraft weight, density altitude, and runway slope are easily understood, we shall find that μ_B—that catch-all phenomenological braking friction coefficient—is extremely problematical. Braking friction coefficient values μ_B vary with runway surface type or contamination, tire inflation and condition, and ground speed. Still, we do the best we can with what we have. The study of aircraft performance, like many applied sciences, is a fiddly business. Landing performance is the fussiest part.

Our theory of maximal landing performance will mirror that for takeoff in having an (only relatively) "exact" version—now to be dubbed "detailed"—with which we shall begin, the possibility of perturbations, and (at least for the last of the five segments) a constant average force version. With thrust absent, equations are somewhat simpler. Almost all of the necessary analytic techniques were introduced in the previous chapter, so our treatment of landing will be somewhat quicker.

Landing involves a complication not shared with takeoff. Taking our standard tricycle gear exemplar, wheels are of two different types—main wheels with brakes, a nose wheel without—so we need to consider not only *forces* on the airplane during the landing roll but also *torques*. Torques directly affect the weight burden on each wheel and thereby the rolling or braking friction forces on it. We will cross that calculation when we come to it.

Because of the myriad complications and confusing side issues in what follows, we will give summaries of what we will have finally calculated, with a comparison to POH "book" values for the base case at the end of each landing segment section. Now on to the quite lengthy "detailed" calculation of landing distance d_L.

First Segment, d_{LA1}, from 50 ft to Just Above the Runway

The small airplane's POH landing performance sheet includes information on the horizontal distance used up in coming to Earth from over a 50-ft obstacle, our $d_{LA1} + d_{LA2}$. For our example Cessna 172, at maximum gross weight (2400 lbf) under standard mean sea level (MSL) conditions, at speed $V_{50} = 62$ KCAS = 62 KTAS = 104.6 ft/s, the POH suggests $d_{LA} = 740$ horizontal ft are used up. Power is off and flaps are deployed 30 deg ($C_{D0} = 0.055$).

We shall find that, if one assumes this first air portion d_{LA1} *stays* at 62 KTAS throughout, there is no way the airplane will be able to touch down (at nominal $V_{TD} = 1.1 V_S = 85.3$ ft/s) in so short a distance. We cannot know that now, so let us run through our first calculation, for drill, as though the (POH) book value were veridical. Later, we will redo and adjust.

We could focus on either rate of descent or glide path angle; we choose the former. The usual bootstrap formula for the (negative) gliding rate of climb is

$$h'_{gl}(V) = \frac{-GV^3 - H/V}{W} \quad (12.1)$$

using composite $G = \frac{1}{2}\rho_0 S C_{D0} = \frac{1}{2}\rho_0(174)(0.055) = 0.011374$ and composite $H = 2W^2/(\rho_0 S \pi e A) = 2(2400)^2/(\rho_0 \times 174 \times \pi \times 0.72 \times 7.378) = 1,668,987$ (in British engineering units) tells us that rate of descent $h' = -12.07$ ft/s. That calculates to 4.14 s to get down 50 ft, during which the airplane has traveled 433 ft along the glide path and (Pythagoras) 430 ft horizontally. The POH, recall, suggests that 740 ft should have been used up before touchdown. That leaves us only 310 ft yet to be consumed during the horizontal flight just off the runway. We charge on. But be prepared to come back and redo the above calculation with a lower initial air speed.

Summary for d_{LA1}, horizontal distance to get from 50 ft to just above runway: immediately after crossing over the 50-ft obstacle, the pilot pulls back from 62 to 57.3 KTAS. To get down to the runway the airplane will traverse 419 ft horizontally. Elapsed time is 4.35 s. This result is finagled in the next section. The Cessna 172 POH does not separate out this distance.

Second Segment, d_{LA2}, from Just Above the Runway to Touchdown

First we find the touchdown speed. For this airplane, $C_{L\max}(\delta_f = 30 \text{ deg}) = 1.93$. This means the stall speed is 77.5 ft/s = 45.9 KCAS. That makes $V_{TD} = 1.1 V_S = 50.5 \text{ KCAS} = 85.3 \text{ ft/s}$.

Second, a calculation using rough averages. Neglecting both any round-out and any ground effect, say we start 2 ft off the tarmac still at 104.6 ft/s. If we bleed speed off to 85.3 ft/s, our average speed, during this flare to land (at sea level), will have been $\langle V \rangle = 95.0$ ft/s. Next we find the associated average horizontal force (drag only), from the standard formula

$$\langle D \rangle = -G \langle V \rangle^2 - H/\langle V \rangle^2 \tag{12.2}$$

This gives $\langle D \rangle = 287.6$ lbf. Rearranging Newton's second law, we have

$$\Delta t = \frac{W \Delta V}{g \langle D \rangle} \tag{12.3}$$

Equation (12.3) evaluates to 5.01 s. During that time interval, the airplane should progress about $d_{LA2} \doteq 476$ ft along and just off the runway. *Considerably* larger than the only 310 ft we have left if our calculation is to agree with the POH. Further adding to our (temporary) embarrassment is the fact that, having neglected ground effect, we overestimated the induced drag component, thereby underestimating the time interval. The corrected figure would be 570 ft.

Does this large discrepancy mean the world of aviation has stopped making sense? No. More likely we have not been given the full story. Once the performance test pilot is over the 50-ft obstacle, he or she has every incentive to slow down. After all, speed is still 23% above that needed for touchdown. Pulling back on the stick both slows the airplane and steepens the glide angle. The former will eventually shorten the ground roll; the latter will immediately shrink the glide, shorten the horizontal air distance. Double duty, and in the commercially correct direction.

Let us not be hasty. If our guess about the performance test pilot's verbal chicanery—while not lying, certainly leaving out information—is correct, we should be able to *show* that there is some (accessible) lower speed, taken just over the obstacle, which will reproduce the POH air-distance figure. Let us find it. We will ignore the transitional maneuver, backstick, from 62 KTAS = 104.6 ft/s, at the obstacle, to some (variable) lower speed we will call V_1. And we will ignore the transitional round-out and flare to the final horizontal flight that will continue from V_1 until $V_{TD} = 85.3$ ft/s.

The geometry from 50 ft down to just above the runway—we pick 2 ft AGL— is used as before. Equation (12.1), trigonometry, and straight-line kinematics will take care of that first portion, $d_{LA1}(V_1)$. This time we shall calculate the second portion, d_{LA2}, more carefully. We need somewhat the same technique we used for calculating $x(V)$ in the Takeoff Performance chapter, but now the force expression is much different:

$$d_{LA2} = \frac{m}{2} \int_{V_1^2}^{V_{TD}^2} \frac{dV'^2}{-GV'^2 - f_{GE}H/V'^2} \tag{12.4}$$

Call the speed squared z, then multiply through on the right by z/z, and one has something of the form of integral I_{10} in the Short List of Integrals (Appendix C). After simplifying and evaluating that integral for this particular problem, one has

$$d_{LA2} = \frac{-W}{4gG} \ln\left(\frac{1 + \dfrac{GV_{TD}^4}{f_{GE}H}}{1 + \dfrac{GV_1^4}{f_{GE}H}}\right) \tag{12.5}$$

The only number missing is the ground effect factor f_{GE}. Peeking back at Chapter 4, Aerodynamic Force, and tacking 2 ft onto the approximate 6 ft the ground-rolling wing is normally off the ground, $f_{GE} = 0.74$. It is important to incorporate ground effect because, at these very low speeds, induced drag dominates parasite drag.

Taking a selection of initial just-over-the-obstacle air speeds V_1, Table 12.1 shows some results of our trial-and-error analysis. This is for the Cessna 172 at MSL, 2400 lbf, flaps 30, coming over a 50-ft obstacle, going immediately to speed V_1, gliding until 2 ft off the runway, then immediately flaring into a level glide, decelerating, until touchdown speed $V_{TD} = 50.5$ KTAS, at which time the ground roll begins. The third, bolded, data line—yielding $d_{LA} = 740$ ft—is the one we are after.

Notice that, for $V_1 = 57.3$ KTAS, the glide path is only steepened, from its 62-KTAS value, 0.2 deg. There is no great mush downwards. On the other hand, only 11 horizontal ft are subtracted in this descent portion of the maneuver. The real gain comes from having started the horizontal portion with lower air speed. Then the bleedoff takes less time, uses up only 321 ft (instead of 570) and puts us on

Table 12.1 Alternative landing path histories for a Cessna 172

V_1, KTAS	V_1, ft/s	$h\dot{}$, ft/s	γ, deg	d_{LA1}, ft	d_{LA2}, ft	$d_{LA} = d_{LA1} + d_{LA2}$
62	104.6	−12.1	−6.6	430	570	1000
60	101.3	−11.8	−6.7	427	461	888
57.3	**96.8**	**−11.5**	**−6.8**	**419**	**321**	**740**
56	94.5	−11.4	−6.9	413	254	667
54	91.1	−11.2	−7.1	403	157	560
52	87.8	−11.1	−7.3	391	65	456
50.5	85.3	−11.1	−7.5	381	0	381

schedule with the POH. So, if we allow the performance test pilot to fudge air speeds a bit, the POH 50-ft to ground distance for this situation is perfectly feasible.

Neglect of a Balloon and Flare

Or is it? Do not forget our own fudges, neglect of the two transitions: 1) just after the obstacle, from 62 to 57.3 KTAS, and 2) near the ground, from a 6.8 deg glide path to horizontal flight. Let us now take a less sanguine look at those two problematical portions of the flight path.

For the pilot to reduce speed 4.7 KTAS, he (or she) pulls back. That will mean a momentary ballooning. Not absolute, upwards, but above the earlier glide path. To compensate, he or she will need to reduce air speed a little more than otherwise, say to around 55 KTAS. Settled at the new lower speed, the glide path will be steeper, crossing and then sinking below the earlier one and even below that path connected to our Table 12.1 solution, 57.3 KTAS. This is not much of a problem. The airplane is stable (we assume no gusts) and still about 11.4 KTAS above the stall.

The problem, if any, comes when it is time to flare. There is some reason to believe—though we will not attempt to calculate it because control calculations are beyond this book's purview—that at 55 to 58 KTAS this airplane may not have enough elevator authority to be able to flare to horizontal. How much flare is needed? Well, taking θ_a to be the airplane's *attitude angle*,

$$\theta_a = \gamma + \alpha \tag{12.6}$$

(here $\gamma < 0$ and $\alpha > 0$), we can simply calculate. To get AOAs α, we need two things. One is the appropriate value of lift coefficient, which we get from

$$L = \tfrac{1}{2}\rho V^2 S C_L(\alpha) = W \cos \gamma \tag{12.7}$$

Assuming the pilot has by now returned to nominal $\gamma = -6.8$ deg at 57.3 KTAS = 96.8 ft/s, Eq. (12.7) tells us that $C_L = 1.230$. Then we need the formula relating lift coefficient and AOA for flaps 30 deg for this airplane:

$$C_L^O(\alpha; \delta_f = 30 \text{ deg}) = 0.082\alpha + 0.64 \tag{12.8}$$

Inverting, we get $\alpha = 7.2$ deg. So, on the gliding portion, $\theta_a = -6.8 + 7.2 = 0.4$ deg, pointed very slightly above the horizon.

On the horizontal portion, going through similar motions, from Eq. (12.7), we first find $C_L^O = 1.24$. But for α we realize that we are now in ground effect; that will make the required α somewhat less than predicted by inverting Eq. (12.8). An easy way to compensate is to find the value of lift coefficient *out of* ground effect, which corresponds to our *in* ground effect value 1.24. Using Eq. (4.63) and some Cessna 172 wing data from the Aerodynamic Force chapter, we find $C_L^O = 1.17$. Not a whole lot different. Then Eq. (12.8) tells us that the actual geometric (body)

AOA is 6.5 deg. So the corresponding level-portion aircraft attitude is also $\theta_a = 6.5$ deg.

To get going horizontally, the pilot must flare through $(6.5 - 0.4) = 6.1$ deg. Not a lot, but he or she is only at 57.3 KTAS. Position of the airplane's c.g. (preferably rearward) will make quite a difference. As mentioned, we will not investigate whether elevator control is efficacious enough for the job. We assume it is.

Our relative dissatisfaction with the calculation of final approach after the 50-ft obstacle is not that the computation cannot be done. It obviously can. It is that it can be done so many different ways. For comparisons with POH performance data, probably not enough information is given. And then there is the practical matter that each airplane, pilot skill level, and approach to landing situation is, especially for these light aircraft, close to unique. We have good days; we have bad days. Still, the airplane eventually touches down, easy or hard. The landing ground roll commences.

Summary for d_{LA2}, from just above runway to touchdown: this segment length is 321 ft. Elapsed time, from averaging initial and final segment speeds, is about 3.53 s. Cumulatively, we have $d_{LA} = 740$ ft, the POH figure and our goal in picking that slower speed once over the obstacle, and a total of 7.9 s.

Third Segment, d_{LG1}, Wheel Spin-Up for 1 s

This (assumed) 1-s interval is just to get the airplane stabilized on the runway with the wheels spun up to roll without further slipping. Only two forces are at work: an impulsive spin to the wheels and a bout of aerodynamic drag. We find the speed loss from each effect, come up with an average speed, and get the segment distance from that. Angle of attack will be assumed to be steady at that value associated with level flight at initial speed $V_i = V_{TD}$.

Conservation of energy during the wheel spin-up—which cannot be precisely correct in view of the screech and the landing skid marks on the runway—is expressed as

$$\Delta E = 0 = \tfrac{1}{2} m (V_f^2 - V_i^2) + \sum_{j=1}^{3} \tfrac{1}{2} I_j \omega_{fj}^2 \qquad (12.9)$$

The sum is over the three wheels. Variable I is the *moment of inertia* of the wheel and ω_f its final angular speed. If we express the final airplane ground speed V_f as $V_i + \Delta V$, where the initial ground speed is $V_i = (V_{TD} - V_{hw})$, and neglect the square of ΔV, we soon get

$$\Delta V \doteq -\frac{g}{W} \left(\frac{I_M}{R_M^2} + \frac{I_N}{2 R_N^2} \right) (V_{TD} - V_{hw}) \qquad (12.10)$$

W is the weight of the airplane, not of the wheel. Now we need some wheel data. Roughly apportioning average radii of the masses of wheel, tube, and tire (weights are given in the POH equipment list) gives the approximate data of Table 12.2.

In our sample Cessna 172 no wind case, Eq. (12.10) gives us $\Delta V \doteq 0.51$ ft/s $= 0.30$ kn. Because this is considerably below the resolution of the airplane's air-speed indicator, the reader can see why we normally ignore this effect, even if slippage doubles the effect.

Experimental Determination of Wheel Moment of Inertia

Take the wheel on a horizontal shaft of radius r. Wrap a light string around the shaft and suspend a (variable) weight from it. First, by trial and error, find the small weight W_0, which will *just* turn the wheel at a steady speed without accelerating it; that is a tare to compensate for kinetic friction. Then hang a somewhat larger weight W from the string and time the interval Δt required for the weight to fall distance h. It turns out, from the laws of uniformly accelerated circular motion, that

$$I = \frac{r^2 \Delta t^2 (W - W_0)}{2h} \tag{12.11}$$

Estimate of Aerodynamic Braking

Because we have the spin-up effect, we might as well use it. Initial speed for aerodynamic braking is then $85.3 - 0.5 = 84.8$ ft/s. Writing Newton's second law in a short-term average sense, we have, for this case

$$\langle F \rangle = D(\langle V \rangle) = k \langle V \rangle^2 = \frac{m \Delta V}{\Delta t} \tag{12.12}$$

Taking $\langle V \rangle = V_i + \frac{1}{2} \Delta V$, V_i the initial speed, and ignoring the term squared in the speed change ΔV (similar shenanigans as in the wheel spin-up case), results in

$$\Delta V \doteq \frac{F(V_i) \Delta t}{m - \frac{F(V_i) \Delta t}{V_i}} \doteq F(V_i) g \Delta t / W \tag{12.13}$$

Table 12.2 Cessna 172 running gear data

Wheel, tire, and tube	Weight, lbf	Radius R, ft	Moment of inertia I, slug-ft^2	I/R^2, slugs
Nose	10.4	0.542	0.0603	0.205
Main	18.0	0.625	0.134	0.343

Landing Performance

The last version is sufficient when $\Delta V / V_i \ll 1$. So we need the initial value of the force, composed of parasite and induced drag components. This will be

$$-D(V_i) = -\tfrac{1}{2}\rho V_i^2 S \left(C_{D0} + \frac{f_{GE} C_L^{O2}}{\pi e A} \right) \quad (12.14)$$

Using similar rules and techniques as before, under slightly different circumstances, $f_{GE} = 0.66$ and $C_L^O = 1.39$. Equation (12.14) evaluates to $F(V_i) = -195.4$ lbf and then the more nearly correct form of Eq. (12.13) gives $\Delta V = -2.7$ ft/s $= -1.6$ KTAS, making the "final" speed at the end of this segment 82.1 ft/s $= 48.6$ KTAS.

The average speed for the segment is then 83.4 ft/s; so $d_{LG1} \doteq 83.4$ ft.

Correction in Case of Runway Slope

We are assuming this runway is level. What if it were not? If the runway is markedly sloped—we *never* bounce on touchdown!—there will be a "weight component" force, $W \sin\theta$. Because this segment occupies only 1 s, this effect can be estimated by assuming an interchange between kinetic energy and gravitational potential energy,

$$\Delta KE + \Delta PE = 0 \quad (12.15)$$

where

$$\Delta PE = W(h_2 - h_1) \quad (12.16)$$

Though we parenthetically consider only the first ground run phase, the same analysis would apply to the second. After wheel spin-up, the airplane was at $V_{TD} - \Delta V^{\text{Spin-Up}} = 84.8$ ft/s; at the end of the segment, level runway, at $V_{TD} - \Delta V^{\text{Spin-Up}} - \Delta V^{\text{Drag}} = 82.1$ ft/s. That makes the average speed (not counting the slope effect we are now considering) $\langle V_i \rangle = 83.4$ ft/s. Say the airplane weighs 2400 lbf and the runway is sloped up 3 deg. In this 1 s, the airplane proceeds about 83.4 ft (actually, a little less!) along the runway. That second position is $83.4 \times \sin 3$ deg $= 4.36$ ft higher than the touchdown point. The associated gain in gravitational potential is $2400 \times 4.36 = 10{,}460$ ft-lbf. Because that potential energy gain comes from the airplane's average unperturbed kinetic energy $\tfrac{1}{2}m(83.4)^2$, its average kinetic energy must be decreased that same amount. Rewriting and rearranging Eq. (12.15) in terms of all these specifics for the final average speed we have

$$\langle V_f \rangle \doteq \sqrt{\langle V_i \rangle^2 - 2g \langle V_i \rangle \Delta t \sin\theta} \quad (12.17)$$

In our no-wind example, the speed decrease due to this uphill 3 deg runway slope would be only 1.7 ft/s $= 1.0$ KTAS, to $\langle V_f \rangle = 81.7$ ft/s. But recall that we are assuming that this extended example involves a *level* runway, so $d_{LG1} \doteq 83.4$ ft still.

Summary for d_{LG1}, wheel spin-up and stabilization: the segment begins with air speed $V = V_{TD} = 1.1 V_S = 50.5$ KTAS $= 85.3$ ft/s. But just for an instant, after which we have $V = 50.2$ KTAS $= 84.8$ ft/s to account for kinetic energy lost to the impulsive wheel spin-up. For the major portion of the assumed 1 s taken by this segment, aerodynamic braking is at work, reducing final segment speed to 48.6 KTAS $= 82.1$ ft/s, horizontal progression 83 ft. The Cessna 172 does not separate out this portion of the ground roll.

Fourth Segment, d_{LG2}, Rotation for 2 s to All Wheels On the Runway

We assume this segment occupies 2 s. In clocking a number of normal landings in airplanes of this type, the author found that this derotation typically took about 3 s (ending at near 4 s posttouchdown). But, we assume the performance test pilot interested in maximal performance will hurry that along by coming in slower, hence, down to 2 s. The airplane will slow a bit. Assuming an average attitude (and lift coefficient) halfway between that for V_{TD} and that for the ground run, of course *in* ground effect, the effect of that deceleration will be calculated. Though brakes often will be applied during the latter portion of this phase, for simplicity we will assume not. No brakes. But there will be rolling friction from the main gear—though not much, the airplane is still very light on its feet—as well as drag. The second phase ends when the nose wheel hits the runway.

It is time to put average speed and impulsive force approximations behind us and get back to a nearly exact formulation and calculation. One moderately sloppy residuum survives: because throughout this segment the airplane's attitude is continually changing, we will use average lift coefficients for describing both lift, $\langle C_L^I \rangle$, and induced drag, $\langle C_L^O \rangle$. A finer-tuned calculation for landing distance (and distance to liftoff) might use a second-order Runge–Kutta numerical procedure with time as the independent variable. Then the sequencing of rotation, flaps retraction, etc., can be described as accurately as one might wish, or as is known. The Runge–Kutta procedure can be carried out, using some @lookup tricks on an ordinary electronic spreadsheet. But we will continue to work one step down, still sufficiently lengthy. The two horizontal forces at issue are rolling friction (which of necessity involves lift) and drag. Because we avoid contaminant drag, that will include only parasite and induced types. The net force is then

$$F_{\text{Net}LG2} = -(W - \tfrac{1}{2}\rho V^2 S \langle C_L^I \rangle)\mu - \tfrac{1}{2}\rho V^2 S \left(C_{DO} + \frac{f_{GE} \langle C_L^O \rangle^2}{\pi e A} \right) \qquad (12.18)$$

There will be additional terms on a sloped or contaminated runway, but Eq. (12.18) is a simple quadratic of form $-(a^2 + b^2 V^2)$. A difference, in comparison with our similar takeoff case, is that here we know the time difference over the segment, assumed 2 s, instead of the final speed. But, multiplying Newton's second law by dt and integrating, we have the true sentence:

$$\Delta t_{LG2} = m \int_{V_1}^{V_2} \frac{dV'}{F_{\text{net}LG2}(V')} \tag{12.19}$$

where we know everything except final speed $V_2 = V_1 + \Delta V$. But to evaluate the integral and get a relation for end-of-segment speed V_2, we first need to evaluate the constants in Eq. (12.18). Collecting information from the Takeoff Performance chapter and from the end of the previous landing segment gives us the information in Table 12.3. It turns out that

$$F_{\text{Net}LG2} = -(a^2 + b^2 V^2) \tag{12.20}$$

with $a = 6.9282$ and $b = 0.1212$. In a detailed calculation, it is necessary to keep track of aircraft attitude, flaps deflection, etc., at all times. Ground control factor f_{GE} is roughly constant at 0.66.

Because the signs of the constants differ from earlier cases, we need integral I_{11} in Appendix C. When the algebraic dust clears, we have

$$\Delta V = \frac{a}{b} \tan\left(\tan^{-1} \frac{bV_1}{a} - \frac{gab\Delta t}{W}\right) - V_1 \tag{12.21}$$

V_1, recall, we have found to be 82.1 ft/s. Equation (12.21) evaluates to $\Delta V = -3.8$ ft/s, so $V_2 = 78.3$ ft/s. All of this (3 s!) and we still have not completely lost the extra 10% over V_S with which we touched down. Next we have to find out about the distance traversed in this 2-s segment.

An approximate figure for d_{LG2} can of course be obtained by taking the average speed $(82.1 + 78.3)/2 = 80.2$ ft/s and letting it run 2 s. That gives us $d_{LG2} \doteq 160.4$ ft. But let us see what the accurate version comes up with.

That calculation is precisely the same as for our main takeoff ground run, using integral I_1:

$$d_{LG2} = \frac{m}{2} \int_{V_1^2}^{V_2^2} \frac{dV'^2}{F_{\text{Net}LG2}(V'^2)} = \frac{-W}{2g} \int_{V_1^2}^{V_2^2} \frac{dz}{A + Bz} \tag{12.22}$$

with $A = a^2 = 48$ and $B = b^2 = 0.01468$. Using our known starting and ending speeds, Eq. (12.22) evaluates to 159.6 ft. What we have really found out, and that

Table 12.3 Aerodynamic coefficients during landing, Cessna 172

Landing stage	C_{D0}	C_L^o	C_L^l	C_{Di}	C_D
Just after wheel spin-up, $\delta_f = 30$ deg	0.055	1.39	1.60	0.0764	0.131
Three-point ground roll, $\delta_f = 0$ deg	0.037	0.75	0.81	0.0222	0.059
Average	0.046	1.07	1.20	0.0493	0.095

is of value, is that our previous "rough" estimate is plenty good enough. Still, because we have it, we will say that $d_{LG2} = 159.6$ ft.

Summary for d_{LG2}, rotation down to all wheels on the runway: initial air speed is 48.6 KTAS = 82.1 ft/s. During the (assumed) 2 s of this segment, air speed is reduced to 46.4 KTAS = 78.3 ft/s. Horizontal motion is 160 ft. The Cessna 172 POH does not separate out this distance.

Fifth Segment, d_{LG3}, Hard Braking to a Full Stop (Air Speed V_{hw})

This segment is to be calculated much as was the first segment of the takeoff roll. The difference is lack of thrust (we ignore any idle thrust or windmilling propeller drag beyond that ordinarily assumed as part of C_{D0}) and there is heavy braking. On bare dry concrete, we normally assume braking coefficient $\mu_B = 0.55$ on a braked wheel. But the effect very much has to be adjudicated according to weight actually placed on those wheels.

The reader who has clocked several landings with a stopwatch may want to alter the early timing assumptions, especially if he or she is calculating for other than short field landings. For example, we found the following reasonable but not sacrosanct numbers for times to get all wheels on the ground for small Piper or Cessna class aircraft: normal landing, 4 s; crosswind landing (smaller AOA, less exposure to wind), 3.4 s; wet runway (more aerodynamic braking), 4.5 s; soft field landing (keeping nose wheel off ground longer), 5.7 s.

Braking Friction and Its Coefficients μ_B

On bare dry concrete, braking coefficient μ_B is usually taken to be somewhere between 0.5 and 0.7. We will ordinarily take that value as $\mu_B = 0.55$ on bare dry concrete or asphalt (0.375 on bare dry grass) and our landing distance calculations will use that value. But there are many other cases, other surfaces, to discuss.

And several ways to discuss them. Pilots in Canada have to deal with snow, slush, ice, and water on their runways quite a bit more than U.S. pilots do. Transport Canada uses a James Brake Index (JBI) to adjust bare dry concrete landing distances for various JBI values. We will not do that—sticking instead with μ_B so we have something with which to calculate—but there is a relationship between those two measures. To add to the confusion, another numerical indication of braking effectiveness is also used: the runway condition reading (RCR). Add to those two verbal indications, our common equivalent braking action (Good, Fair, Poor, or Nil) and a longer-winded more descriptive alternative, runway surface condition (RSC). Not to worry. Here are the approximate numerical relations involved.

$$\text{RCR} = -2.30 + 30.8 \times \text{JBI} \tag{12.23}$$

$$\mu_B = 0.208 + 0.00860 \times \text{RCR} \tag{12.24}$$

or

$$\mu_B = 0.188 + 0.265 \times \text{JBI} \tag{12.25}$$

Loose as they are, the R^2 measures of goodness of fit of these relations hover around 0.985. Not too shabby. Table 12.4, taken from the *Canada Flight Supplement*[3] gives some operational guidance. A rightmost entry 1.46 means, for instance, that for RCR 12, barely "poor" braking, one should expect landing distances about 46% greater than on the same (almost always concrete) runway surface when bare and dry.

Table 12.5 provides rough numerical braking measures corresponding to the more prolix RSC descriptions. This extract from the *Canada Flight Supplement* gives us a ballpark estimate of braking friction coefficients for various messy

Table 12.4 Runway condition reading

RCR	Median μ_B	Equivalent braking action	d_{LG}/d_{LG0}
02 to 05	0.238	Nil	≥ 2.00
06 to 12	0.286	Poor	1.46 to 1.99
13 to 18	0.342	Fair	1.16 to 1.45
19 to 25	0.397	Good	1.00 to 1.15

Table 12.5 Runway surface conditions

RSC	JBI range	Low μ_B	High μ_B	Median μ_B
Bare and dry	≥0.80	0.4		
Damp, <0.01 in. water	0.60 to 0.70	0.347	0.374	0.36
Very light snow patches	0.55 to 0.60	0.334	0.347	0.34
Wet concrete, 0.01 in. to 0.03 in. water	0.40 to 0.55	0.294	0.334	0.314
Wet asphalt, 0.01 in. to 0.03 in. water	0.30 to 0.60	0.268	0.347	0.307
Sanded packed snow or ice	0.4			0.294
Compacted snow, $T < 5°F$	0.40 to 0.50	0.294	0.321	0.307
Heavy rain, 0.03 in. to 0.10 in. water	0.28 to 0.30	0.262	0.268	0.265
Snow covered	0.25 to 0.30	0.254	0.268	0.261
Compacted snow, $T > 5°F$	0.20 to 0.25	0.241	0.254	0.248
Cold ice, $T < 14 °F$	0.10 to 0.20	0.215	0.241	0.228
Wet ice, $T \geq 32°F$	0.05 to 0.10	0.201	0.215	0.208
Hydroplaning, standing water ≥0.1 in.	0.05			0.201

runway types. For the last entry, involving hydroplaning, the contaminant drag formula in the Takeoff Performance chapter should be applied.

How Much Load is on the Various Wheels?

The braking force on a wheel we take to be a "steady"

$$F_B = \mu_B N \tag{12.26}$$

where μ_B is the braking friction coefficient and N is the "normal" force pushing the two surfaces together. The rolling friction force F_f has the same form, but we use simple μ for its coefficient.

When both occur together, we assume the rolling friction coefficient is absorbed into the braking friction coefficient. Fine. But the question here is "How big are the normal forces N_N on the nose wheel and N_M on the main wheels (considered as a unit)?"

For our sample Cessna 172 sitting on the tarmac, $N_N \doteq 0.23W$ and $N_M \doteq 0.77W$; weight on the nose wheel is only 30% of the weight on the two main wheels, 60% of the weight on either main wheel. One can get those figures from measurements on the scale drawing of the airplane in the weight and balance section of the POH. But, at the start of this final braking segment, applying the brakes dumps *extra* weight onto the nose wheel and *relieves* weight (less is the pity!) on the mains. Backstick—which we will *not* consider (control considerations again!)—alleviates this problem to some extent. But how will we get an (approximate) handle on this weight apportionment? Newton's first law, extended.

While forces on the airplane in the horizontal direction (drag at or near the c.g., rolling friction on the nose wheel, braking friction on the mains) are certainly *not* balanced (the airplane is decelerating to a stop); they *are* balanced in the vertical direction. The sum of N_N and N_M and lift L (which we are assuming acts at the c.g., for simplicity ignoring the wing's pitching moment) is still equal and opposite to weight W. Moreover, the airplane is not rotating. So Newton, extended, says that torques on the airplane, about any axis (we will choose a spanwise axis through the c.g.) must add to 0 (see Fig. 12.2).

Calling the c.g. height h and the horizontal lever arms to the nose and main wheels l_N and l_M, respectively, the torques-about-the-c.g. condition, counting counterclockwise torques as positive, is

$$l_N N_N - (h\mu_B N_M + h\mu N_N + l_M N_M) = 0 \tag{12.27}$$

Plugging in our sample values for the lever arms and coefficients (0.02 and 0.55) tells us that $N_N = 0.70 N_M$. Weight on the nose wheel is now 70% of the weight on the mains! No wonder that up-elevator/backstick during hard braking

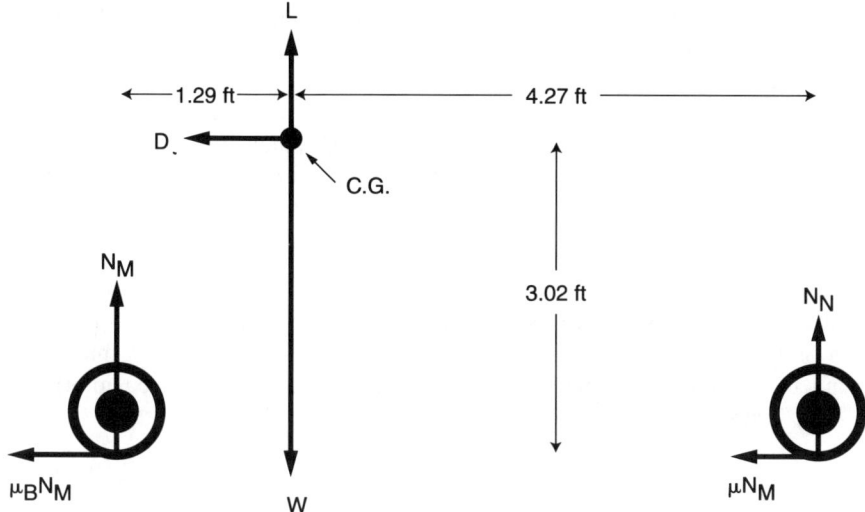

Figure 12.2 Forces and lever arms (from c.g.) on the rolling braking airplane.

(which we are ignoring) is so important. We need as much weight as possible back on the braking wheels. Combining this knowledge with the force relation,

$$L + N_N + N_M - W = 0 \qquad (12.28)$$

tells us that the total force of friction, rolling together with braking, is conveniently expressible as

$$F_{f/B} = \frac{0.7}{1.7}(W-L)\mu + \frac{1}{1.7}(W-L)\mu_B = 0.332(W-L) = \langle \mu_B \rangle (W-L) \qquad (12.29)$$

In other words, this airplane acts as though an average friction/braking coefficient $\langle \mu_B \rangle = 0.332$ were being used with the standard normal force $N = W - L$.

The average friction/braking coefficient and the weight distribution differ on different surfaces. On short grass, for instance, $\mu = 0.05$ and $\mu_B = 0.375$ are typical values. The general formula for the average effective coefficient is

$$\langle \mu_B \rangle = \frac{l_M \mu + l_N \mu_B}{l_M + l_N + h(\mu_B - \mu)} \qquad (12.30)$$

and so on grass we find $\langle \mu_B \rangle = 0.255$. The general formula for the wheel load ratio is

$$\frac{N_N}{N_M} = \frac{l_M + h\mu_B}{l_N - h\mu} \qquad (12.31)$$

This ratio, using our short grass figures for this airplane, is 0.588. As the airplane is being braked to a stop the nose wheel supports 37% of the weight while the mains support 63%.

With Eqs. (12.29) and (12.30), we now have simpler expressions for the braking/rolling frictional force, even though we may not like what they say. So we are finally ready to move on and complete our landing distance calculation.

Tag End of the Story

This hard-braking ground roll segment, in a sense the main dish, is by this time almost anticlimactic. The horizontal forces, as in the immediately preceding segment, are drag (only parasite and induced, in our sample calculation) and rolling/braking friction. In coefficient form,

$$F_{\text{Net}LG3} = -\tfrac{1}{2}\rho V^2 S C_{DGR} - \langle \mu_B \rangle (W - \tfrac{1}{2}\rho V^2 S C_{LGR}^l) \quad (12.32)$$

where, moreover, we have secured values of all of the coefficients (see Table 12.3): $C_{DGR} = 0.059$, $C_{LGR}^l = 0.81$, and $\langle \mu_B \rangle = 0.332$. Evaluating with the rest of the known constants, we have

$$F_{\text{Net}LG3} = -(A + BV^2) \quad (12.33)$$

with $A = 796.8$ and $B = -0.04337$. Evaluation of the landing segment distance proceeds as before

$$d_{LG3} = \frac{m}{2} \int_{V_1^2}^{0} \frac{dV'^2}{F_{\text{Net}LG3}(V'^2)} = \frac{-W}{2g} \int_{V_1^2}^{0} \frac{dz}{A + Bz} \quad (12.34)$$

Evaluating this I_1 integral with starting speed $V_1 = 78.3$ ft/s, we find that $d_{LG3} = 349.2$ ft.

Collecting previous results for the landing roll, we finally have

$$d_{LG} = d_{LG1} + d_{LG2} + d_{LG3} = 83.4 + 159.6 + 349.2 = 592 \text{ ft} \quad (12.35)$$

This is 52 ft, almost 10%, longer than the POH landing ground roll value 540 ft. But, if nothing else, the reader has seen the large number of fairly tenuous assumptions going into such a detailed calculation. A 0.5 s here, some backstick there... without expert pilots, instrumentation and computer work, ±10% is about what one can expect. This author, for instance, doubts his landing rolls under identical circumstances are more tightly clustered than about 10%. Reality sets in. Quicker approximations are hoped for (earnestly desired!) and will be located.

Before we get to them, one further point. We have been keeping easy track of landing roll segment time intervals because we have specified them, but what about this last segment? A rewritten Eq. (12.19) does that job:

$$\Delta t_{LG3} = m \int_{V_1}^{0} \frac{dV'}{F_{\text{Net}LG3}(V')} \tag{12.36}$$

and evaluates (integral type I_2) to $\Delta t_{LG3} = 8.36$ s.

Summary for d_{LG3}, hard braking to a full stop, $V = V_{hw}$, taken here to be zero: this last segment starts with $V = 46.4$ KTAS $= 78.3$ ft/s and ends with the airplane at a full stop. The segment length is 349 ft, which makes the total ground roll $d_{LG} = 592$ ft compared to 540 ft from the POH. And it makes the overall landing distance $d_L = d_{LA} + d_{LG} = 740 + 592 = 1332$ ft compared to 1280 ft in the Cessna 172 POH. The last segment took 8.36 s, the landing roll took 11.4 s, and the entire landing maneuver occupied only 19.3 s. Landing is mandatory, but that mandate is relatively short-lived.

"Constant Average Force" Approximation for the Braked Landing Ground Roll

Something must be done. This entire chapter, so far, has been devoted to calculating time and distance consumed in a single landing maneuver. Even that lengthy calculation was somewhat simplified—an impulsive force assumption, nonoverlapping rotation and braking segments, ignored pitching moment coefficients, etc.—and much circumscribed. We only considered one surface type, usually no wind and no slope, never any contaminant drag, only one gross weight, and one density altitude. Twenty seconds to land, ten pages to figure it out. Not economical.

Can the "constant average force" approximation give sufficiently accurate landing ground roll results? Let us take a look.

Though there was much insight to be gained in analyzing air segments—with plenty of flaps, with sufficient elevator authority, and with fairly calm air, one should get the speed down on short final; even a few knots lower helps a lot—let us now ignore the landing air portion. After all, we found that the first two ground roll segments, through rotation to all wheels on the runway, involved only small speed losses. And because a stopwatch is easy to carry and use, why not use this empiricism? Time several landings from touchdown on the mains until the nose wheel touches the runway. (In our case, we assumed this interval was 3 s.) Speed was a little above stall, but there were some retardation effects (mostly aerodynamic drag). So let us provisionally assume we touch down *at* the stall speed (45.9 KCAS = 45.9 KTAS = 77.5 ft/s, in our case) and let that speed continue undiminished through wheel spin-up and airframe rotation. That would have made the first two ground roll segments add to 232 ft. We calculated, in our earlier and overly assiduous mode, $d_{LG1} + d_{LG2} = 243$ ft. Negligible difference. So, to highlight this *ansatz*:

$$d_{LG1} + d_{LG2} \doteq V_{TD}(\Delta t_{LG1} + \Delta t_{LG2}) \qquad (12.37)$$

We are left with only the final hard-braking segment of the landing ground roll.

Constant average force approximation to the rescue. To use it, we shall need to know the forces at the beginning (at air speed $V = V_{TD} = V_S$ for the in-air flaps setting) and at the end of the segment (at $V = V_{hw}$) and use the appropriate uniform acceleration formula. Easily done. From the uniform acceleration formulary in the Takeoff Performance chapter, we have (using our situation, $d_0 = 0$, $V_0 = V_{TD} - V_{hw}$, $V = V_g = 0$, $a = \langle F_{Net} \rangle / m < 0$):

$$d_{LG3} = d(V_g = 0) \doteq \frac{-W(V_{TD} - V_{hw})^2}{2g\langle F_{Net}\rangle} = \frac{-WV_{TD}^2}{2g\langle F_{Net}\rangle}\left(1 - \frac{V_{hw}}{V_{TD}}\right)^2 \qquad (12.38)$$

Because timing plays such an important role in the landing maneuver, we might also need the uniform acceleration elapsed time formula:

$$\Delta t_{LG3} \doteq \frac{-W(V_{TD} - V_{hw})}{g\langle F_{Net}\rangle} \qquad (12.39)$$

Now let us redo our extended standard-conditions example and compare results. The decelerating force at the beginning of the three-point ground roll was made up of combined rolling and braking friction (with average coefficient $\langle \mu_B \rangle = 0.332$), mitigated by some lift, plus drag. Just before stopping, only the composite frictional force was operating. If there had been (constant) runway slope, that would have provided an additional constant force $-W \sin \theta$ throughout. For our sample calculation, we need only recall Eq. (12.33) and the constants ($A = 796.8$, $B = -0.04337$) appearing therein.

$$F_{NetLG3} = -(A + BV^2) \qquad (12.40)$$

Using our "maintained" $V_{TD} = V_S(\delta_f = 30 \deg) = 77.5$ ft/s, $\langle F_{NetLG3}\rangle = (-796.8 - 536.3)/2 = -666.6$ lbf. The (no-wind) distance, from Eq. (12.38), is 336 ft; the "detailed" calculation gave 349 ft. For the time intervals, the uniform acceleration formula in Eq. (12.39) gives 8.7 s compared to the detailed calculation's 8.4 s.

Will this good agreement hold up under all circumstances? Perhaps not. But the underlying idea behind the approximation is so simple and robust, we should use it—with occasional checks, either by detailed calculation or, even better, by experiment—unless it leads us astray.

A simple even though approximate theory, in the practical realm, has several advantages. First, it is more likely to be used than its complex cousin. Second, it can be used piecemeal. One example might be a runway that first slopes up, say 2 deg for 100 yd, then down 1 deg. Granted one would have to use the $V(d)$

relation in the uniform acceleration formulary to find the corresponding approximate ground speed at the point where the slope changes. A second example might involve hydroplaning. The force law starts off with one expression, then (speed having slowed below hydroplaning speed) shifts to another. Inhomogeneous runways (different partial surfaces, standing water or other contamination over only part of the runway length) provide further examples.

Wind Effect on Landing Ground Roll: Detailed and Constant Average Force Methods

Because wind effects on landing distance are large, and because they somewhat confusingly force a change of reference frame from the air to the Earth, we need to consider wind in explicit detail. With a few changes of notation, the problem is precisely as considered in the Takeoff Performance chapter. Here are the explicit changes:

- For takeoff, $F_{\text{Net}} = a^2 - b^2 V^2$; for landing $F_{\text{Net}} = -(a^2 - b^2 V^2)$.
- In the pivotal distance to liftoff example, $a = 22.00$ and $b = 0.1202$; in our braked landing ground roll example, we had $a = 28.23$ and $b = 0.2083$.
- In the takeoff example, the integral was from ground speed zero up to ground speed $(V_R - V_w)$, where V_R was the rotation air speed and $V_w = V_{hw}$ was the headwind speed (negative for tailwinds); in the landing example, the integral is from some ground speed $(V_1 - V_{hw})$ to ground speed zero.

The change in integral limits precisely compensates the landing case force's prefactor -1. The upshot is that the long formula for "exact" distance to rotation with a wind, in the Takeoff Performance chapter, with a simple notational switch from its V_R to our current V_1, applies to the landing case as well. Here is the rewrite:

$$d_{LGw} = \frac{W}{2gb}\left[\frac{1}{b}\ln\left(\frac{(a-bV_w)(a+bV_w)}{(a-bV_1)(a+bV_1)}\right) + \frac{V_w}{a}\ln\left(\frac{(a-bV_1)(a+bV_w)}{(a+bV_1)(a-bV_w)}\right)\right] \quad (12.41)$$

Taking the air speed at the beginning of the braked landing ground roll segment to be $V_1 = 78.3$ ft/s (and choosing, for apt comparison, the same no-wind distance value for the constant average force approximation and a variant of that last with exponent 1.85 instead of 2.00), our previous example's constants and conditions give the results shown in Fig. 12.3. Just as in the takeoff case (compare with Fig. 11.6 in the Takeoff Performance chapter), there is little practical distinction among the "detailed" braked landing roll theory and the two very much simpler constant average force approximations with various exponents close to 2.00.

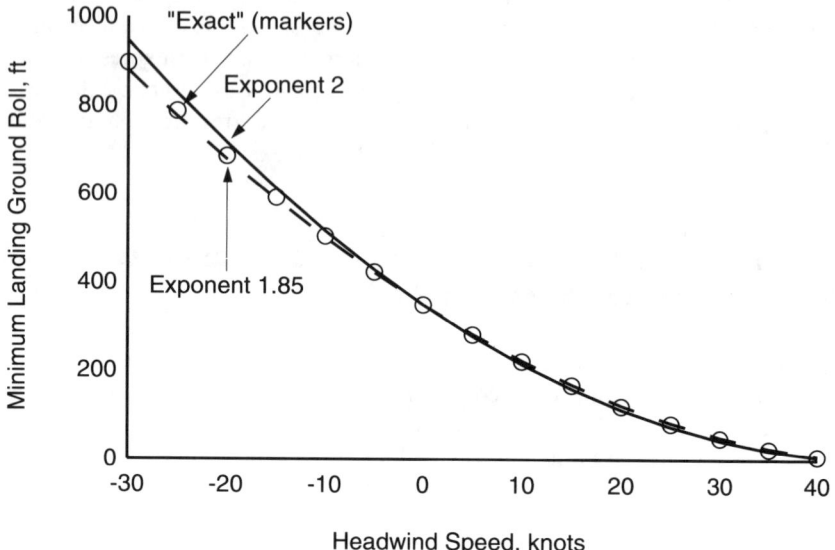

Figure 12.3 Braking ground roll vs headwind speed, three models.

The takeoff formula for time required to get up to rotation air speed can similarly be brought over. The braked landing ground roll version is

$$\Delta t_{LGw} = \frac{W}{2gab} \ln\left(\frac{(a+bV_1)(a-bV_w)}{(a-bV_1)(a+bV_w)}\right) \quad (12.42)$$

Again, this works for the no-wind case simply by setting $V_w = 0$. With $V_1 = 78.3$ ft/s, and the above sample values of W, a, and b, Eq. (12.42) evaluates to the same $\Delta t_{LG3} = 8.36$ s as before.

Perturbation Approach for Landing

The perturbation approach, recall, is laid on top on the constant average force approximation. Comparing the landing constant average force formula, Eq. (12.38), with its takeoff counterpart, Eq. (11.77) in the Takeoff Performance chapter, one first sees a difference in overall sign. Secondly, the landing average net force is negative, opposite the positive direction of travel, whereas for takeoff the average net force is positive. When one goes through the algebraic manipulations entirely similar to the takeoff case, signs cancel and one ends up with the identical formula (except for changed nomenclature) for a "perturbed" landing distance:

$$d_{LG1} = \frac{d_{LGB}}{1 + \frac{\Delta F_1}{\langle F_B \rangle}} \qquad (12.43)$$

The figure 1 in the landing distance subscript now identifies the perturbing force and no longer means the first ground roll segment. Equation (12.43) shows us how to find the new "perturbed" braked landing ground roll distance in terms of the unperturbed base case distance d_{LGB}, the base case average net force $\langle F_B \rangle$, and the perturbing force ΔF_1. A more practical version of Eq. (12.43), which does have a changed sign relative to its takeoff counterpart, is

$$d_{LG1} = \frac{d_{LGB}}{1 - \frac{2g d_{LGB} \Delta F_1}{W V_{TD}^2}} = \frac{1}{\frac{1}{d_{LGB}} - \frac{2g \Delta F_1}{W V_{TD}^2}} \qquad (12.44)$$

Let us now take a look at examples of the perturbation approach.

Example 12.1 Consider a downhill landing. The unperturbed (flat dry concrete) case involved standard conditions, maximum gross weight for the Cessna 172 sample aircraft, initial touchdown speed $V_{TD} = 77.5$ ft/s, and average net force $\langle F_{Net} \rangle = -666.6$ lbf. The base case landing roll was then found to be $d_{LGB} = 336$ ft.

Now we move to the same runway surface but sloped down 2 deg. That means an additional force $\Delta F_1 = W \sin 2 \deg = 83.8$ lbf. We have firmly decided, to avoid confusion, to always treat runway slopes as positive angles. Plugging these values into either Eq. (12.43) or (12.44) tells us that the braked landing roll has now increased to 384 ft. That is an increase of 48 ft.

Redoing our previous exact calculation (which had a slightly different initial segment speed, 78.3 ft/s, instead of 77.5 ft/s) with this modified force shows an increased braked landing roll from 349 to 401 ft, an increase of 52 ft. In this example, the perturbation approach is sufficiently accurate.

Example 12.2 Consider a landing on short grass. Taking the same unperturbed base case as before, now consider landing on short grass instead of on concrete. Where the base situation involved $F_{f/B} = 0.332 (W - L)$, this new surface changes that to $F_{f/B} = 0.255 (W - L)$. That is not the whole story, however. At the start of the braked landing roll, because the lift coefficient (in ground effect) is approximately 0.81, lift (for both cases) is

$$L_i = \tfrac{1}{2} \rho V_{TD}^2 S C_{LGR}^l = 1006 \text{ lbf} \qquad (12.45)$$

That makes $W - L = 1394$ lbf. At the end of the run there is no lift (in the no-wind case) and $W - L = 2400$ lbf. The average is $\langle W - L \rangle = 1897$ lbf and the perturbing force is

$$\Delta F_2 = -(0.255 - 0.332)\langle W - L \rangle = 146.1 \text{ lbf} \tag{12.46}$$

Equation (12.43) turns these facts into $d_{LG2} = 430$ ft, 94 ft or 28% longer than on concrete. The Cessna 172 POH suggests that landing on grass will increase the ground roll by 45%. Perhaps our assumption that μ_B on grass is 0.375 is overly generous.

When to Land Downhill with a Headwind Instead of Uphill with a Tailwind

The takeoff version of this problem was treated at length in the previous chapter. Because the analysis was a variation on the constant-average-force-cum-sloped-runway-perturbation (plus approximate headwind/tailwind correction), there is little to add for the landing case except to rewrite the pivotal formulas with appropriately changed signs. The more precise version for the break-even landing headwind is

$$V_{hw}^{LBE} = V_{TD} \times \frac{\sqrt{1 - \dfrac{W \sin\theta}{\langle F_B \rangle}} - \sqrt{1 + \dfrac{W \sin\theta}{\langle F_B \rangle}}}{\sqrt{1 - \dfrac{W \sin\theta}{\langle F_B \rangle}} + \sqrt{1 + \dfrac{W \sin\theta}{\langle F_B \rangle}}} \tag{12.47}$$

where we reiterate that runway slope angle θ is always to be taken as positive. As in the takeoff case, one will probably want to make use of the fact that

$$\frac{W \sin\theta}{\langle F_B \rangle} = \frac{-2g d_{LGB} \sin\theta}{V_{TD}^2} \tag{12.48}$$

The Break-Even Headwind Rule for Landing

So we have this Break-Even Headwind Rule: If the actual wind coming up the sloped runway has greater speed than is given by the right-hand side (RHS) of Eq. (12.47)—or given approximately by the RHS of Eq. (12.49)—land downhill into the wind. Otherwise, land uphill with the wind.

Approximation to Break-Even Landing Headwind Formula

A pattern as complicated as Eq. (12.47) is hard to interpret. But if one is confronted only with a relatively small slope (or a relatively small wind speed), a binomial expansion gives the much simpler result

$$V_{hw}^{LBE} \doteq \frac{d_{LGB} g \sin \theta}{V_{TD}} \qquad (12.49)$$

An even more practical version, possibly suitable for cockpit prelanding checks, comes from converting wind speed to (ground) knots and air speed to calibrated knots using the close linearity of the sine function for small angles and approximating a combination of factors equal to 0.1971 as 1/5:

$$V_{hw}^{LBE} \doteq \frac{d_{LGB} \theta \deg \left(1 - \dfrac{h_\rho}{70{,}000}\right)}{5 \, \text{KCAS}(V_{TD})} \, \text{kn} \qquad (12.50)$$

What is the break-even headwind speed for our base case Cessna 172 on a 2 deg dry concrete runway? The RHS of Eq. (12.48) evaluates to -0.1256. The RHS of Eq. (12.49) is then 4.89 ft/s or 2.90 kn. The cockpit-ready approximate version, Eq. (12.50), gives 2.92 kn. Perhaps that close agreement is only fortuitous. An important fact to keep in mind is that, for a given actual wind and slope, different airplanes may do better to land in opposite directions. Of course terrain clearance and possible wind shear are practical factors that must be considered even though they were ignored in this calculation.

As with the takeoff case, a pilot following the break-even landing headwind rule can *almost* be assured that he or she will not land longer than on the same runway were it flat and calm.

How Gross Weight and Density Altitude Affect the Landing Roll

A glance at the typical POH landing performance page shows very little discussion of alternative surfaces, alternative flaps settings, headwinds or tailwinds, or etc. In the Cessna 172 POH, the possibility of landing lighter than at maximum gross weight (2400 lbf for sample model) is not even mentioned. That page is preoccupied with landing at various density altitudes (split up, as usual, into separated values of pressure altitude and outside air temperature). The POH authors are right about that emphasis—density altitude *is* a very important landing performance variable. But so also is gross weight.

In the takeoff case, we were able to find an (admittedly complicated) way to take weight and density altitude into simultaneous consideration: the takeoff power parameter. The reader will be pleased to know that a similar analysis can be conducted for landing. And may be overjoyed to learn that, this time, the result is excruciatingly simple:

$$d_L(w/\sigma) \doteq d_L(w/\sigma = 1) \times \left(\frac{w}{\sigma}\right) \quad (12.51)$$

Say for example your landing (ground) distance is 540 ft at maximum gross weight at MSL. And say you are interested in predicting the corresponding distance when your airplane lands at $h_p = 7000$ ft weighing only 2200 lbf. Then $w \equiv W/W_0 = 2200/2400 = 0.917$ and $\sigma = 0.811$ (0.810 by our approximate rule). So $w/\sigma = 0.917/0.810 = 1.13$; you will land about 13% longer than under standard conditions at standard (maximum gross) weight.

Why is this rule true? And how good is it? We take the second question first, comparing Eq. (12.51) with the Cessna 172 POH landing performance table ground roll data (see Figs. 12.4 and 12.5). Why is agreement between Eq. (12.51) and the POH so good? Because, at least for the density altitude (or relative air density) dependence, they used the same theory. Unfortunately that POH considers only a single gross weight value.

Why Landing Roll Distance d_{LG} is Proportional to W/σ

Or approximately so. Although we have so far kept the first two brief ground roll segments separated from the third and longer one, from this point on we lump all

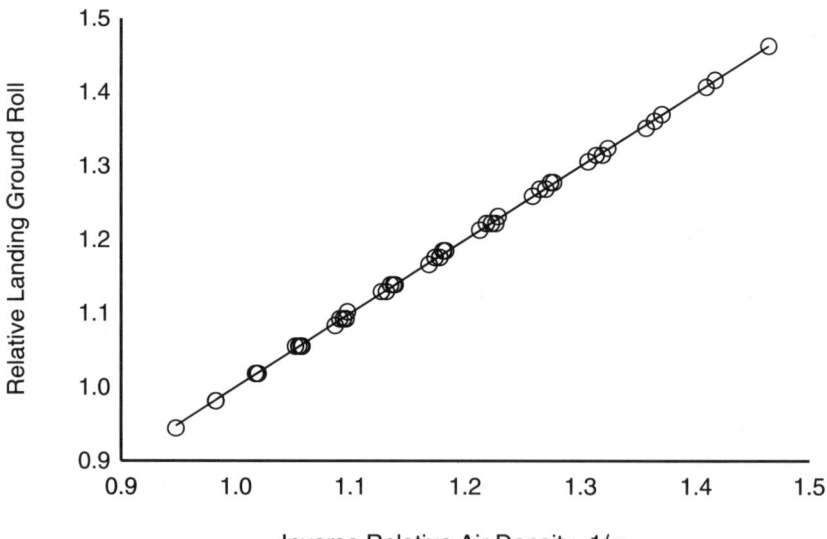

Figure 12.4 Cessna 172 POH landing ground roll figures are inversely proportional (within rounding errors) to relative air density σ.

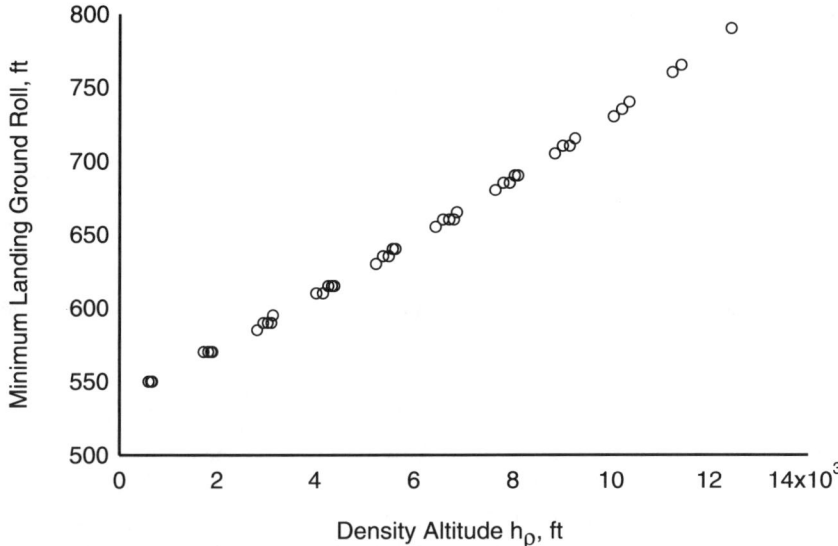

Figure 12.5 Though displayed in more practical fashion, this is the same Cessna 172 POH landing roll data as shown in Fig. 12.4.

three together into landing (ground roll) distance d_L. Next we recall the constant average force approximation:

$$d_L = \frac{-WV_{TD}^2}{2g\langle F_{\text{Net}}\rangle} \qquad (12.52)$$

We have neglected the wind dependence factor [see Eq. (12.38)] because it is an unchanging feature to be tacked on at the end. To derive Eq. (12.51), all we have to do is look into the weight and air density dependence of each of the factors on the RHS of Eq. (12.52).

- Gross weight $W = wW_0$ by definition of the relative weight w.
- Touch down speed $V_{TD} = kV_S$, where k is some constant close to unity and V_S is the stall speed. As we have seen in Chapters 4 and 8, Aerodynamic Force and Maneuvering Performance, respectively,

$$V_S(w/\sigma) = \sqrt{\frac{w}{\sigma}}\, V_S(w/\sigma = 1) \qquad (12.53)$$

so

$$V_{TD}(w/\sigma) = kV_S(w/\sigma) = \sqrt{\frac{w}{\sigma}}\, V_{TD0} \qquad (12.54)$$

where we have abbreviated the condition $w/\sigma = 1$ with an additional subscript 0.

- The average "constant" net force on the airplane is

$$\langle F_{\text{Net}} \rangle = \tfrac{1}{2}(F_{\text{Net}}(V_{TD}) + F_{\text{Net}}(0)) \tag{12.55}$$

and because

$$F_{\text{Net}} = -\langle \mu_B \rangle W_0 w + \tfrac{1}{2}\rho_0 S(C^l_{LGR} - C_{DGR})\sigma V^2 \tag{12.56}$$

we see that

$$\langle F_{\text{net}} \rangle = w \langle F_{\text{Net0}} \rangle \tag{12.57}$$

Putting these results together into Eq. (12.52), and canceling factors w, we have our desired result:

$$d_L(w/\sigma) \doteq d_{L0} \times \left(\frac{w}{\sigma}\right) \tag{12.58}$$

While one might get a slightly more accurate result by separating out the two early landing ground roll segments (whose combined length, assuming the two Δt's do not vary, goes as $(w/\sigma)^{1/2}$ because V_{TD} does), applying Eq. (12.58) to only the third segment then adding back in the differently modified early segment lengths, it is almost certainly not worth the effort. At least at this level, landing defies precise analysis. To obtain much more accuracy predicting landing ground runs requires numerical analysis.

Comparisons with the Cessna 172 POH can only use the single gross weight 2400 lbf used there. Figure 12.6, however, uses Eq. (12.58) to construct a cockpit-

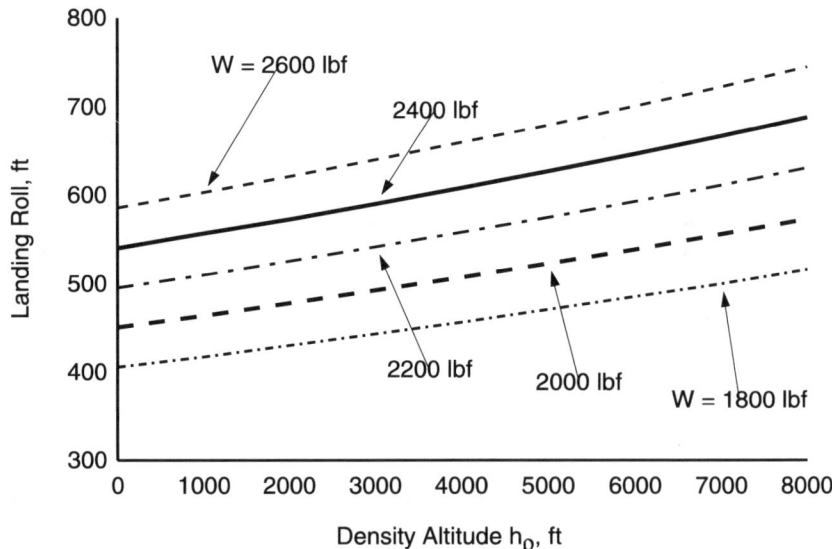

Figure 12.6 Landing roll vs density altitude for several gross weights, Cessna 172.

ready graph for landing ground roll as a function of density altitude for several different gross weights.

Conclusions

Calculating landing ground roll has been our most complicated (and probably least accurate!) performance analysis. One starts with a detailed time sequence of assumed speeds, flaps configurations, and braking techniques, adds in pertinent information about the landing environment (type surface, slope, contamination, density altitude, and wind), and appends aircraft weight and balance details. Then come various approximations to make the problem tractable—for instance, in our example, we assumed flaps were being retracted only during the 2-s rotation down onto the runway. To avoid having a large number of landing roll segments clutter the analysis, some details must be ignored. But one should be careful to not fudge consistently in only one direction, always lengthening or always shortening the ground roll.

With the stage set, one is then ready for action—a suitable method of calculation. In most cases the constant average force technique, augmented by perturbations, will be sufficiently accurate. Our goal, recall, was only to obtain a *respectable* analysis of (predominately) quasi-steady-state aircraft performance.

The final act is to compare the calculated result with experiment or, as we did, with a finer-honed calculation. Ultimately, experimental landing roll measurements have final say on whether chosen approximations and calculational techniques are up to the job. If "identical" landings under "identical" circumstances give experimental landing roll numbers no more tightly clustered than say 5%, a theory calculating only about that close is acceptable "for all practical purposes."

References

1. FAA, "Flight Test Guide for Certification of Part 23 Airplanes," Advisory Circular 23-8A, Change 1, 1993.
2. Roskam, J., and Lan, C.-T. E., *Airplane Aerodynamics and Performance*, DARcorporation, Lawrence, KS, 1997, p. 479.
3. Transport Canada, *Canada Flight Supplement*, 1992, pp. A46, E14-15.

Appendix A

How Big Are the Error Bars?

Introduction

Normally it is best to avoid buzz words, but this subject drives us to display a whole bunch:

Error; accuracy; precision; significant figures; rounding; uncertainty; tolerance limit; bias, or systematic error; statistical or stochastic or random error; uncertainty; correction; adjustment; operational definition; true value; estimator; normal distribution; direct vs indirect measurement; degrees of freedom; dispersion; standard error; normal or Gaussian distributions; Student t distributions; confidence limit; correlation coefficient; least squares; outlier; and propagation of error.

You get the point: reality is one thing and numerical description of it is another. "There is many a slip twixt the (theoretical) cup and the (practical) lip." We want the world's "true values" to mesh well with our estimates of them. While we can not hope for a perfect fit, we would like a simple, consistent, informative accommodation. We are willing to endure an occasional bent epistemological cog, but the disaster of stripped descriptive gears we want to avoid at all costs. Good statistics are more crucial with airplanes than elsewhere in engineering. Overbuilding an airplane the way many overbuild furniture or staircases, it could not fly! This appendix is designed to help you successfully run the gauntlet between files of credulity and cynicism.

When some client (or professor, or fellow researcher, etc.) asks you to please measure the variable corresponding to a particular physical attribute of a particular system, your first question should be "How precisely do you want to know it?" Or, "How much money do you want to spend?" Usually amounts to the same thing.

Your naive petitioner often responds to the first question with "I want the true value, perfectly." That embarrassing (for the questioner) situation happens more frequently than one might think. And, if so, then you have to decide how best to respond, without hurting either facts or feelings, that *none of us can know whether any offered candidate number is the true value or not*. In almost all cases the concept of "true value" has no operational meaning. Yet statistical discussions assume there *is* such a thing. Practically speaking, we *think* we are close to that apocryphal "true value" when, having eliminated or corrected for all *discernible* systematic errors (bias), our repeated measurements cluster tightly around some "central" value. Especially when various parts of that cluster come to us through different measurement techniques or by using different instruments. This situation is unresolved, and unsatisfying philosophically, but tight clustering is as good as it gets. Measurements involve errors. Measuring instruments have finite resolutions. We have to live with uncertainty. But there are ways to reduce the sizes of those error bars.

Sample Measurement Job

Suppose your job is to find the rate of climb (feet per minute) at 70 KCAS for a particular airplane and operational situation—assume flaps and gear up, full throttle, leaning for best power, gross weight 2200 lb, density altitude 5000 ft. Seems simple enough.

As neophyte performance test pilot, you load on enough gasoline or ballast (guided by the airplane's official weight-and-balance record) to get the weight to 2200 lbf, crank in the altimeter setting (broadcast on the ATIS), take off, and make a stabilized and timed climb, at 70 KIAS, from 4500 to 5500 ft indicated altitude. Elapsed time, by the stopwatch, is 93.4 s. One thousand ft in 93.4 s means $1000/93.4 = 10.7066$ ft/s or 642.3983 ft/min. (Your calculator is set to four decimal places.) You may be a neophyte, but you were not born yesterday; you report the asked-for rate of climb as a rounded 642.4 ft/min.

Back down, your boss is not fully satisfied with your reported 642.4 ft/min at 70 KCAS and mentions **some possible sources of bias, or systematic error**, along with actions you might take to mitigate them and how well those efforts are likely to work. Here is the list:

1) The weight-and-balance record might be faulty. Weigh the airplane (calibrated scales) to get a good new empty weight (± 20 lbf).
2) The weight density of the fuel (avgas) added might not be the nominal 6.0 lbf/gal. And the quantity registered by the pump might be faulty. Calibrate the pump with containers of accurately known capacity ($\pm 0.5\%$) and correct the fuel density using known volume expansion coefficients (± 0.0001). Also, of course, get accurate weights of the pilot, observer, and whatever else is taken on board (± 3 lbf).
3) Fuel burn between engine start and the climb through 5000 ft diminished gross weight. Allow for that by keeping track of time intervals for various

How Big Are the Error Bars? 419

phases of the flight and using gallons per hour information from the engine manual (±2 lbf). Adjust the raw rate-of-climb figure to desired weight 2200 lb.

4) Your altimeter use was ill-conceived. Set the Kollsman window to 29.921 in. Hg to have it read pressure altitude. (This is not strictly necessary. Any altimeter setting and altimeter reading, if the instrument is well calibrated, will give you the corresponding ambient pressure, and thereby the pressure altitude, if you know the formula in Chapter 2.) Measure temperature at several pressure altitudes and calculate the corresponding density altitude for each to compute the pressure altitude corresponding, in "today's atmosphere," to density altitude 5000 ft. Reduce the pressure altitude climb band to say 500 ft to reduce averaging effects of nonlinearity of rate of climb as a function of density altitude. Then use a temperature ratio correction (or a numerical integration with respect to pressure altitude) to find the elapsed vertical tapeline (geopotential) altitude difference represented by that pressure altitude band (±35 ft).

5) The outside air thermometer suggested just above needs to be calibrated (±0.25°F) and have its reading corrected for ram temperature rise (±1°F). This requires knowing the temperature recovery factor which, assuming it does not differ with air speed, can be found (Chapter 2) from flights at two different air speeds in atmospheres of the same temperature.

6) Uncalibrated altimeter. Calibrate the altimeter at setting 29.921 in. Hg (±10 ft).

7) Altimeter lag. From the length and diameter of associated tubing, calculate the lag. Make an argument (because of cancellation) as to the small size of net effect (±2 ft) of this error source.

8) Unknown pitot-static position error. Ascertain position error, with altimeter setting 29.921, to get the corresponding altitude correction (±10 ft) at the angle of attack (AOA) arguably expected for this climb maneuver.

9) Uncalibrated air-speed indicator. Calibrate this particular instrument to get the appropriate corrections (±0.5 KCAS), including that for position error.

10) "Kinetic energy" effect. A climb at constant calibrated air speed is in fact accelerating (in true terms) because of decreasing air density. This small error impacts the space-interval averaging used to estimate the tapeline vertical climb band. But, if that band is appropriately small, not much (±2 ft).

11) Uncalibrated stopwatch and thumb. Calibrate the watch against a known standard time interval of about 100 s (±0.1 s). Also test the observer's stopwatch technique by having him repeatedly time accurately known intervals (±0.1 s).

12) Propagation of errors due to arithmetic (here, dividing). Even ignoring uncertainties in gross weight and midpoint density altitude, errors in true

altitude band height $\Delta h(\pm 50$ ft) and elapsed time $\Delta t(\pm 0.14$ s) cause uncertainty in the calculated rate of climb $\Delta h/\Delta t(\pm 30$ ft/min). This in spite of the statistical "independence" of errors in the two variables Δh and Δt.

Should this be called "nit-picking on a grand scale"? Perhaps. Strenuous exercise of the critical faculty is, *on occasion*, required. You should be able to ask, when offered a measure number, "What reasons are there to doubt this claim?" You temporarily become a "nattering nabob of negativity," but in a controlled and positive fashion. You should know quite a bit about possible sources of error and how to reduce them. You should know how to properly express measure numbers and their uncertainties and how to calculate the consequences those first-line measurement errors generate onto calculated quantities down the line. This appendix goes over much of that necessary background.

Having implemented each of the above dozen suggestions, a few days later you run the climb test again. This time you get 724.1 ft/min. But the boss *still* is not happy, this time asking: "but how big are the error bars?"

Time out! This is turning from a 2-min climb into a 1-week-long research project! We second the sentiment. Back to the white board, for a short course on theory of errors. Let us start low and slow.

The Short (Necessary but Insufficient) Course on Error Theory

Here, without proof and with only little discussion, are some precepts to consider.

1) Get in the habit of assigning error bars (estimates of the mean standard error, the standard deviation, 90% tolerance limits, etc.) to any measure number you encounter. And particularly those you come up with yourself. Do not wait to be asked. Try to make the error bar estimates as much a part of the quantity as its physical units (or its dimensionlessness).

2) A matter of nomenclature. Considering the (apocryphal!) true value as the starting point, the standard, here is how best to define "error"
   Error = Measured value − true value.
   A "correction," on the other hand, is what is applied to the measured value to get the true value:
   Correction= −error = true value − measured value.

2) Do whatever you can to reduce or eliminate bias or systematic errors. These errors are consistent from one measurement to another and can generally be recognized only by using a second measuring instrument or method.

3) Random errors can never be wholly eliminated. These are due to one or more stochastic (chance) inputs. Random errors are generally as likely to

give negative errors as to give positive ones. Unless convinced otherwise, assume that random errors for individual measurements are normally distributed according to $N(0, \sigma)$, the normal distribution with mean zero and standard deviation σ. To estimate that standard deviation, one needs to run repeated sample measurements.

4) Significant figures are a rough-and-ready way of assigning error bars. So 93.4 (three significant figures) means "93.4 plus or minus 0.05," an implicit error bar as big as one unit at the rightmost significant digit, one-half above and one-half below. The usual rules for operating with significant figures are

a) When adding or subtracting two figures, first round each to the same (least) number of decimal places: 93.4 and 55.36 become 93.4 and 55.4; 3.0014 and 7.02 become 3.00 and 7.02.

b) When rounding a rightmost digit 5, round down if the preceding digit is even, round up if the preceding digit is odd: 44.625, to two places, is 44.62; 44.615, to two places, is also 44.62. This rule tends to make the terminal midpoint digit 5, in a long string of addends, contribute evenly up and down.

c) When multiplying or dividing two figures, keep only as many significant figures in the product or quotient as contained in the contributor (factor, divisor, or dividend) with the fewer number of significant figures: $44.625 \times 7.01 = 313$ (three significant figures); $0.0069/3.04 = 0.0023$ (two significant figures).

The "significant figure" concept is a blunt tool, useful in casual measurement discussions but not to be taken as the last word.

6) It is not reasonable to discard a measurement simply because it is far from the average of its companion measurements. On the other hand, it is reasonable to throw out a measurement because, when that single measurement was performed, something appeared "not quite right." Especially when that suspicious measurement has in fact resulted in an outlier.

The longer course starts with a discussion of the normal distribution.

The Normal Distribution $N(\mu, \sigma)$

The *normal probability density* for variable x, with mean μ and standard deviation σ, is given by the familiar bell-shaped curve:

$$N(x; \mu, \sigma) = \frac{e^{-(x-\mu)^2/2\sigma^2}}{\sigma\sqrt{2\pi}} \qquad (A.1)$$

In almost all of the material to follow, we will assume sampling from a normal distribution. There are two good reasons for this assumption. First, Gauss, almost two hundred years ago, discovered that concatenation of many small random errors resulted in measured values distributed according to a normal distribution. Second, it is an interesting and useful fact of mathematical statistics (the Central Limit Theorem) that *samples*, even those drawn from an underlying distribution that is *far* from normal, are distributed very close to normal.

There are as many normal distributions as there are pairs of values (μ, σ), but any of them can easily be described in terms of the so-called "standard normal" or Z distribution with mean zero and standard deviation unity, $N(0, 1)$. Because the origin (zero point) and scale (units) of most measurements are arbitrary, we can redefine a variable X, distributed according to $N(\mu, \sigma)$, by considering instead variable Z defined by

$$Z = \frac{X - \mu}{\sigma} \quad \text{(A.2)}$$

This redefinition process is known as "standardization" of the primordial variable X. It is clear that the standardized variable Z will follow the standard normal distribution $N(0, 1)$. Figure A.1 is the graph of $N(0, 1)$ with the graph of its integral, the cumulative distribution function (c.d.f.) $F_Z(x)$, the probability that a single measured value (distributed according to $N[0, 1]$) will be less than or equal to x.

Continuous distributions such as $N(0, 1)$ do not quite give you a probability, only a "probability density." There is only a vanishingly small chance that you

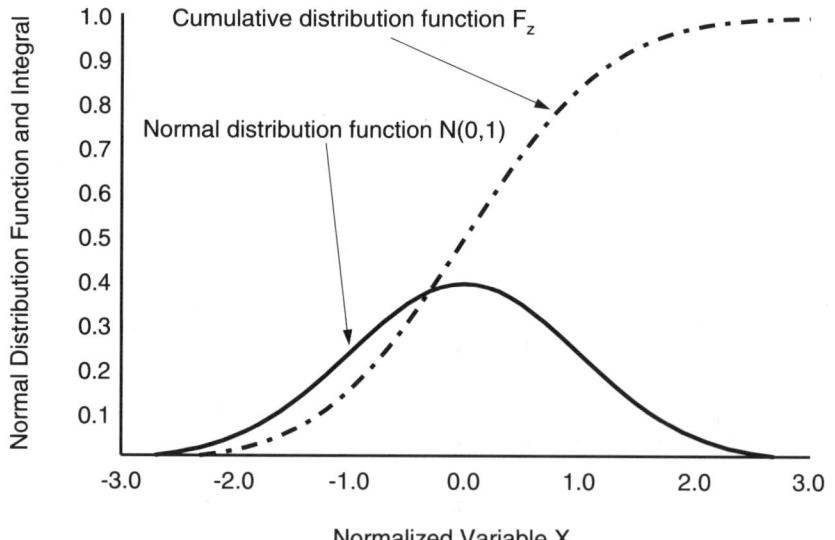

Figure A.1   Graph of (bell-shaped) $N(z; 0, 1)$ vs z and the sigmoid graph of the integral of N from $(-\infty)$ to z.

will find say $z = 0.3$, no matter how many times you sample in $N(0, 1)$; there is an infinite set of possible measure values and, while you will always hit one of them, there is essentially no chance you will hit a particular one specified in advance. That is one reason the c.d.f. F_Z is so useful; it *does* give you a probability number. When sampling $N(0, 1)$, the probability of coming out with a value less than or equal to 0.3, $F_Z(0.3)$, is a definite (though approximate) 0.61791. There is no exact formula for $F_Z(z)$, as z ranges over the entire z axis, but almost all elementary statistics books (and serious spreadsheet programs) have tables (and @-functions) giving its values. See Table A.1 for an excerpt of both functions. Of course, some particularly simple values are known: $F_Z(0) = 0.5$ (because the underlying $N[0, 1]$ is symmetric about $z = 0$), and the limits of $F_Z(z)$ as z approaches $\pm\infty$, are, as for any c.d.f., 1 and 0.

In using F_Z, we commonly want information slightly backwards (inverse) from that presented in Table A.1. For example, we might want to know how far out on the Z axis one has to go before 95% of the area under the $N(0, 1)$ curve is to the

Table A.1 Density $N(z; 0, 1)$ and its cumulative distribution $F_Z(z)$

z	$N(z; 0, 1)$	$F_Z(z)$
−3.00	0.0044	0.0014
−2.75	0.0091	0.0030
−2.50	0.0175	0.0062
−2.25	0.0317	0.0122
−2.00	0.0540	0.0228
−1.75	0.0863	0.0401
−1.50	0.1295	0.0668
−1.25	0.1826	0.1057
−1.00	0.2420	0.1587
−0.75	0.3011	0.2266
−0.50	0.3521	0.3085
−0.25	0.3867	0.4013
0.00	0.3989	0.5000
0.25	0.3867	0.5987
0.50	0.3521	0.6915
0.75	0.3011	0.7734
1.00	0.2420	0.8413
1.25	0.1826	0.8944
1.50	0.1295	0.9332
1.75	0.0863	0.9599
2.00	0.0540	0.9773
2.25	0.0317	0.9878
2.50	0.0175	0.9938
2.75	0.0091	0.9970
3.00	0.0044	0.9987

Table A.2 Selected values from the cumulative distribution function of $N(0, 1)$

z	1.282	1.645	1.960	2.326
$F_Z(z)$	0.90	0.95	0.975	0.99

left. From Table A.1 you can see that $z(0.95)$ is somewhat less than 1.75 standard deviation units above the mean. Because most of our questions of this sort revolve around whether we have such round figures as 80, 90, 95%, etc., of the area to the left, Table A.2 is useful.

There is one additional wrinkle to keep in mind. Say you want to know which value of z corresponds to having 90% of the area under the standard normal curve less far out, from the mean, than it is. This is a "two-tailed" requirement because the far-out values can be either larger or smaller than the mean. So the appropriate number to look up in Table A.2 is above 0.95, 1.645, and not above 0.90, 1.282. *Twenty* percent of the area under the $N(0, 1)$ curve is more than 1.282 units away from mean zero.

Important Questions

We are finally ready to get to the heart of matter, how to properly describe and quantify errors. To keep what we might call "mental altitude excursions" under control, we will use concrete examples to flesh out the math-speak. The main questions we need to answer are

1) How do we use sampling results to estimate true values?

In repeated measurements of the same quantity under the same conditions, the best estimator of "true value" is the arithmetic mean (ordinary average) of those repeated good measurements. This is the result you would have expected.

When you reran the rate-of-climb determination seven times—under identical circumstances, at the same weight, and over the same altitude band, though not on a day with particularly smooth air—you got these $n = 8$ values: {724.10, 738.44, 721.15, 745.39, 713.40, 731.52, 722.80, 695.70}. The average value, 724.064 ft/min; is your best estimate (on the basis of information so far available) of the airplane's, rate of climb (ROC) at 70 KCAS.

Using μ for the true value, and the symbol $\approx$ to mean "is estimated by," the sample mean "x-bar," is defined by:

$$\mu \approx \bar{x} = \frac{\sum_{i=1}^{n} x_i}{n} \quad \text{(A.3)}$$

When averaging, keep only one further decimal place, in the sample mean, than there is in individual measurements.

2) What certainty can we attach to these estimates of the true value?

Certainly not perfect certainty. After all, there was quite a bit of scatter among your eight measurement results. If the extent of scatter is large, you very reasonably feel less sure that a ninth measurement will be close to the average of your earlier eight. With a given amount of scatter, you can either feel quite sure that you know the true value within wide limits, or alternatively you can feel not very sure that you know the true value within narrow limits. As you can see, this is a much more sophisticated business than simply estimating the true value itself. The first job is to get a numerical handle on measuring the sample's scatter.

All nonpathological distributions have a standard deviation σ (or its square σ^2, the variance, which we will emphasize to avoid large radical signs). The best estimator of the true value of the population variance is the sample variance (notice the $[n-1]$ weighting):

$$\sigma^2 \approx S^2 = \frac{1}{n-1}\sum_{i=1}^{n}(x_i - \bar{x})^2 \tag{A.4}$$

where there are n individual measurements x_i and $\bar{x}$ is their mean. The divisor $(n-1)$ occurs because, once the mean is known, not all n individual measurements are independent; knowing "only" $(n-1)$ of them, the nth can easily be reconstructed. An estimate of dispersion can also be gotten from the range (highest value minus lowest value) of the measurements x_i, but it is not as "efficient" as S^2; more on that technique later. As in the case of the mean, keep only one more decimal place in your estimate of S than there is in individual measurements. For your eight ROC measurements, $S = 15.331$.

You are closing in on an answer to the question "How big are the error bars?" The sample variance itself, Eq. (A.4), is often used to answer this question, but for small samples (and for somewhat technical reasons) the Student t distribution—which approaches the normal distribution as the number in the sample grows—is preferred. Let us say we want error bars that give us 80% assurance (in much of aviation, our statistical standards are not very high!) that the ROC at 70 KCAS is within their confines. The shortest two-sided 80% confidence interval (CI)—described by specifying $\alpha = 0.2$ or 20%—is given by

$$L, U = \bar{X} \pm S_{\bar{X}} t_{n-1; 1-\alpha/2} \tag{A.5}$$

Here L and U stand for lower and upper bounds, arrived at by using the minus and plus signs respectively. $\bar{X}$ is the sample average, here 724.06. $S_{\bar{X}}$ measures the variability to be expected in averages of repeated samples of this size (here, 8) and is related to the sample variance S by:

$$S_{\bar{X}} = \frac{S}{\sqrt{n}} \tag{A.6}$$

If you go through the arithmetic, using Eq. (A.4), you will find $S = 15.331$. So $S_{\bar{X}} = 15.331/2.828 = 5.421$.

So far so good, but what is that strange t with all of the subscripts; how do we find its numerical value? From a statistical table of values of the Student t distribution (Table A.3), which we now reproduce in part.

In our case, $n - 1 = 7$ and $1 - \alpha/2 = 1 - 0.2/2 = 0.90$; $t_{7;0.9} = 1.4149$. This is not terribly far from the two-tailed 80% figure one would get from the $N(0, 1)$ distribution, 1.2816. Because the c.d.f. of Student t is much less often tabulated than the standard normal distribution's c.d.f. F_Z, the following approximation may be useful to you:

$$t_{f;\gamma} \doteq z_\gamma + \frac{(z_\gamma^3 + z_\gamma)}{4f} \qquad (A.7)$$

Letter f stands for degrees of freedom, the number of independent numbers that go into making up the statistic being considered. In our case, $f = n - 1 = 7$ and $z_\gamma = z_{0.9} = 1.2816$ (either from the bottom of Table A.3 or Table A.2.

Table A.3 Percentiles of Student t distribution

v, deg of freedom	$\gamma = 0.90$	$\gamma = 0.95$	$\gamma = 0.975$
1	3.0777	6.3138	12.7062
2	1.8856	2.9200	4.3027
3	1.6377	2.3534	3.1824
4	1.5332	2.1318	2.7764
5	1.4759	2.0150	2.5706
6	1.4398	1.9432	2.4469
7	1.4149	1.8946	2.3646
8	1.3968	1.8595	2.3060
9	1.3830	1.8331	2.2622
10	1.3722	1.8125	2.2281
11	1.3634	1.7959	2.2010
12	1.3562	1.7823	2.1788
13	1.3502	1.7709	2.1604
14	1.3450	1.7613	2.1448
15	1.3406	1.7531	2.1315
16	1.3368	1.7459	2.1199
17	1.3334	1.7396	2.1098
18	1.3304	1.7341	2.1009
19	1.3277	1.7291	2.0930
20	1.3253	1.7247	2.0860
25	1.3163	1.7081	2.0595
30	1.3104	1.6973	2.0423
40	1.3031	1.6839	2.0211
Limit (∞)	1.2816	1.6449	1.9600

Equation (A.7) then gives us 1.4026, a relative error of only 0.9%. The end result is that (using the correct percentile T value):

$$\text{ROC} = 724.1 \pm 7.7 \text{ ft/min} \tag{A.8}$$

or that the full-throttle average ROC at 70 KCAS probably (80%, remember) lies somewhere in the interval [716, 732]. The (80% CI) error bars are about 16 ft/min in size.

Example A.1 The eight raw ROC numbers above were actually arrived at by asking the Quattro Pro spreadsheet program for eight random numbers from the normal distribution with mean 720 and standard deviation 25. The fact that that particular sample was skewed a bit to the high side and had less scatter than the underlying population is "just the breaks." When we made the same request of the spreadsheet program a second time, the sample average came out 712.071, with a sample standard deviation of 32.606. That is the reality of statistics, especially with small samples. In flight testing, because of time and money constraints, we usually do have small samples.

In this example, we back up a moment to consider and exemplify what may be a confusing aspect of statistics on statistics. We were sampling from a normal distribution with mean $\mu = 720$ and standard deviation $\sigma = 25$. We took a first sample of size $n = 8$ and got sample mean $\bar{X} = 724.06$. This was a bit high, knowing what we know, though in the real world we seldom know what the true population mean actually is. That sample's standard deviation (using the $n - 1$ weighting) was $S = 15.331$. From our Olympian height of foreknowledge, that was quite low. Now if further samples are taken, we expect to find sample means and sample standard deviations that differ, from those first values, in almost all cases. Those statistics, $\bar{X}$ and S, are also so-called "random variables," with their own means, standard deviations, and estimates of those. Here are the several questions in order:

1) What is the mean of $\bar{X}$ over many trials? And how do we estimate, having made just one trial, what $\mu_{\bar{X}}$ is?

The mean of $\bar{X}$ is (simply, again) the population mean μ. We estimate it the same way we estimated μ, with the sample $\bar{X}$ itself. So after that first sample, our best estimate of the long-term average of many $\bar{X}$ also is 724.06. Of course, we know that this estimate is not exact, but (knowing only what we have seen realized) what better candidate number could there be? This brings us to further questions.

2) What is the average standard deviation of $\bar{X}$ over very many trials? And how do we estimate that mean having made just one trial?

The standard deviation of $\bar{X}$ is the population standard deviation σ divided by the square root of the sample size n. Small samples' means vary more than do the means of large samples. We estimate this by Eq. (A.6) which, in turn, uses Eq. (A.4).

3) Turning to the sample standard deviation S (the first sample value we obtained was 15.331), what would be the average of many such sample standard deviations after many trials and how would we estimate it?

The average sample standard deviation is simply the population standard deviation σ. We estimate σ, after one sample, by that sample's standard deviation S.

4) What is the standard deviation of sample standard deviations S over many trials? And how do we estimate that (super) standard deviation having made just one trial?

The standard deviation of sample values S is the population standard deviation σ divided by the square root of $2(n-1)$. And (now we are getting repetitive!) we estimate this by

$$\frac{S}{\sqrt{2(n-1)}} = \sqrt{\frac{\sum_{i=1}^{n}(x_i - \bar{x})^2}{2(n-1)^2}} \qquad (A.9)$$

To give the above discussion (repetitive and semicircular as it was) some semblance of reality, consider the following example involving six samples of eight ROCs each. Table A.4 gives the raw data and the two summary statistics ($\bar{X}$ and S) for each sample.

The mean of the six sample means is 720.186, very close to the population mean 720. The standard deviation of the six sample means is 9.351. The

Table A.4 Six samples, with summary statistics, of eight ROC figures each

	Sample[a]	Sample	Sample	Sample	Sample	Sample
	724.10	670.39	688.63	755.82	706.84	684.94
	738.44	729.51	697.55	739.94	713.57	735.51
	721.15	696.35	721.69	702.99	695.73	723.88
	745.39	662.59	729.74	731.28	725.17	681.53
	713.40	738.32	732.32	761.17	743.06	721.59
	731.52	715.20	739.05	713.70	730.91	716.07
	722.80	752.17	754.09	699.07	712.60	695.01
	695.70	732.04	754.49	755.01	712.52	704.40
Mean	724.06	712.07	727.19	732.37	717.55	707.87
STDS[b]	15.33	32.61	24.02	24.70	14.85	19.54

[a] Insider code for $(n-1)$ weighting.
[b] STDS = standard deviation sample.

theoretical value of that latter quantity is $\sigma/8^{1/2} = 25/2.828 = 8.839$, a bit smaller.

With the statistics of the sample standard deviation, S, things are really getting rarified. But let us push on. The six S values average to 21.841. The theoretical value is $\sigma = 25$. Not bad. The standard deviation of the six S values is 6.714. The theoretical value is $\sigma/(2[n-1])^{0.5} = 6.682$. Again, fairly close.

If you want to work out your own extended example and have no obvious easy way to get a random sample from $N(\mu, \sigma)$, here is how you can do that. First, find a random number in the interval [0,1]; any reasonable spreadsheet program will do that much for you. (Or, use a table of random numbers.) Take that random number to be a value of F_Z, the c.d.f. for the standard normal $N(0, 1)$. Functionally invert, using the $(z, F_Z[z])$ table, to find z. That result is a standardized value. So, to find the original value x, you have to multiply z by σ, then add μ. Say for example you wanted random values from $N(720, 25)$. Assume your random number from [0,1] is 0.412. The corresponding value of z is about -0.22. Multiplying by 25 and adding 720, you get 714.5. That is your first random number from the $N(720, 25)$ density. It is not so slow once you get organized.

Confidence Intervals for Estimation of Mean with Dispersion Measured Through the Range

First of all, we need to know how to estimate dispersion (σ) using the range. For this purpose we need short Table A.5. One uses Table A.5 by simply picking up the factor d_n for his or her sample size n and measuring the range (high value minus low value) in the sample. The estimate of dispersion is then

$$\sigma \approx d_n \times r \qquad (A.10)$$

In our first sample of eight purported ROC values, the range was $r = 745.39 - 695.70 = 49.69$. For $n = 8$, our estimate of σ is 17.441. Recall that $\sigma = 25$ and our earlier more elaborate estimate of it, using the $n - 1$ weighted sample standard deviation, was 15.331. Both estimates are on the low side, for this particular sample, but they are not far from each other.

The third column in Table A.1, relative efficiency is a factor telling us how much smaller a sample we could use, for the same efficiency of estimation, if we employed the sample standard deviation estimate. So, in this example, we could get by with a sample size of $0.89 \times 8 = 7.12$, instead of 8, and maintain efficiency of estimation, if we had used the more elaborate standard deviation estimator instead of our simpler range one.

To make us of this range-estimation technique for confidence intervals, we have to have yet another small table (Table A.6), one for the t^* distribution.

Table A.5 Range estimation of dispersion for normal populations

n	d_n	Relative efficiency
2	0.886	1.000
3	0.591	0.992
4	0.486	0.975
5	0.430	0.955
6	0.395	0.933
7	0.370	0.912
8	0.351	0.890
9	0.337	0.869
10	0.325	0.850
11	0.315	0.832
12	0.307	0.815
13	0.300	0.798
14	0.294	0.783
15	0.288	0.768
16	0.283	0.753

Because this table does not contain the $\gamma = 0.90$ column we need for 80% confidence, we are thrown into using either a 90 or 95% shortest two-sided CI. It is aviation, remember, so we will use the looser alternative; then we will have the 90% CI or (area under the distribution curve) $\gamma = 0.95$ or (area under either of the two tails) $\alpha = 0.10$.

Table A.6 Percentiles of t^* distributions

n	$\gamma = 0.95$	$\gamma = 0.975$
2	3.157	6.353
3	0.885	1.304
4	0.529	0.717
5	0.388	0.507
6	0.312	0.399
7	0.263	0.333
8	0.230	0.288
9	0.205	0.255
10	0.186	0.230
11	0.170	0.210
12	0.158	0.194
13	0.147	0.181
14	0.138	0.170
15	0.131	0.160
16	0.124	0.151

The formula for the $100(1-\alpha)\%$ CI is then

$$L, U = \bar{X} \pm R t^*_{n;1-\alpha/2} \qquad (A.11)$$

Redoing the ROC confidence interval problem with this tighter interval and looser dispersion estimate, we find $t^*_{8;0.95} = 0.230$, and hence

$$\text{ROC} = 724.06 \pm 11.43 \qquad (A.12)$$

Larger error bars go along with greater confidence. Using the Student t technique for assessing error bars, with this tighter 90% CI, gives error bars of size $2 \times 5.421 \times 1.8946 = 2 \times 10.27 = 20.54$. Not significantly different than the $2 \times 11.43 = 22.86$ of Eq. (A.12). But then one isolated example proves little. The important point is that you use one or another of the calculational strategies. Get into the habit of associating, with estimated experimental quantities, definite and supportable error bars.

Estimates of Dispersion from Several Shorter Series of Measurements

In aircraft performance flight testing, especially in the general aviation realm, there is something a bit unrealistic about our example in which eight "identical" climbs were performed, at a single weight, density altitude, and air speed, to assess rate of climb. It is more likely one would do two to four climbs at each of several air speeds. Then the flight test engineer would draw a smooth curve through those estimates to find the maximum (located at V_y), and perhaps draw a tangent line from the origin to that curve (V, ROC[V]) to find V_x. That sort of thing.

Now performing only (say) three climbs at each air speed certainly tends to expand the error bars. $S_{\bar{X}}$ goes up [smaller samples, see Eq. (A.6), have greater scatter], and so does the Student t percentile (for the two-tailed 80% CI, from 1.4149 for $n = 8$ to 1.8856 for $n = 3$). But one can argue as follows. Similar climbs at different air speeds, while they of course differ as to means μ_i, can be counted on to have just about the same standard deviations σ_i. We would be surprised to find that at 70 KCAS the ROC error bars were 16 ft/min tall and then suddenly, at 80 KCAS, the error bars were as large as 50 ft/min or as small as 5 ft/min. So is not there some way we can lump all the scatter measurements, for different air speeds, together? Yes there is.

You have m short series of measurements, indexed by i, each of which consists of n_i measurements. (Fleshing out the example above, assume $m = 7$ and that

Table A.7 Sample ROC data at seven different air speeds

i	1	2	3	4	5	6	7
n_i	3	3	3	3	4	3	3
KCAS	60.0	65.0	70.0	72.4	75.0	80.0	60.0
	585.4	634.6	752.2	744.4	713.4	655.6	637.7
	644.5	710.3	732.0	727.1	731.5	664.6	640.3
	611.3	687.2	724.1	751.4	722.8	688.7	647.1
					695.7		
Mean	613.75	677.37	736.10	740.99	715.86	669.62	641.70
STDS[a]	29.63	38.81	14.47	12.48	15.34	17.10	4.81

[a] STDS = standard deviation sample.

each $n_i = 3$ except, for some reason, $n_5 = 4$.) The lumped estimate of σ is given by

$$\sigma \approx s \equiv \sqrt{\frac{(n_1 - 1)s_1^2 + \ldots + (n_m - 1)s_m^2}{(n_1 - 1) + \ldots + (n_m - 1)}} \qquad (A.13)$$

Note: 1) the primacy of the degrees-of-freedom figures; 2) the denominator is simply $n - m$; and 3) Eq. (A.13) reduces to (the square root of) Eq. (A.4) when $m = 1$. Let us take an example.

Plugging the numbers from Table A.7 into Eq. (A.13), we get $\sigma \approx s = 21.35$, close to the $\sigma = 25$ used to construct each of the seven short samples.

With short samples of only three or four climbs, Table A.5 reminds us that it is almost as efficient to use ranges, rather than individual values of s_i, to estimate individual sample dispersions. For $n_i = 3$ the efficiency is 0.992; for $n_i = 4$ the efficiency is 0.975. The set of ranges r_i is {59.12, 75.73, 28.07, 24.24, 35.82, 33.06, 9.31}. When multiplied by their respective values of d_{ni}, the estimates of σ_i are {34.94, 44.76, 16.59, 14.36, 17.41, 19.54, 5.50}. These are each a little larger than their corresponding standard deviation sample (STDS) estimates in the bottom row of Table A.7, but quite comparable. When used in place of s_i values, in Eq. (A.13), the lumped result is $\sigma \approx s = 24.69$, very close to our "undercover" value $\sigma = 25$.

The moral is: with very small samples, use sample ranges, instead of standard deviations, to estimate dispersion. And do not be afraid to lump information from comparable samples with different means.

Indirect Measurements

It is not very often that we merely make measurements, record the results, and that is an end to it. Our hard-earned measurement results generally end up being used to calculate "downstream" quantities of further interest. Here, for example, is the "instant drag polar" formula for the parasite drag coefficient:

$$C_{D0} = \frac{W \sin \gamma_{bg}}{\rho_0 V_{Cbg}^2 S} \tag{A.14}$$

W is gross weight, γ_{bg} is the best glide angle, ρ_0 is the standard mean sea level (MSL) value of atmospheric density, V_{Cbg} is the calibrated best glide air speed in feet per sec, and S is the reference wing area. Weight, angle, and air speed are the genuine variables in terms of which C_{D0} is to be calculated; the standard density and wing area (for a given airplane) are constants.

Without going into details on just where we got these figures, the numbers of repeated trial measurements, and so forth, assume for the moment that each of the three variables has sample mean and sample standard deviation, $(n - 1)$ weighted, as in Table A.8. In addition, we know $\rho_0 = 0.002377$ slugs/ft³ and $S = 174$ ft². On this basis, what can we infer about the most probable value, and relative dispersion, of C_{D0}?

Obtaining our best estimate of C_{D0} is no problem. Plugging the constants and mean values of variables from the second column of Table A.8 into Eq. (A.14), we find $C_{D0} = 0.0381$.

Getting the estimate of scatter of C_{D0} is a problem and the one we now consider. The rule for getting an estimate of scatter for a function $f(x_1, x_2, x_3, \ldots)$, in terms of the sample variances s_i of the variables on which f depends, is:

$$\sigma^2 \approx s^2 = \left(\frac{\partial f}{\partial x_1}\right)^2 s_1^2 + \left(\frac{\partial f}{\partial x_2}\right)^2 s_2^2 + \left(\frac{\partial f}{\partial x_3}\right)^2 s_3^2 + \cdots \tag{A.15}$$

The partial derivatives are to be evaluated at the variables' mean values. Taking $x_1 = W$, $x_2 = \gamma_{bg}$, and $x_3 = V_{Cbg}$, and remembering to use radian measure for the standard deviation of the angle, we take the derivatives and soon have

$$s^2 = \left(\frac{C_{D0}}{W}\right)^2 21^2 + \left(\frac{C_{D0}}{\tan \gamma_{bg}}\right)^2 0.0157^2 + \left(2\frac{C_{D0}}{V_{Cbg}}\right)^2 1.8^2 = 0.00554^2 \tag{A.16}$$

Notice that we substituted, after taking partial derivatives, to make it convenient to express scatter of the parasite drag coefficient as *relative dispersion*,

Table A.8 Sample data from repeated weighing and glide tests

Variable	Mean	STDS
Weight W	2209 lbf	21 lbf
Best glide angle γ_{bg}	6.3 deg	0.9 deg = 0.0157 rad
Best glide air speed V_{Cbg}	124 ft/s	1.8 ft/s

dispersion divided by mean value. This is a commonly used term even in ordinary parlance, in the guise "We know the weight within 1%." Meaning we know it within 1% of its estimated value

$$\frac{s^2}{C_{D0}^2} = \left(\frac{21}{W}\right)^2 + \left(\frac{0.0157}{\tan \gamma_{bg}}\right)^2 + \left(2\frac{1.8}{V_{Cbg}}\right)^2 = 0.1454^2 \quad (A.17)$$

Hence, we only know C_{D0} to within about 14.5% of its estimated value 0.0381. Not very well. The middle term is the large one; better go do some more glides. We here deftly skipped over the fact that the best glide angle is nowhere near an "elementary sense datum." It came out of a calculation involving air speed, vertical glide interval, and elapsed time. Perhaps the atmosphere was too unstable. Needs to be traced back.

One good use of Eq. (A.15) in flight testing is constructive rather than critical. You start by saying you want to know important parameter C_{D0} to within say 4%. What does that imply about how much scatter you can tolerate in the numbers going into making up C_{D0}? Making various assumptions and playing with Eq. (A.15), you can find out.

Do not be dissuaded from using Eqs. (A.15) or (A.13) because those are not featured in your current flight test engineering manual or handbook. Part of your job is to expand the envelope of current practice by doing even better. It is usually more cost effective to throw more brain power at a problem than to throw only more instrumentation at it.

Simple Linear Regression

Now we treat this simplest case of "curve fitting," getting a formula that summarizes experimental data. We will only deal here with the simple linear case, fitting data with only one independent variable and only to a straight line. There are lots more complicated variants and those are useful. The general aviation general propeller chart, for example, was constructed by fitting propeller data to sixth-order polynomials with Jandel's TableCurve program. That company also has a two-independent-variable version called TableCurve 3D. Jandel's competitors have similar programs. A common aviation place to first run into curve fitting is in the older "noninstant" business of fitting lots of glide data, taken at various speeds, to a rearranged power-required curve to get drag parameters C_{D0} and e. Another, familiar to readers of this book, is the linearization of the propeller polar diagram.

Some curves that do not plot as straight lines can make use of linear regression after being "linearized." Take, for instance, the power law relation:

$$y = ax^b \quad (A.18)$$

where parameters a and b are to be "fit," determined, from experimental ordered pairs (x, y). The "secret" is to take logarithms of both sides, and then redefine variables to get

$$Y \equiv \ln(y) = \ln(a) + b \ln(x) = A + BX \tag{A.19}$$

where of course X is $\ln(x)$ and $B = b$. After you find A and B via simple linear regression, you use such facts as $a = \exp(A)$ to unravel the necessary complication. Similar manipulations can be used on simple exponential functions. But beware! Curve fitting works by finding parameters that minimize the sum of the squares of all "errors," vertical differences between y values and the fit curve. Parameters that minimize this sum in the transformed relation [e.g., Eq. (A.19)] are extremely unlikely to correspond to the choice of parameters that minimizes the sum in the original relation [Eq. (A.18)]. Hence, here is one place computer work is to be preferred over hand calculation. Good computer programs use complicated trial-and-error optimization techniques to find the genuine "best" fit parameters in the least-squares sense.

There are all sorts of traps one can fall into in this business of fitting curves to experimental data. We do not have space to do more than simply mention, without explanation, one further trap: it is particularly hazardous to curve fit, and especially to extrapolate, using polynomials. Unless, of course, you have good reason to believe the data actually does follow a polynomial curve.

We are going to stick to one straight line, where the theory is simple and the practice is sound. Even so, there will be enough complications to go around. Our (artificially constructed) raw data will come from another close-to-straight line relationship common to general aviation: the variation of maximum rate of climb, for a given airplane, with density altitude. Figure A.2 shows the scatter diagram of the raw data and the curve fit line Quattro Pro came up with; Table A.9 shows the raw data and the Quattro Pro regression output.

Nothing could be simpler. You get your raw data into a spreadsheet, call up the linear regression facility, hit a few buttons, and there you have it. In this case, the best fit line is

$$y_{\text{fit}}(x) = -0.04315x + 690.7556 \tag{A.20}$$

Our next job—and a fairly number-consuming one it will be—is to explain (though not derive) from where each of the spreadsheet regression output numbers came. To get maximum benefit from this, you will need to crank the raw data into a spreadsheet of your own and follow along as we explore the various statistics comprising the linear regression process. To cut down on the number of summation signs and subscripts, we are going to use a bracket notation that works like this

$$[x] \equiv \sum_{i=1}^{n} x_i; \quad [xy] \equiv \sum_{i=1}^{n} x_i y_i; \quad [xx] \equiv \sum_{i=1}^{n} x_i^2 \tag{A.21}$$

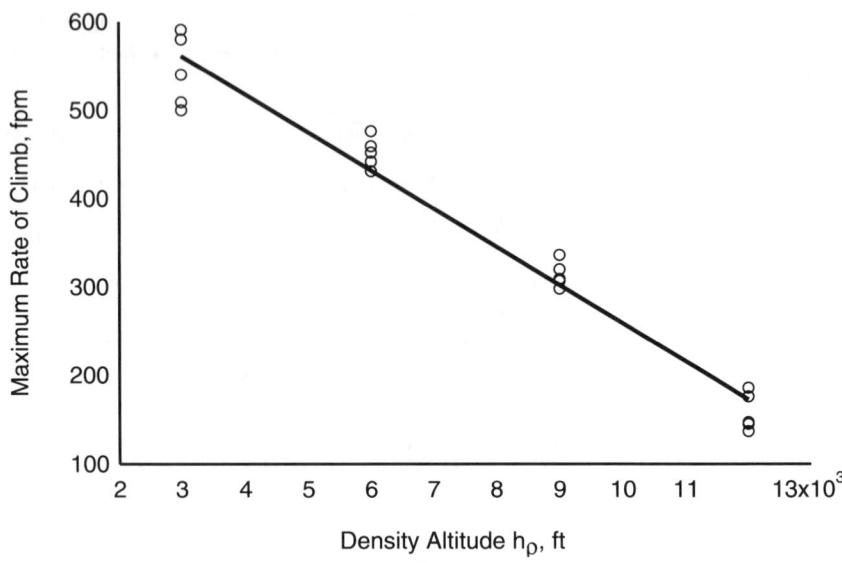

Figure A.2 Constructed maximum ROC data vs density altitude, with best fit line.

Table A.9 Constructed maximum ROC data as a function of density altitude, with regression output statistics

$x = h_\rho$	$y = h'_{max}$	Regression output	
3000	509	Constant	690.8
3000	580	Standard error of Y estimate	29.6
3000	540	R squared	0.96
3000	500	No. of observations	20
3000	591	Degrees of freedom	18
6000	431		
6000	476	X coefficient	−0.04315
6000	452	Standard error of coefficient	0.00198
6000	442		
6000	459		
9000	307		
9000	336		
9000	298		
9000	320		
9000	309		
12000	145		
12000	137		
12000	147		
12000	176		
12000	186		

How Big Are the Error Bars? 437

This will not be confusing, once you get the hang of it, because virtually all of our sums will be over the (in our case) 20 pieces of raw data, $n = 20$. We also will round figures off when that makes sense. Let us get started.

1) The Major Curve Fit Parameters (m and b)

By these we mean X Coefficient -0.04315 and constant 690.8, respectively, in Table A.9. The formulas for these ultimately come from a calculus treatment of the least-squares criterion mentioned above. The results, using CP for cross-products and SS for sum of the squares, are:

$$m = \frac{CPxy}{SSx} \equiv \frac{[xy] - [x][y]/n}{[xx] - [x]^2/n} = -0.04135 \qquad (A.22)$$

and

$$b = [y]/n - m[x]/n = 690.8 \qquad (A.23)$$

Time out for two quick comments. First, CPxy and SSx, as closer befits their names, were actually originally defined somewhat differently, as

$$CPxy \equiv [(x - \bar{x})(y - \bar{y})] \qquad (A.24)$$

and

$$SSx \equiv [(x - \bar{x})^2] \qquad (A.25)$$

The bars denote averages, $[x]/n$, etc., and deviations from those averages are "understood" in the plain language names of CPxy and SSx. It is a good exercise, and a check on understanding, to get the earlier forms of these two definitions (which are easier to compute with), from the latter ones, by expanding those latter terms. Second, it is an interesting fact that, if you consider an x that is the average of all of the x's, $[x]/n$, then the definitions of m and b, Eqs. (A.22) and (A.23), will tell you that the fit y value corresponding to that average x is simply the average of all y values, $[y]/n$. The regression line passes through the c.g. End of digression.

2) Standard Error (S_e)

By this we mean standard error of Y estimate 29.6. This is an estimate, from the data itself, of the scatter of datapoints about the fit line. This estimate is arrived at via

$$S_e \equiv [(y - y_{fit})(y - y_{fit})]/(n - 2) = (SSy - (CPxy)^2/SSx)/(n - 2) \qquad (A.26)$$

The $(n - 2)$ weighting will be discussed below. It is a fundamental assumption, in the linear regression procedure, that points at any x have the same scatter about the fit line as those at any other x. The normal curve centered on the regression line has the same width everywhere. Of course, in any practical application this may or may not actually be the case. In any event, this is the kind of scattter measured by S_e. There are others.

3) Coefficient of Determination (R^2)

R squared 0.96. This is the square of the so-called "correlation coefficient." An R^2 number close to unity says that the hypothesized linear regression relationship explains a great deal of the variation among y values. In a sense, R^2 is getting at the same thing as S_e, but from the other end and in a relative sense. The definition is

$$R^2 \equiv \frac{\text{SS due to regression}}{\text{SS}y \text{ ignoring regression}} = \frac{(\text{CP}xy)^2}{\text{SS}x \text{SS}_y} \qquad (A.27)$$

4) Number of Observations and Degrees of Freedom

In our example, there are of course 20 observations, $n = 20$. The reason for the number of degrees of freedom being less by two is that this curve fit has two parameters, m and b. Under this circumstance, not all of the datapoints are independent; if we had only 18 of them, we could find the other two from knowledge of m and b.

5) Scatter in the Estimates of Slope (m) and of Intercept (b)

This involves standard error of coefficient 0.00198 for an estimate of dispersion of estimate m. For some reason the spreadsheet regression output does not include a similar number for estimating scatter in intercept b. We will add that figure. The important idea here is that the particular values of m and b we came up with [Eqs. (A.22) and (A.23)] may be not quite correct because they came from a sample of (here) 20 datapoints. If we ran another 20 climbs, even though at exactly the same density altitudes, we would not expect to get precisely those same values. This added wrinkle, it turns out—and we will discuss it below—contains a rarely appreciated implication for aircraft operations safety.

The formula for estimation of the variance of regression slope m is:

$$S_m^2 = \frac{S_e^2}{\text{SS}x} \qquad (A.28)$$

and, for estimation of the variance of regression, intercept b is

$$S_b^2 = S_e^2 \left(1/n + \frac{(\bar{x})^2}{\text{SS}x} \right) \qquad (A.29)$$

Lower and upper limits on $100(1-\alpha)\%$ CIs are of the same form as any similarly scattered parameters (with somewhat different degrees of freedom):

$$L, U = m \pm S_m t_{n-2; 1-\alpha/2} \tag{A.30}$$

and

$$L, U = b \pm S_b t_{n-2; 1-\alpha/2} \tag{A.31}$$

Plugging our values into Eqs. (A.28) and (A.29), we find $S_m = 0.00198$ ft/min/ft, as on the spreadsheet, and $S_b = 16.23$ ft/min. Following through, for a 90% CI, and looking up the fact that $t_{18; 0.95} = 1.7341$, we get estimates of regression line parameters complete with error bars:

$$m = -0.04315 \pm 0.00343 \tag{A.32}$$

and

$$b = 690.8 \pm 28.1 \tag{A.33}$$

For our final linear regression subject, we turn to the problems of predicting mean values and individual performance. There is a difference between these. Predicting a mean value (for example predicting the average rate of climb at say 5000 ft) connotes taking samples of size n that work to reduce variation within the population of all climbs done at 5000 ft. The individual climb is the world's smallest sample, of size one, hence has to bear up under additional scatter.

The estimate of variance for the mean value at x is given by

$$S_{\bar{y}}^2 = S_e^2 \left(1/n + \frac{(x - \bar{x})^2}{SSx} \right) \tag{A.34}$$

Notice that the scatter is greater the farther x is from the overall average of all values of x. This is why error bands around regression lines are curved away from the line, farther from the line near its ends.

The estimate of variance for an individual performance at x has an additional addend S_e^2. This is connected to the (uniform) width of the normal curve describing variation within that subpopulation, the estimate of "standard error." The formula is

$$S_{\bar{y}}^2 = S_e^2 \left(1 + 1/n + \frac{(x - \bar{x})^2}{SSx} \right) \tag{A.35}$$

Let us compute these two estimates of standard deviation for climbs, and for an individual climb, at 5000 ft. Because the overall average x is 7500 ft, and SSx is $15,000^2$, the addend on the far right in Eq. (A.35) is $1/36 = 0.0278$. Recalling that $S_e = 29.6$ ft/min and $n = 20$, we get

$$S_{\bar{y}(5000)} = 4.52; \quad S_{y(5000)} = 30.7 \text{ ft/min} \tag{A.36}$$

With these values at hand we can go on to calculate the corresponding say 90% CIs for ROC. In both cases, those are centered about ROC(5000 ft) = $-0.04315 \times 5000 + 690.8 = 475.0$ ft/min. Looking up the fact that $t_{18;0.95} = 1.7341$, we find the following prediction for the average ROC:

$$\overline{\text{ROC}}(5000) = 475.0 \pm 4.52 \times 1.7341 = 475.0 \pm 7.8 \quad (A.37)$$

The prediction of best ROC for one individual effort at 5000 ft, on the other hand, is

$$\text{ROC}(5000) = 475.0 \pm 30.7 \times 1.7341 = 475.0 \pm 53.2 \quad (A.38)$$

There is a factor of not quite seven between sizes of the error bars for the average and individual predictions. Charter and air tanker operators in mountainous terrain, take note! Getting over a rocky ridge "on the average" is not good enough.

Conclusions

This concludes our discussion of those aspects of the theory of errors we think most important to general aviation flight operations and performance flight testing. The interested reader will find more information on the theory of errors in standard texts such as those written by Arley and Buch[1] and by Lentner.[2]

References

1. Arley, N., and Buch, K. R., *Introduction to the Theory of Probability and Statistics*, Wiley, New York, 1966.
2. Lentner, M., *Elementary Applied Statistics*, Bogden & Quigley, Inc., Tarrytown-on-Hudson, New York, 1972.

Appendix B

Bootstrap Approach Inputs and Outputs

Some of the inputs and outputs depend on whether one has a fixed-pitch or a constant-speed, propeller-driven airplane. And also on the flight regime level—1) full throttle, 2) partial throttle, or 3) takeoff and landing—in which one is interested. Discussion of inputs and outputs is therefore organized accordingly.

Bootstrap Approach Level 1

Full throttle (or gliding) quasi-steady flight with wings either level or banked. The operational (pilot-chosen) variables are 1) W, gross weight; 2) σ, relative air density, or h_ρ, density altitude; 3) ϕ, bank angle; and, except for V speed determinations; 4) V, true air speed. Also the flaps/gear configuration and, for constant-speed airplanes, propeller circular speed (revolutions per minute), N.

Inputs for Fixed-Pitch Airplanes (Bootstrap Data Plate)

Airframe inputs:

1) S, reference wing area (ft^2),
2) A, wing aspect ratio (span$^2/S$),
3) C_{D0}, parasite drag coefficient (depends on flaps/gear configuration), and
4) e, airplane efficiency factor (possibly depends on flaps configuration).

Inputs 1 and 2 come from the airplane's pilots operating handbook (POH). Inputs 3 and 4 come out of glide tests in which speed for best glide V_{bg} and angle for best glide (in calm air) γ_{bg} are found by trial and error for (only) one known density altitude and gross weight and for each flaps/gear configuration of interest.

Engine inputs:

5) M_0, rated full-throttle engine torque (ft-lbf) and
6) C, engine power altitude dropoff parameter, the proportion of indicated power that goes to engine friction losses (close to 0.12).

Input 5 comes from the airplane's POH values of rated (full-throttle) power and revolutions per minute. Input 6 comes from the airplane's engine manual.

Propeller inputs:

7) d, propeller diameter (ft),
8) m, slope of the linearized propeller polar diagram, and
9) b, intercept of the linearized propeller polar diagram.

Input 7 comes from the POH. Inputs 8 and 9 come out of 1) a climb test for best angle of climb V_x, and 2) *either* a test for maximum level flight speed V_M or a climb test for speed for best rate of climb V_y. These tests need only be performed at any one known density altitude and known gross weight.

Inputs for Constant-Speed Airplanes (Bootstrap Data Plate)

Items 1 through 7, same as for the fixed-pitch airplanes. In place of the linearized propeller polar parameters m and b, the constant-speed airplane requires

8) TAF total activity factor and
9) Z, ratio of fuselage diameter (taken one propeller diameter behind the propeller) to propeller diameter.

Item 8 comes from measuring propeller blade width at several stations along its length. Item 9 comes from measuring the circumference of the fuselage and from item 7. A copy of the general aviation general propeller chart or its computer version is required for the calculations.

Outputs for Either Fixed-Pitch or Constant-Speed Airplanes

A) V speeds (at any weight, density altitude, or bank angle)

1) V_M, maximum level flight speed,
2) V_m, minimum level flight speed (only when greater than V_S, stall speed),
3) V_x, speed for best angle of climb (in calm air),
4) V_y, speed for best rate of climb,
5) V_{bg}, speed for best angle of climb (in calm air), and
6) V_{md}, speed for minimum descent rate.

B) Rates or angles of climb or descent (including for climbing or descending turns) at any chosen air speed V, any bank angle ϕ, any weight W, any density altitude h_ρ.

C) Service and absolute ceilings (and speeds required there).

D) Power available, power required, thrust, drag (parasite and induced) at any V, ϕ, W, and h_p. Also speeds and bank angles required for various "safety" maneuvers such as the engine-out (return-to-airport) gliding turn with minimum loss of altitude or speed and bank angle required for minimum turn radius (or rate) for given descent rate.

Bootstrap Approach Level 2

Partial-throttle quasi-steady flight with wings either level or banked. The operational (pilot-chosen) variables are 1) W, gross weight; 2) σ, relative air density or h_p, density altitude; 3) ϕ, bank angle; 4) N, engine revolutions per minute; and, except for V speed determinations; 5) V, true air speed. Also the flaps/gear configuration.

Inputs for Fixed-Pitch Airplanes

Same as for level 1 (items 1 through 9) with these additions:

10) Several pairs (level cruise speed V_{cr}, engine revolutions per minute N) for (only) one known density altitude and gross weight (and for any flaps/gear configurations of interest) and
11) $c(P)$, a graph or formula of brake specific fuel consumption rate (bsfc) as a function of engine shaft power.

Item 10 comes from a single level cruise flight test made at known weight and density altitude and at various air speeds. Item 11 can be gleaned from the airplane engine manual.

Inputs for Constant-Speed Airplanes

Same as for level 1 (items 1 through 9) with these additions:

10) Cruise performance data giving data pairs (manifold absolute pressure, %power) for a selection of density altitudes and engine revolutions per minute and
11) $c(P)$, a graph or formula of bsfc as a function of engine power.

Item 10 is essentially the airplane's POH cruise performance section, but without using its often-optimistic air speeds. Item 11 can be taken either from that same cruise performance section or (better) from the airplane's engine manual.

Outputs for Fixed-Pitch or Constant-Speed Airplanes

All outputs of level 1 plus these additional V speeds:

1) V_{br}, speed for best range (at any given W, h_ρ) and
2) V_{be}, speed for best endurance (at any given W, h_ρ).

For fixed-pitch airplanes, revolutions per minute needed to make climbing or descending turns under given conditions of weight, altitude, turn rate, air speed, and climb or descent rate.

Bootstrap Approach Level 3

Takeoff and landing maneuvers. The operational (pilot-chosen) variables are 1) W, gross weight; 2) σ, relative air density, or h_ρ, density altitude; and 3) various air speeds (for takeoff V_R, air speed at rotation; for landing V_{50}, at 50 ft and V_1 just below 50 ft). Also flaps configuration and several rotation rate and flaps-retraction/braking timing decisions.

Inputs for Fixed-Pitch and Constant-Speed Airplanes

Much depends on whether one carries out A) a "detailed" calculation or a "relative" one, using the takeoff power parameter (TOPP), either B) calculated or C) experimentally determined.

A) In the detailed case, one needs quite a few additional data items such as 1) height of wing above runway (to determine ground effect); 2) graphs of $C_L(\alpha; \delta_f)$ for any desired flaps configuration (for induced drag contributions); 3) weight and balance information (for landing, to determine how much weight is on the nose wheel and on the main wheels); and 4) several facts about the runway environment (surface type, contamination type and depth, runway slope, wind conditions). If there is contamination, then even more data are required about the airplane's running gear (number of wheels of each type, tire width, tire deflection, and pressure). For the "detailed" method, we refer the reader to Chapters 11 and 12.

B) For takeoff calculations, the TOPP method is considerably simpler. Referring to Eqs. (11.121) and (11.122), "all" that is needed to calculate TOPP is eight of the nine items of the bootstrap data plate (C is not required because of an approximation we made) plus:

1) μ, the rolling friction coefficient for the tire/runway interface;
2) f_{GE}, the ground effect factor for the airplane's average wing height above the runway;
3) C_{LGR}^O, ground roll lift coefficient out of ground effect;

4) C_{LGR}^I, ground roll lift coefficient in ground effect;
5) x, the proportion that rotation speed V_R is greater than stall speed V_S;
6) $C_{L\text{Max}}$, the maximum (critical) lift coefficient for the takeoff flaps configuration;
7) W_0, the airplane's standard weight (normally maximum certificated gross weight);
8) W, the airplane's actual gross weight;
9) σ, relative air density on the runway;
10) Headwind/tailwind speed; and
11) Any other runway peculiarities (e.g., runway slope) to be used in a further (perturbation) calculation.

So even in the simpler formulation, calculation of distance to liftoff is a relatively complex enterprise.

C) Using the purely relative experimental method, all one needs for takeoff is two good distances to liftoff at fairly widely separated values of w/σ, where $w = W/W_0$. For landing, only one good landing roll distance is needed, at a known value of w/σ.

Outputs for Fixed-Pitch and Constant-Speed Airplanes

1) For takeoff, ground roll distance d_{LO} and distance to 50 ft AGL, d_{50}.
2) For landing, distance from 50 ft AGL onto the runway, d_{LA}, and landing ground roll distance d_{LG}.

Another bootstrap requirement is that one have a few calibrated cockpit instruments.

Calibrated Instruments Required

- Air-speed indicator. Using the open-box, at-altitude method requires a GPS unit, but the air-speed indicator can also be calibrated, somewhat more laboriously, by running back and forth along a ground course of known length. The needed output is a graph of KCAS as a function of KIAS.
- Altimeter. This will necessitate shop work. Concentrate on ascertaining accuracy when the Kollsman window is at 29.92 in. Hg.
- Tachometer. Borrow an optical tachometer running off the propeller blade flashes. Construct a curve of actual engine revolutions per minute as a function of indicated (cockpit tachometer) revolutions per minute.
- Stopwatch. For climb and glide tests.
- Thermometer. To ascertain density altitude. The installed aircraft OAT thermometer is slow to equilibrate. One may want to purchase a rapid-

reading type with a small probe that can be hung out a (closed) cockpit window.

Conclusions

In summary, implementing the bootstrap approach requires 1) calibrated instruments; 2) about 1.5 h of performance flight tests (climbs, glides, and level flight) to ascertain the nine numbers comprising the bootstrap data plate; 3) two takeoffs; 4) one landing; and 5) 1 or 2 h of calculation massaging test data into useful parameters and graphs. For your first aircraft, you might not get all this done in one day. On subsequent aircraft, you should be able to. Once the above numbers and graphs are in hand, predicting the airplane's performance—for a pivotal example see the calculation (example 10.4 in Chapter 10) of revolutions per minute needed for a completely specified descending turn under given ambient and aircraft conditions—is only a matter of substituting into formulas.

Appendix C

A Short List of Integrals

For full citations of the referenced integrals tables, see the references at the end of this appendix.

Integral I_1—Dwight[1] (90.1) or Peirce[2] (26)

$$\int \frac{dz}{A + Bz} = \frac{1}{B}\ln|A + Bz| \tag{C.1}$$

Integral I_2—Dwight[1] (140.02) or Peirce[2] (50)

$$\int \frac{dz}{A^2 - B^2 z^2} = \frac{1}{2AB}\ln\left|\frac{A + Bz}{A - Bz}\right| = \frac{1}{AB}\tanh^{-1}\left(\frac{Bz}{A}\right) \tag{C.2}$$

The last form is only correct when $A^2 > B^2 z^2$. The hyperbolic functions are related to each other as are the corresponding ordinary circular functions; for instance

$$\tanh z \equiv \frac{\sinh z}{\cosh z} \tag{C.3}$$

where

$$\sinh z \equiv \tfrac{1}{2}(e^z - e^{-z}), \quad \cosh z \equiv \tfrac{1}{2}(e^z + e^{-z}) \tag{C.4}$$

Integral I_3—Dwight[1] (691.01) or Peirce[2] (448)

$$\int \tanh(Az)dz = \frac{1}{A}\ln\cosh(Az) \tag{C.5}$$

Integral I_4—Dwight,[1] (160.01) Peirce,[2] (68) or Petit Bois[3] (p. 3)

$$\int \frac{dz}{Az^2 + Bz + C} = \frac{1}{\sqrt{B^2 - 4AC}} \ln \left| \frac{2Az + B - \sqrt{B^2 - 4AC}}{2Az + B + \sqrt{B^2 - 4AC}} \right| \quad (C.6)$$

Integral I_s—Dwight,[1] (160.11) Peirce,[2] (72) or Petit Bois[3] (p. 4)

$$\int \frac{z \, dz}{Az^2 + Bz + C} = \frac{1}{2A} \ln|Az^2 + Bz + C| - \frac{B}{2A} \int \frac{dz}{Az^2 + Bz + C} \quad (C.7)$$

Integral I_6—Petit Bois[3] (p. 10)

$$\int \frac{dz}{Az^3 + Bz^2 + Cz + D} = \sum_{i=1}^{3} \frac{\ln(z - z_i)}{3Az_i^2 + 2Bz_i + C} \quad (C.8)$$

The three numbers z_i are roots of the integrand's denominator. For instructions on solving that cubic, see Abramowitz and Stegun[4] or Birkhoff and MacLane.[5]

Integral I_7—Petit Bois[3] (p. 11)

$$\int \frac{z \, dz}{Az^3 + Bz^2 + Cz + D} = \sum_{i=1}^{3} \frac{z_i \ln(z - z_i)}{3Az_i^2 + 2Bz_i + C} \quad (C.9)$$

Again the three z_i are roots of the denominator. For instructions on solving that cubic, see citations for I_6 just above.

Integral I_8—Dwight[1] (565.1) or Peirce[2] (401)

$$\int e^{Az} \, dz = \frac{1}{A} e^{Az} \quad (C.10)$$

Integral I_9—Dwight[1] (566) or Peirce[2] (401)

$$\int f(e^{Az}) \, dz = \frac{1}{A} \int \frac{f(y)}{y} \, dy \quad (C.11)$$

where

$$y \equiv e^{Az} \quad (C.12)$$

Integral I_{10}—Dwight,[1] (121.1) Peirce,[2] (53) or Petit Bois[3] (p. 4)

$$\int \frac{z \, dz}{a^2 + z^2} = \tfrac{1}{2} \ln(a^2 + z^2) \quad (C.13)$$

Integral I_{11}—Dwight,[1] (120.01) Peirce,[2] (49,50) or Petit Bois[3] (p. 2)

$$\int \frac{dz}{a^2 + b^2 z^2} = \frac{1}{ab} \tan^{-1} \frac{bz}{a} \quad (C.14)$$

References

1. Dwight, H. B., *Tables of Integrals and Other Mathematical Data*, 4th ed., Macmillan, New York, 1966.
2. Peirce, B. O., *A Short Table of Integrals*, 3rd rev. ed., Ginn, New York, 1929.
3. Petit Bois, G., *Tables of Indefinite Integrals*, Dover, New York, 1961.
4. Abramowitz, M., and Stegun, I. A., *Handbook of Mathematical Functions*, Dover, New York, 1965, p. 17, 20.
5. Birkhoff, G., and MacLane, S., *A Survey of Modern Algebra*, Rev. ed., Macmillan, New York, 1956, p. 112.

Appendix D

Numerical Integration

In both the trapezoidal rule and in Simpson's rule, intervals Δx along the horizontal axis are (in any given application of the formulas) all the same size. For fine points such as approximations of errors, and for other techniques, see Abramowitz and Stegun.[1]

The trapezoidal rule:

$$\int_{x_0}^{x_n} f(x)\, dx \doteq \Delta x \sum_{i=0}^{n} f_i - (f_0 + f_n)(\Delta x/2) \tag{D.1}$$

Considering the subtracted portions on the right of Eq. (D.1) as coming out of the initial and final centered bars' widths, rather than heights, makes the trapezoidal rule easy to remember.

Simpson's rule:

$$\int_{x_0}^{x_{2n}} f(x)\, dx \doteq \frac{\Delta x}{3}[f_0 + 4(f_1 + f_3 + \cdots + f_{2n-1}) + 2(f_2 + f_4 + \cdots + f_{2n-2}) + f_{2n}]$$

$$\tag{D.2}$$

Simpson's Rule demands the x-interval be broken into an even number ($2n$) of subintervals.

Reference

1. Abramowitz, M., and Stegun, I. A., *Handbook of Mathematical Functions*, Dover, New York, 1965, pp. 885–886.

Appendix E

Derivation of Propeller Master Equation

Our goal is to derive, from Eqns. (6.16) for dT^{MT}, (6.25) for dQ^{MT}, (6.35) for dT^{BE}, (6.36) for dQ^{BE}, (6.40) for W and a, and (6.41) for W and a', using subsidiary relations given by Eq. (6.42) for local solidity σ and several others, the propeller master Eq. (6.43) or (E.1), below. This is done by combining the axial momentum theory and blade element theory along with the Prandtl momentum loss factor F.[1] The practical advantage of the master equation is that it is a *single* equation and therefore much easier to solve (still by trial and error) than pairs or larger sets of linked equations used by earlier investigators.

The propeller master equation is

$$J = \frac{\pi x (4F \sin^2 \phi - \lambda_T \sigma)}{(4F \sin \phi \cos \phi + \lambda_P \sigma)} \quad \text{(E.1)}$$

where propeller blade relative station x and advance ratio J are taken as parameters and where $F = F(x, \phi)$ is the Prandtl momentum loss factor, to be described. During practical calculation with Eq. (E.1), the question is "What angle of attack α will give flow angle ϕ, Prandtl factor F, and aerodynamic coefficient values λ_T and λ_P, to make Eq. (E.1) true. During any single trial-and-error process, J, x, and $\sigma(x)$ are known and fixed. Once a solution is found for a given pair of parameters (x, J), one must move on to consider other values of relative station x and, when the blade length has been exhausted, then move to other values of J. So, of course, this is, by hand, a laborious procedure. Even with modern spreadsheet tools designed for iterative solution of equations (Back-Solver, SolveFor, Solver) it is still somewhat laborious. Such is the reality of propeller theory.

But where does Eq. (E.1) come from in the first place? Before starting on the road to its derivation, we must discuss the Prandtl momentum loss function

$F(x, \phi)$. That is a somewhat difficult subject, for this relatively light correction (needed because of slightly radial flow), so we proceed as lightly yet definitely as we can. The paper by Adkins and Liebeck[2] is our major reference. The loss function is defined as

$$F \equiv \frac{2}{\pi} \cos^{-1}(e^{-f}) \tag{E.2}$$

where

$$f \equiv \frac{B(1-x)}{2 \sin \phi_t} \tag{E.3}$$

where ϕ_t is the flow angle at the blade tip. Adkins and Liebeck argue that for any blade position x with flow angle ϕ, this flow angle at the tip obeys

$$\tan \phi_t = x \tan \phi \tag{E.4}$$

which means that

$$\frac{1}{\sin \phi_t} = \sqrt{1 + \frac{1}{x^2 \tan^2 \phi}} \tag{E.5}$$

Example E.1 When $B = 2$, $x = 0.6$, and $\phi = 10$ deg, one will find $\sin \phi_t = 0.1052$ and hence $f = 3.802$ and $F = 0.9858$.

At this point, the reader should refurbish his or her acquaintance with definitions given in Chapter 6 for relative blade station x, axial interference factor a, rotational interference factor a', propeller angular speed Ω, the relation between power and torque, advance ratio J, the two propeller coefficients C_P and C_T, local solidity ratio

$$\sigma(x) \equiv \frac{Bc(x)}{\pi x d} \tag{E.6}$$

and Eq. (6.34) [here Eq. (E.7)] for the properly rotated coefficients

$$\begin{bmatrix} \lambda_T \\ \lambda_P \end{bmatrix} = \begin{bmatrix} c_l \cos \phi - c_d \sin \phi \\ c_l \sin \phi + c_d \cos \phi \end{bmatrix} \tag{E.7}$$

Figure E.1 sets the physical picture.

Slightly rewriting Eqs. (6.16) and (6.25), appending factors F previously ignored, the axial momentum half of our theory gives us

$$\frac{dT}{dr} = 4\pi r \rho (V + v) v F \tag{E.8}$$

$$\frac{dQ}{dr} = 4\pi r^3 \rho (V + v) \omega F \tag{E.9}$$

Derivation of Propeller Master Equation

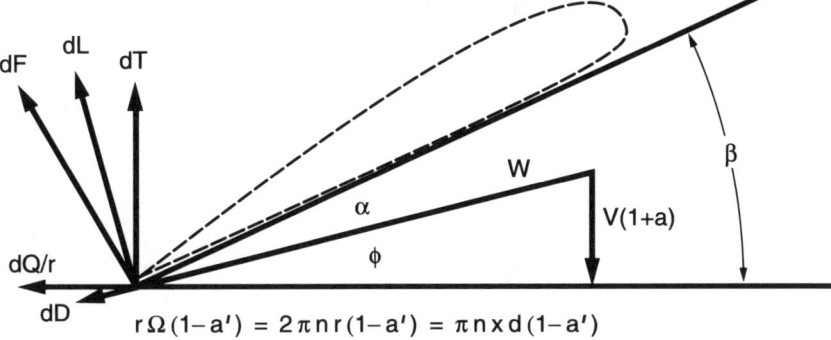

Figure E.1 Flow geometry for propeller blade element at $r = xR$.

The blade element half of our theory as seen through Eqs. (6.35) and (6.36), using Eq. (6.30), gives us

$$\frac{dT}{dr} = \tfrac{1}{2}\rho W^2 \lambda_T cB \tag{E.10}$$

$$\frac{dQ}{dr} = \tfrac{1}{2}\rho W^2 \lambda_P rcB \tag{E.11}$$

In Eqs. (E.8–E.11), the apparent unknowns are v, ω, ϕ, W, λ_T, λ_P, T', and Q', where the primes stand for derivatives with respect to blade station r. Eight variables and only four equations. If we exchange v for a, ω for a', and the two λ coefficients for the assumed known two functions $c_d(\alpha)$, $c_l(\alpha)$ and ϕ, α and ϕ unknown, we have four equations in these seven variables: a, a', W, α, ϕ, T', and Q' which are functions of parameters x and J. But our physical picture, Fig. E.1, also gives us three further subsidiary conditions:

$$\phi = \beta - \alpha \tag{E.12}$$

$$W = \pi x n d (1 - a') \sec\phi = V(1 + a)\csc\phi \tag{E.13}$$

$$\tan\phi = \frac{V(1+a)}{r\Omega(1-a')} = \frac{J}{\pi x}\frac{(1+a)}{(1-a')} \tag{E.14}$$

With seven equations for seven unknowns, at long last we can set out to solve the system of the seven Eqs. (E.8) through (E.14). First, equate the right-hand sides (RHSs) of Eqs. (E.8) and (E.10) to get

$$4\pi r\rho V^2(1+a)aF = \tfrac{1}{2}\rho W^2 \lambda_T 2\pi r\sigma \tag{E.15}$$

and then solve for

$$\frac{V(1+a)F}{W^2 \sigma} = \frac{\lambda_T}{4Va} \tag{E.16}$$

Second, equate the RHSs of Eqs. (E.9) and (E.11) to get

$$\frac{V(1+a)F}{W^2\sigma} = \frac{\lambda_P}{4\pi nxda'} \tag{E.17}$$

Third, equate the RHSs of Eqs. (E.16) and (E.17) to get

$$\frac{\lambda_T}{4Va} = \frac{\lambda_P}{4\pi nxda'} \tag{E.18}$$

and then solve for

$$J \equiv \frac{V}{nd} = \frac{\pi x \lambda_T a'}{\lambda_P a} \tag{E.19}$$

Next we start to work on formulas for axial and rotational interference factors a and a'. Our fourth step is to divide the RHSs of Eqs. (E.8) and (E.10) to get

$$1 = \frac{4\pi r \rho (V+v) vF}{\frac{1}{2}\rho W^2 \lambda_T cB} \tag{E.20}$$

and simplify, using Eq. (E.13), to get

$$1 = \frac{4aF \sin^2 \phi}{\lambda_T \sigma (1+a)} \tag{E.21}$$

Sixth, divide the RHSs of Eqs. (E.9) and (E.11) to get

$$1 = \frac{4\pi r^3 \rho (V+v) \omega F}{\frac{1}{2}\rho W^2 \lambda_P rcB} \tag{E.22}$$

and then use both forms of Eq. (E.13) to obtain

$$1 = \frac{4a' \sin \phi \cos \phi F}{\lambda_P \sigma (1-a')} \tag{E.23}$$

Seventh, solve Eq. (E.21) for a:

$$a = \frac{1}{\dfrac{4 \sin^2 \phi F}{\lambda_T \sigma} - 1} \tag{E.24}$$

and solve Eq. (E.23) for a':

$$a' = \frac{1}{\dfrac{4 \sin \phi \cos \phi F}{\lambda_P \sigma} + 1} \tag{E.25}$$

Eighth and finally, put Eqs. (E.24) and (E.25) into Eq. (E.19) and simplify to get the combined axial momentum and blade element propeller master equation:

$$J = \frac{\pi x(4F(x, \phi)\sin^2 \phi - \lambda_T \sigma)}{(4F(x, \phi)\sin \phi \, \cos \phi + \lambda_P \sigma)} \quad (E.26)$$

References

1. Betz, A., with appendix by Prandtl, L., "Screw Propellers with Minimum Energy Loss," Gottingen Repts., 1919, pp. 193–213.
2. Adkins, C. N., and Liebeck, R. H., "Design of Optimum Propellers," *Journal of Propulsion and Power*, Vol. 10, No. 5, 1994, pp. 676–682.

Appendix F

Flight Test for Drag Parameters

The formulas for getting to the needed drag polar parameters—parasite drag coefficient C_{D0} and airplane efficiency factor e—from known speed and angle for best glide, V_{bg} and γ_{bg}, are given in the Glide Performance chapter as Eqs. (9.41) and (9.43). So our job here is to describe how to fly the airplane and to do the postflight calculations for obtaining V_{bg} and γ_{bg}.

The idea, in a nutshell, is to repeatedly glide down over the *same* vertical interval (range of altitudes), at different air speeds, until the *product* $V_C \times \Delta t$ is a maximum. To calculate that maximum, you will need to take a pocket calculator up with you. V_C is calibrated air speed. In the cockpit, one can simply use KCAS or calibrated miles per hour, whichever your air-speed indicator uses. For example, you might find $V_{Cbg} = 70.5$ KCAS. Time interval Δt is the elapsed time for the glide, in seconds, from a stopwatch. For purposes of illustration, assume that the maximizing value is $\Delta t = 39.10$ s. Once that maximizing air speed is located accurately, you can return to base to complete the exercise.

Here is a more detailed cockpit activity sequence for one run. Your altimeter setting should be 29.92 in. Hg, so that your altimeter will directly read *pressure* altitude h_p. See Figure F.1 for nomenclature.

- Climb well above h_{p2}. Leave yourself plenty of room. Start your glide and be well stabilized at your trial V_I before you hit h_{p2}.
- At h_{p2}, start the stopwatch.
- Keep your wings level, maintain a predetermined heading, and keep your indicated air speed within one unit (1 kn, or 1 mph) of your trial V_I. If actual V_I varies during the course of the run, keep a mental note of its average value.
- Recording propeller revolutions per minute as you glide down is optional but always a good idea. That data can help you later construct accurate

propeller charts, as outlined in the Cruise and Partial-Throttle Performance chapter (Chapter 10).
- At h_{p1}, stop the watch. Your altimeter will lag a bit, of course, but it will lag by essentially the same amount at both ends of the run; the errors cancel.
- Record your actual (or average) V_I, use your calibration curve to find V_C, then compute and record $V_C \Delta t$ for that run.
- Power up and climb back up for another go. (In fact, you will probably do a climb test, as described in Appendix G, on the way back up.)

The observer—it is hard, though not impossible, for a lone pilot to do all the work alone—might want a data collection sheet set up as follows.

For the header: 1) date and page number; 2) aircraft tail number; 3) airplane make, model, and year; 4) airplane owner; 5) name of pilot; 6) name of observer; 7) starting gross weight; 8) Lower pressure altitude h_{p1}; 9) upper pressure altitude h_{p2}; 10) average OAT (temperature halfway between h_{p1} and h_{p2}; and 11) Note (such as "speeds in knots").

For the column headings: 1) run number (sequential); 2) clock time near beginning of run; 3) revolutions per minute; 4) gross weight (figured later); 5) target indicated air speed (IAS); 6) actual IAS (average during that run); 7) V_C (after application of ASI calibration curve); 8) Δt; and 9) $V_C \Delta t$.

It is also a good idea for the observer to graph products $V_C \Delta t$, against V_C, to better locate the maximum. And, having provisionally done so, it is only prudent to take a last final glide run to check that speed.

Getting the true air speed for best glide (with respect to the air mass) V_{bg} is almost immediate. We assumed you found $V_{Cbg} = 70.5$ KCAS. (The corresponding *indicated* air speed V_{Ibg} might have been a little different; correct V_I to the calibrated value V_C with your air-speed indicator's calibration curve.) Because we need to work in British engineering units, our sample value would be $V_{Cbg} = 70.5/0.592468 = 119.0$ ft/s. The corresponding true air speed value is given by

$$V_{bg} = \frac{V_{Cbg}}{\sqrt{\sigma}} \doteq \frac{V_{Cbg}}{\sqrt{\langle\sigma\rangle}} \tag{F.1}$$

$\langle\sigma\rangle$ is the *average* relative air density, halfway down the vertical glide interval. If you are gliding down repeatedly from say pressure altitude $h_{p2} = 6000$ ft to pressure altitude $h_{p1} = 5500$ ft, simply go to $\langle h_p \rangle = 5750$ ft, fly level, and record OAT $= \langle T \rangle$. Figure F.1 depicts the various temperatures, pressure altitudes, etc. Be sure to let your outside air thermometer equilibrate, and make sure it is located where it is not washed over by hot exhaust gases. One converts pressure altitude and OAT to relative air density by means of

$$\sigma = \frac{518.7}{(\text{OAT}°\text{F} + 459.7)}(1 - 6.8752 \times 10^{-6} \, h_p \text{ ft}) \tag{F.2}$$

Flight Test for Drag Parameters

	TEMPERATURES		PRESSURES	DENSITY
Top of Glide	T_2	$T_s(h_{p2})$	h_{p2}	
Midpoint	$\langle T \rangle$	$T_s(\langle h_p \rangle)$	$\langle h_p \rangle$	$\langle \sigma \rangle$
Bottom of Glide	T_1	$T_s(h_{p1})$	h_{p1}	

Figure F.1 Atmospheric variables in the glide test for drag parameters.

If, for instance, your measured OAT is $\langle T \rangle = 45°F$, then you would have $\langle \sigma \rangle = 0.9871$. That would make our sample $V_{bg} = 119.0/0.9935 = 119.8$ ft/s.

Getting the best glide *angle* is a bit more work. Look at Fig. F.2, showing the gliding space triangle with reference to the air mass. While it is certainly true that a *practical* best glide, say in case of losing your engine, would be with respect to the Earth, we do not care about that aspect for present purposes. Headwinds or tailwinds are all right as long as they are steady; horizontal wind shear not allowed. We do, however, care about updrafts or downdrafts. Because those are not reproducible from run to run (or even steady within a given run), make sure that you are *not* gliding in any vertical air currents.

From Fig. F.2, it is clear that

$$\gamma_{bg} = \sin^{-1} \frac{\Delta h}{V_{bg} \Delta t} \tag{F.3}$$

All that is left is to get Δh, the tapeline geometrical distance you glided down through. A sufficiently accurate approximation for it [derived in Chapter 1, Eq. (1.35)] is

$$\Delta h \doteq \frac{\langle T \rangle}{T_s(\langle h_p \rangle)} \Delta h_p \tag{F.4}$$

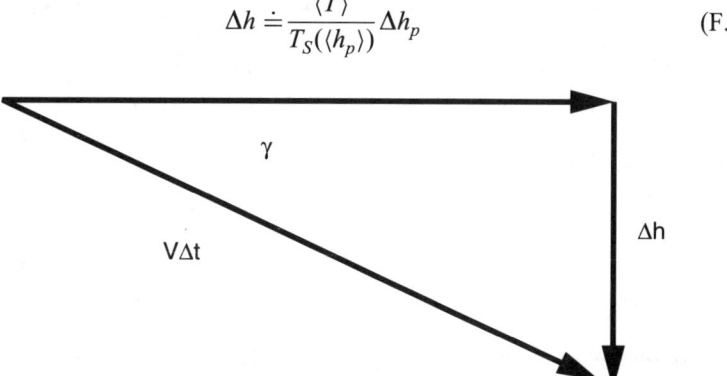

Figure F.2 Glide test space triangle with respect to the air mass.

Both temperatures in Eq. (F.4) must be expressed in the Rankine (absolute Fahrenheit) or in the Kelvin scale. Because standard temperature, in degrees Fahrenheit, for a given pressure altitude is

$$T_S(h_p) = 59 - 0.003566 h_p \tag{F.5}$$

we would have, for our sample problem, $T_S(5750 \text{ ft}) = 38.50°\text{F} = (38.50 + 459.7)°\text{R} = 498.2°\text{R}$. And our measured average temperature would be $(45 + 459.7)°\text{R} = 504.7°\text{R}$. Tapeline Δh would then be $(504.7/498.2) \times 500 = 506.5$ ft, a little larger than indicated.

Collecting these sample values, you would find

$$\gamma_{bg} = \sin^{-1} \frac{506.5 \text{ ft}}{119.8 \text{ ft/s} \times 39.1 \text{ s}} = 6.21 \text{ deg} \tag{F.6}$$

As you can see from Eq. (9.41) for C_{D0} and Eq. (9.43) for airplane efficiency factor e, there is some additional background data needed to find C_{D0} and e. You will need an estimate of gross weight W—keep track of the clock time so that you can approximate your fuel burnup until the maximizing glide—and air density that $\rho = \rho_0 \sigma$ (actually, you could have gotten by with just your calibrated air speed in feet per second), as well as reference wing area S and wing aspect ratio A from pilots operating handbook data. If we assume $W = 2209$ lbf, $S = 174$ ft^2, and $A = 7.38$, then Eqs. (9.41) and (9.43) would tell us that, for this flaps/gear configuration, $C_{D0} = 0.0408$ and $e = 0.595$. The final calculation, for e, is shorter if one uses Eq. (9.42) in lieu of Eq. (9.43).

If you will sometime want various performance figures not only for flaps 0 but also for flaps 10, flaps 20, and flaps 30, the entire experiment and calculations must be repeated for each of those additional flaps/gear configurations. Drag figures into almost everything.

There is also the question of whether to do the drag glide tests with the engine at idle power (the safest procedure), with the engine turned off and the propeller windmilling, or with the engine turned off and the propeller stopped. If you choose either of the latter two alternatives, be sure your starter is functioning flawlessly and that you have a landing strip within glide range. The experiment does not require any special altitude, so think high. Here is our suggested procedure for stopping the propeller.

- Just before beginning your glide, pull the mixture knob out to full lean (idle cutoff), turn off the ignition, and pull up, as though to stall (but do not!), until the propeller stops turning.
- At this point (thinking ahead!) switch the ignition back to "both" and (unless you are very high) push the mixture to full rich.

- When you have finished gliding, crank the engine to start. Steepen your glide, if necessary. Once the engine starts, lean the mixture.
- Again, do not try this procedure unless you are sure your starter works and your engine will get going again readily. Or else have a wide choice of emergency strips close at hand.

What if you have a biplane? Does not matter. The bootstrap approach does not care how many wings you have. Lots of ink has been spilt on biplane aerodynamics: gap, stagger, decalage, Munk span factor K, square tip correction τ, equivalent monoplane aspect ratio, etc. If you are combining two known monoplane wings to *compute* the aerodynamics of a certain geometric combination from individual monoplane numbers, those details would be necessary. But if you *have* the airplane you just run the flight tests. Remember the bootstrap approach axiom: Fly the airplane to see how the airplane flies. Another way to say that is: The airplane is its own best instrument.

Appendix G

Determination of Propeller Polar Constants *m* and *b*

Five of the nine bootstrap data plate (BDP) numbers for the fixed-pitch, propeller-driven airplane—reference wing area S, wing aspect ratio A, engine rated torque M_B, engine power/torque altitude dropoff constant C, and propeller diameter d—can easily be obtained from the airplane's pilots operating handbook and its engine manual. The remaining four BDP parameters are the so-called "harder to get" ones. Appendix F dealt with the performance glide test, which allows calculation of the airframe drag parameters C_{D0} (parasite drag coefficient) and e (airplane efficiency factor). The final two parameters needed are the slope m and y intercept b appearing in the linearized propeller polar

$$\frac{C_T}{J^2} = m\frac{C_P}{J^2} + b \tag{G.1}$$

Constant b is obtained first, from full-throttle climb tests designed to find V_x, speed for best angle of climb (in calm air). Constant m is then obtained *either* from the climb test results plus a single full-throttle level speed run (giving V_M, maximum speed for level flight) *or* from the climb test results plus a further series of climb tests which locate V_y, speed for greatest rate of climb.

Let us start with b. We do not need to go over all of the details because the experimental situation closely mirrors the glide tests in Appendix F. The difference is that now the airplane climbs, at full throttle, up through the vertical interval from h_{p1} to h_{p2}. And this time it is the *minimum* of the product $V_C \times \Delta t$ (calibrated air speed times elapsed time), which tells the pilot and observer that V_x has been found. So the recordkeeping details and various tips in Appendix F can be used—"with the necessary changes having been made," as philosophers often say—to run the climb tests for V_x.

So let us say we have V_x. How is b related? Start with the definition of bootstrap composite parameter F, the only one that involves b.

$$F \equiv \rho d^2 b \tag{G.2}$$

Because composite $K \equiv F - G$, we then have

$$b = \frac{K + G}{\rho d^2} \tag{G.3}$$

Because $K = -H/V_x^4$ [see Eq. (7.27)], and then referring to the definitions of composites G and H [Eqs. (7.7) and (7.8)], we soon have

$$b = \frac{SC_{D0}}{2d^2} - \frac{2W^2}{\rho^2 d^2 S\pi e A V_x^4} \tag{G.4}$$

We see, of course, that the experimental conditions under which the climb test was performed—the flaps/gear configuration, gross weight W and air density ρ—make a difference.

On to m. Let us assume the level speed run for V_M method was chosen. The mathematical manipulations needed if the V_y method is chosen are entirely similar. Start with the bootstrap formula for V_M, Eq. (7.19):

$$V_M = \sqrt{\frac{-E - \sqrt{E^2 + 4KH}}{2K}} \tag{G.5}$$

Square this and multiply by $2K$, then add E, to get

$$2KV_M^2 + E = -\sqrt{E^2 + 4KH} \tag{G.6}$$

Square again, simplify, and solve for E:

$$E = \frac{H - KV_M^4}{V_M^2} = H\left(\frac{1}{V_M^2} - \frac{KV_M^2}{H}\right) \tag{G.7}$$

Use the fact that

$$-K/H = V_x^{-4}$$

and the definition of composites H [Eq. (7.8)] and E [Eq. (7.15)] to get

$$m = \frac{2n_0 dW^2}{\Phi(\sigma) P_0 \rho S\pi e A} \left(\frac{1}{V_M^2} + \frac{V_M^2}{V_x^4}\right) \tag{G.9}$$

If you decide, instead, to start from the bootstrap equation for V_y, you will end up with the quite similar relation

$$m = \frac{2n_0 dW^2}{\Phi(\sigma) P_0 \rho S\pi e A} \left(\frac{3V_y^2}{V_x^4} - \frac{1}{V_y^2}\right) \tag{G.10}$$

Index

A

Absolute ceiling, 190
   bootstrap formula, 216-217
   partial-throttle, 339-341
Acceleration vector, 54
Advance ratio. *See* Propeller, advance ratio
Adverse yaw, 240, 246
Aerodynamic braking, 396
Aerodynamic center, 75, 82, 85, 87, 111-112
Aerodynamic force, 85
Air speed indicator (ASI), 23
   calibration, 324
   calibration techniques, 26-32
   high speed, 24
   low speed, 24
   position error, 26
Airplane efficiency factor, 72, 87, 96, 115, 117, 193
Allometric formula, 120
Altimeter, 37
   consistency check, 39
Altimeter setting formula, 37
   derivation, 37-38
Altitude
   determining differences, 16-18
   indicated, 40
Angle of attack (AOA), 63, 66, 71, 89-91
   body, 67, 72, 100, 239, 352, 395
   induced, 70, 85, 88, 90, 95, 100
   propeller blade, 165, 168-169
   stall, 89, 92-93, 102-103, 106
   zero lift, 89-90, 92-93
Angle of climb, 190
   bootstrap formula, 199, 202-203
Angle of descent, 190
Angle of incidence, 70, 97, 101
Angular acceleration, 58-59
Angular position, 58-59
Angular velocity, 58-59
Anhedral, 97
Aspect ratio, 85, 88-95
Assumptions, performance, 237-238
Atmosphere
   density, 7, 10
   layers, 5 (table)
   modeling, 15, 40-44
   molecular composition, 6
   relative density, 10, 15
   relative pressure, 10
   relative temperature, 10
Attitude angle, 66, 239, 394

B

Bank angle, 66, 239
   maximum for level flight, 257-258
Banked absolute ceiling, 252, 255-258, 267
Barometer, altimeter as, 39
Bernoulli's law, 22, 70, 84, 156-157
   compressible, 23
   incompressible, 22-23, 78
Best climb angle
   bootstrap formula, 202-203
   variation with weight or altitude, 210-211

Best climb rate, variation with weight or altitude, 213-216
Best glide, headwind/tailwind graphical solution, 308-310
Best glide angle, 189-192, 198-199, 281-283, 334
  bootstrap formula, 204
  headwind/tailwind, weight or altitude effect, 312-314
  headwind/tailwind with updraft/downdraft, 314-315
  small flight path angle approximation, 292
  variation with weight or altitude, 210-211
Best glide speed, 189-192, 198-199, 283, 334
  bootstrap formula, 205, 292
  headwind/tailwind with updraft/downdraft, 314-315
Binomial expansion, 84, 291
Biot-Savart law, 87
Body axis, 58
Bootstrap approach, advantages, 233-234
Bootstrap composite parameters, 205-207, 256, 298, 321, 324
Bootstrap data plate, 188-190, 221, 256, 271, 298, 320
  gliding, 284
Brake mean effective pressure (bmep), 132-133
Brake specific fuel consumption rate (bsfc), 123, 136, 326
Braking friction, 400-404
Braking friction coefficient, 390, 400-404
Break-even headwind
  landing, 410-411
  takeoff, 375-379
Breguet formula, 317
British engineering units, 50, 150
Buckingham's pi theorem, 150-151

C

Calibrated air speed (CAS), 15, 27, 29
Camber line height, 76
Center of gravity (c.g.), 60, 110-111
  aft, 111
  forward, 111
Center of mass (c.m.), 60
Center of pressure, 69, 81-82
Centrifugal force, 61, 242-245
Centripetal force, 61, 240
Chord, wing, 71, 97, 101
Circular dynamics, 59-61
Circular kinematics, 58-59
Circular mechanics, correspondences with linear mechanics, 60 (table)
Circulation, 81, 83-84, 87
Confluent alternant matrix, 73
Constant average force approximation, 363-364, 379, 405-407, 413-414
Coriolis acceleration, 62
Coriolis force, 61-63, 244
Cosine series, 290
Cross product, 52
Crossflow, 100
Cruise performance, scaling rules, 336-337
Cruise performance table, bootstrap, 332-336
Cylinder head temperature, 123

D

Density altitude, 10
Descent rate, 272
  bootstrap formula, 308, 313
  gliding, 285-286
Dewpoint temperature, 12, 127
Dihedral, 97
Dimensional analysis, 150-152
Dimensionless group, 150-152
Dimensionless number, 50
Dimensions, 49
Displacement vector, 54
Doghouse plot, 250

Dot product, 52
Downwash, 85-88
Downwash angle, 92, 94
Downwind turn, 273-274
Drag, 65-66, 114-117, 272
   biplane interference, 94
   bootstrap formula, 199-201
   contaminant, 353-355
   cooling, 70
   due to flaps, 117
   due to lift, 70
   during takeoff, 351-352
   gliding, 287
   ground effect, 351-352
   induced, 70, 72, 85, 190, 251, 392-393
   interference, 70
   minimum, 116, 283-284
   parasite, 69-70, 190
   pressure or profile, 70
   skin friction, 70
   trim, 70
Drag coefficient, 114, 279
   parasite, 70-71, 115, 193
Drag polar, 192, 279
   determined from best glide, 288-289
Dynamic pressure, 77, 82

E

Energy, 56-57, 59-60
Engine
   accessory bleed efficiency, 125
   brake horsepower, 120-121, 129, 194
   brake moment, 150, 164
   clearance volume, 125
   compression stroke, 130-131
   controls and instruments, 134-135
   cylinder bore, 120-121
   displacement volume, 120-121
   efficiency, 123
   excess air, 134
   friction power, 129
   indicated power, 126, 128
   intake stroke, 130
   manifold pressure, 140
   mechanical efficiency, 125
   normally aspirated, 119, 124
   piston speed, 120-121
   power, 164
   power stroke, 131, 133
   products of combustion, 123, 133
   rotation speed, 120-121
   standard cycle (table), 132-134
   stroke, 120-121
   torque. *See* Torque, engine
   volumetric efficiency, 124
Engine operator's manual, 135-137
Engine performance chart, 138-142
Envelope definition of graph, 312-313
Equations of motion
   gliding, 279
   quasi-static, 239-240
   small flight path angle, 290-291
Equivalent air speed (EAS), 36
Equivalent braking action, 400, 401 (table)
Excess power, 191-192
   bootstrap formula, 195
Excess thrust, 192
Exhaust backpressure, 129, 141-142
Exhaust gas temperature, 123

F

Federal Aviation Administration (FAA), 61, 233, 242-245, 353, 388
Flap effectiveness factor, 104-106
Flapped area, 104
Flaps, 103-110
Flaps deflection angle, 101
Flare, landing, 392, 394-395
Flight path angle, 66, 239, 272
   banked or unbanked glides, 280-282
   bootstrap formula, 322
   small angle approximation, 291
Force, 55-56
Frame of reference, 53
   inertial, 61, 244-245

non-inertial, 61-62, 244
Fuel
 higher heating value, 122
 lower heating value, 122
 specific weight, 122
 stoichiometric ratio, 122, 125, 134
Fuel consumption rate, 328
Fuel-air mixture
 lean, 123
 rich, 123
Fuel-to-air ratio, 122
Fuselage, 87

G

Gagg-Farrar altitude power dropoff factor, 141, 195, 217, 228, 231, 319
 derivation, 126-129
Gasoline, aviation, 122. *See also* Fuel
General aviation general propeller chart (GAGPC), 177-182, 187, 349, 361, 381
General Aviation Manufacturers Association (GAMA), 317-318, 327-328, 332
General propeller chart, 158-159, 177-180
Geopotential altitude, 9
Glide path angle, headwind/tailwind, 307-308
Glide ratio, 285-286
Glide sink rate, 203
Global positioning system (GPS) unit, 28
Gravitational potential energy, 56
Ground effect, 70, 94-97, 392-394
Ground effect factor, 393
Ground speed, 29
Gust load factor, 250

H

Headwind/tailwind effects, graphical analysis, 221-223
Headwind/tailwind with updraft/downdraft, 223-225

Helical flight path, 241-241, 280
Hodograph, 277
Horizontal tail, 87, 92, 112
Horizontal wind effect, wings-level glide, 303, 306-314
Horsepower, 58
Horseshoe vortex, 87
Humidity, 11
 engine effect, 13
Hydroplaning, 407, 353-355

I

Ideal gas, 7
Ideal gas law, 7, 132
Inclinometer, 245-249
Indicated air speed (IAS), 27
International standard atmosphere (ISA), 9-11
Invariant, banking, 283, 289

J

James brake index (JBI), 400, 401 (table)

K

Kinetic energy, 56-57, 274-276
Kinetic energy effect, 32-33, 238, 280
Knot, 50
Kutta-Joukowski theorem, 70, 83-84

L

Landing
 downhill with headwind, 410
 perturbation approach, 408-410
 segments, 389-390
 uphill with tailwind, 410
 weight and altitude effect, 411-414
Lapse rate, atmosphere, 5
Law of cosines, 219, 307
Lift, 65-66
 factors influencing, 114

gliding, 287
ground effect, 355
Lift coefficient, 70-71, 89-90, 102, 114, 279, 394
   maximum, 92-93, 102, 107, 110
   pressure coefficient relation, 81-82, 84
   section, 104
   slope, 89, 91-95, 103, 106
   slope in ground effect, 352
Linear dynamics, 55-58
Linear kinematics, 54-55
Linearized propeller polar, 194, 231, 329
   assumptions, 231
   intercept, 188, 197, 323
   slope, 189, 196, 323
Load factor, 249-250
Load factor limit, 266

M

Mach number (M), 34, 71, 91, 153, 172
Manifold absolute pressure (MAP), 64, 174-175
Maximum camber, 76
Maximum level flight speed
   bootstrap formula, 196-198
   variation with weight or altitude, 212-213
Maximum thickness, 76
Mean aerodynamic chord (mac), 99-100
Mean geometric chord (mgc), 98-100
Minimum descent rate
   bootstrap formula, 203
   small flight path angle, 293
   variation with weight or altitude, 209-210
Minimum level flight speed
   bootstrap formula, 196-198
   variation with weight or altitude, 212-213

Mixture control, 122
Moment, 60
   engine, 121
Moment of inertia, 60
   wheel, 396
Momentum, 56

N

NACA airfoil profile, 76
   2412, 89-93
   4412, 76
   4-digit, 76
National Advisory Committee on Aeronautics (NACA), 76
National Aeronautics and Space Administration (NASA), 353
Nautical mile, 51
Newton's laws of motion, 55, 61, 244, 348

O

Obstacle clearance speed, 346-347
Offset angle, 67
Otto cycle, 124, 129-133
Outside air temperature (OAT), 29
Outside air thermometer, 34

P

Partial-throttle operation, 338-341
Pilots operating handbook (POH), 232-233, 317-319, 327-328, 412-414
Pitch angle. *See* Attitude angle
Pitching moment, 64, 69, 74
Pitching moment coefficient, 70-71, 81-83, 89, 92-93, 109-110, 112
Position error, 44
   altimeter and ASI errors, 47
   determination, 44-48
Position vector, 54
Potential energy, 57-58
Power, 58-60
Power adjustment factor, 272

Power available, 134, 190-192, 194, 320
  bootstrap formula, 195
Power coefficient grading, propeller, 164-166, 169
Power drop factor (p.d.f.), 137
Power dropoff parameter, 323
Power picture, assumptions, 230
Power required, 190-192, 194, 320
Power setting parameter, 319-321
Prandtl momentum loss factor, 168 171
Pressure
  dynamic, 22
  total, 22
Pressure altitude, 10, 29, 40
Pressure coefficient, 70, 77-79, 81-82
Profile thickness, 76
Propeller
  actuator disk, 154-156
  advance ratio, 64, 151, 194, 272, 322
  axial interference factor, 155-158, 161
  blade activity factor (BAF), 178-180
  blade chord, 65
  blade element theory, 162-174
  blade tip losses, 168
  combined momentum and blade element theory, 166-173
  constant-speed, 158, 176-182
  flow angle, 162
  geometry, 146-150
  hub or boss drag, 172
  ideal efficiency, 157-159, 161, 177
  momentum theory, 154-161
  nominal pitch, 148
  power adjustment factor, 177-181
  representative blade element theory, 176, 231
  resistive torque, 150
  rotational interference factor, 155, 160-161
  slip stream wake contraction, 172-173
  slowdown efficiency factor (SDEF), 172, 181-182, 272
  static thrust, 155, 170-171
  total activity factor (TAF), 178-180
  twist angle, 65
  ultimate wake interference factor, 156
  vortices, 160-161
  windmilling, 199, 278
Propeller efficiency, 151-152
  constant-speed, 177-179, 181-182, 272
  fixed-pitch, 194, 325, 328-333
  rated, 190, 229-230
Propeller master equation, 171
Propeller polar diagram, 174-176
Propeller power coefficient, 64, 152-154, 272, 322, 325, 328-329
  momentum theory, 158
Propeller section profile
  Clark Y, 147
  drag coefficient, 163
  lift coefficient, 163
  RAF6, 147-148, 163-164, 170
Propeller theory approximations, 171-174
Propeller thrust coefficient, 64, 152, 164-165, 325, 328-329
Pseudoforce, 61-63
P-factor, 237, 328

Q

Quadratic drag polar, 114
Quarter-chord point, 74, 82, 111-112
Quasi-static equations of motion, 65

R

Radian, 58
Radius of curvature, 240
Ram temperature rise, 34
Rate of climb, 190
  bootstrap formula, 199-202, 257, 299, 322
  correcting, 15

Rate of descent, 190
   banked or unbanked glides, 280, 282
   bootstrap formula, 391
Ratio of specific heats, 8, 125
Reference wing area, 64, 101
Relative motion, 219, 306
Return to airport maneuver, 294-305
   best bank angle, 295-296
   best turn speed, 295
   minimum altitude loss, 296
Reynold's number (Re), 71, 153, 173
Rolling friction, 356
Rolling friction coefficient, 356 (table)
Root chord, 97
Round-out, landing, 392
Runge-Kutta numerical solution, 359, 398
Runway condition reading (RCR), 400, 401 (table)
Runway reaction force, 356
Runway slope, landing, 397-398
Runway surface condition (RSC), 401 (table)

S

Saturation vapor pressure, 12, 127
Scalar, 51
Scalar product, 52
Service ceiling, 216, 257
Sine series, 290
Slug, 50
Small flight path angle approximation, 115, 220-221, 285, 293-295
Sonic speed, 14, 25
Specific endurance, 330-331, 333-334
Specific gas constant, 7, 127
Specific range, 330-331, 333-336
Speed at absolute ceiling, bootstrap formula, 216-217
Speed for best climb angle
   bootstrap formula, 197
   variation with weight or altitude, 209-210
Speed for best climb rate
   bootstrap formula, 197-198, 200
   variation with weight or altitude, 212-213
Speed for best glide, 115-116, 283
   headwind/tailwind, infinite series solution, 310-312
   small flight path angle, 292
   variation with weight or altitude, 209-210
Speed for minimum descent rate
   gliding, 287-288
   small flight path angle, 293
   variation with weight or altitude, 209-210
Stagnation point, 79
Stall, 101
Stall speed, 107-108, 189-191, 197-198, 209, 413
   banked, 265, 300
Static thrust, 349-351
Steady maneuvering chart, 258-268
Structural load limit, 249-251
Sweepback angle, 91, 97

T

Tail force, 110, 112-113
Takeoff
   70/50 rule, 369-370
   assumptions, 371
   downhill with wind, 375
   factors, 371
   forces, 348 (table), 357-359
   headwind, 345, 365-368
   order of approximations, 381-382
   perturbation approach, 371-375
   phases, 346
   safety factor, 388
   tailwind, 345
   uphill into wind, 375
Takeoff power parameter (TOPP), 345, 379-387
Taylor series, two-dimensional, 213-214, 372
Temperature recovery factor, 34-35

Terminal velocity dive, 303
Terrain clearance, 377, 411
Thin airfoil theory, 79
Thrust, 65, 190
    bootstrap formula, 194-195, 199-200
Thrust axis, 58
Thrust coefficient grading, propeller, 164-166, 169
Thrust offset angle, 239
Time to climb, determining, 19-20
Tip chord, 97
Tire and wheel runup, 356
Tire and wheel spinup, 395-396
Torque, 59-60
    engine, 121, 132-133, 137, 194-195, 317-322, 327
    landing roll, 390, 402-404
Trapezoidal rule, 80, 165, 180
Troposphere, 4-5
True air speed (TAS), 15, 27, 66
Turn
    coordinated level, 240-245
    maximum rate, 268-270
    minimum radius, 268-270
    slipped or skidded, 245-249
Turn performance, constant-speed airplane, 270-272
Turn radius, 265
    banked or unbanked glides, 280, 282
    minimum. *See* Turn, minimum radius
    nonlevel turn, 282
    small flight path angle approximation, 292
Turn rate, 66, 239, 265, 272
Turn-and-bank indicator, 245
Twist, wing, 70

U

Uniform acceleration, 406
Uniform acceleration approximation, 368-369
Units, 50

V

V speed
    absolute ceiling, 189-191
    banked, 251-255
    best climb angle, 189-192, 202-203, 325, 341-342
    best climb rate, 189-191, 213-216
    best endurance, 317, 330-331, 333-334
    best glide, 189-192, 198-199, 283, 334
    best range, 317, 330-331, 333-334
    design cruising, 250
    design diving, 250
    liftoff, 346
    maneuvering, 250
    maximum level flight, 189-191, 321, 325
    minimum descent rate, 189-191, 198-199, 334
    minimum level flight, 189-191, 321, 325
    never exceed, 250
    rotation, 386
    touchdown, 389-391
Vector, 51
Vector product, 52
Velocity vector, 54
Viscosity, 14, 157
    absolute or dynamic, 14
    kinematic, 14, 71
Viscosity correction factor, flaps, 104-107
V-n diagram, 249-251, 265

W

Washout, 100-101
Weight and balance, 112
Wind correction angle, 218
Wind effect
    landing, 407-408
    climbing, 217-229
    speed for best climb angle, 225-229
Wind shear, 273, 378, 411

Wing, elliptically loaded, 87, 92
Wing aspect ratio, 64
Wing span, 72
Work, 55-56, 59-60

Z

Zero-thrust glide, 199
Zooming, 300-301